# FOUNDATIONS OF
# MATHEMATICAL LOGIC

## HASKELL B. CURRY

*Evan Pugh Research Professor in Mathematics*
*The Pennsylvania State University*

DOVER PUBLICATIONS, INC., NEW YORK

# PREFACE TO THE DOVER EDITION

In this reprinting it has been possible to correct a number of errors of certain sorts. These comprise purely typographical errors, and also certain cases where an essential word or two was inadvertently omitted or misplaced. The only corrections which affect the sense are the following two.

The original proof of (ii) under Theorem 4D7, page 157, contained an error which was first noticed by E. C. W. Krabbe, then a student at the University of Amsterdam. The theorem, however, is correct as stated. That incorrect part of the proof is here replaced by a new one, which is actually simpler than the original.

In Exercise 7C6, page 349, the student was asked to prove something false. For the discovery and published correction of the error see my article in *Contributions to Mathematical Logic*, edited by H. A. Schmidt, K. Schütte, and H. J. Thiele, Amsterdam, 1968, p. 101. The exercise is here replaced by another one.

Besides these corrections a rather large number of changes have been suggested by various persons or by my own further study. One of the most interesting of these is the following. The rules Px on page 193 and Nx on page 262 are not strictly analogous. If we were to modify Px so as to be analogous to Nx, or Nx so as to be analogous to Px, certain simplifications in various parts of the work would ensue. This would require extensive changes. Although such changes would doubtless improve the work, they are not suitable to be made here.

My thanks are due to the rather considerable number of persons who have suggested corrections, and to Dover Publications, who have made this reprint possible.

HASKELL B. CURRY

*State College, Pennsylvania*
*July 10, 1976*

Published in Canada by General Publishing Company, Ltd., 30 Lesmill Road, Don Mills, Toronto, Ontario.
Published in the United Kingdom by Constable and Company, Ltd., 10 Orange Street, London WC2H 7EG.
This Dover edition, first published in 1977, is an unabridged and corrected republication of the work first published by the McGraw-Hill Book Company, Inc., in 1963.
*International Standard Book Number: 0-486-63462-0*
*Library of Congress Catalog Card Number: 76-55956*
Manufactured in the United States of America
Dover Publications, Inc.
180 Varick Street
New York, N.Y. 10014

# PREFACE TO THE FIRST EDITION

For several years I have given a general course in logic for graduate students at The Pennsylvania State University. At first this was an introductory course, intended for persons who might be mature mathematically but had no previous experience with logic; however, it soon developed that the students would benefit much more by the course if they had a comparatively elementary one first. This book is a development of that course.

In view of this origin the reader will not expect to find here a book for absolute beginners, nor for persons who do not have the degree of maturity regarded as normal in a graduate student. But for students to whom it is not necessary to explain the most elementary facts about logical symbolism, nor to give extensive training in the translation of notions of ordinary language into that symbolism or vice versa, this book is intended to be self-contained. It aims to give a thorough account of a part of mathematical logic which is truly fundamental, not in a theoretical or philosophical sense, but from the standpoint of a student; a part which needs to be thoroughly understood, not only by those who will later become specialists in logic, but by all mathematicians, philosophers, and scientists whose work impinges upon logic.

The part of mathematical logic which is selected for treatment may be described as the constructive theory of the first-order predicate calculus. That this calculus is central in modern mathematical logic does not need to be argued. Likewise, the constructive aspects of this calculus are fundamental for its higher study. Furthermore, it is becoming increasingly apparent that mathematicians in general need to be aware of the difference between the constructive and the nonconstructive, and there is hardly any better way of increasing this awareness than by giving a separate treatment of the former. Thus there seems to be a need for a graduate-level exposition of this fundamental domain.

The exposition here given differs in its point of view from that generally current, particularly in the textbooks designed for beginners. The traditional approach to the logical calculus is that it is a formal system like any other; it is peculiar only in that it must be formalized more rigorously, since we cannot take "logic" for granted, and in that it can be interpreted in the statements of ordinary discourse. Here the point of view is taken that we may interpret our systems in the more circumscribed set of statements which we form in dealing with some other (unspecified) formal system. The origin of the point of view so far as I am concerned is described in Sec. 5S1. This is a point of view which I share with Lorenzen, who arrived at a similar position independently and on the basis of a quite different philosophy. Since in the study of a formal system we can form statements which cannot be decided by the devices of that system, this brings in possibilities which did not arise, or seemed only pathological, in the traditional theories. The result is that various different systems of logical calculus stand here more or less on a par, and there are a number of alternatives, for example, in the definition of negation.

From this point of view logical study is a part of the methodology of formal systems. In order to develop it, we have to begin with the study of formal systems themselves. Thus, after Chapter 1, which is an introduction to the whole subject, the next two chapters are devoted to the study of formal methods as such. This is done here with some care, for unfortunately there has been a great deal of misunderstanding of this matter even on the part of experts. A brief treatment of the Markov theory of algorithms is included. After these two chapters there is a rather easy chapter, Chapter 4, concerned with the elementary facts about lattices and similar algebraic systems.

After these preliminaries the proper business of this book begins with Chapter 5. The general method of the inquiry is to seek for a formulation which expresses the meaning we wish to attach to the logical connectives, and then to develop the properties which follow from the assumptions so motivated. The connectives are not taken up all at once; the positive connectives—implication, conjunction, and alternation—are taken up in Chapter 5, negation in Chapter 6, quantification in Chapter 7, and modal operations in Chapter 8. Of these, Chapter 5 is fuller and more detailed than the others, not only because its principal connective, implication (conjunction and alternation are relatively trivial, and are carried along as by-products) is in a sense the central connective of logic, but because the general principles of the inquiry are established in that chapter; in the later chapters we merely extend to the new operations the results already obtained for the old. On the other hand, the treatment of modality is very brief, for the reasons which are explained in the introduction to that chapter.

This mode of approach is adapted to the semantical situation—each chapter is concerned with a factor in that situation which was not present in the previous one. However, the arrangement has one disadvantage, viz., that when one extends the proofs of theorems to cases not contemplated when those theorems were originally proved, one tends to forget certain important details and to assume too readily that everything is going to be all right in the new situation. Anyone with experience can cite cases of errors which have arisen in just this way. This is, however, a price which has to be paid for the advantages of the semantical approach. Besides, it is typical of the way in which mathematics grows; we are continually extending results, often with suitable modifications, to situations similar to, but not exactly the same as, those originally encountered.

The last four chapters in this book make extensive use of the methods of Gentzen. These have been presented hitherto as having their essential significance in showing that a proof can be put in a certain standard form, and from this fact important necessary conditions for deducibility follow. Important as these considerations are, they do not exhaust the possibilities of the method. As the reader proceeds through this book, it will become clear to him that the Gentzen rules have a natural interpretation directly in terms of the semantical situation we are attempting to formalize. They are thus a tool of the semantical analysis. Further, the significance of the principal theorem is that it shows that the formulation obtained from the semantical situation is adequate for that situation, or to put it the other way around, failure of the theorem is evidence that there is some factor inherent in the meaning of the operations which has not been adequately taken account of. Thus formulations for which the theorem fails are to be regarded with suspicion. This is a conviction which does not lie on the surface where it can be seen and explained at the very beginning, but it will become evident as one proceeds.

This being a book for graduate students, it has certain features which would be out of place in an elementary text. In the first place, it is thoroughly documented. Graduate students are expected to become proficient in research; for that purpose they need to know something about the technical literature and to go there on occasion for additional information. Thus there are copious references for supplementary purposes. I have not hesitated to include material in foreign languages, some of which are relatively unfamiliar to English-speaking students. Even if a student cannot read the material in question, or for some other reason it is inaccessible to him, he should know what he is missing.

Each of the eight chapters of this book is divided into from three to five sections, indicated by the letters A to E; in addition, there is a supplementary section S. As its name suggests, the last treats rather informally topics which supplement the text in various ways. Here there are historical and bibliographical comment and other supplementary references to the literature, discussions of aspects of the subject which are too specialized, or perhaps controversial, to be included in the main text, etc. Two precautions should be kept in mind about this material. The first is that it may require more background of the reader than is supposed in the main text. The second is that it is intended to be suggestive rather than definitive. The last remark applies particularly to the historical comments; I have attempted to state completely the material used in the preparation of the main text, but beyond that point to give only such information as I happen to possess.

At the end of most sections will be found a list of exercises. These vary greatly in difficulty, but on the whole there is relatively little of the kind of practice material one finds in an undergraduate textbook. Problems of which I do not know the solution myself are marked with a star; however, one should not infer that these are the most difficult ones. For the benefit of readers who may use this book for self-instruction, references are given to places in the literature—sometimes later in this same book—where information bearing on the problem, perhaps a solution, may be found; however, the solutions found in the places cited may often be improved.

It is a pleasure to acknowledge here the help which I have received from various sources. To my secretary, Veronica P. Zerbey, I am indebted for the somewhat arduous task of preparing the manuscript. My research assistants, Josiah P. Alford, Franklin S. Brenneman, and Frederick C. Zerbey, have helped a great deal with the details, including the preparation of the bibliography. For financial support which rendered this assistance possible, I am indebted to the U.S. National Science Foundation. Suggestions have also been received from my students William Craven and Herman J. Biesterfeldt. The illustrations in this book are used by permission of Gauthier-Villars of Paris and originally appeared in my book "Leçons de logique algébrique." In the later stages of preparation I have received valuable assistance from my colleague, Professor Hugo Ribeiro, and my student Luis E. Sanchis.

*Haskell B. Curry*

# CONTENTS

# EXPLANATION OF CONVENTIONS

Here are collected for reference certain conventions in regard to the format of this book.

**Cross references.** The system used for designating parts of the book and making cross references from one part to the other is essentially the same as that used in the author's book, "Combinatory Logic."

The chapters of this book are designated by Arabic numerals. References to whole chapters are made thus: Chap. 3.

The major divisions of the chapters are called sections and are designated by Roman capital letters. A reference to a section in the same chapter is made by the section letter preceded by the abbreviation Sec.; e.g., Sec. B. For a section in a different chapter the chapter number precedes the section letter; thus Sec. 5B indicates Chapter 5, Section B.

Sections are divided into subsections designated by Arabic numerals. Subsections in the same section are referred to by this numeral preceded by 'Sec.'; thus, Sec. 5. Subsections of a different section are cited by giving the number of the subsection preceded by the section designation and, if necessary, the chapter number. Thus, Sec. 5C3 and Sec. C3 are subsections 3 of Sec. 5C or Sec. C, respectively.

Occasionally subsections are divided into still smaller subdivisions designated $a$, $b$, $c$, which are cited in an analogous fashion.

Theorems and formulas are numbered consecutively throughout the sections. When a theorem or formula is cited without a section designation, the reference is to a theorem or formula in the same section. Otherwise the theorems are identified by section number thus: Theorem D2 means Theorem 2 of Sec. D in the same chapter; Theorem 5D2 is Theorem 2 of Sec. 5D. Formula numbers can be distinguished from subsection numbers by the fact that they are enclosed in parentheses. If a formula number is given without further identification the reference is to a formula in the same section. Otherwise it is identified by section; for example, (5) in Sec. 5D. Corollaries are numbered by adding a digit, in the form of a decimal fraction, to the number of the theorem.

The numbering of lemmas, remarks, and examples is less formal. Unless the situation is clear these will be cited explicitly in connection with a subsection; e.g., Remark 2 of Sec. 7A4.

In this book definitions are generally made informally in the main text, with the new term being defined in italics; such definitions may then be found through the index. In the few cases where it seems necessary to be more formal they are treated like lemmas and remarks.

**Citations in brackets.** References to the Bibliography are made by abbreviations consisting of Roman letters in brackets (e.g. [CLg]), with or without the author's name. For a fuller explanation see the beginning of the Bibliography.

**Use of letters.** With few exceptions the use of letters in formulas conforms to established mathematical practice. Only certain special letters, which are used with a fixed meaning throughout the book, will be commented on here. The letters 'B', 'C', 'F', 'I', 'K', 'S', 'W' will be used to designate special notions of combinatory logic. These special letters are used only in those few occasions where a relationship with combinatory logic is under discussion; notions derived from them are symbolized in ways which suggest this origin, but do not require special type. Roman letters are used in the main text and explanations accompanying formulas as fixed abbreviations, and the Greek letters 'Λ', 'Π', and 'Σ' are used in the same way. In Chapter 6 the letters $F$ and $\mathfrak{J}$ are given fixed meanings.

**Special symbols.** Single quotation marks are used throughout this book to indicate that the expression enclosed in them is being mentioned, not used. This is a technical usage; for explanations in connection with it see Sec. 2A3. Double quotation marks are used in their ordinary sense.

For conventions connected with symbols used as functors, dot notation, etc., see Secs. 2A3 and 2A4.

# Chapter 1

# INTRODUCTION

It is appropriate to start the study of mathematical logic by inquiring what mathematical logic is. This question will be answered in a preliminary manner in Sec. A. We shall then turn to a somewhat deeper consideration of the nature of mathematics and its relation to logic. This consideration will begin, in Sec. B, with a discussion of the paradoxes of logic and of the lessons that are to be drawn from them in regard to the logic of mathematics. We shall then proceed, in Sec. C, to a critique of the various views as to the nature of mathematics. Finally, in Sec. D, we shall return to the relation of mathematics and logic. The chapter is intended to give some background for the formal developments which begin in Chap. 2.

## A. THE NATURE OF MATHEMATICAL LOGIC

We open this inquiry by examining three senses which the word 'logic' has in ordinary discourse.

The first sense is that intended when we say that "logic is the analysis and criticism of thought."[1] We observe that we reason, in the sense that we draw conclusions from our data; that sometimes these conclusions are correct, sometimes not; and that sometimes these errors are explained by the fact that some of our data were mistaken, but not always; and gradually we become aware that reasonings conducted according to certain norms can be depended on if the data are correct. The study of these norms, or principles of valid reasoning, has always been regarded as a branch of philosophy. In order to distinguish logic in this sense from other senses introduced later, we shall call it *philosophical logic*.

In the study of philosophical logic it has been found fruitful to use mathematical methods, i.e., to construct mathematical systems having some connection therewith. What such a system is, and the nature of the connection, are questions which will concern us later. The systems so created are naturally a proper subject for study in themselves, and it is customary to apply the term 'logic' to such a study. Logic in this sense is a branch of mathematics. To distinguish it from other senses, it will be called *mathematical logic*.

[1] See Johnson [Lgc, part I, p. xiii].
For an explanation of the convention that has been followed in this and all subsequent citations, see the introduction to the Bibliography, at the end of the book.

1

In both of its preceding senses 'logic' was used as a proper name. The word is also frequently used as a common noun, and this usage is a third sense of the word distinct from the first two. In this sense a logic is a system, or theory, such as one considers in mathematical or philosophical logic. Thus we may have classical logics, modal logics, matrix logics, Aristotelian logics, Kantian logics, etc.

We may clarify somewhat the relation between these three senses of 'logic' if we consider the corresponding senses of 'geometry'. In the first sense geometry is the science of space. Etymologically the word means the measurement of the earth, and the oldest geometry is said to have been the rules of measurement of the ancient Egyptian surveyors. So conceived, geometry is a branch of physics. But alongside this there is geometry as a branch of mathematics. In this one considers mathematical systems which have some connection with the study of space. Finally, we have many sorts of geometries: we can speak of a projective geometry, a differential geometry, a nonarchimedean or a nondesarguesian geometry, a four-dimensional geometry, and so on.

*Mathematical logic, then, is a branch of mathematics which has much the same relation to the analysis and criticism of thought as geometry does to the science of space.*

This is as far as it is desirable to go, at present, in defining 'mathematical logic'. As a matter of fact, it is futile to attempt to define any branch of science by delimiting precisely its boundaries; rather, one states the central idea or purpose of the subject and leaves the boundaries to fall where they may. It is an advantage that the definition of logic is broad enough to admit different shades of opinion. Furthermore, it will be permissible to speak of "logical systems," "logical algebras," without giving a precise criterion for deciding whether a given system is such; it suffices that such systems have a connection of one sort or another with the analysis of thought.

There are, however, several remarks which it is appropriate to make now to amplify and clarify the above discussion.

In the first place, the connection between a geometry and actual space may be quite remote, as, for example, in the case of a finite geometry, a nondesarguesian geometry, or a geometry with infinitely many dimensions. Indeed, it is not clear just when a system is a geometry and when it is not. In this respect the situation in regard to a logic is analogous. We can and do consider logics as formal structures, whose interest from the standpoint of philosophical logic may lie in some formal analogy with other systems which are more directly applicable.

In the second place, current usage restricts 'geometry' to the mathematical aspect of the subject. Indeed, this mathematical aspect has developed to such an extent that, if one wishes to speak of the physical aspect, one is forced to use some other term. An analogous development in the case of logic has not yet taken place—whether or not it will do so in the future, as some maintain, it is not our business to decide. It is quite in accord with current usage to speak of philosophical and mathematical logic in the way it has been done here.

In the third place, although the distinction between the different senses of 'logic' has been stressed here as a means of clarifying our thinking, it would be a mistake to suppose that philosophical and mathematical logic are completely separate subjects. Actually, there is a unity between them. Mathematical logic, as has been said, is fruitful as a means of studying philosophical logic. Any sharp line between the two aspects would be arbitrary.

Finally, mathematical logic has a peculiar relation to the rest of mathematics. For mathematics is a deductive science, at least in the sense that a concept of rigorous proof is fundamental to all parts of it. The question of what constitutes a rigorous proof is a logical question in the sense of the preceding discussion. The question therefore falls within the province of logic; since it is relevant to mathematics, it is expedient to consider it in mathematical logic. Thus the task of explaining the nature of mathematical rigor falls to mathematical logic, and indeed may be regarded as its most essential problem. We understand this task as including the explanation of mathematical truth and the nature of mathematics generally. We express this by saying that *mathematical logic includes the study of the foundations of mathematics*.

## B. THE LOGICAL ANTINOMIES

We now proceed with the program of discussing, on an intuitive basis, the nature of mathematics and its relation to logic as ordinarily conceived. We shall begin with considerations affecting the nature of mathematical rigor as it was understood at the close of the nineteenth century.

To the mathematicians of that time a mathematical proof was rigorous when it was "strictly logical." Take, for example, the theorem that if the real function $f(x)$ is continuous for $a \leq x \leq b$, and if further $f(a) < 0 < f(b)$, then there is a value $c$ such that $a < c < b$ and $f(c) = 0$. Before the era of arithmetization, one could only "see" that this was true from the fact that the graph of $f(x)$, being a continuous curve which was above the $x$ axis at one end and below it at the other, must cross the $x$ axis at some point. Experience with contradictions derived from such reasoning (and especially from arguments regarding infinite series without adequate investigation of convergence) showed the need for a more precise treatment. This was achieved, as is well known, by conceiving a function as a set of ordered pairs, by an arithmetical definition of continuity, and by a "strictly logical" proof of the above theorem in terms of these definitions.[1]

But now what was the logic in terms of which such a logical proof could be defined? Certainly it was not the traditional logic, for that was inadequate to express reasonings in terms of relations (such as inequality) which were the very soul of such a proof. Indeed, it seems that the mathematicians of that time carried out their reasoning in terms of logical intuitions which were never formulated as explicit principles. Apparently it was tacitly assumed that everyone had such intuitions and that they formed an absolutely reliable criterion of rigor.

Into this situation the discovery, about the beginning of the twentieth century, that there are arguments which, although perfectly sound from the intuitive point of view, nevertheless lead to contradictions, fell like a

---

[1] On this point cf. Black [RMP, especially pp. 156–157].

bombshell. Such arguments are now called paradoxes, or antinomies.[1] Some of them had been known since antiquity, but their relevance in this particular connection was not appreciated. Because of their importance for mathematical logic, it will be expedient to consider some of them here.

**Russell's paradox.**[2]  Our intuition tells us that we can consider classes of objects as forming new objects. Thus we can consider the class of all chairs in this room, the class of all men, of all houses, of natural numbers. Likewise we can consider classes of classes, and even such notions as the class of all classes, or the class of all ideas. Among these classes there will be two sorts, which we shall call proper and improper classes. Proper classes are those, like men, houses, numbers, which are not members of themselves; improper classes are those which, like the class of all classes or the class of all ideas, are members of themselves. Now let $R$ (the Russell class) be the class of all proper classes. If $R$ is a proper class, then, since $R$ is the class of all such classes, $R$ is a member of $R$, and hence $R$ is not a proper class. On the other hand, if $R$ is not a proper class, then $R$ is not a member of $R$, and therefore $R$ is a proper class. Either assumption leads to a contradiction.

It is instructive to express the paradox in symbols. Let the statement that $x$ is a member of the class $y$ be symbolized by the notation

$$x \epsilon y$$

'$x$' and '$y$' being variables for which names of arbitrary notions can be substituted, and let $\rceil$ and $\rightleftarrows$ be symbols for negation and logical equivalence, respectively. Then, by the definition of $R$, we have, for arbitrary $x$,

$$x \epsilon R \rightleftarrows \rceil(x \epsilon x)$$

and hence

$$R \epsilon R \rightleftarrows \rceil(R \epsilon R)$$

Thus the statement that $R \epsilon R$ is equivalent to its own falsehood, and hence if it is true it is also false, and vice versa.

The following two arguments, although not paradoxical, are of a nature similar to that of the Russell paradox and shed some light upon it.

**Barber pseudoparadox.**[3]  The council of a certain village is said to have given orders that the village barber (supposedly unique) was to shave all the

---

[1] These terms are equivalent in meaning. Some persons have, to be sure, attempted a hairsplitting distinction between them, but I think such persons have been misled by false etymology. 'Paradox' means something which is not in agreement with prevailing opinion; it is formed from παρά + δόξα, where δόξα means 'opinion', or 'expectation', and παρά means generally 'beside', but often has the connotation of missing the target and going beyond it, as in English 'beside the point'. 'Antinomy' on the other hand, means contrary (ἀντί) to law or custom (νόμος). Although the metaphor in the two words is slightly different, I do not see any etymological justification for the claim that one of them is to be preferred over the other for designating the logical contradictions.

[2] That is, Bertrand Russell (1872–    ), English philosopher, coauthor with A. N. Whitehead of [PMt]. For a popular biography, see Leggett [BRP]; for a critical account of his philosophy, see Schilpp [PBR]. The paradox appeared in his [PMt], pp. 79ff. It is first treated as related to predicates rather than classes; then later (pp. 101ff.) a statement in terms of classes is given. The paradox was communicated by letter to Frege (see his [GGA. II], pp. 253ff.; cf. the footnote to Sec. C2 below).

[3] This variant appears to have been due to Russell. The subject index in Church [BSL] refers to an item (number 111.25) which is not available to me. My source is some lectures delivered by M. Geiger at Göttingen in 1928.

men in the village who did not shave themselves, and only those men. Who shaved the barber?

**Catalogue pseudoparadox.**[1] A certain library undertook to compile a bibliographic catalogue listing all bibliographic catalogues, and only those catalogues, which did not list themselves. Did the catalogue list itself?

These arguments are here called pseudoparadoxes because there is no actual contradiction. In the first case the village barber could not obey the law, which was therefore ridiculous, like that said to have been passed by an American state legislature to the effect that, when two trains approach a crossing at right angles, each one must wait until the other one passes by. Likewise, the library simply could not make a catalogue satisfying the stated requirements. But such explanations as these do not apply to the Russell paradox. In terms of logic as it was known in the nineteenth century, the situation is simply inexplicable. This is true in spite of the fact that in the greater sophistication of the present time one may see, or think he sees, wherein the fallacy consists.

**Burali-Forti[2] paradox.** This, the first of the mathematical paradoxes to be published, is of a more technical nature. It involves the theory of transfinite ordinal numbers. For readers acquainted with that theory the paradox can be stated as follows. It is shown in that theory that (1) every well-ordered set has a (unique) ordinal number; (2) every segment of ordinals (i.e., any set of ordinals arranged in natural order which contains all predecessors of each of its elements) has an ordinal number which is greater than any ordinal in the segment; and (3) the set $B$ of all ordinals in natural order is well ordered. Then, by statements (3) and (1), $B$ has an ordinal $\beta$; since $\beta$ is in $B$, we have $\beta < \beta$ by statement (2), which is a contradiction.[3]

**Cantor[4] paradox.** This paradox, although not published until 1932, was known to Cantor as early as 1899; it had a great influence in leading Russell to construct his paradox, in fact rather more than did the earlier paradox of Burali-Forti.[5] It is based on the theory of cardinal numbers. According to that theory the set of all subsets of a set $M$ has a cardinal number higher than that of $M$. This is a contradiction if $M$ is the set of all sets.

We now consider paradoxes of a different sort, involving a notion of description and definition.

**Liar paradox.** This paradox takes several forms. The simplest is that of the man who says "I am lying"; if he lies, he is speaking the truth, and vice versa. Another version is that of Epimenides the Cretan, who is alleged to have stated that all statements made by Cretans were lies, it being

---

[1] See Gonseth [MRl, p. 253].

[2] C. Burali-Forti (1861–1931), Italian mathematician, one of the collaborators with G. Peano (1858–1932) in the production of the *Formulaire de mathématiques,* which has had a great influence on mathematical logic, particularly in matters of notation. Burali-Forti's [LMt] is the most readable systematic presentation of the Peanese logic.

[3] For a study of the history of this paradox see Copi [BFP]. Copi argues that historical statements made by Fraenkel in regard to Cantor's anticipation of the paradox are not established with certainty. In particular, attention should be called to his footnote 7 regarding the allegation that the paradox was known to Cantor in 1895.

[4] Georg Cantor (1845–1918), German mathematician, founder of the theory of transfinite numbers and of the theory of point sets.

[5] There is abundant evidence of this in Russell [PMt]. See also Copi, *op. cit.*

understood that all other statements made by Cretans were certainly false. Modern versions give a statement to the effect that a proposition described in such and such a way is false, the description being constructed so as to apply uniquely to the statement itself. The paradox caused a great commotion in antiquity, and is said to have caused the death of a certain Philites of Cos. The first-mentioned form of the paradox seems to be due to Eubulides of Miletus; any actual connection of Epimenides with any form of it is doubtful.[1]

**Richard paradox.**[2] The following argument is often used to prove that the set of all numerical functions is nonenumerable. Suppose there were an enumeration; let $f_m(n)$ be the value, for the argument $n$, of the $m$th function in the enumeration. Form the function $g$ such that, for any $n$,

$$g(n) = f_n(n) + 1$$

Let $p$ be the index of $g$ in the enumeration, so that

$$g(n) = f_p(n)$$

Then

$$f_p(p) = g(p) = f_p(p) + 1$$

Since this is a contradiction, we conclude that the numerical functions are not enumerable.

Now suppose that we consider, not the set of all numerical functions, but the set of all definable ones. By 'definable' is meant, of course, definable in some fixed language, such as (mathematical) English, with a fixed dictionary and grammar. Since the number of words in the language is finite, the number of expressions is enumerable, and hence the expressions which constitute definitions of numerical functions, and thus the definable functions themselves, must also be enumerable. Now such a language can be so constituted that the above argument can be carried out in it. This again is an insoluble contradiction.

**Berry paradox.**[3] The number of natural integers which can be named in English in less than a fixed number of syllables (or letters) is certainly finite; hence there must be a least number which cannot be so named. But "the least integer which cannot be named in English in less than 50 syllables" is an English name of less than 50 syllables. Various modifications exist.

**Grelling paradox.** Among English adjectives there are some, such as 'short', 'polysyllabic', 'English', which apply to themselves. Let us call such adjectives autological; all others heterological. Thus 'long', 'monosyllabic', 'green' are heterological. Then if 'heterological' is heterological, it is autological, and vice versa.

**Skolem paradox.** This argument, although not strictly a contradiction, and therefore not relevant to the main point of this section, is appropriate for consideration here because it is frequently associated with the paradoxes and, in the broad sense, has a paradoxical character. It requires some anticipation of matters which we shall take up later. According to a celebrated theorem of Löwenheim and Skolem, any system which can be

---

[1] For references on this paradox see Sec. 1S6.

[2] For references see Church [BSL].

[3] Whitehead and Russell [PMt.I, p. 60].

formalized in the first-order predicate calculus will be such that, if it has a model at all, it will have an enumerable model.  Now various systems of higher logic and set theory, within which standard proofs of nonenumerability can be formalized, come under this theorem.  Of course, this means simply that, when a nonenumerable set has an enumeration obtained from the model, that enumeration cannot be obtained, so to speak, within the system, and therefore one can conclude that it is impossible to characterize the situation by means of a system of the kind considered.  This conclusion appears so counterintuitive to many people, that the application of the name 'paradox' seems justified.

These examples are typical of a number of paradoxes.  I shall not pause to consider more of them here, nor to discuss the great variety of attempted explanations.  There are, however, a few general remarks which it is expedient to make because of their bearing on our future program.

The English logician Ramsey[1] proposed in 1925 that paradoxes could conveniently be divided into two groups, which he called Group A and Group B. The Russell, Burali-Forti, and Cantor paradoxes are examples of his Group A; the other paradoxes here mentioned, from the liar to Grelling's, are examples of his Group B; Skolem's paradox does not fit into the classification because it is not a paradox in the same sense as the others.  Ramsey maintained that paradoxes of Group A contain only notions such as one would expect to find in a logical or mathematical system, whereas those of Group B contain notions of naming, defining, truth, etc., which are not strictly mathematical, but belong rather to epistemology, linguistics, or what not, and can therefore be ignored.  It is now customary to call the paradoxes of Ramsey's Group A *logical* paradoxes, those of Group B *semantical* (sometimes "epistemological") paradoxes.  Ramsey was not quite correct in his view that mathematics did not have to take account of the semantical paradoxes, and some of the most significant results of modern logic have come from a deeper study of them.  Inasmuch as the two kinds have been defined only by examples, the distinction is a little vague and tends to be obliterated in modern logic; nevertheless it is of some importance.

These paradoxes show incontrovertibly that logic, as it was taken intuitively in the nineteenth century, is inadequate as a final criterion of mathematical proof.  The absolute rigor, which was then thought to have been attained by the arithmetization of analysis, turns out to be a delusion. This, to be sure, does not mean that mathematics is in danger of utter collapse.  There have been crises before, and this has not happened.  Although the paradoxes have caused something of a furor, yet the number of mathematicians who concern themselves with them, and with foundational questions generally, is relatively small.  This is as it should be, for mathematicians have other things to do.  As a matter of fact, the inexact methods of the eighteenth century were used by mathematicians for deriving significant results in certain areas until quite recent times; they are still used by most engineers without affecting their ability to build Grand Coulee dams and nuclear reactors.  But the problem of explaining the paradoxes remains an important open problem.  Although there is a vast literature devoted to

[1] Died in January, 1930, when he was not quite twenty-seven years old.  See his [FML], p. 20.

them and a great variety of explanations have been offered, yet there is at present no one explanation which is universally accepted. It seems that a complete reform of logic is called for, and mathematical logic can be the principal instrument for bringing this about.

## C. THE NATURE OF MATHEMATICS

The fact that mathematical logic is to be studied by mathematical methods makes it pertinent to examine current views on the nature of mathematics. Let it be said at once that there is no unanimity on the subject. This is a healthy situation, for each point of view suggests problems and methods which the others do not. This discussion begins, in Sec. 1, by listing the different types of opinion now current in regard to the nature of mathematics. In Sec. 2, two examples will be discussed which illustrate some of the different views. Finally, in Sec. 3, there will be some critical comments.

**1. Opinions as to the nature of mathematics.** There are two main types of opinion in regard to the nature of mathematics. We shall call these contensivism and formalism. According to *contensivism*,[1] mathematics has a definite subject matter or content; the objects with which its statements deal, as they are ordinarily understood in mathematical discourse— its numbers, sets, relations, functions, etc.—exist in some sense, and those statements are true just in so far as they agree with the facts. On the other hand, from the point of view of *formalism*, mathematics is characterized more by its method than its subject matter; its objects either are unspecified or, if they are specified, are such that their exact nature is irrelevant, so that certain sorts of changes can be made in them without affecting the truth of the theorems. We must, for example, assign to formalism any view which asserts that mathematics deals with symbols, for even though a unique symbolism may be specified, no one would seriously maintain that any particular symbolism is essential. In contrast, it is characteristic of contensivism that the mathematical objects are unique.

Contensivism may be further divided into two principal species. One species, known as *platonism*[2], affirms that essentially all the notions of number and set have a real existence apart from our knowledge of them, and that classical mathematics, though it needs a more secure foundation, is not actually unsound. On the other hand, there are those who maintain that there is something rotten in the state of mathematics and that large parts of classical analysis must be discarded. It is appropriate to call this second species *critical contensivism*. Each of these two species has several varieties.

One might think that the paradoxes make some of the more extreme forms of platonism untenable. As that term has been defined here, that is not so. For in each of these paradoxes there is formulated a sentence purporting to make a statement equivalent to its own negation. If this statement is either true or false, then it is both, and we have indeed a contradiction; but we get no contradiction if we suppose that it is not a statement at all, and thus is

---

[1] The word 'contensive' was coined in [APM] to stand as translation for the German '*inhaltlich*'.

[2] This term is not to be confused with 'Platonism' as used by philosophers. The term, as used here, comes from Bernays [PMt].

neither true nor false. That this is not too preposterous a conclusion appears evident if we change the sentence in the liar paradox to "this statement is true"; there is now no contradiction, but neither is there any information in the supposed statement. Thus the contradiction does not arise from the "self-contradictory concepts" themselves, but from the properties which one affirms of them.[1] Still, the paradoxes show that, if one takes this view, one cannot take logic in a naive fashion also.

Probably platonism is the view adopted, more or less subconsciously, by most mathematicians who do not concern themselves explicitly with foundational questions. It is also the position of the pioneers in mathematical logic, Frege (see Sec. 2a) and Russell, and it is defended today by some of the ablest logicians.

The presently leading variety of critical contensivism is called *intuitionism*. This is the doctrine espoused by Brouwer[2] beginning with 1907; the name is sometimes also extended to include some similar doctrines, especially those of his predecessors. The doctrine presupposes a primordial intuition, in terms of which the human mind "constructs" the natural numbers and the continuum. Only those mathematical objects exist which the human mind so constructs. The construction is necessarily finite, so that such notions as the totality of all natural numbers cannot be considered a completed construction, but only as something which is in process of growth. Likewise, an infinite sequence is replaced by a sequence of choices, which may be entirely free and unpredictable or restricted by some law. Thus the theory has two main characteristics: (1) its constructive character, and (2) a certain metaphysical background in terms of which the ontology of mathematical entities is to be explained. It will be convenient to discuss these two aspects separately.

So far as the constructive character is concerned, the following example, due to Heyting,[3] illustrates the point. Consider the following two definitions:

1. $p$ is the greatest prime such that $p - 1$ is also prime, or $p = 1$ if such a number does not exist.

2. $q$ is the greatest prime such that $q - 2$ is also prime, or $q = 1$ if such a number does not exist.

Here it is clear that statement (1) defines a unique number, namely, $p = 3$. On the other hand, there is no way known at present of calculating the number $q$. Of course, if the "twin prime" problem were solved, the construction of the number $q$ would be complete, but until that time no construction depending on the calculation of $q$ can actually be carried out. Even an argument of the form "If $q = 1$, $A$ is true, if $q > 1$, $A$ is also true"

---

[1] Cf. [GKL], p. 515. It is worthwhile to notice that the same argument applies to void descriptions. For the notion expressed by 'the king of France' is not meaningless, since we can always tell whether a given object is the king of France or not. We get no contradiction (with the real world) if we suppose that 'the king of France' denotes an object of some kind; we only get a contradiction if we suppose that object is really the king of France (cf. [rev. Rosser]). In the language of Frege, such expressions would be said to have a sense (*Sinn*) but no denotation (*Bedeutung*).

[2] L. E. J. Brouwer (1881–    ), noted Dutch mathematician, formerly professor at the University of Amsterdam. He is distinguished also for his contributions to topology.

[3] [Int, pp. 1–2].

cannot be admitted as a constructive proof that $A$ is true, because until we know the value of $q$ we cannot decide which of the alternatives is correct, and the construction is blocked at that point.[1] It should be clear from this why the intuitionists deny the law of excluded middle for constructions involving an infinite totality. The effect of the restriction to constructive methods is to destroy a large portion of modern mathematics, and some of what remains is so changed as to be almost unrecognizable. Intuitionistic set theory, for example, is so different as to be a new subject; this effect is heightened by the fact that the terminology used is such as to intensify rather than to minimize the differences.

The intuitionist metaphysics, on the other hand, is very obscure. In Brouwer's thesis,[2] the primordial intuition is linked to an a priori intuition of time,[3] but I do not know to what extent present intuitionists regard this idea as essential. However, they still appear to postulate the following characteristics for their basic intuition: (1) It is a thinking activity of the human mind. (2) It is independent of language; the intuitionist construction has no need of being bound to any linguistic expression, and although language is necessary to communicate the results, this language can give only an imperfect reproduction of the pure thought, which alone is exact. (3) It cannot be adequately described by any prescribed rules: a proof is valid when it is a construction the individual steps of which are immediately evident; no matter what rules are given, a valid proof can be found which does not conform to them. (4) It has an a priori character, in the sense that it is independent of experience. (5) It has objective reality, in that it is the same in all thinking beings.

Before passing to the discussion of formalism, it is necessary to guard against a possible misconception. Even in the nineteenth century[4] many persons realized that mathematics often deals with "postulate systems." In such cases a branch of mathematics can be characterized by giving a set of postulates, such as those for a group, a field, a betweenness relation, a euclidean geometry, etc.; then the theorems of that branch are the statements obtained from the postulates by "logical deduction." Although the primitive ideas of such a postulate system are unspecified, yet the view in question is not formalism as here conceived. For each theorem of such a postulate system really asserts that a certain property of (or relation between) the primitive ideas, viz., the property (or relation) expressed in the statement of the theorem, is a logical consequence of those stated in the postulates.[5] The crux of the matter is then the definition of the term 'logical consequence'. Until this term has been explained, one does not have an opinion as to the nature of mathematics at all. The explanation may be

---

[1] From the constructive point of view the 'if ——$_1$, then ——$_2$' connective is to be understood (cf. Sec. 3A2) as meaning that there is an effective process (Sec. 2A5) of obtaining the conclusion from the premise. This effective construction is indeed blocked if we cannot determine which of the two premises is true.

[2] Brouwer [GLW].

[3] Perhaps akin to Kant's a priori intuition of space and time. Noneuclidean geometry upset the a priori conception of space.

[4] One might say even in antiquity. But the ideas of the ancients were different. For the development of the nineteenth-century ideas see Nagel [FMC].

[5] The "primitive ideas" are thus the bound variables in a statement of pure logic.

given from any of the points of view here sketched, and the view of mathematics is to be classified accordingly.

The best-known formalist view—by many persons considered the only form of formalism—is that of Hilbert.[1] His basic idea was that the transfinite notions of mathematics were ideal constructions of the human mind. He admitted that there were certain "finitary" intuitive reasonings which had a priori absolute certainty; the transfinite notions, which went beyond these, he considered mental creations bearing a relation to the finitary intuitive processes similar to that which imaginary numbers have to the reals. We can form such ideal creations freely, subject to only one basic limitation, namely, that we be consistent. He proposed to establish this consistency for ordinary mathematics by examining the language in which this mathematics is expressed. This language was to be formulated so completely and so precisely that its reasonings could be regarded as derivations according to precisely stated rules—rules which were mechanical in the sense that the correctness of their application could be seen by inspection of the symbols themselves as concrete physical objects, without regard to any meaning which they might or might not have. These so formalized reasonings were to be the subject of a new mathematical investigation which he called metamathematics. In metamathematics he admitted only the finitary, absolutely certain methods of reasoning. His program was to establish the consistency of ordinary mathematics by this means. Its realization would then guarantee the absolute safety of mathematics for all time.

This program received a severe setback when, in 1931, Gödel[2] showed that the consistency of a sufficiently powerful theory could not be established by means which could be formalized in the theory itself. Thus either the theory is inconsistent or it is inadequate to formalize any proof of its own consistency. This circumstance has led to a difference of opinion among modern formalists, or rather, it strengthened a difference of opinion which already existed. Some think that the consistency of mathematics cannot be established on a priori grounds alone and that mathematics must be justified some other way. Others maintain that there are forms of reasoning which are a priori and constructive in a wider sense and that in terms of these the Hilbert program can be carried out. Still others regard the Gödel argument as showing that the whole formalistic position, at least in its extreme form, is untenable. There is also a large group of logicians who argue that formalism must be supplemented by "semantical" considerations of platonistic character.[3]

**2. Examples.** In order to illustrate these notions I shall treat here two concepts of mathematics from several points of view. It will not be feasible to discuss all points of view in connection with each, but the discussion of certain typical ones will help to illuminate the above discussion.

*a. The Natural Numbers.* From a certain platonist standpoint the natural

---

[1] David Hilbert (1862–1943), German mathematician, professor at Göttingen, considered by many to be the greatest mathematician of his time.

[2] Kurt Gödel (1906–     ), Austrian mathematician (born in Czechoslovakia), professor at the Institute for Advanced Study, Princeton, N. J. For references and discussion see Secs. 3A1*h* and 3S1.

[3] For references supporting this paragraph see Sec. S4*b*.

numbers are defined as classes of mutually equivalent classes. Here two classes are equivalent just when there is a one-to-one mapping of one of them in the other. The number 1 is the class of all classes such that there is an element having the property that membership in the class is interdeducible with being equal to that element.[1] Given a number $n$, the number $n + 1$ is the class of classes such that by removing an element we form a class belonging to the class $n$. In this sense all the natural numbers are uniquely defined,[2] except for two considerations. In the first place the paradoxes force the platonist to define natural numbers of different "types," and in the second place he is worried about the possibility that there may be no classes with $n$ elements[3] if $n$ is sufficiently large, and he feels obliged to postulate an "axiom of infinity" to the effect that this is not the case.[4]

The intuitionists regard the natural numbers as objects of pure thought which are generated by the primordial intuition. This is sometimes explained as follows. We take some object of perception, abstract from its nature, and so form the idea of a unit. We conceive that this unit can divide in two, and so spawn a new unit; this new unit may in turn divide, and so on indefinitely. The natural numbers in their totality do not form a class.

A formalist would not speak of "the natural numbers" but of a set or system of natural numbers. Any system of objects, no matter what, which is generated from a certain initial object by a certain unary operation in such a way that each newly generated object is distinct from all those previously formed and that the process can be continued indefinitely, will do as a set of natural numbers. He may, and usually does, objectify this process by representing the numbers in terms of symbols; he chooses some symbol, let us say a vertical stroke '|', for the initial object, and regards the operation as the affixing of another '|' to the right of the given expression. But he realizes there are other interpretations; in particular, if one accepts the platonist or intuitionist metaphysics, their systems will do perfectly well.

Certain platonists[5] object to the formalist conception in that it does not give any explanation of the process of counting, i.e., that it gives no explanation of a statement saying that such and such a class has $n$ objects. To this a formalist—and probably also an intuitionist—would retort that it is easy to give such an explanation, viz., that the class has $n$ elements precisely

---

[1] This is frequently expressed in symbols, thus:

$$\alpha \; \epsilon \; 1 \; \rightleftarrows \; (\exists x)(\forall y)(y \; \epsilon \; \alpha \; \backsim \; y = x)$$

[2] The theory so far presented is essentially that of the German logician Gottlob Frege (1848–1925), as presented in his [GGA]. Frege had an extraordinarily keen and subtle mind, and the marks of his genius are still left deep in the heart of our subject. Unfortunately, he had a caustic disposition, and was said to have been scathing in his criticism of his contemporaries. Perhaps on that account he was slow in getting recognition. He was just beginning to be recognized when Russell wrote that Frege's system admitted the Russell paradox. The second volume of [GGA] was then about to appear. The remarks with which it closes have a tragic touch. Although he was then only fifty-five and lived for over twenty years more, he published nothing further of major logical interest. (See Church [BSL], item 49] and the biography by Church in Runes [DPh].)

[3] That is, belonging to the class $n$.

[4] Cf. Sec. D below.

[5] For example, Russell [PMt$_2$, p. vi] and Ramsey [FMt] (in his [FML], p. 2).

when it is in one-to-one correspondence with the class formed by cutting off the number series at the number $n$ (or $n-1$, if 0 is included).

*b. The Axiom of Choice.* This is the axiom of set theory to the effect that, given a class of mutually exclusive nonempty classes, there is a class containing one and only one element from each of the member classes. From the platonist point of view this is a question of fact, and the platonists are, indeed, divided into opposing camps, in which some accept the axiom of choice, others deny it, and still others accept it under certain restrictions. (The last two groups may, of course, be reckoned among the critical contensivists.) An intuitionist cannot even formulate the question and pretends he does not understand it. A formalist would say, perhaps, this: "I can formulate systems in which an analogue of the axiom of choice holds, with or without various restrictions; I can also formulate systems in which it is not assumed, and also, under certain restrictions, systems in which the axiom fails. As to which of these is the most useful system, that is a question for you to decide, possibly with reference to the purpose for which you want to use it; but they are all mathematics, and should be considered as such."

**3. Critical remarks.** The discussion in Sec. 1 was intended to set forth the facts concerning the various current views as to the nature of mathematics. Here I shall make some critical comments which are more matters of opinion.

All forms of contensivism are open to two basic objections, viz., that the criterion of rigor is vague at best, and that it rests on metaphysical assumptions of dubious character. These two objections are interrelated. It will be convenient to discuss them first in regard to intuitionism, then in regard to platonism.

Let us recall the five properties of the primordial intuition which were listed in Sec. 1. According to the third of these properties it is impossible to give an exact description of a rigorous proof. A construction is valid when it is a finite series of steps each of which is directly evident to the mind. If two minds do not agree as to what is directly evident, there is no objective criterion by which their difference can be resolved. This shows that the criterion of rigorous proof is vague. As to its metaphysical character, it suffices to note that, from certain philosophical standpoints, the five characteristics are unacceptable, at least in conjunction. An empiricist, for example, would maintain that there is no a priori knowledge, and therefore no a priori intuition which can be used as a criterion of truth. Likewise, some thinkers have defended the thesis that there is no thought without language. Finally, even if we were to grant the existence of an intuition having the first four characteristics, it seems very doubtful if it would have the fifth, and without that fifth characteristic intuitionist mathematics is not an objective science at all. Thus the intuitionist requires the existence of an a priori intuition which is objective and prelinguistic, and although this ontological assumption is agreeable to certain types of philosophy, yet it is an assumption for all that, and one which from other philosophical viewpoints is highly dubious and metaphysical.[1]

In regard to platonism, it hardly needs saying that to ascribe reality to all the infinitistic notions of mathematics is a metaphysical assumption which is

[1] Cf. [OFP], p. 6.

highly repugnant to certain types of minds—including those of the intuitionists. Thus the situation in regard to metaphysical character is rather worse for platonism than it is for intuitionism. In regard to the criterion of rigor, however, the situation is rather better, because most platonists admit the possibility of formalization (in a sense to be explained presently). Indeed, most platonistic theories have no other criterion of rigor than that derived from the formalization, and the criterion is precise insofar as the formalization is strict and thorough.

The detailed study of a certain species of formalism will concern us in the next chapter. There we shall see that formalism can give a conception of rigorous proof which is objective and precise and that this conception is free from metaphysical assumptions of the sort just discussed for platonism and intuitionism. Since it is desirable to conceive of mathematics as an objective science free from metaphysical assumptions, a formalist point of view is adopted in this volume.

It is, however, necessary to make some further remarks in order to clarify the position. Naturally, these remarks may need further clarification later.

In the first place, the incompleteness theorems of Gödel, mentioned at the end of Sec. 1, show that no single formal theory can exhaust mathematics. Accordingly, formalism is not to be identified with the view that such a theory exists. Rather, the species of formalism here adopted maintains that the essence of mathematics lies in the formal method as such, and that it admits all sorts of formal theories as well as general and comparative discussions regarding the relations of formal theories to one another and to other doctrines. In this sense *mathematics is the science of formal methods*.

A formalist theory contains by definition certain unspecified elements or parameters, or, as already said, elements which can be changed. If these elements are determined in a particular contensive way, then the interpreted theory becomes a contensive theory. Let us call such a theory a formalized contensive theory. Furthermore, if the true statements of this formalized theory are acceptable from a certain contensive viewpoint, let us say that the formal theory is compatible with that viewpoint. Thus we may have formal theories which are compatible with platonism; others which are compatible with intuitionism. Now there is a fundamental difference between such a formalized contensive theory and a contensive theory which is primary. For a proof or other argument arrived at in the formal theory remains valid regardless of the interpretation. If rigor is defined by the formal theory—as it is for many forms of platonism—that rigor remains an objective fact whether one accepts the presuppositions of the contensive theory or not. Moreover, formalisms which differ in these respects may be considered simultaneously. Thus formalism is not a doctrine which excludes or denies the doctrines of this or that sort of contensivism; it is equally independent of, and compatible with, any reasonable form of contensivism.

It follows from this that what was said above in criticism of the metaphysical assumptions of various forms of contensivism does not refute the positions of those doctrines. It was not intended to. It was simply maintained that a definition of mathematical truth ought not to depend on such metaphysical assumptions. Those who accept these assumptions will naturally prefer formal theories which are compatible with them. In any event such formal theories show what follows from adopting the assumptions in question.  Since formal systems compatible with different points of view can be considered simultaneously, it is conceivable that one might adopt different theories for different purposes.  Thus platonistically oriented formalizations for mathematics appear to be simpler and more in agreement with ordinary mathematical procedures, and thus more suitable for ordinary theoretical purposes, whereas intuitionistically oriented ones are appropriate where the actual carrying out of operations is important, as for example in recursive arithmetic and in operational theories of physics.  Formalism does not prejudge such questions; it admits the possibility of various kinds of mathematics existing side by side.

It was said above that intuitionism had two aspects, its metaphysical aspect and its constructivism.  The latter is independent of the former.  The notion of constructibility is of great importance for mathematics.  The intuitionists are quite right in emphasizing this importance, and they have certainly had a hand in developing it.  However, they are not solely responsible for it.[1]  The conception of finitary construction which is basic to Hilbert's metamathematics is actually more stringent.  This is shown by the following circumstances.  About 1930 Heyting gave a formalization of arithmetic which was compatible with intuitionism; somewhat later Gödel[2] showed that classical arithmetic could be interpreted in intuitionistic arithmetic; this gives an intuitionistic proof of the consistency of classical arithmetic, whereas a strictly finitary proof would contradict one of Gödel's incompleteness theorems.  Thus there is, from the finitary standpoint, a nonconstructive element in the intuitionistic arithmetic; just where this enters I do not know.  Modern improvements of the constructivity notion have been made by persons who are primarily formalists.  Thus the development of the notion of constructibility is due to an interaction between intuitionists, formalists, and possibly others, in which all parties have played a part.

Many of the theses advanced by the intuitionists, when stripped of their metaphysical accompaniment, are acceptable to formalism.  That this is true for constructivism was the theme of the preceding paragraph.  As for the notion that the concept of intuitively valid proof cannot be exhausted by any single formalization, we have noted already that the Gödel theorem shows just this, and thus that mathematical proof is precisely that sort of growing thing which the intuitionists have postulated for certain infinite sets.  We can admit also that intuition of some sort enters into mathematics, provided that we are allowed to consider this intuition as essentially one of linguistic nature, or a natural development of experience without regard to a priori truth.  Only the thesis that mathematics with another sort of motivation is necessarily worthless is not acceptable.

Again, it is not claimed that formalism is absolutely free from all assumptions which might be called metaphysical.  In the theory presented here, one may conceive such assumptions as entering in certain abstractions.  The first of these is involved in the use of such terms as 'symbol' and 'expression'; these denote, not individual marks on paper or the blackboard—which

---

[1] For a claim on their behalf see Brouwer [IBF].

[2] In his [IAZ].

are called *inscriptions*—but classes of such inscriptions which are "equiform." Thus the same expression may have several "occurrences." The doctrine that only concrete individuals can be taken as objects is known as *nominalism*. It is probable that a nominalistic account of formalism can be given, but that is not attempted here. The second abstraction is with respect to the limitations of time and space. We assume that any process is constructive for which the number of operations is finite, even though there might not be space enough in the universe or time enough in the cosmic order to actually do so. Thus we have an idealization of experience such as we have in almost all branches of science. A third abstraction, by virtue of which infinitely many acts may be conceived of as completed, is characteristic of a nonconstructive point of view. This abstraction enters in modern "semantical" discussions about formal theories. Although such mixtures of formalism and platonism play an important role in modern logic, yet in this volume, which is devoted to the foundations of mathematical logic, they are not considered.

Finally, we take up the question of to what extent absolute certainty attaches to mathematics. The search for absolute certainty was evidently a principal motivation for both Brouwer and Hilbert. But does mathematics need absolute certainty for its justification? In particular, why do we need to be sure that a theory is consistent, or that it can be derived by an absolutely certain intuition of pure time, before we use it? In no other science do we make such demands. In physics all theories are hypothetical; we adopt a theory so long as it makes useful predictions and modify or discard it as soon as it does not. This is what has happened to mathematical theories in the past, where the discovery of contradictions has led to modifications in the mathematical doctrines accepted up to the time of that discovery. Why should we not do the same in the future? Using formalistic conceptions to explain what a theory is, we accept a theory as long as it is useful, satisfies such conditions of naturalness and simplicity as are reasonable at that time, and is not known to lead us into error. We must keep our theories under surveillance to see that these conditions are fulfilled and to get all the presumptive evidence of adequacy that we can. The Gödel theorem suggests that this is all we can do; an empirical philosophy of science suggests it is all we should do. Moreover, since usefulness is relative to a purpose, differently constituted theories may be accepted for different purposes, so that an intuitionistic and a classical mathematics may stand side by side.

### D. MATHEMATICS AND LOGIC

A reader with some acquaintance with the recent expository literature on the foundations of mathematics will perhaps be surprised at the absence of any mention of "logicism"[1] in the foregoing discussion. Logicism, which is reputed to be the view that mathematics is reducible to "pure logic," has been frequently taken, along with formalism and intuitionism, as one of the three leading schools of thought in regard to the nature of mathematics.[2] On closer examination, however, one sees that it is vague, inasmuch as the

---

[1] Also called "logisticism." The accompanying adjective used here will be 'logicistic'.
[2] Especially during the thirties; see, for example, Hardy [MPr].

term "pure logic" is undefined.[1]  In fact, logicism is not a unified view as to the nature of mathematics at all, but a special thesis as to the relation of logic to mathematics.  That thesis merits some discussion; this will bring us back to the theme of Sec. A.

First let us be clear in regard to the historical facts in regard to the term 'logicism'.  The term was originally applied to the systems of Frege and Russell.  Those authors perceived the fact (which we have already noticed in Sec. C1) that a mathematical theorem in an axiomatic system can be regarded as a statement of logical consequence.  They observed, however, that besides the terms which function as bound variables in such a statement, there are also constants; e.g., the word 'two' in "a quadratic equation has two roots" has a fixed meaning.  They advanced the thesis that all the constants occurring in mathematics could be defined in terms of certain "logical constants"; in particular, the natural numbers were defined as properties of classes as explained in the discussion of the first example in Sec. C2.  In this way any mathematical theorem could be explained as a statement of "logic."  The logic in terms of which this explaining was to be done was definitely platonistic, and thus their theories are to be so classified.[2]  In the course of time the term 'logicism' came to be applied to the work of other persons who drew their inspiration from them, such as Ramsey, Wittgenstein, Lewis, Carnap, and Quine.  The theories of the later authors in this list are distinctly formalistic.  Thus 'logicism' is the name for a group of theories which are historically connected, rather than for a type of view parallel to those considered in Sec. C.  There are, moreover, certain systems whose status as related to logicism is doubtful, and it is not clear that different writers using this term include under it exactly the same doctrines.

The various theories just mentioned do, however, have certain features in common.  The logic to which mathematics reduces is not philosophical logic in the sense of Sec. A, but a special kind of mathematical system.  They accept the definition of natural number à la Frege; in this respect they contrast with systems which take the natural numbers as primitives (e.g., Hilbert's), as well as with those which define them in other ways (e.g., set theory,[3] combinatory logic[4]).  Again the logic is essentially classical; i.e., it is based on two-valued propositional algebra and avoids the paradoxes by denying significance to certain constructions violating a "theory of types" or some modification of it.  Finally, most of them have the conviction that logic is, in principle, unique.  These will be taken here as characteristic traits of logicism.[5]

---

[1] We have already noticed a similar point at the beginning of the discussion of formalism in Sec. C1.

[2] In regard to Frege, this is the prevailing view.  For a dissenting opinion, see Bergman [FHN].  In regard to Russell this is very clear, e.g., in the introduction to the second edition of his [PMt].

[3] In abstract set theory (cf. Sec. S4b), one can define 0 as the null set and $n + 1$ as either the set whose only element is the number $n$ (Zermelo) or the set consisting of all natural numbers $\leq n$ (Von Neumann).  The latter has the advantage that the set representing $n$ has always exactly $n$ elements.  Usage seems to vary as to whether abstract set theory is regarded as logicistic.

[4] Cf. [CLg], pp. 6ff. and 174, and the references cited in the former place; the subject is to be treated more extensively in volume II of [CLg].

[5] This removes some, but not all, of the vagueness noted at the end of the preceding paragraph.

Early objections to logicism were that it required—at least in Russell's version—axioms which it was not natural to classify as "logical." A notable example was the "axiom of infinity," which asserts that there are infinitely many "individuals," i.e., objects of lowest type.[1] Later the theorem of Gödel showed that no single formalized system of logic could be adequate for mathematics, so that, if the thesis of logicism is to be upheld at all, it must be with respect to some sort of a hierarchy of logics rather than to a single one. Finally, one can significantly consider systems which differ from the classical ones in that they can be so interpreted that certain laws of the classical logic are invalid; the interpretation of these from the standpoint of logicism is, to say the least, highly unnatural.

Now, from the standpoint of formalism—and incidentally from some other points of view as well—one can characterize a mathematical system objectively without presupposing anything which it would be natural to call "logic." This will, indeed, be the business of the next chapter. It is this fact which makes mathematical logic possible, since otherwise it would be circular to apply mathematics for the purposes mentioned in Sec. A. The systems proposed by the logicists (at least if they are made sufficiently precise) are special cases of such mathematical systems. Thus what may well be regarded as a part of the logicist thesis—viz., that mathematics can be significantly applied to logic, and thus that certain mathematical systems are logical in nature—is in agreement with the position taken here. Whether one should go further and identify mathematics and logic is a matter of how the latter subject is to be defined. The position of this book is that the somewhat vague statements made in Sec. A are sufficient for a working definition, and from that point of view mathematics and logic are not to be identified.

The related question of how mathematical logic is to be characterized within mathematics can be answered best after we have finished our work.[2] Clearly, one necessary characteristic of a logical system is that it be so explicitly formulated that it does not take logic for granted; how this may be done we shall learn in due course. Ordinary mathematics may be based on some logic which it does take for granted, and there may, conceivably, be different such logics for different purposes. The task of mathematical logic should be to investigate these logics and their relations to one another; in particular, to develop techniques which the mathematician can use with confidence and with awareness of the special nature of the logic that is the basis (e.g., whether an axiom of choice is involved). This is one way in which mathematical logic can contribute to the progress of mathematics as a whole.

[1] The term 'individual' means here an object which, so far as the system is concerned, is not a set; the totality of these constitutes the lowest type. In abstract set theory there is also an "axiom of infinity," but it has a different significance. As that theory is ordinarily presented, there are no individuals, let alone infinitely many; there are infinitely many sets, and the axiom of infinity says that a certain infinite totality of such sets constitutes a set. In either case the axiom affirms the existence of sets (or classes) with infinitely many elements.

[2] Various attempts have been made to characterize more precisely those mathematical systems which are logical, but none of these has met with sufficiently general acceptance to be adopted here.

## S. SUPPLEMENTARY TOPICS

The following indications are intended to assist the reader in finding material in the current literature which supplements this book. References are made to the Bibliography at the end of the book according to the explanations made there. It will be convenient to classify the citations under the following heads: (1) systematic general treatments of mathematical logic which are still of current interest; (2) bibliographic aids, including journals, sources of reviews, bibliographies, etc.; (3) historical material, including classics and material not now current, although there may still be much to be learned from them; (4) material related to special parts of mathematical logic which are not treated in this book and often are not treated in the general works cited in Sec. 1; (5) source material for this book in general and alternative or amplifying treatments of the same subject matter; and (6) material related specifically to this chapter. The reader should bear in mind the natural limitations of a list of this kind, and should supplement it by reference to general sources of bibliographic information. The citations given here pertain to this book as a whole or to this chapter; references pertaining solely to one of the later chapters will be given with that chapter.

**1. General systematic works.** These will be presented under several subheads, as follows:

*a.* The following are the more important general treatises, each of which is a standard presentation of some viewpoint: Church [$IML_2$]; Hilbert and Bernays [GLM]; Kleene [IMM]; Lorenzen [EOL]; Quine [MLg]; Schütte [BTh]. In the same list should be included works by Carnap, but to get a complete outline of his more strictly logical contributions, one would need several titles, viz., [LSS], [ISm], [FLg], [MNc]. The monumental work of Whitehead and Russell [PMt] belongs now under Sec. 3. Some items here classed under Secs. *b* and *e* may also belong in this group.

*b.* Somewhat more elementary and textbooklike, but still containing original work, are Fitch [SLg]; Ladrière [LIF] (an extensive survey); Prior [FLg]; Quine [MeL]; Rosenbloom [EML]; Rosser [LMt].

*c.* There are certain elementary works, chiefly remarkable for their expository achievements, which nevertheless present some features of interest for advanced students. Those available in English (some of them appeared originally in or have been translated into other languages) are Carnap [ISL]; Goodstein [MLg]; Hilbert and Ackermann [PML]; Leblanc [IDL]; Nagel and Newman [GPr]; Quine [ELg], [MeL]; Rosenbloom [EML]; Suppes [ILg]; Tarski [ILM]; Wilder [IFM]. In German there are Carnap [ESL]; Hilbert and Ackermann [GZT]; Lorenzen [FLg]; Scholz [VGZ] (rather more voluminous). In other languages there are Blanché [ILC], Feys [Lgs], Mostowski [LMt], Novikov [EML]. Tarski [ILM] is an expansion of a work which appeared originally in Polish; it has been translated into several languages. This list does not exhaust the class of elementary texts, but is a selection of what seem to me the most appropriate items. I have not listed items about which I have inadequate information.

*d.* The following are shorter surveys or summaries: Hermes [GLM], Hermes and Markwald [GLM], Hermes and Scholz [MLg], Schmidt [MGL], Mostowski [PSI]. The first four of these are in German; the last is available in

English, German, Polish, and Russian. There are articles by competent reporters in the Encyclopaedia Britannica (see review in *J. Symb. Logic*, **23**: 22–29), Encyclopedia Americana (see review in *J. Symb. Logic*, **23**:207–209), and (mostly by A. Church) in Runes [DPh]. Bochenski's [PLM], now available in English, French, and German, is a handbook of formulas and definitions, with references.

*e.* In the nature of critical reviews, or more inclined to a philosophical viewpoint, are the following: Beth [FMt], Black [NMt], Dubislav [PMG], Fraenkel and Bar-Hillel [FST], Ladrière [LIF], Strawson [ILT], Waisman [EMD], Weyl [PMN], Wilder [IFM]. Some of these may well have been included under Sec. *a.* The work of Wilder is intended for mathematical students at the level of an American baccalaureate major.

**2. Bibliographic aids.** There are at the present time six journals primarily devoted to publishing original articles and reviews in the field of mathematical logic, viz., the following: *Archiv für mathematische Logik und Grundlagenforschung*, West Germany; *Journal of Symbolic Logic*, United States; *Logique et analyse*, Belgium; *Notre Dame Journal of Formal Logic*, United States; *Studia Logica*, Poland; *Zeitschrift für mathematische Logik und Grundlagen der Mathematik*, East Germany. A great variety of other mathematical and philosophical journals also publish work in the field; there are publications of one sort or another in over twenty languages.

The *Journal of Symbolic Logic* also publishes reviews of practically all the current literature. The reviews are indexed by authors every two years and by subjects every five years. Reviews of the more mathematical items are also published in *Mathematical Reviews* and *Zentralblatt für Mathematik und ihre Grenzgebiete*. The reviews in the *Journal of Symbolic Logic* are apt to be more critical and fuller, but they often appear later, and related fields of mathematics are not always adequately covered. Reviews, mostly in the Polish language, are also published by *Studia Logica*. There is a review journal *Referativny Zhurnal Matematika* published in the Soviet Union.

Book reviews, of course, appear in a number of mathematical, philosophical, and scientific journals.

An extensive bibliography compiled by A. Church [BSL] appeared in 1936. With supplements published later, it is virtually complete from 1666 to the end of 1935. The review section of the *Journal of Symbolic Logic* keeps this bibliography up to date. Other extensive bibliographies are found in Fraenkel [EML₃], [AST]; Fraenkel and Bar-Hillel [FST]; Ladrière [LIF]. An older bibliography which may possibly be still useful is in Lewis [SSL]. More selective and classified bibliographies appear in Beth [SLG], Bochenski [PLM], Church [BBF], Hermes and Scholz [MLg], Leblanc [IDL], Schmidt [MGL].

**3. Historical material.** It will be convenient to consider separately the recent history and that before, let us say, 1920. In the former case one must consult recent writers who have published incidental accounts of it; in the latter case enough time has elapsed so that systematic historical works exist. Naturally there is no sharp line of demarcation between these two.

*a.* There is a great deal of historical information of a semisystematical sort in Beth [FMt]. This relates particularly to the work of various schools, their interactions with one another and with philosophical ideas. Somewhat the same thing can be said for writings of Fraenkel, especially Fraenkel and Bar-Hillel [FST], but the general tenor is more mathematical and the bibliographies are very nearly exhaustive. In a different sense Church [IML$_2$] is a rich source of information, largely in the form of very extensive footnotes and remarks relating to the original contributors of mathematical ideas. Scholz [MJL], discussing the work of Łukasiewicz, also gives incidentally a lot of information of this kind. One may also mention Kotarbinski [LPO] (for developments in Poland); Yanovskaya [OMM], [MLO] (for developments in Russia). See also the general surveys mentioned in Sec. 1e.

b. For the older period there are several general treatments of the history of logic. Those which I have found most interesting are Bochenski [FLg], [AFL]; Jørgensen [TFL]; Lewis [SSL] (for the period from Leibniz to 1918); Moody [TCM]; Boehner [MLg] (the last two for the Middle Ages). A brief bibliography of works on the history of logic is given in Church [BBF] and in Hermes and Scholz [MLg]; a more extensive one in Bochenski [FLg].

c. Selected lists of important works which are too old to be included in Sec. 1 are found in Beth [FMth], [SLG]; Church [BBF]; Hermes and Scholz [MLg]; Leblanc [IDL]. In Church [BSL] a small number of works which the compiler thought particularly important are marked with a star. Some of these classics have been made more accessible by recent publication of editions of Frege's works in German and English, and by that of the second and third volumes of the works of Peano, containing most of his early writings on logic.

**4. Special topics.** In order to give an idea of the place of the subject of this book in mathematical logic as a whole, I shall mention here some of the ways in which mathematical logic goes beyond the foundations. The discussion and references given are, in general, intended to give an approximate idea of the direction of the investigation, and not a full and technically correct account of achievement; furthermore, not every possible direction has been reported.

This book relates to the first-order "predicate calculus," in which we consider propositional functions ranging over a fixed range of "individuals," such that no proposition or propositional function is ever considered as itself belonging to the range; furthermore, the methods used are strictly constructive and limited as to their complexity. One can go beyond this in two directions. On the one hand, one can introduce calcululuses of higher order, in which propositions or propositional functions (and therefore sets, etc.) can appear as arguments to other functions. On the other hand, one can use methods which go beyond those used here. In some ways intermediate between these are systems in which numbers are explicitly introduced into the domain of arguments. In Secs. a to c below I shall discuss briefly the three principal varieties of higher-order logicistic calculus; in Sec. d the direction in which natural numbers are taken as primitives; and in Sec. e some miscellaneous matters. The discussion of higher methods will be deferred to Sec. 3S3.

a. *Theory of Types.* The basic principle of this theory is that the logical notions (individuals, propositions, propositional functions) are classified into a hierarchy of "types" and that a function can take as arguments only notions which precede it in the hierarchy. There are many different forms.

The classic in this direction is "Principia mathematica" by Whitehead and Russell; it was preceded by a number of writings (see Church, [BSL, under Russell], also, for an anticipation, Schröder [VAL] and Church [SAS]), and was based on the work of Frege and Peano. For an elementary exposition of its underlying philosophy see Russell [IMP]. The form of the theory proposed in Whitehead and Russell [PMt] is known as the "ramified" theory of types; Ramsey [FMt] proposed to reduce it to the "simplified" theory, and this idea has been followed in most later theories. There has been no comprehensive treatise since Whitehead and Russell [PMt]. For critical reviews see Fraenkel and Bar-Hillel [FST] and Beth [FMt]; a brief sketch is given in Hilbert and Ackerman [GZT] and in Wang [SAT, pp. 11–15].[1] Church [FST] gives an exact formulation of one part of it. Henkin [CTT] made a study of completeness. In some forms of the theory the hierarchy may be transfinite.

There have been several recent variants. To be reckoned as such, even though they have departed considerably from the original position, are those of Lorenzen [EOL], Wang [FMt], and Schütte (see, for example, his [BTh], chap. 9); even the theories of Quine, although here classified separately (see Sec. c), were motivated by it. The Japanese mathematician G. Takeuti has proposed a Gentzen-like formulation of the theory which he calls the GLC (general logical calculus). His "fundamental conjecture" is to the effect that the elimination theorem (i.e., the theorem on the eliminability of cuts, Sec. 5D) holds for this calculus. Somewhat similar ideas have been advanced by Kuroda. This is an area in which research at the present time is very active. In view of Gödel's theorem, one would not expect to establish the fundamental conjecture by constructive methods.

*b. Axiomatic Set Theory.* The basic characteristics of this theory are as follows: (1) propositional functions are taken extensionally, i.e., they are identified when they have always the same truth values for the same arguments; (2) propositional functions of more than one argument can be reduced to functions of one argument, i.e., to classes; (3) there is a class, whose elements are called sets, such that a class can be an element of another class if and only if the former is a set; (4) sets are characterized genetically, according to their construction, in such a way that classes which are too extensive, e.g., the class of all sets, cannot (assuming consistency) be shown to be sets. Typical axioms are the power axiom, to the effect that the class of all subsets of a set is a set, and the selection axiom (*Aussonderungsaxiom*), to the effect that, given any set, any subclass of that set is a set.

Axiomatic set theory appears to go back to the naive set theory which Cantor developed in the last decades of the nineteenth century, just as the theory of types appears to go back to Frege—both the work of Cantor and that of Frege admitted inconsistencies. Cantor's principal papers have been collected and published (see Cantor [GAb]). The first axiomatic theory was proposed in Zermelo [UGL]. This was later extended, improved, and modified by Fraenkel, Skolem, von Neumann, Bernays, Gödel, Ackermann, and Azriel Levy. For a brief summary of this development see Wang

[1] There is also some elaboration of a form of the theory in Carnap [ISL].

[SAT]; for a full historical and critical account see Fraenkel and Bar-Hillel [FST], cf. also Beth [FMt, chap. 14]; for the latest forms of the theory see Gödel [CAC], Bernays and Fraenkel [AST], and various publications of Levy; for a textbook see Suppes [AST]. For the higher technical developments of the theory without regard to the axiomatic foundations, including the theory of transfinite numbers, see Hausdorff [GZM] (the first edition contains material which was left out of the revisions); Kamke [MLh]; Sierpinski [LNT], [CON]; Bachmann [TZh]. A large part of the current discussion in the theory is concerned with the role of the axiom of choice, the continuum hypothesis, and assumptions relating to the existence of very large numbers; the relative consistency of the first two assumptions was the theme of Gödel's [CAC].

c. *Quine's Theories.* The third sort of higher calculus is that proposed by Quine. This may be regarded as a synthesis of the theory of types and axiomatic set theory. After making a study of the theory of types (see, for example, his [SLg]) and of set theory (in a series of papers in the *Journal of Symbolic Logic*) he made a fundamental suggestion in his [NFM]. This was that in the most essential axiom scheme of set theory, the selection axiom, we replace the principal requirement, viz., that the class to be shown to be a set be a subclass of a preexisting set, by the requirement that its defining property satisfy the restrictions of the theory of types for some possible assignment of types to its constituents. Quine calls a property satisfying this requirement a "stratified" property; then his form of the selection axiom is, roughly, to the effect that every stratified property determines a set. Quine did not develop the theory proposed in his [NFM] in detail; this was only done much later in Rosser [LMt]. Quine himself adjoined further assumptions, and on this basis developed his treatise [MLg]. The first edition of this work was found to be inconsistent by Lyndon and Rosser (see Rosser [BFP]); but it was found possible to repair this, and the second edition is not known to be inconsistent.

The Quine systems have a number of strange features. Unlike any of the systems of axiomatic set theory, the systems contain a universal set, and the complement of any set is also a set. On the other hand, we are saved from the Cantor paradox (Sec. 1B) by the fact that the class of all unit subsets of a set cannot be shown to be a set which is in $1 = 1$ correspondence with the given set. Specker, a Swiss mathematician, has studied the system; in particular, in his [ACQ], he disproved the axiom of choice for the system of Quine [NFM].

d. *Arithmetical Systems.* In the preceding directions the natural numbers and other arithmetical notions are usually defined in terms of sets (or some equivalent notion). This is one of the characteristics of logicism as noted in Sec. D. We shall now turn attention to a group of directions in which the natural numbers, for one reason or another, are regarded as primitives.

The directions corresponding to this specification include some strange bedfellows. On the one hand, the intuitionistic theories answer to the description; so do the theories of Hilbert, leading to the system Z of Hilbert and Bernays [GLM] and to its generalizations. For both of these directions, in their present state, a report is deferred to Sec. 6. Here it will be appropriate to make a brief report about background studies, which, although made at an early date, have still an appreciable effect on present-day work in the foundations of mathematics.

The principal background study referred to is that of Dedekind [WSW]. This is the grandfather of present-day recursive arithmetic (see Sec. 3S3). Peano, in his "Arithmetices principia," developed a symbolic theory of the natural numbers; in this he acknowledged indebtedness to Dedekind and also to a work of H. Grassman which appeared in 1861. The postulates stated by Peano in his [APN] and [CNm] have become a part of mathematical tradition under the name of "Peano postulates." These postulates form the basis, for instance, of the development of arithmetic for the purposes of ordinary mathematics (without formalizing the logic) which is contained in Landau [GLA]. Hilbert's system $Z$ is essentially the addition of these postulates, together with recursive definitions of the sum and product, to the first-order predicate calculus (with equality).

After Peano the next important step was Skolem [BEA]. This paper introduced the idea that constructive arithmetic could be developed without quantification. From this idea modern recursive arithmetic took its start (cf. Sec. 3S3).

The work of Lorenzen and Schütte, mentioned under Sec. $a$ because they use types, or "*Schichten*," may also be considered as belonging to the present direction.

  *e. Other Directions.* Not all the directions of higher logical investigation can be made to fit under the preceding heads. Indeed, we have just noticed that the work of Lorenzen and Schütte can be included under type theory only with some forcing. Here we shall be concerned with some theories which are even more difficult to classify.

Combinatory logic is one of these theories. It is, in fact, a comparative treatment of fundamental matters pertaining to all directions. Church's theory of $\lambda$ conversion is regarded in this book as a part of combinatory logic. For a brief description and references see Secs. 3D4 and 3D5.

Behmann, in his [WLM], proposed an explanation of the paradoxes as resulting from definitions which could not be eliminated. A system based on this idea appeared only years later, in his [PKL]. In spite of apparent differences, this seems closely related to Church's $\lambda$ conversion. In the meantime, Ackermann and Schütte proposed systems of type-free logic which appeared to have been influenced, initially at least, by Behmann. For a report on these see Schütte [BTh] and references there given.

Another system of type-free logic is the "basic logic" of Fitch. The latest publications on this system are his [QCF] and [EVE]. These refer to previous publications.

The Polish logician Leśniewski (1886–1939) developed a system which was evidently one of great subtlety. It seems to have had a nominalistic tendency. This system, however, is very difficult of access. The only accessible publication of Leśniewski (his [GZN]) is scarcely intelligible. No one except those who have had direct contact with the author professes to understand his work, and a presentation of the views of the master which satisfies all his pupils has not yet been made. For such information as is available, see Słupecki [SLP], [SLC], [GML]; Grzegorczyk [SLR]. Kotarbinski [LPO] gives a general report on Polish logic, with emphasis on the philosophical side and with almost no technical details. See also Jordan [DML].

Mathematical logic is beginning to be applied directly in several fields.

The application to linguistics will be mentioned later (Sec. 2S1) in the discussion of semiotics. There are also applications to the design of electrical-circuit networks, automata, cybernetics, etc.

**5. Source material.** This book is based on a lecture course at the Pennsylvania State University. The part of it which concerns the Gentzen methods is a revision of [TFD]; the purely algebraic part is based on [LLA]; and the part relating to formal methods in general on [CLg], chaps. 1 and 2. These works contain references to sources; so far as formal methods are concerned, a rather full statement of source materials was given in [CLg], sec. 1S1. I shall therefore confine myself to the most important sources, to material too recent to be included in the previous bibliographies, and to important sources of collateral information (see also Sec. 5S1).

In regard to formal methods in general, one should consult Smullyan [TFS] and Lorenzen [EOL]. The ideas given here may also be compared with the definition of a "*Kodifikat*" in Schmidt [VAL]. The articles [CFS], [IFI], [DFS] contain discussion of special points which will be cited in the proper place.

In regard to the inferential methods, the ultimate source is Gentzen [ULS]. This has recently been translated into French with brief notes by Feys and Ladrière. A bibliography and brief historical statement are given in [IAL]. Works containing important information about it are Kleene [IMM, chap. 15 and to a lesser extent, chaps. 4–8]; Kleene [PIG]; Quine [MeL], [NDd]. Bernays [LC1] gives a brief informal presentation of the T rules. Fitch [SLg] uses a technique of similar nature. Beth's semantic tableaux constitute, in some respects, a refinement of the Gentzen rules. They are described in his [FMt], chaps, 8, 11, and 15; for the sources see his [SEF], [SCI]. The German writers tend to shy away from the Gentzen technique and to devise ways of modifying the ordinary formulations so as to obtain its advantages without its formal machinery. This is much the same as if one attempted to develop group theory without introducing the abstract group operation, on the ground that in any practical case the group operation could be defined in terms of the other operations of the system with which one is dealing. In other words, they insist on interpreting what are here, and in Gentzen, taken as primitives, and they must, of course, have separate interpretations for the classical and nonclassical theories. With this understanding the works of Schütte (see his [SVS], [SWK]) and H. A. Schmidt (see his [VAL]) are references on the Gentzen methods and make some contributions even to the abstract theory. Connection with Gentzen methods can be seen in Schütte [BTh] but is less obvious. The work of Lorenzen, although he hardly mentions Gentzen's work and appears not to be well acquainted with it, is suggestive. Church [IML$_2$] devotes only a few exercises to inferential methods.

In regard to the more conventional deductive treatment of propositional algebra and predicate calculus, see especially Church [IML], Hilbert and Bernays [GLM], Kleene [IMM], Schmidt [VAL]. The first of these treats one form of deductive development and is rich in historical and critical comment about other developments down to 1951. The second and third are concerned with what are here (from Chap. 5 on) called H systems; Hilbert and Bernays [GLM] is, indeed, a primary source for much of their theory, but contains relatively few references; Kleene contains some modern improvements, showing the influence of Gentzen, and pays attention to demonstrability in an intuitionistic system. ˙ Schmidt [VAL] is a rather encyclopedic treatment of the propositional algebra (a second volume is to treat the predicate calculus); it is perhaps too much concerned about being readable without the use of pencil and scratch paper and is rather meager in its historical material, but it contains a lot of information and detailed proofs which are not easily available elsewhere.   Quine treats propositional calculus from the standpoint of tautologies, and his [MeL] is a standard reference for that approach; his quantification theory has peculiar features which will come up for consideration later (Chap. 7).   For the standard epitheorems of predicate calculus see Church [IML$_2$], Kleene [IMM], Hilbert and Bernays [GLM]; they are treated also from a little more elementary point of view in Hilbert and Ackermann [GZT].

**6. References for this chapter.**   The discussion of Sec. A has been used for years as an introduction to the course mentioned in Sec. 5; it appears also at the head of [LLA].

The literature in regard to the paradoxes is enormous.   For general and comprehensive discussions, see Beth [FMt, chap. 17], Fraenkel and Bar-Hillel [FST, chap. 1].   The latter gives (pp. 16–18) an almost exhaustive and classified bibliography.   Of the critical literature I have found Stenius [PLA] and Specker [AML] especially stimulating.   Kleene [IMM, sec. 11] discusses a few paradoxes with some insight.   There is also a brief discussion, with emphasis on the liar paradox, in Church [PLg].   Kempner [PCS] gives a semipopular treatment.   Wilder [IFM] discusses the Russell paradox and its significance for mathematics; this is interesting because the author is not primarily a logician.   For the liar paradox, in particular, see Beth [FMt, p. 485]; Fraenkel and Bar-Hillel [FST, p. 11]; Church [PLg]; Tarski [SCT], [WBF]; Bochenski [FLg, secs. 23, 25]; Prior [ECr]; Kleene [IMM p. 39]; Weyl [PMN, p. 228].

For the discussion of the nature of mathematics in general see [OFP], Kleene [IMM, chap. 3], Weyl [PMN], Beth [FMt, parts V and IX], Fraenkel and Bar-Hillel [FST], Wilder [IFM, part II], Black [NMt].

For the platonistic conception of mathematics, Russell [IMP] is an elementary exposition; for further details see the references in Sec. 4$a$.   For critical comment see also Bernays [PMt]; Gödel [RML], [WIC].

For intuitionism the authoritative recent work is Heyting [Int.]   This is, however, chiefly an exposition of intuitionist mathematics, with little regard to either the deeper intuitionist philosophy or its history.   To supplement it see Heyting [CIL], [FMI] (gives history and references and includes discussion of the predecessors of intuitionism, including Kronecker, Poincaré, Borel, etc.); Brouwer [HBP]; Beth [FMt, chap. 15]; Fraenkel and Bar-Hillel [FST, chap. 4].   Wilder [IFM] gives a good account of intuitionist mathematics from the viewpoint of an outsider.   For the deeper philosophy it is probably necessary to read Dutch; thus the notion of a time intuition is explicit in Brouwer [GLW], but appears not to be mentioned in versions in English and German.   (See, however, Brouwer [CPM] and the quotation from it in Beth [FMt, p. 618].)   See also Weyl [PMN]; his position is intermediate between intuitionism and Hilbertism, and he had a profound understanding of both schools.

For brief summaries of the work and position of Hilbert see Bernays [HGG], Weyl [DHM, pp. 635–645]; cf. also Heyting [FMI, pp. 37–60]. The standard work, completed by Bernays, is Hilbert and Bernays [GLM]. Some of Hilbert's logical papers have been reprinted in the third volume of Hilbert [GAb] and also in his [GLG$_7$]; in some cases passages have been deleted in the process of reprinting. Schütte [BTh] may be regarded as a continuation of Hilbert's program in one of the several possible directions. See also Kreisel [HPr].

Somewhat divergent views of formalism are taken in Goodstein [CFr], [NMS]; Lorenzen [EOL], [LRF].

For the Gödel theorem, and the semantic developments mentioned at the end of Sec. C, see Sec. 3S1.

The discussion of the three levels of abstraction at the end of Sec. C3 is based on Shanin [LPA]. For nominalistic views see Beth [FMt, chap. 16] and the references cited there; also Gilmore [AST].

The account of the relations between mathematics and logic in Sec. D is a revision of that in [OFP], chap. 12. Collateral references are Russell [PMt, especially the introduction to the second edition]. For the possibility of alternative logics see, for example, Church [LEM], Lewis [ASL].

The term 'finitary', used in Sec. C1 in describing constructive methods, was introduced by Kleene [IMM, p. 63] as translation of German 'finit'.

# Chapter 2
# FORMAL SYSTEMS

In the preceding chapter the discussion was intended to be intuitive. Accordingly, such terms as 'formal theory', 'mathematical system', 'acceptable', etc., were used in nontechnical senses which were expected to be self-explanatory. In this chapter formalism is considered more technically. These and other terms will be given technical meanings. With few exceptions (as stated later), they will be used in these technical senses from now on.

We shall begin, in Sec. A, with general considerations of a preliminary nature. In Sec. B we shall define a "theory" as a class of statements and discuss some matters related to this notion in its most general form. In Sec. C we define a system as a theory whose elementary statements concern certain "formal objects" in a specified way. Specialization of these notions will concern us in Sec. D. Finally, in Sec. E, there will be a discussion of the notion of algorithm due to Markov; this will enable us to make more precise certain notions introduced in Sec. A and for that purpose is more suitable, from the present point of view, than other notions which are equivalent to it.

## A. PRELIMINARIES

In this section we shall be concerned with certain matters concerning the use of language and with the explanation of technical terminology which will be used throughout the rest of the work. Some of these terms have already been used in nontechnical senses which do not always agree exactly with the technical senses. The latter are to be regarded as a refinement of the former. Other terms, such as 'theory', 'system', will be refined later.

**1. The U language.** Every investigation, including the present one, has to be communicated from one person to another by means of language. It is expedient to begin our study by calling attention to this obvious fact, by giving a name to the language being used, and by being explicit about a few of its features. We shall call the language being used the *U language*.

It is impossible to describe this U language exhaustively. All we can say is that it contains the totality of linguistic conventions which, at the moment, we understand. This may seem vague, but in that vagueness we are no worse off than in any other field of study. Every investigation, in any subject whatsoever, must presuppose that same datum. Thus there would be no point in calling attention to it, if it were not for the fact that language is more intimately related to our job than to most others. We do not postulate that thought is possible without language, and therefore we are obliged to pay some attention to the U language from the very beginning. But since we cannot describe it exhaustively, we can only notice certain of its features and be explicit concerning usages about which there is danger of misunderstanding.

The U language has the following features: (1) It is specific. If we were to speak about several different U languages, then at most one of these would be used, and this one is the U language; the others are only talked about, and therefore none of them is the U language. (2) It contains the technical terminology and other linguistic devices—such as the use of letters for variables—which are generally understood by mathematicians of an appropriate degree of maturity. (3) It is not immutable, but is continually in process of growth: from time to time we may introduce new technical terms and new symbolism; likewise we may agree to use old terms in new senses, or to abandon them altogether. In this way the use of the language becomes more precise as we proceed. (4) Although it is necessarily rather vague to begin with, yet by careful use we can attain any reasonable degree of precision, just as we can measure to millionths of an inch using tools which are ultimately the creation of human hands and brains. Absolute precision is not attainable in the field of measurement; there is no need to suppose that it can be reached in the use of the U language either.

The presentation of the paradoxes in Sec. 1B, which was made in the U language, has led many persons to assert that the U language is inconsistent. So it is, if carelessly used, and carelessness would be expected to cause trouble in any kind of activity. The discussion of platonism in Sec. 1C1 shows that the paradoxes can be avoided if certain precautions are used with the word 'statement'. The exact nature of these precautions will concern us later. For the present, suffice it to say that we use the U language for describing certain more or less concrete activities and for drawing certain conclusions from these descriptions by finitary constructive processes, such as those admitted by Hilbert. The intuitive evidence that no contradictions can enter in this way is very strong indeed; the question of whether this evidence amounts to absolute certainty will be left to the reader's philosophy.[1]

**2. Languages and expressions.** In recent years a certain school of thinkers has maintained that many problems can be solved, or at least illuminated, by examining critically the language in which they are expressed. This has led to a whole science of symbolism which has been called *semiotics*. Certain terminology arising from (or suggested by) that discipline will be introduced here for use in what follows. Further terminology of this kind will be introduced later; here attention will be confined to some items which will be useful at once.

The basic concept of semiotics is that of a *language*. A language, in its most general sense, is defined by two sorts of conventions. First, there is specified an *alphabet*, i.e., a certain stock of objects, called *symbols* (or *letters*), which can be produced in unlimited quantity like the letters of ordinary print

---

[1] Cf. the remarks at the end of Sec. 1C3.

or the phonemes of speech.   Secondly, there are rules specifying how certain combinations, called *expressions*, or *words*,[1] can be formed from the letters.   Generally the expressions are arbitrary linear series of letters in which repetitions may occur; in that case the language will be called a *linear language*.   Thus in the linear language whose alphabet consists of the three letters

$$a, b, c$$

the following are examples of expressions:

$$a, abcba, bcccaa, abcbabcccaa$$

In this semiotical concept of a language it is irrelevant whether or not the language is used for human communication; if it is so used, we call it a *communicative language*.   A language in the sense of linguistics will be called a *natural language*.

The terms 'letter', 'symbol', and 'expression' make sense for any language, natural or not.   In particular, they make sense for the U language. For these and other semiotical terms we shall use the letter 'U' to indicate reference to the U language; thus we can have U symbols and U expressions. Evidently, the U language is a communicative language which is mutually understood by author and reader; it is the one which is actually used.

It is evident from the above discussion that expressions (including symbols) are not single concrete physical objects, but types or kinds of such objects, such as apples, trees, men, which may occur in numerous instances. The individual marks on paper which constitute the instances of expressions will be called *inscriptions*.[2]   In a more nominalistic account, the notion of inscription would be fundamental and we should speak of *equiform* inscriptions instead of inscriptions which are instances of the same expression. But it is more convenient to introduce expressions and to speak of sameness of expressions, instead of equiformity of inscriptions.   We are thus making the first of the two abstractions mentioned in Sec. 1C3; we are also making the second abstraction, since we suppose that expressions can be of any length. Presumably it would not be difficult to transform what is said here into a manner of speech which does not use these abstractions, but we shall not attempt it here.

Sometimes we wish to talk about expressions as objects.   For such purposes we need names for the expressions.   Of course, one could use the expression itself as its own name; this practice, which is known as the *autonymous mode of speech*, is particularly suitable when the expression mentioned is not used in the U language, as when one says that the Greek word σῆμα means a sign.   On the other hand, if this is not the case, there is some danger of confusion between mention of an expression and its ordinary use. Thus from the sentences

John is a redheaded man

John is a name with four letters

---

[1] In certain contexts 'word' will be used instead of 'expression'.   However this cannot always be done, as 'word' has nontechnical uses which sometimes conflict.

[2] The term has primary reference to written languages, but could be used with some awkwardness for spoken ones.

one might be tempted to conclude that a certain name with four letters is redheaded. In the first sentence, a common given name is used to refer to a man who was baptized with it, whereas in the second that same name is mentioned. To avoid this danger it is customary to use a specimen of an expression enclosed in quotation marks as a name for that expression. Since this is a rather technical use of quotation marks, we reserve *single quotes* for that purpose, using *double quotes* for all other uses of quotation marks in the U language. With this understanding it is 'John', not John, which is a name with four letters. This convention is of some help; it has been used several times already in this work, and will be used generally in the future when the context or other conventions of the U language do not make it superfluous.[1]

The notation just described is not foolproof.[2] For the quoted expression is an indivisible unit in the U language—a new symbol, if you will—and substitutions ordinarily made for certain letters or expressions in the U language cannot be made inside the quotes. Where such "quotation functions" are needed, further devices are required. This will sometimes occur in connection with explanations of notation. Then the difficulty can generally be avoided by using the idiom of the U language whereby, when we wish to introduce the name 'John Doe' (which has not previously occurred in the same context) for a man described in some way, we say "John Doe is the man who . . . ," rather than " 'John Doe' will be the name of the man who . . . ." This brings the new name in without quotes, so that the usual conventions of the U language in regard to variables will apply.[3]

Sometimes we wish to talk about one language $L_1$ within another language $L_2$. In such a case it is customary to call $L_1$ the *object language;* $L_2$, the *metalanguage.* It is not excluded that $L_1$ and $L_2$ may overlap. Ordinarily the object language will be a certain portion of the U language which it is agreed to remove from it, while the metalanguage is the U language after the removal. But sometimes we may wish to talk about languages $L_1$ and $L_2$ which are related to each other as object language and metalanguage, respectively; in that case we use a third language, $L_3$, customarily called the *metametalanguage.* In this way we can continue to form hierarchies of languages with any number of levels. However, no matter how many levels there are, the U language will be the highest level: if there are two levels, it

---

[1] This will be the case, in particular, when the expression being mentioned is displayed on a separate line. In certain instances, connected with verbs like 'call', I have made no attempt to make the usage strictly uniform.

[2] See Exercise 2 at the end of this section.

[3] Cf. Quine's "quasi-quotation" [MLg, sec. 6]. Quine needed this device because he had to mention expressions in his formal developments. Here this does not occur, and the usage in the text is usually clear enough. For those rare cases where greater explicitness is desired, we shall use the following form of speech:

expression of the form $A$,      where $a_1, a_2, \ldots$ are . . .

where '$A$' is replaced by the name of some expression $B$, and '$a_1$', '$a_2$', . . . , by names of letters $b_1, b_2$, occurring in $B$. After the replacements, the speech form is to be construed as designating the expressions obtained by substituting expressions satisfying the stated conditions for the letters $b_1, b_2, \ldots$ in $B$. Certain details and modifications can safely be left to the reader's common sense.

will be the metalanguage; if there are three levels, it will be the metameta-language; and so on.  Thus the terms 'U language' and 'metalanguage' must be kept distinct.[1]

**3. Grammatics.**  If we study a communicative language from the standpoint of the meaning which is conveyed, the expressions of the language do not form a natural class of symbol combinations.  Of much more significance is another class of combinations which form units in the rules for determining how the sentences are constituted.  We shall call the study of the rules for determining the sentences of a language its *grammatics*, and the symbol combinations which form the units in grammatics its *phrases*.  Thus in the sentence

John has both green and purple pajamas

the following are samples of expressions:

has both
paj
hn has both gr

while the following are phrases:

John
both green and purple . . .
both – – – and . . .

where the dots and dashes in the last two examples indicate blanks to be filled.  The first set of examples shows that not all expressions are phrases; the last example shows that not all phrases are expressions.

There are three main classes of phrases, viz., nouns, sentences, and functors.  A *noun* names some object (real or imaginary); a *sentence* expresses a statement; and a *functor* is a means of combining phrases to form other phrases.  We shall call nouns and sentences *closed phrases* to distinguish them from functors.  As to functors, the phrases combined by a functor we shall call its *arguments;* the result of the combination we shall call its *value*.  Functors can evidently be classified according to the number and kind of the arguments and the nature of the value.  Thus if a certain functor combines $m$ arguments of types $X_1, X_2, \ldots, X_m$, respectively, to form a value belonging to type $Y$, we shall say that it belongs to the type

$$\mathsf{F}_m X_1 \cdots X_m Y$$

These "types," which are formed from the basic types such as $n$ (noun) and $s$ (sentence) by the "functionality operators" $\mathsf{F}_n$, will be called *grammatical categories*.[2]

In order to attain complete generality we must admit functors with other functors as arguments; there are numerous examples of such notions in linguistics.  As to the value, note that

$$\mathsf{F}_1 X(\mathsf{F}_1 YZ), \quad \mathsf{F}_3 UVW(\mathsf{F}_2 XYZ)$$

---

[1] Cf. [LMF], sec. II; also Sec. S4.

[2] For examples of phrases from ordinary English and mathematics, see [LAG], sec. 6; also [CLg], pp. 264 and 274, and [TFD], sec. I5.

are the same categories, respectively, as

$$F_2 XYZ, \quad F_5 UVWXYZ$$

and so on. We may therefore do either of the following: (1) require that $m = 1$, but allow the value to be a functor, so that we in effect define all $F_m$ in terms of $F_1$, or (2) allow functors of any number of arguments but require the value to be closed, in which case we speak of the value as the *closure*. The former procedure is usually followed in linguistics and is the main idea back of the reduction to one operation discussed later (in Sec. 2D2); the latter is in the spirit of ordinary mathematics and will be adopted here. Accordingly, the value of a functor will be called henceforth its *closure*, and the number of arguments, which is now uniquely determined, will be called its *degree*. Functors of degree $m$ will be called $m$-place or $m$-ary; for $m = 1$ and $m = 2$ these terms, as here used, are '*unary*'[1] and '*binary*', respectively.

The foregoing account of grammatics is, of course, very sketchy. For an adequate grammatical theory of a natural language we should undoubtedly need several kinds of nouns and probably of sentences also. But the categories of $n$ (noun) and $s$ (sentence) give us a basis which is adequate for most purposes; refinements, if necessary, may be introduced later.

The principal kinds of functors which occur in our work are the following: *operators*, which combine nouns to form other nouns; *verbs*, or *predicators*, which combine nouns to form sentences; *connectors*, which combine sentences to form other sentences; and *subnectors*, which form nouns out of sentences. It is also convenient to have terms for referring to the meanings of these phrases, and to say that such and such a phrase *designates* such and such a meaning. The terms used for this purpose are shown in the following table.

| *Phrase* | *Designatum* |
|---|---|
| Phrase | Designatum, element |
| Noun | Object, ob |
| Sentence | Statement |
| Clause† | Proposition† |
| Functor | Function, functive |
| Operator | Operation |
| Verb (predicator) | Statement or verbal‡ function (predicate) |
| Connector | Connection |
| Subnector | Subnexus |

The use of this dual terminology does not imply a commitment to a philosophy which postulates the existence of these meanings as esoteric entities of a mysterious sort. Actually, one can regard the usage as purely rhetorical. One could, for example, consider a statement as a class of sentences which are mutually equivalent in some sense, just as expressions are classes of equiform inscriptions. However, since I give no way of judging when the sentences are equivalent—in other words, no criterion for the identity of

---

[1] Some authors prefer the term 'singulary'. For a statement of reasons for preferring 'unary' see [rev. Church, end of article] as corrected in [DFS], footnote 4.

† For these see Chap. 5.

‡ See Sec. 7A3.

statements—the reader is free to interpret the designation relation in any way he likes, in particular to make no distinction between a sentence and a statement. The duplication of terms has the advantage that it does not commit us to a particular type of philosophy; it introduces, so to speak, a free parameter which one can adjust to suit different philosophical prejudices. The vague indication which it gives of the level of abstraction seems to be better than none at all, and thus is an aid to thinking.

It is a general convention of the U language that a phrase which is grammatically not a noun may sometimes be used in a nominal context as name for its designatum. This convention will be used occasionally in the sequel.

**4. Technique for functors.** Here there will be stated certain conventions of a rather special nature which are concerned with the handling of functors. The discussion is technical and may be passed over until it is required.

There is need for a notation for mentioning functors which is as efficient as quotation marks are for mentioning expressions. Since functors are by definition modes of combination, the specification of a functor is not complete until it is shown how the closure is obtained from the arguments. This requires that there be blanks or similar devices to indicate where the arguments are to be put, and also some method of indicating the order in which the blanks are to be filled. Dashes with numerical subscripts will be used for this purpose. Parentheses to indicate the extent of the arguments must be considered a part of the functor. Thus a complete notation for the addition operator would be $‘(\text{———}_1) + (\text{———}_2)’$.

However, this is more elaborate than is necessary for most purposes. Many functors are of one or the other of the following three types: *prefixes*, which are written before the arguments; *infixes*, which are binary functors written between the arguments; and *suffixes*, which are written after the arguments. Prefixes and suffixes can have any degree, and it has been shown that parentheses are not necessary when either kind is used exclusively with functors of known degree.[1] Infixes require parentheses; the rules for these parentheses, including the omission of superfluous ones according to standard practice, will be regarded as known. Accordingly, it is sufficient to consider functors of these three types as simple symbols. Such a simple symbol will be called an *affix*. When this affix is used as a noun without quotation marks, it is to be understood as a name for the function; with quotation marks it is, of course, a name for the functor itself.

Another case where abbreviated notation is appropriate is the case of an ordinary mathematical function. Here the full notation would be an expression of the form $‘f(\text{———}_1, \text{———}_2, \ldots, \text{———}_n)’$, where $‘f’$ is a letter (or word acting as a letter). Unless there is danger of confusion (due to special circumstances), it will suffice to use the function letter (i.e., what replaces the

---

[1] This is a result of J. Łukasiewicz, a Polish logician (1878–1956), for a long time professor of philosophy at the University of Warsaw. He is noted for his work on multiple-valued logics and on the history of logic. For an account of his work, see Mostowski [OSJ]; also (less readable, but giving further details) Borkowski and Słupecki [LWJ], Kotarbinski [JLW], Scholz [MJL], Sobociński [MJL]. A bibliography of his writings also appears in *Studia Logica*, 5:9–11 and 8:63. For the origin of the notation see Borkowski and Słupecki [LWJ, p. 24]. The notation is discussed more fully later.

'$f$') without quotes as a name for the function and in quotes as a name for the functor.[1]

*Special Functors.* The following table lists for reference some special functors which will be used in technical senses. Here the first column lists the affix; the second column, its meaning (or a translation scheme for its closure); and the third column, the places later in this book where it is introduced or where further information can be found.

*Binary infixed connectors*

| | | |
|---|---|---|
| $\rightarrow$ | If ——$_1$, then ——$_2$ (or ——$_1$ only if ——$_2$) | §3A2[2] |
| $\rightleftarrows$ | ——$_1$ if and only if ——$_2$ | |
| or | ——$_1$ or ——$_2$ | §3A2 |
| & | ——$_1$ and ——$_2$ | §3A2 |

*Binary infixed verbs*

| | | |
|---|---|---|
| $\equiv$ | ——$_1$ is the same (by definition) as ——$_2$ | §3C3 |
| $=$ | ——$_1$ equals ——$_2$ | |
| $\leq$ | ——$_1$ precedes ——$_2$ | |
| $\subseteq$ | ——$_1$ is included in ——$_2$ | §2B1 |

*Unary verbs*

| | | |
|---|---|---|
| $\vdash$ | —— is asserted | §2D1 |
| $\dashv$ | —— is refuted | §6A3 |

*Binary infixed operators*

| | | | |
|---|---|---|---|
| $\supset$ | ——$_1$ ply ——$_2$ | (implication operation) | §4C1 |
| $\mathsf{s}$ | (——$_1$ $\supset$ ——$_2$) $\wedge$ (——$_2$ $\supset$ ——$_1$) | (equivalence operation) | §4D3 |
| $\vee$ | ——$_1$ ad ——$_2$ | (union) | §4A1 |
| $\wedge$ | ——$_1$ con ——$_2$ | (meet, or conjunction) | §4A1 |

*Unary operation*

| | | |
|---|---|---|
| $\neg$ | (negation operation) | §6A3 |
| $\square$ | (necessity operation) | §8A2 |

*Dot Notation.* When complex phrases are built up by iterated application of such functors, there may be so many parentheses that the resulting expression is difficult to read. Various devices—of which we have already noticed the elimination of superfluous parentheses—have been proposed to aid in such reading. One such device, known as the dot notation, which is in rather common use, is explained as follows.

The principle of this device is that one replaces parentheses by groups of

---

[1] This convention and the analogous one at the end of the preceding paragraph are special cases of the general convention at the end of Sec. 3.

[2] This infix is used in connection with algorithms (Sec. 2E) in a different sense.

dots in such a way that the more inclusive parentheses have a larger number of dots and that one distinguishes between left and right parentheses by their position relative to their associated affixes.   One can then see at a glance where the main breaks occur, and the structure of the phrase can be more readily perceived.   Unfortunately, when one attempts to state precisely the rules for such a notation, one finds that it is rather a complicated business, and on this account many writers avoid the notation altogether.[1]   However, it does have certain advantages, and consequently it is used, somewhat sparingly, in the more technical portions of this book.

It is advisable to modify the technique so as to include the standard rules for omission of parentheses and to be consistent with common sense.   The modified conventions are as follows.

A group of dots on either side of a binary infix and on the right of a unary prefix will be called a *point;* we include the possibility that the number of dots may be zero.   A point on the right of an affix will be known as a *right* (*-facing*) *point;* one on the left as a *left*(*-facing*) *point;* the beginning and the end of the entire expression will also be right and left points, respectively. These points will be ranked in order of seniority according to rules to be given presently.   Then a point indicates that the argument on that side of the affix is to extend from that point in the indicated direction until the first senior point facing in the opposite direction.   The expression so determined will be called the *scope* of the point.   The relation of seniority is to be a transitive relation generated by the following rules, it being understood that a rule stated earlier in the list takes precedence over one stated later: (1) the beginning or end of the entire expression is senior to any point in its interior; (2) a point attached to a connector is senior to one attached to a verb; (3) a point attached to a verb is senior to one attached to an operator; (4) a point with a larger number of dots is senior to one with a smaller number; (5) a point attached to a functor appearing earlier in the table of Sec. 3 is senior to one appearing later; (6) a left point is senior to a right point (rule of association to the left).

In applying these rules in cases where parentheses are present, the parenthesized expressions are to be treated as units.   Given a pair of corresponding parentheses, the expression between them is the "entire expression" for any point within it, and the scope of a point outside of such a pair of parentheses includes either the entire unit or none of it.

These conventions are stated under the assumption that no point will contain within its scope a senior point facing in the same direction.[2]   It would be possible to state more complicated conventions allowing this possibility, but the convenience of the dot notation would be lost.   In practice, we obtain maximum perspicuity by using more dots than are strictly necessary

---

[1] The Łukasiewicz notation (see footnote, p. 34) is much easier to describe and to treat theoretically, but it is by no means perspicuous and requires some practice before one can read it.   Dots could be used to increase perspicuousness in connection with that notation, but this has not been done (cf., however, the use of spacing in Bochenski [NLL]).

[2] However, the rules do admit the possibility that there may be a point of the same seniority.   Thus we can interpret

$$A_1 \supset_. A_2 \supset \cdots \supset_. A_n \supset B$$

for any value of $n$.

and combining dots judiciously with explicit parentheses, so that the senior point in a formula can be seen at a glance.

It is easy to modify the rules so as to include the case where simple juxtaposition is used as a binary functor, as in the case of multiplication in ordinary algebra. There is then no affix. We can suppose, however, that a fictitious affix is supplied and that the points on either side of it have the same number of dots; then the affix and one of the points can be removed. The remaining point has to be considered as both a left and a right point simultaneously.

**5. Processes and classes.** Besides the more or less symbolic conventions which we have just discussed, it is necessary to be clear about certain terms in the more discursive part of the U language. These are terms with which the reader is already familiar; the purpose of this discussion is not to define them formally, but to sharpen their use. We shall discuss the notion of effective process, of definite question, and of a class or totality.

*Effective Process.* Suppose that we have certain transformations which can actually be carried out on certain elements. Suppose that we have specifications determining a sequence of transformations to be applied successively to an element. These specifications will be said to define an effective process for attaining a certain goal relative to an element if, given that element, the specifications determine uniquely a sequence of transformations such that the goal is reached after a finite number of steps. There must never be any ambiguity whose solution requires examination of an infinite number of possibilities. Thus the notion of an effective process is akin to that of a construction as understood by an intuitionist, but it has all the stringency of Hilbert's finitary standpoint, and it does not depend on any idealistic intuition, temporal or otherwise.

Ordinarily, the specifications define an effective process simultaneously for many elements; i.e., the process is defined in general terms involving parameters. In such cases the elements for which the transformations are defined will be called the *admissible elements*. Then the description of the process must be such that we know without equivocation the following: (1) whether a given element is admissible; (2) given an admissible element, exactly what transformation is to be applied and what its result will be; (3) when the goal is attained. Then, given an admissible element, the process will be effective for that element just when we know in addition that a finite number of steps will actually reach the goal. In such cases we shall say that the specifications define an *effective process* and that the process is *applicable* to those elements for which it actually is effective in the sense of the preceding paragraph.

Examples of such processes are the Markov algorithms, which we shall study in Sec. 2E. Here the admissible elements are the expressions in a language with a finite alphabet; these evidently satisfy condition (1). Finally, the attainment of the goal is signalized by the carrying out of one of certain specially indicated transformations, or, in some cases, by reaching an element for which no further transformation is defined. Very general sorts of effective processes can be specified by such algorithms; there is heuristic evidence to the effect that every effective process can be.

In the foregoing we have tacitly supposed that a "transformation" was a function of one argument.  The notion of effective process may be extended to the case where two or more elements determine the result of a transformation.  In fact, this case can be reduced to the foregoing by the device of taking ordered sequences of elements as new elements.

*Definite Questions.*  A question will be said to be definite if it can be answered yes or no and there is an effective process for finding the answer. Thus the question is as to the truth of a certain statement, this statement being then the element with which the process begins.

This notion can be extended to simultaneous consideration of admissible statements, just as in the case of effective process.  We have to do with effective processes for which the admissible elements are certain statements and the goal a judgment of their truth or falsity.  The question is *definite* if there is such an effective process which is applicable to every admissible statement.  If there is an effective process which is applicable whenever the admissible statement is true, then the question is called *semidefinite*.[1]

It will be seen that there seems to be a certain circularity about these definitions.  We shall return to discuss this point later.  However, we note in passing that the question of whether an effective process is applicable to a given admissible element is always semidefinite and may or may not be definite.

*Conceptual Classes.*  We shall often have to formulate, by means of the U language, properties (or relations) which define, in a strictly intuitive (or contensive) way, a totality of elements or notions.  In order to distinguish such intuitive totalities from the "sets" or "classes" formed later[2] (and conceived rather as objects of some theoretical study than as intuitive notions), we shall call them *conceptual classes* (or relations).  The elements eligible for consideration for membership in such classes will be called admissible elements.

A conceptual class will be said to be definite just when the question of whether an admissible element belongs to the class is definite.  Similarly, we shall speak of a conceptual class as semidefinite if the corresponding question concerning membership is semidefinite.

An *inductive class* is a conceptual class which is generated from certain initial elements by certain specified modes of combination.[3]  More precisely, this means the following.  Let $\mathfrak{X}$ be the class in question.  Then $\mathfrak{X}$ is defined by two sorts of specifications, called the *initial specifications* (I) and the *generating specifications* (II).  The initial specifications define the *initial elements;* the latter constitute a definite class, say $\mathfrak{B}$, often called the *basis* of $\mathfrak{X}$.  The generating specifications define a definite—usually, but not necessarily, finite—class, say $\mathfrak{M}$, of modes of combination; with each such mode $\mu$ there is associated a fixed number called its degree; it is then understood that the application of any such $\mu$, of degree $n$, to a sequence of $n$

---

[1] For example, if $\alpha$ is a recursive set of natural numbers, then the question of whether a natural number $n$ belongs to $\alpha$ is a definite question; if $\alpha$ is a recursively enumerable class, the same question is semidefinite.  The admissible elements are statements to the effect that an explicitly given number is a member of $\alpha$.

[2] In the higher parts of logic, for example, set theory, but not in this book.

[3] The term 'mode of combination' is intended to suggest that we have to do with functions of any number of arguments.  The possibility that there may be only one argument is not excluded.

arguments, each an element of $\mathfrak{X}$, produces an element of $\mathfrak{X}$; further, that the question of whether an element is so produced from the given arguments is definite. It is further understood that every element of $\mathfrak{X}$ can be reached by an effective process (in the generalized sense) which starts with certain initial elements and at each later step applies a mode of combination of $\mathfrak{M}$ to arguments already constructed; this requirement, often called the *closure specification*, is to be understood as part of the definition of an inductive class, and it is therefore not necessary to state it explicitly in the definition of a particular $\mathfrak{X}$.

The definition of an inductive class $\mathfrak{X}$ is sometimes expressed by saying that $\mathfrak{X}$ is the class defined by the following three properties: (1) $\mathfrak{X}$ includes the basis; (2) $\mathfrak{X}$ is closed under the modes of combination; (3) $\mathfrak{X}$ is included in every class which satisfies properties 1 and 2. This definition does not spell out the fact that $\mathfrak{X}$ is the totality of all elements which can be reached from the initial elements by iterated application of the modes of combination. It therefore has a platonistic character, which is objectionable from our point of view.

Nothing in the foregoing requires that there be at most one element produced by a mode of combination from given arguments $Y_1, \ldots, Y_n$. This condition is, however, fulfilled in many interesting cases. Just in case a mode of combination satisfies this condition, it will be called *determinative*.

The notion of inductive class may apply in either of the following two cases (and perhaps in others): (1) the elements are objects, and the modes of combination are operations; (2) the elements are statements, and the modes of combination are connections.

Under the restrictions made it will appear later (in Sec. 6) that an inductive class is semidefinite. Under special circumstances it is definite, but there are many interesting cases in which it is not.

For some purposes (e.g., where it is not necessary to retain a strictly constructive viewpoint), it may be advantageous to relax somewhat the definiteness requirements in the definition of an inductive class. In that case we may speak of a "generalized inductive class," or perhaps a "semi-inductive class."

It remains to comment on the apparent circularity in the above discussion. This arises since in condition (1) for an effective process we have in effect said that the admissible elements form a definite (conceptual) class, whereas the notion of definite class depends on that of effective process. This brings us back to the remark that these terms are not being formally defined. We must, in fact, begin with some initial class of admissible elements whose definiteness is not open to doubt. As such a class we may take, for example, the expressions of some language with a finite alphabet, let us say, for example, the U language or some portion of it. (Note that, in view of the remarks in Sec. 3, one may identify a concept or notion with the U expression which designates it.) In terms of these elements we may generate effective processes, other definite classes, etc. If we accept Markov's (or Church's) thesis (Sec. E1), one may define the notion of effective process with a high degree of precision.

**6. Constructions.** A process for reaching an element $X$ of an inductive class $\mathfrak{X}$ by iterated application of the modes of combination will be called a *construction* of $X$ (relative to $\mathfrak{X}$). We shall here study some technical matters connected with such constructions. It will be convenient to treat primarily the first of the two cases mentioned toward the close of Sec. 5, viz., that in which the elements are objects and the modes of combination are operations. Appropriate changes of terminology, such as replacements of 'name of an element' by 'sentence designating a statement', will then automatically extend our conclusions to other cases. However, we shall continue to use '$\mu$' and '$\mathfrak{M}$' as in Sec. 5.

In connection with such constructions we consider certain diagrams called *tree diagrams*. A tree diagram $\mathfrak{D}$ will consist of *nodes* joined together in the following way. There will be a unique bottom node; every node other than the bottom node will be joined to a unique node below it, and there will be no other junctions; further, an operation $\mu$ of $\mathfrak{M}$ will be assigned to each node not a top node, and the number of nodes joined to that node from above will be precisely equal to the degree of $\mu$. Let $\mathfrak{C}$ be a construction of $X$. Then a tree diagram $\mathfrak{D}$ will be said to be associated with $\mathfrak{C}$ just when there is a one-to-one correspondence between the nodes of $\mathfrak{D}$ and the occurrences of elements of $\mathfrak{X}$ appearing in $\mathfrak{C}$, so that the following conditions are satisfied: the bottom node corresponds to $X$; and if $Y$ is formed in $\mathfrak{C}$ by applying an operation $\mu$ to arguments $Y_1, Y_2, \ldots, Y_n$, in that order, then the node corresponding to $Y$ has that same operation $\mu$ assigned to it, and the nodes joined to it from above, in order from left to right, are precisely those corresponding to $Y_1, Y_2, \ldots, Y_n$. In such a case the top nodes of $\mathfrak{D}$ will correspond to initial elements. We shall say that a tree diagram $\mathfrak{D}$ is *labeled* (relative to a construction $\mathfrak{C}$) just when each node of $\mathfrak{D}$ is marked with a name of the corresponding element in $\mathfrak{C}$. In practice we can realize such a labeled $\mathfrak{D}$ as follows. We take as nodes instances of the names of the various elements of $\mathfrak{X}$ to which they correspond; over each node not a top node we draw a horizontal line, with the name of the operation used in forming that node at the extreme right; above this line we write in order the nodes corresponding to the arguments to which that operation was applied. Thus, if we take addition, multiplication, and squaring as operations and use the notation of elementary algebra, the labeled tree diagram for $a^2b + ac$ would be as follows:

$$\frac{\dfrac{a}{a^2}\,\mathrm{sq}\; b}{\underline{\quad a^2b \quad}}\times \frac{a\; c}{ac}\times$$
$$\frac{a^2b + ac}{}+$$

Here 'sq' is used for the name of the squaring operation; the other names are self-explanatory.

It is clear that an element belongs to an inductive class $\mathfrak{X}$ just when there exists a construction of it. Since the question of whether a tree diagram corresponds to a construction of a given $\mathfrak{X}$ is a definite question, an inductive class is always semidefinite.

Constructions, tree diagrams, and labeled tree diagrams are like expressions in that they may have innumerable instances. Given two instances of a construction, we say that they are instances of the same construction just when they have the same associated tree diagram and the same elements corresponding to the same nodes, i.e., the same labeled tree diagram. (If the element produced by an operation from given arguments is unique, it would be sufficient to have, in addition to identity of the tree diagrams, the same initial elements at all the top nodes.) Generally, the same element $X$ may have several different constructions. Just when this construction is unique for any $X$ in $\mathfrak{X}$, we shall say that the inductive class $\mathfrak{X}$ is *monotectonic*[1]; when there may be more than one construction, we shall say that it is *polytectonic*. We shall meet examples of both kinds later.

We now consider a notion related to a construction, viz., that of a *construction sequence*. This is a sequence of elements such that each term either is an initial element or is constructed from some of its predecessors by an operation. Evidently, a construction sequence may be formed from a construction in various ways. The sole requirement is that $X$ must precede $Y$ in the sequence whenever $Y$ is below $X$ in the associated tree diagram, i.e., whenever, starting from $X$,[2] we can reach $Y$ by a succession of steps each of which carries us from a node to the one joined to it from below. The sequence may even contain extraneous elements, i.e., ones not needed in the construction. One particular such sequence, the *normal construction sequence* of $\mathfrak{C}$, is the one obtained from the labeled tree diagram by enumerating the nodes from the top down and from left to right; more precisely, it is the unique construction sequence which satisfies the following additional conditions: (1) its elements are precisely the elements of $\mathfrak{C}$, with repetitions taken as they occur as if they were distinct elements; (2) if $X$ and $Y$ are used as arguments in the formation of any element (i.e., if the nodes for $X$ and $Y$ are both joined from above to the same node immediately below) and $X$ is to the left of $Y$, then $X$ precedes $Y$ and also precedes any $Z$ such that $Y$ is below $Z$. For example, the sequences

$$a, b, c, a^2, ac, a^2b, a^2b + ac$$

$$a, b, c, c, a^2, b^2, ac, ad, a^2b, abc, a^2b + ac$$

$$a, a^2, b, a^2b, a, c, ac, a^2b + ac$$

are construction sequences for the above construction of $a^2b + ac$; the last of the three sequences is the normal construction sequence.

Evidently $X$ will be in $\mathfrak{X}$ just when it is the last term in a construction sequence. But in order to reestablish the construction from the construction sequence, it is necessary to give some additional information, viz., the operation and the arguments used in forming any term. This additional information may be called an *analysis*[3] of the construction sequence. In order to make it a definite question whether an arbitrary sequence of elements is or is not a construction sequence, we must strengthen somewhat the

[1] This is the same as the term 'tectonic' in [CFS]. It was introduced in [LAG] because the term 'polytectonic' is useful for the opposite property. If there is an effective process for obtaining the construction for admissible $X$, then $\mathfrak{X}$ is *effectively monotectonic*. In this book, 'effectively' in this context will generally be understood.

[2] Strictly speaking, the node corresponding to $X$. For the purpose of this discussion it is permissible to identify the elements with their names.

[3] This term is due to Kleene [IMM, p. 87].

definiteness requirements with regard to operations; viz., it must be definite, not only whether—given $Y_1, \ldots, Y_n, X$, and $\mu$—$X$ is obtained from $Y_1, \ldots, Y_n$ by $\mu$, but also whether—given $Y_1, \ldots, Y_n, X$—$X$ is obtained by some $\mu$ from some subsequence of $Y_1, \ldots, Y_n,$† and if so, the number of possibilities is finite.

Further techniques related to constructions will be taken up in Sec. 3B1.

**7. Natural numbers.** That we have to use notions connected with natural number in the U language is already evident from the fact that we have had to speak of the degree of a functor as the number of its arguments. It is therefore appropriate to pause to consider just what is involved in this intuitive usage of numerical notions.

We use numbers as counters. The essence of this process has already been set forth in Sec. 1C2a. We can select certain words forming a sequence in the U language; from these we can form standard (conceptual) classes corresponding to each number; and a class has $n$ members just when there is a pairing (one-to-one correspondence) between its members and those of the appropriate standard class. More than this we have not needed.

Later on we may formalize the notion of number in ways which will be explained in due course. Then we can introduce further arithmetical ideas and use these in our study. This will concern us in Sec. D and Chap. 3.

In no case is it necessary that any sort of idealistic meaning—platonistic, based on temporal intuition, or what not—be associated with the number. Neither is it necessary that numbers be introduced as formal primitives, so that a formalized arithmetic is not necessarily prior to any other sort of logical study.

**EXERCISES**

**1.** Suppose we agree that the U language is to be ordinary English, except that we can form names of expressions in either the U language or the Greek language by single quotes as above explained, that '$a$' is a U noun designating the first letter of the Greek alphabet, and that autonymous usage is not permitted. Consider the expressions written on the following line.

$$a, \ \alpha, \ \text{alpha}, \ \ \text{ἄλφα}, \ \ 'a', \ \ '\alpha', \ \ '\text{alpha}', \ \ '\text{ἄλφα}', \ \ ''a'', \ \ ''\alpha''$$

Write down all true U sentences formed by substituting these expressions for the blank(s) in the following:

  *a.* —— is a U expression.
  *b.* —— is a Greek expression.
  *c.* —— is a letter.
  *d.* —— is a word of more than one letter.
  *e.* ——$_1$ designates ——$_2$.
  *f.* ——$_1$ is a part of ——$_2$.
  *g.* —— contains quotation marks.
  *h.* —— is an expression of ordinary English.
  *i.* —— contains Greek letters.

**2.** The following statements appear—with some slight changes—in reputable logical publications of the last thirty years. Bearing in mind that in all cases except

---

† Cf. Church [IML$_2$, footnote 121].

case $e$ the authors were intending to state general principles in which substitutions could be made for the letters, criticize the use of quotation marks in these statements.

($a$) Consider two statements, '$P$' and '$Q$', of symbolic logic which are translations of the English sentences '$A$' and '$B$'. Then '$(P \& Q)$' is the statement which is a translation of '$A$ and $B$'.

($b$) A statement such as "If $x$ and $y$ are numbers, then $x + y = y + x$" violates the rule about using names of things when speaking of those things. It should properly be written as "If '$x$' and '$y$' are numbers, then '$x + y$' = '$y + x$'."

($c$) If '$P$' is a translation of a statement, then the negation of the statement is translated '$\neg P$'.

($d$) If '$A$' and '$B$' are true, then '$A \& B$' is true.

($e$) If we wanted to state that Chicago lies between New York and Denver, we might well use $C$ for Chicago and $D$ for Denver, but to use $NY$ for New York would be confusing.

($f$) The conjunctive proposition '$p$ and $q$' will be symbolized by '$p \cdot q$'. The dot expresses that both propositions are asserted together. Hence '$p \cdot q$' may be read 'both $p$ and $q$'.

($g$) For 'not-$p$' we shall write '$\neg p$'.

**3.** Assuming that the basic grammatical categories are $n$ (noun) and $s$ (sentence) and the other categories are formed by the $F_n$, determine the grammatical categories of the following phrases:

| | | |
|---|---|---|
| '——$_1$ like ——$_2$' | in | 'Horses like oats'. |
| '——$_1$ give ——$_2$——$_3$' | in | 'Dogs give their masters much affection'. |
| '——'s' | in | 'John is Henry's brother'. |
| 'that ——' | in | 'I know that Anne is happy' |
| 'too ——' | in | 'Edward's shirt is too large'. |
| 'much ——' | in | 'Edward's shirt is much too large'. |

(Cf. [CLg], pp. 274–275.)

**4.** Assuming that the basic categories are $N$ (number) and $S$ (statement or proposition) and that one takes a naive set-theoretic point of view (whereby sets, relations, functions are conceived as objects), determine the categories of the following:

($a$) 2

($b$) The factorial function

($c$) Primeness, i.e., the property of being prime

($d$) Divisibility of one number by another

($e$) The greatest common divisor

($f$) The minimum value of a unary numerical function

($g$) The monotonic property of a numerical function

($h$) The finite difference operation

($i$) Finiteness of a set of numbers

($j$) Dominance of one function by another

(Cf. [CLg], pp. 264–265.)

**5.** If '$n$' and '$s$' abbreviate 'noun' and 'sentence', respectively, give examples from ordinary or mathematical language of the following:

($a$) $F_2 nnn$

($b$) $F_1 ss$

($c$) $F_1 s n$

($d$) $F_2 sss$

($e$) $F_1(F_2 nns)(F_2 nns)$

(Cf. references to Exercises 3 and 4.)

**6.** Assuming that '$\neg$' is a unary prefix, and that '$\supset$', '$\wedge$', '$\vee$' are binary infixes, express the following unambiguously without parentheses, using the conventions regarding dots described in Sec. 4, the binary infixes being senior to the prefix and the order of seniority among the infixes being '$\supset$', '$\vee$', '$\wedge$'.

(a) $\qquad (A \supset (B \supset C)) \supset ((A \supset B) \supset (A \supset C))$

(b) $\qquad (A \supset B) \supset ((A \supset C) \supset (A \supset (B \wedge C)))$

(c) $\qquad A \supset (B \supset (A \wedge B))$

(d) $\qquad (A \supset C) \supset ((B \supset C) \supset ((A \vee B) \supset C))$

(e) $\qquad (A \wedge (B \vee C)) \supset ((A \wedge B) \vee (A \wedge C))$

(f) $\qquad (A \supset (B \supset C)) \supset (\neg(A \supset C) \supset \neg(A \supset B))$

(g) $\qquad (A_1 \supset B_1) \supset ((A_2 \supset B_2) \supset (\cdots \supset (A_n \supset B_n)\cdots))$

(h) $\qquad \vdash \neg(\neg(\neg(\neg A)) \supset (\neg(\neg B))) \rightarrow \vdash A \supset B$

('$\rightarrow$' defined as in Sec. 4).  (For (a) to (d), see [TFD], p. 43; for (g), cf. [UDB], (3). For further such exercises, see Rosser [LMt, p. 23, example II2.1].)

**7.** Express the following in the ordinary notation of elementary algebra (the slant '/' is used for division):

(a) $\qquad a -. b + c :-: a + c .—. b + c$

(b) $\qquad a - b ./. a + b :-: a + b ./. a - b = -. 2ab/. a^2 - b^2$

(c) $\qquad a_1 +. 1/. a_2 +. 1/. \cdots +. 1/. a_n$

(Cf. Rosser [LMt, p. 23, example II2.2].)

**8.** Show that if we start with a finite alphabet, the words in that alphabet form an inductive class; further give an argument to support the thesis that that class is definite by exhibiting an effective process for deciding the question of being such a word.

**9.** Let $\mathfrak{X}$ be the class of words in the infinite alphabet

$$C, \, N, \, a_1, \, a_2, \, a_3, \, \ldots$$

defined by the inductive specifications (in the autonymous mode of speech):

(a) $a_1, a_2, \ldots$ are all in $\mathfrak{X}$.

(b) If $X$ and $Y$ are in $\mathfrak{X}$, so are $NX$ and $CXY$.

Show that $\mathfrak{X}$ is a definite class.  (This is a special case of the Łukasiewicz notation mentioned in Sec. 4.  Several solutions have been published; see, for example, [CFS], sec. 6; [LLA], appendix sec. 2; Rosenbloom [EML, sec. IV I and the references given there on p. 205].  Cf. below, Exercise E5.)

**10.** Write a labeled tree diagram and the corresponding normal construction sequence for the construction by the operations of addition, subtraction, and division of the formula $A$ of elementary algebra defined as follows:

$$A \equiv \cfrac{\cfrac{a^2 - b^2}{a^2 + b^2} - \cfrac{a^2 + b^2}{a^2 - b^2}}{\cfrac{a + b}{a - b} + \cfrac{a - b}{a + b}}$$

(Adopt and use suitable abbreviations for the various component phrases.)

**11.** Suppose a sequence is given, together with the information that it is a construction sequence.  What further information is necessary in order to determine the construction uniquely?

**12.** Show that the following conditions are sufficient so that every element $X$ of an inductive class $\mathfrak{X}$ has a unique construction: (a) an element obtained by an operation is distinct from every atom; (b) elements obtained by distinct operations, or by the same operation from different arguments, are always distinct.  Further discuss the analogy between these two properties, the specifications I, II, and the extremal clause, on the one hand, and the Peano postulates for natural numbers, on the other hand. ([CLg], sec. 2E7.)

## B. THEORIES

In this section a theory will be defined as a class of statements. We shall consider here the formulation of this definition, and consequences following from it that do not require any assumptions concerning the objects which the statements of the theory refer to.

**1. Theories in general.** We begin by postulating a certain nonvoid, definite class $\mathfrak{E}$ of statements, which we call *elementary statements*. As explained in Sec. A5, this means that the question of whether a given U expression does or does not express a statement of $\mathfrak{E}$ is definite. The statements of $\mathfrak{E}$ are called elementary statements to distinguish them from other statements which we may form from them or about them in the U language; later on we shall call some of these latter statements "epistatements," but for the moment we do not need this term.

A *theory* (over $\mathfrak{E}$) is defined as a conceptual class of these elementary statements. Let $\mathfrak{T}$ be such a theory. Then the elementary statements which belong to $\mathfrak{T}$ we shall call the *elementary theorems* of $\mathfrak{T}$; we also say that these elementary statements are *true* for $\mathfrak{T}$. Thus, given $\mathfrak{T}$, an elementary theorem is an elementary statement which is true. A theory is thus a way of picking out from the statements of $\mathfrak{E}$ a certain subclass of true statements. We shall then say that the statements of $\mathfrak{E}$ constitute the elementary statements for (or of) the theory $\mathfrak{T}$.[†]

The terminology which has just been used implies that the elementary statements are not such that their truth and falsity are known to us without reference to $\mathfrak{T}$. The U sentences which express them must therefore contain some undetermined constituents or parameters whose meaning is not fixed until $\mathfrak{T}$ is defined. In other words, they are *formal statements*, and they stand, in this respect, in contrast to the *contensive statements* whose truth and falsity are known to us completely beforehand. Of course, one may argue that this is improper usage; that the elements of $\mathfrak{E}$ are not statements until the meaning of these undetermined constituents is fixed; and that therefore we must postulate a separate $\mathfrak{E}$ for each $\mathfrak{T}$. This is, however, a matter of usage of terms. There are two arguments in favor of the usage here adopted. In the first place, it is convenient, in that it enables us to speak of two or more theories with the same $\mathfrak{E}$. In the second place, it agrees with the ordinary usage of the word 'sentence'[1]; for the English expression

he is a jackass

is certainly a sentence, and one which my readers must have heard, yet it is not possible to judge of it as true or false until it is embedded in a context which will tell us what 'he' stands for and in which particular sense the word 'jackass' is intended. Later on we shall consider ways in which these

[†] It is not excluded that we may have theories with different classes $\mathfrak{E}$.
[1] It will be recalled that 'sentence' and 'statement' may be identified.

constituents may enter; for the present we are concerned with matters which do not require such specification.

Although this notion of theory is very general, yet certain notions relating to theories can be defined in terms of it. In the first place, we can define a theory $\mathfrak{T}_1$ to be a *subtheory* of another theory $\mathfrak{T}_2$, or $\mathfrak{T}_2$ to be an *extension* (or supertheory) of $\mathfrak{T}_1$, which relationships we can express by the notation (borrowed from set theory)

$$\mathfrak{T}_1 \subseteq \mathfrak{T}_2$$

just when every elementary theorem of $\mathfrak{T}_1$ is also one of $\mathfrak{T}_2$.† Again we can define a *consistent* theory as one which does not exhaust the whole of $\mathfrak{E}$, and a *decidable* theory as one which is a definite class.

The definition of consistency may seem a bit strange. It is motivated as follows.[1] In theories based on the ordinary predicate calculus, the elementary statements are of the form

$$\vdash P \tag{1}$$

where $P$ is a "formula" and '$\vdash$' designates a unary predicate of assertibility. There is in that calculus a negation operation. Such a theory would be considered inconsistent if both a formula and its negation were assertible, and from such a "contradiction" it follows by the principles of the calculus that every formula is assertible. Hence a theory containing a contradiction would be inconsistent according to the definition given here. Conversely, if such a theory were inconsistent according to the present definition, then any formula $P$ and its negation would both be assertible, and so there would be a contradiction. Thus, for such a theory, the two definitions would be equivalent. The present definition, however, applies under much more general conditions. It has the same damning connotation, for whether negation is present or not, an inconsistent theory is useless.

The term 'primitive frame', introduced later (in Sec. C1) for systems, makes sense for theories and may, on occasion, be so applied.

**2. Deductive theories.** A theory $\mathfrak{T}$ is called *deductive* just when $\mathfrak{T}$ is an inductive class (of elementary statements, of course). From the definitions in Secs. A5 and 1, this implies that the initial elements constitute a decidable theory $\mathfrak{A}$. The elements of $\mathfrak{A}$ will be called *axiomatic statements*, or *axioms*.[2] The modes of combination are then a set, say, $\mathfrak{R}$, of *deductive rules*, or rules of inference; each of these produces an elementary theorem when a suitable number of elementary theorems are given as *premises*. The rules are called *determinative* (cf. Sec. A5) when the elementary statement produced is uniquely determined by the premises. Sometimes it is convenient to use the term '*postulates*' for the rules and axioms together.

A construction under these conditions is called a (*formal*)[3] *demonstration*.

---

† In this it is supposed that $\mathfrak{T}_1$ and $\mathfrak{T}_2$ have the same $\mathfrak{E}$, but it may be extended to the case where they have different ones, $\mathfrak{E}_1$ and $\mathfrak{E}_2$, such that

$$\mathfrak{E}_1 \subseteq \mathfrak{E}_2$$

[1] This argument is given in Post [IGT].

[2] For the present we shall use the shorter term, but later, when we have to talk of axiomatic formulas, the longer term will be useful.

[3] The term 'formal' will be used when there is a possibility of confusion with other uses of 'demonstration'. The terms 'deduction', 'proof', 'derivation' will be used as synonyms for 'demonstration' when there is no conflict with other uses of these words

The elementary theorems are precisely those elementary statements for which a demonstration exists.

Sometimes one considers theories in which one has rules with premises which are not elementary, but are nevertheless such that the conclusion can be regarded as obtained constructively. We cannot take account of such theories until we develop the notion of epitheory later. However, such theories have many of the characteristics of deductive theories as defined here. We shall say that they are deductive in a generalized sense.[1] Much of what we say will apply to them.

Another generalization of a deductive theory arises if we relax the restrictions on definiteness so that $\mathfrak{X}$ is a semi-inductive class. Such a theory will be called *semideductive*. In a semideductive theory the notion of demonstration may not be effective.

For a deductive theory—and also for some of its generalizations—we can define a concept of *completeness* as follows. A deductive theory $\mathfrak{X}$ is complete just when the adjunction to its axioms, leaving the rules unchanged, of an elementary statement which is not an elementary theorem makes the theory inconsistent—in other words, when the theory is incapable of a consistent proper axiomatic extension. (This definition makes an inconsistent theory complete; there is some difference of usage on that point.) This kind of completeness is called *Post completeness*.[2] It is a rather strong property and fails for most systems of any importance, but it holds for the classical two-valued propositional calculus when it is formulated with a substitution rule.

**3. Consequence relations.** Let $\mathfrak{X}$ be a fixed (semi)deductive theory with axioms $\mathfrak{A}$ and rules $\mathfrak{R}$. A semideductive theory formed by adjoining additional axioms to $\mathfrak{A}$, leaving the rules and $\mathfrak{E}$[†] unchanged, is called an *axiomatic extension* of $\mathfrak{X}$. This will be a deductive theory if the additional axioms constitute a decidable theory and if the definiteness of the rules is not disturbed by the extension. This definition makes sense even when $\mathfrak{A}$, and hence $\mathfrak{X}$ (i.e., the class of elementary theorems), is void.

Suppose the additional axioms form a theory $\mathfrak{B}$. Then the axiomatic extension of $\mathfrak{X}$ will be called the (deductive) closure of $\mathfrak{B}$ and designated $\mathrm{Cn}(\mathfrak{B})$.[‡] Further, a statement $X$ will be said to be a *consequence* of $\mathfrak{B}$ relative to $\mathfrak{X}$ just when $X$ is in $\mathrm{Cn}(\mathfrak{B})$.

The operation of passing from $\mathfrak{B}$ to $\mathrm{Cn}(\mathfrak{B})$ is a closure operation in the sense in which it is ordinarily understood in mathematics. It has the following properties:

I $$\mathfrak{B} \subseteq \mathrm{Cn}(\mathfrak{B})$$

II $$\mathrm{Cn}(\mathrm{Cn}(\mathfrak{B})) \subseteq \mathrm{Cn}(\mathfrak{B})$$

III $$\mathfrak{B}_1 \subseteq \mathfrak{B}_2 \rightarrow \mathrm{Cn}(\mathfrak{B}_1) \subseteq \mathrm{Cn}(\mathfrak{B}_2)$$

---

[1] The more restricted kind of deductive theory can then be called an *elementary deductive theory*. Cf. Sec. 2D3.

[2] It was introduced by Post [IGT].

[†] The new axioms must then belong to the original $\mathfrak{E}$.

[‡] 'Cn' here is an abbreviation of 'consequence'. For its introduction see Tarski and Woodger [LSM, p. 63]. (This is a translation from Tarski [FBM].)

These characterize a closure relation in general. The deductive closure has the following further property:

IV   If $X$ is in Cn($\mathfrak{B}$), then there is a finite subtheory $\mathfrak{C}$ of $\mathfrak{B}$ such that $X$ is in Cn($\mathfrak{C}$).

The further study of this closure operation belongs to the part of our subject known as epitheory, which will be considered in Chap. 3. But it is mentioned here because certain logical doctrines take the consequence relation, rather than the notion of deductive theory, as fundamental. From that point of view the rules $\mathfrak{R}$ establish a relation of *direct consequence;* then Cn($\mathfrak{B}$) is, essentially, the least class containing ($\mathfrak{A}$ and) $\mathfrak{B}$, which is closed with respect to the direct consequence (i.e., with respect to $\mathfrak{R}$). In view of the discussion of Sec. A5, this is essentially the same as that here given. It is clear that Cn($\mathfrak{B}$) coincides with $\mathfrak{T}$ if $\mathfrak{B}$ is void.

In terms of the consequence relation, one can characterize Post *completeness* (Sec. 2) as follows. $\mathfrak{T}$ is complete just when every statement of $\mathfrak{E}$ is a consequence (relative to $\mathfrak{T}$) of any statement $X$ not in $\mathfrak{T}$.

**4. Interpretation of theories.** Up to the present we have been studying a theory purely as a conceptual class of statements, without regard to its relation to other notions. Evidently a theory, conceived of as a class of statements which may be defined in any conceivable way, is of interest to us only in so far as there is some relation between the theory and some contensive subject matter, i.e., some subject matter which is known to us independently of the theory. A theory is useful to us in so far as it enables us to make predictions concerning the subject matter.

Such a relationship between a theory and a contensive subject matter will exist when there is a many-to-one correspondence between certain elementary statements of the theory and certain contensive statements related to the subject matter. In such a case we say we have an *interpretation* of the theory in the subject matter. This interpretation will be said to be *full* if there is such a contensive correspondent to every elementary statement; otherwise it will be *partial.* The correspondent contensive statement will also be said to be the *interpretant* of the original elementary statement; when it is not necessary to be too exact we shall call it the interpretation also. Thus an elementary statement of a physical theory may have as interpretant some statement which can be subjected to an experimental test; most physical theories contain statements which are not capable of direct experimental test, so that the interpretation is only partial.

An interpretation is *valid* just when the interpretant of every elementary theorem (i.e., of every true elementary statement) is true. It is *adequate* (or relatively complete) if every elementary statement whose interpretant is true is a theorem. These terms are thus the analogues of consistency and completeness of an uninterpreted theory. They are relative to the interpretation, and hence to the subject matter, but when the latter is understood, they may also be applied to the theory. If the subject matter is empirical, these notions are empirical too.

As already said, we generally study a theory because we wish to use it for some purpose. Just when the theory is suitable for that use we say it is *acceptable* for that purpose. Acceptability thus may involve all the information which we have on hand in regard to validity; but of two theories equally valid, so far as known, one may be more acceptable because it is simpler, more natural, more aesthetically or philosophically satisfying, etc. Acceptability is relative to our knowledge at a given moment as well as to the purpose or intended use; a theory may be acceptable today and unacceptable tomorrow, or it may be acceptable for one purpose but not for another.

As here conceived, interpretation is a correspondence between statements, each of the two corresponding members being an actual statement with its own criterion of truth. This seems a preferable mode of speech to that in which we think of the same statement as having two kinds of truth, formal and contensive. Likewise, interpretation does not involve assigning "meaning" to certain constituents in the elementary statement and determining the truth from their meanings. We have an interpretation whenever we have a correspondence, no matter how the correspondence is set up.[1]

## EXERCISES

In the following it is supposed that $\mathfrak{B}$, $\mathfrak{C}$, etc., are extensions of a fixed theory $\mathfrak{T}$. The symbols '$=$', '$\subseteq$', '$\cup$', '$\cap$' are used in their ordinary set-theoretic senses, viz., as denoting set equality, inclusion, union, and intersection, respectively. The citations beginning with 'Th.' are to theorems in Tarski [FBM], where a great variety of more complex such statements may be found.

**1.** Show that the deductive closure has the properties I to IV. (Cf. Th. 1.)

**2.** Show that
$$\mathrm{Cn}(\mathfrak{B}) \cup \mathrm{Cn}(\mathfrak{C}) \subseteq \mathrm{Cn}(\mathfrak{B} \cup \mathfrak{C}) = \mathrm{Cn}(\mathfrak{B} \cup \mathrm{Cn}(\mathfrak{C}))$$
$$= \mathrm{Cn}(\mathrm{Cn}(\mathfrak{B}) \cup \mathrm{Cn}(\mathfrak{C}))$$
(Th. 2.)

**3.** Let $\mathfrak{B}$ be called deductively closed just when
$$\mathrm{Cn}(\mathfrak{B}) = \mathfrak{B}$$
Show that if $\mathfrak{B}$ and $\mathfrak{C}$ are deductively closed, so also is $\mathfrak{B} \cap \mathfrak{C}$, but in general $\mathfrak{B} \cup \mathfrak{C}$ is not.

**4.** Let $\mathfrak{B}$ be called axiomatizable[2] just when it has the same deductive closure as some finite $\mathfrak{C} \subseteq \mathfrak{B}$. Show that the union of any finite number of axiomatizable theories is axiomatizable. (Cf. Th. 20.)

**5.** Let a theory $\mathfrak{B}$ be called independent when no one of its statements is a consequence of the rest. Formulate this condition in terms of the Cn operation; further show that each of the following conditions is necessary and sufficient for it:

(a) $\qquad \mathfrak{B} \cap \mathrm{Cn}(\mathfrak{C}) \subseteq \mathfrak{C} \qquad\qquad\qquad$ for all $\mathfrak{C} \subseteq \mathfrak{B}$

(b) $\qquad \mathfrak{C}_1 \cup \mathfrak{C}_2 \subseteq \mathfrak{B} \ \& \ \mathrm{Cn}(\mathfrak{C}_1) = \mathrm{Cn}(\mathfrak{C}_2) \to \mathfrak{C}_1 = \mathfrak{C}_2 \qquad$ for all $\mathfrak{C}_1, \mathfrak{C}_2$

(Th. 31.)

**6.** Show that the properties in the preceding exercises follow abstractly from I to IV—i.e., they hold if a theory is regarded as a subset of some set of objects $\mathfrak{E}$, not necessarily elementary statements—and that Cn is a unary operation on such subsets such that I to IV hold. What role, if any, does IV play in this proof? Show further that if the $\mathfrak{B}_1$ in III is restricted to be finite, III becomes the converse of IV, and that with this converse (i.e., with IV as an equivalence, and I, II), III is superfluous. (Th. 1.)

---

[1] In this respect usage differs from that of Carnap [ISm].

[2] The word 'axiomatizable' has other senses in the literature. In such cases one uses 'finitely axiomatizable' for the present sense.

## C. SYSTEMS

It was pointed out in Sec. B1 that the elementary statements on which a theory is based necessarily contain certain unspecified constituents or parameters; i.e., they are formal statements. We were not concerned there with the way in which these parameters were introduced. We shall now specify that the parameters enter as unspecified objects about which the elementary statements assert that they have certain properties or that certain relations hold. A theory whose statements are formed in this way will be called a *system*. We shall consider two main sorts of systems and discuss their natures and relations to one another.

**1. Systems in general.** As explained in the introduction to this section, we here postulate a certain conceptual class of objects, called the *formal objects*, and a conceptual class of predicates, called the *basic predicates*, each of the latter having associated with it a natural number called its *degree*. The elementary statements are then precisely all those statements which assert that a basic predicate holds of an ordered sequence of formal objects in number equal to its degree. We can, for the moment, symbolize such a statement by an expression of the form

$$\Phi(a_1, \ldots, a_n) \tag{1}$$

where '$a_1$', ..., '$a_n$' are abbreviations for names of specific formal objects, '$\Phi$' is an abbreviation for an $n$-argument verb designating a basic predicate of degree $n$, and the parentheses and commas indicate in the usual fashion the fact that $\Phi$ is to apply to the arguments $a_1, \ldots, a_n$. Under the assumption that the formal objects form a definite class, and the basic predicates do likewise (with the determination of degree being understood to be definite also), it is clear that the elementary statements form a definite (conceptual) class.

There are, as we shall see in Secs. 2 and 3, two principal variants of the notion of system which differ in regard to the nature of the formal objects. Before discussing these, it is expedient to take up here certain preliminary matters which apply to both types of system.

In order to present such a system in the U language, it is necessary to decide on a notation for naming the formal objects and designating the basic predicates, and also on devices for combining these to form the U sentences expressing the elementary statements. This notation, in its totality, forms a language in the semiotical sense; this language is here called the *A language*. Its nouns, the names of the formal objects, will be called *A nouns;* its verbs, which designate the basic predicates, will be called *A verbs*, and each will be said to have the same degree as the predicate it designates; and its sentences, which express the elementary statements, will be called *A sentences*. Thus the A language contains those linguistic devices which suffice for the expression of the elementary statements.

It cannot, however, be too strongly emphasized that the A language is not a language being talked about; it is adjoined to the U language to be used therein. The A nouns are a special kind of U noun; the A verbs a special kind of U verb; and the A sentences a special kind of U sentence. The adjunction of this new terminology to the U language does not differ in any essential point from any other procedure where we introduce technical expressions into the U language.

In the following it will be convenient to have a certain standard terminology for use in the A language of a system (or family of systems) whenever particular conditions do not make some other usage desirable. For each $n = 1, 2, 3, \ldots$, let the system contain $m_n$ basic predicates of degree $n$. If $m_n > 0$, then for $k = 1, 2, \ldots, m_n$, let the $k$th predicate of degree $n$ be $\phi_n^k$,[1] and its closure for arguments $X_1, \ldots, X_n$ be

$$\phi_n^k X_1 X_2 \cdots X_n$$

If $m_n = 0$, there is no predicate $\phi_n^k$ for any $k$.[2]

The considerations relating to the formal objects and elementary statements are conveniently referred to as the *morphology* of the system. This contrasts with the *theory proper*, which relates to the theory built upon that morphology. The notion of system is to be understood as admitting the possibility that there may be no basic predicates, in which case we speak of it as a *pure morphology*. The conventions defining a system will be called its *primitive frame*.

**2. Syntactical systems.** The first of our two kinds of system is the syntactical system. In such a system the formal objects are taken to be the expressions of some object language. Let us call this language the *O language*. Then there is a certain stock of *O symbols*, or *letters*, constituting the *O alphabet;* the formal objects are the finite strings of these letters.

One way of conceiving the formal objects as an inductive class is to visualize them as formed, one letter at a time, from left to right. If the void expression is admissible, it can be taken as the single initial element; otherwise we must have an initial element for each letter. As operations, we must have one for each letter, viz., affixing it on the right. Then the expressions form an inductive class; indeed, a monotectonic one. A syntactical system conceived in this way will be called an *affixative system*. It is not usually satisfactory, as we shall see.

A second way is to conceive the expressions as an inductive class in which the letters are the initial elements and there is a single binary operation called *concatenation*. We sometimes symbolize this in the A language (hence in the U language) by '$\wedge$' used as a binary infix (i.e., placed between its argument like '$+$' in elementary algebra), sometimes by simple juxtaposition (like multiplication in algebra). Its meaning is as follows. If $X$ and $Y$ are expressions, $X \wedge Y$ is formed by writing $Y$ immediately after (to the right of) $X$. Thus if '$\alpha$', '$\beta$' are letters, $X$ is '$\alpha\alpha\beta$', $Y$ is '$\beta\beta\alpha$', then $X \wedge Y$ is '$\alpha\alpha\beta\beta\beta\alpha$'.

---

[1] For some purposes it is preferable to have a single sequence of predicates $\phi_k$ and to let the degree of $\phi_k$ be $n_k$.

[2] Note that we are here using the idiom of the U language mentioned in connection with "quotation functions" in Sec. A3. In the more explicit language of the footnote we should say: "The A verbs will be prefixes of the form '$\phi_n^k$', where '$n$' and '$k$' are numerals, of which '$n$' designates the degree and '$k$' is the index in an enumeration of those, if any, of that degree; the A sentences are of the form '$\phi_n^k X_1 X_2 \cdots X_n$', where '$n$' and '$k$' are as before and '$X_1$', $\ldots$, '$X_n$' are A nouns." Note that the ellipses and the indication of the arguments have still to be left to common sense.

It is clear that the expressions are now an inductive class; but since concatenation is associative, it is polytectonic. Thus '$\alpha\beta\alpha$' has the two constructions

$$\frac{\dfrac{\text{`}\alpha\text{'}\quad\text{`}\beta\text{'}}{\text{`}\alpha\beta\text{'}}\quad\text{`}\alpha\text{'}}{\text{`}\alpha\beta\alpha\text{'}}\qquad\qquad\frac{\text{`}\alpha\text{'}\quad\dfrac{\text{`}\beta\text{'}\quad\text{`}\alpha\text{'}}{\text{`}\beta\alpha\text{'}}}{\text{`}\alpha\beta\alpha\text{'}}$$

This kind of syntactical system will be called *concatenative*.

Let us now look at two examples. In both of these we shall suppose the alphabet consists of three letters $a$, $b$, $c$† (where '$a$' is the name of '$\alpha$', '$b$' of '$\beta$', etc.), and we shall indicate concatenation by simple juxtaposition in the A language.

The letters '$X$', '$Y$', '$Z$' are used as "U variables" (Sec. 3D1), i.e., as pronouns for unspecified O expressions.

*Example* 1. (Sams, first form.)

ELEMENTARY STATEMENTS.    Three unary predicates:

—— is a sam.

—— is a tettle.

—— is a tantet.

AXIOMS.    $a$ is a sam.

$aca$ is a tantet.

RULES.    $X$ is a sam $\rightarrow$ $Xb$ is a sam.

$X$ is a sam, and $Y$ is a sam $\rightarrow$ $XcY$ is a tettle.

$XcY$ is a tantet $\rightarrow$ $XbcYb$ is a tantet.

*Example* 2. (Sams, second form.)

ELEMENTARY STATEMENTS.    One binary, one unary predicate:

—— is a sam.

——$_1$ = ——$_2$.

AXIOMS.    $a$ is a sam.

$a = a$.

RULES.    $X$ is a sam $\rightarrow$ $Xb$ is a sam.

$X = Y \rightarrow Xb = Yb$.

*Remarks on These Examples.* One can easily convince oneself that the sams in both examples are the expressions in the list

$a$    (that is, '$\alpha$')

$ab$    (that is, '$\alpha\beta$')

$abb$    (that is, '$\alpha\beta\beta$')

and that $X = Y$ in Example 2 exactly when $XcY$ is a tantet in Example 1, viz., when $X$ and $Y$ are the same sam.

The systems have been formulated as concatenative systems. If we were to attempt to formulate them as affixative systems, there would be no difficulty about Example 2; but there would be serious difficulty about Example 1, for we could not state the rules without bringing in the concatenation operation (see below, Sec. D3).

† Example 2 does not use $c$.

An at present widely accepted form of syntactical system,[1] often known as a *calculus*, has the following characteristics. The formal objects are taken in the concatenative sense. The deductive theory based on this contains two sorts of rules, called *formation rules* and *transformation rules*, respectively. The formation rules state what are the sentences in the O language, so that the formation rules involve having the notion of being an O sentence as a basic predicate. The transformation rules define a consequence relation among the O sentences which is analogous to that described for the elementary statements (or U sentences) in Sec. 2B3. As the discussion given there shows, this is equivalent to defining a family of systems depending on a class of initial sentences, the axiomatic O sentences, as a parameter; in each of these the main basic predicate is being an O theorem. Some authors do not insist on the definiteness restrictions insisted on here, so that their systems are only semideductive.

The system of Example 1 is a calculus in this sense if we regard the tettles as O sentences and the tantets as O theorems. As this example shows, one may need additional (sometimes called "auxiliary") basic predicates, such as that of being a sam, in order to formulate the system. Systems like Example 2, with a binary basic predicate, are not admitted as calculuses (although such predicates may doubtless occur as auxiliaries). The Church theory of $\lambda$ conversion, in the form in which he himself presents it, is a less trivial example of a system with a binary basic predicate;[2] it has nothing analogous, at least directly, to notions like O sentence and O theorem.

In connection with such calculuses one should guard against confusing the meaning of 'sentence' as it occurs in 'O sentence' and 'U sentence'. In the latter case the word 'sentence' is itself part of the U language, and is supposed to be understood as such. In the former case it denotes simply an inductive class of expressions conceived as shapes. Whether or not these expressions are actually sentences as ordinarily understood is just as irrelevant as whether they are associated in someone's mind with monkeys in a zoo. For precisely this reason I have translated these terms into the Hungarian language, where 'sam' (i.e., '*szám*') means 'number', 'tettle' (i.e., '*tétel*') means 'sentence', and 'tantet' ('*tantét*') means 'theorem'. By so doing I hope to strip off the semantical (i.e., meaning-related) connotation which is frequently sneaked in by the use of the word 'sentence'. At the same time I hope it will bring out more clearly the fundamental distinction between those notions which we express in the A language—and hence in the U language—by verbs and sentences, and those which we simply name.

Again, even though we say we are talking about an O language, it is not necessary to trot such a language out explicitly. One can make remarks about the President of the United States without having that dignitary actually present. If one does exhibit the O language, it is for illustrative purposes only. Not only that, but the precise nature of the O letters is entirely irrelevant. Thus, I have said above that the O letters for the system of sams consisted of the Greek letters $a$ and $b$ (i.e., '$\alpha$' and '$\beta$'), but

---

[1] For example, in Carnap [LSL]. A calculus in the sense of Lorenzen [EOL] is slightly different. Cf. [CFS].

[2] For a modified and strict formulation as a calculus in the sense of Lorenzen, see [CFS], example 7.

actually, it is irrelevant whether $a$ and $b$ are Greek letters, Egyptian hieroglyphs, two distinguishable and reproducible kinds of bricks, or two kinds of noises. All we need to exhibit are the A names of those letters; all that we need to know about those letters is that they act like links of different kinds which can be forged into chains. In what follows, the O language is not exhibited; the symbols which appear when we are talking about O letters are the A nouns which name them.

As a standard terminology, to be used when something else is not suitable, we agree that the O letters shall be $a_0, a_1, \ldots$, and that concatenation shall be indicated, as above, by juxtaposition or by an infixed '$\wedge$'. There may be any number of letters, but the number of letters can always be reduced to two, since we can replace the original letters by sams.

We shall use the sams as natural numbers in the fashion described in Sec. A7. The natural notation

$$0, 1, 2, \ldots$$

will be preferred to

$$a, ab, abb, \ldots$$

for such use, and the successor function will be denoted by priming.

**3. Ob systems.** In the second type of deductive system, here called an *ob system*, the formal objects form a monotectonic inductive class. The elements of this inductive class are called *obs*, its initial elements *atoms*, and its modes of combination (*primitive*)[1] *operations*. Every ob is thus the result of a construction from the atoms by the primitive operations; by the monotectonic property this construction is unique. Thus an ob can be identified with such a construction, objectified, if you will, by means of a tree diagram (or a normal construction sequence); in this respect it stands in contrast to an O expression which can be objectified as a linear series. Beyond this specification it is irrelevant what the obs are; this colorless word 'ob' has been deliberately chosen to emphasize this irrelevance.

The system of Example 2, and indeed any affixative syntactical system, is an example of an ob system. This follows at once from the above remark that in these systems the formal objects are monotectonic.

Other examples of ob systems are found in those syntactical systems in which there is a special conceptual class of "well-formed expressions," here called *wefs*,[2] such that this class is monotectonic and exhausts all the expressions which play any actual role in the system. Practically all the systems considered in modern mathematical logic and mathematics are of this character.

We shall now look at some other examples of ob systems.

*Example 3.* (Generalized sams.)

ATOMS. One, namely, $a$.

OPERATIONS. $n$ unary operations: the application of the $k$th one to the argument $X$ is $Xb_k$. One binary operation: its application to arguments $X$, $Y$ in that order is $(X \wedge Y)$.

---

[1] The term 'primitive' will be used when it is desired to distinguish the operations specified in the primitive frame from others introduced, e.g., by definitions; otherwise it may be omitted.

[2] The usual term is 'wff' (for 'well-formed formula'), but 'wef' has the advantage of being pronounceable.

ELEMENTARY STATEMENTS.   One binary predicate, equality, giving rise to elementary statements of the form

$$X = Y$$

AXIOMS.   If $X$, $Y$, $Z$ are obs, the following are axiomatic statements:

$$X \wedge a = X$$
$$X \wedge Yb_k = (X \wedge Y)b_k$$

RULES

1. If $X = Y$, then $Y = X$.
2. If $X = Y$ and $Y = Z$, then $X = Z$.
3. If $X = Y$, then $Xb_k = Yb_k$ ($k = 1, 2, \ldots, n$).
4. If $X = Y$ and $U = V$, then $X \wedge U = Y \wedge V$.†

In this example there are infinitely many axioms, for each separate determination of $X$ and $Y$ gives rise to an axiom.[1]  Such a formulation of a whole infinity of axioms, using U variables like '$X$', '$Y$', is called an *axiom scheme*. It would probably be instructive for the reader, at this stage, to show that the binary operation is associative.[2]

The following example is a variant of Example 3.

*Example* 4. (Associative system.)
ATOMS.   $b_1, b_2, \ldots, b_n$.
OPERATIONS.   A single binary one, like the last one in Example 3.
ELEMENTARY STATEMENTS.   Same as in Example 3.
AXIOMS.   If $X$, $Y$, $Z$ are any obs,

(1)                                        $X = X$
(2)                        $(X \wedge (Y \wedge Z)) = ((X \wedge Y) \wedge Z)$

RULES.   Same as Rules 1, 2, 4 in Example 3.
Examples 3 and 4 may be regarded as formulating a syntactical system as an ob system (cf. Secs. 5 and 6).

*Example* 5. (Propositional algebra.)
ATOMS.   An infinite sequence $p_1, p_2, \ldots$.
OPERATIONS.   One unary, one binary.   The closures of these for argument $X$ and arguments $X$, $Y$, respectively, are

$$\neg X \qquad (X \supset Y)$$

(The latter should be read "$X$ ply $Y$" since $\supset$ is an operation and '$p_1 \supset p_2$', for instance, is a noun.)

---

† This rule, as stated, is determinative.   If determinativeness did not interest us, we could split this rule into the simpler rules:
If $X = Y$, then $X \wedge Z = Y \wedge Z$.
If $X = Y$, then $Z \wedge X = Z \wedge Y$.
But to make these determinative we should need a premise such as "$Z = Z$"; then the rules would be no simpler than the above Rule 4.

[1] The following are axioms of Example 3:
$$ab_2b_1 \wedge a = ab_2b_1$$
$$ab_2b_1 \wedge ab_2b_3b_5 = (ab_2b_1 \wedge ab_2b_3)b_5$$

[2] Cf. Exercise 2.

ELEMENTARY STATEMENTS.    One unary basic predicate forming elementary statements of the form

$$\vdash X$$

where $X$ is an ob.

AXIOMS.    If $X$, $Y$, $Z$ are any obs

$$\vdash (X \supset (Y \supset X))$$
$$\vdash ((X \supset (Y \supset Z)) \supset ((X \supset Y) \supset (X \supset Z)))$$
$$\vdash ((\neg Y \supset \neg X) \supset (X \supset Y))$$

RULES.    If $\vdash (X \supset Y)$ and $\vdash X$, then $\vdash Y$.

In this example there is an infinite sequence of different atoms, the "propositional variables." No properties of these atoms are used except that they form an infinite sequence. Consequently we could get a system of strictly finite morphology by taking the sams themselves in the role of propositional variables, as follows.

*Example 6.*    (Finite form of propositional algebra.)

ATOMS.    One, namely, $a$.

OPERATIONS.    Two unary, one binary, with closures (for arguments $X$, $Y$)

$$Xb, \quad \neg X, \quad (X \supset Y)$$

ELEMENTARY STATEMENTS.    Three unary predicates giving rise to elementary statements of the following form:

| | |
|---|---|
| $S(X)$ | ($X$ is a sam) |
| $P(X)$ | ($X$ is a proposition) |
| $\vdash X$ | ($X$ is asserted) |

AXIOMS.    $S(a)$.

RULES

If $S(X)$, then $S(Xb)$.

If $S(X)$, then $P(X)$.

If $P(X)$, then $P(\neg X)$.

If $P(X)$ and $P(Y)$, then $P(X \supset Y)$.

If $P(X)$, $P(Y)$, then $\vdash (X \supset (Y \supset X))$.

If $P(X)$, $P(Y)$, $P(Z)$, then $\vdash ((X \supset (Y \supset Z)) \supset ((X \supset Y) \supset (X \supset Z)))$.

If $P(X)$, $P(Y)$, then $\vdash ((\neg Y \supset \neg X) \supset (X \supset Y))$.

If $\vdash (X \supset Y)$ and $\vdash X$, then $\vdash Y$.

In order to define an ob system we must, of course, choose some systematic way of assigning an A noun to each ob. A particular way of doing this will be called a *presentation* of the system. The following standard presentation for an arbitrary formal system will be adopted, like that of Secs. 1 and 2, for use when some special consideration does not dictate otherwise: the atoms will be $a_0, a_1, a_2, \ldots$ , forming a finite or infinite sequence; the $k$th operation of degree $n$ will be $\omega_n^{(k)}$, and its closure for arguments $X_1, \ldots, X_n$ will be

$$\omega_n^{(k)} X_1 \cdots X_n \dagger$$

† Compare the explanation of the $\phi_n^k$ in Sec. 1. Note that I am explaining the use of a symbol in the U language by saying what the designatum is, and in doing so, the symbol is used, not mentioned. This idiom is more natural than the circumlocution used in Sec. 1. As in Sec. 1 it is also sometimes convenient to enumerate the operations in a single sequence, the $k$th operation being $\omega_k$ and its degree $n_k$.

Parentheses are then unnecessary; in fact, the notation can be shown to be monotectonic in the sense that every construction is uniquely described. The presentation will be called the *Łukasiewicz standard presentation*. A presentation which differs from it in the choice of symbols, but preserves its fundamental idea of using prefixed functions of fixed degree without parentheses, will also be called a Łukasiewicz presentation (but not the standard one).

The concept of an ob formal system is to be understood as including certain degenerate cases. Thus we may have a pure ob morphology with no basic predicates, and hence no overlying theory; we may consider ob systems without any operators; etc. For some purposes it is expedient to regard the atoms as operations of degree 0.

**4. Representation of a system.** It has been stressed that the exact nature of the formal objects of a formal system of either type is irrelevant. Any way of regarding the formal objects as specified objects given from experience will be called a *representation* of the system, provided the contensive objects retain the structure of the formal objects. Thus, when we said, in Example 1, that $a$ was '$\alpha$' and $b$ was '$\beta$', and made the convention that concatenation of $A$ nouns was to indicate concatenation of the designated O expressions, we were making a representation of the system.

The restriction that the contensive objects retain the structure of the formal objects is important. It means that there is a separate contensive object for each formal object, and in the case of an ob system, this means a separate object for each construction. It means further that the operations must be reflected in some way as modes of combination of the contensive objects. In technical terms there must be a one-to-one correspondence, isomorphic with respect to the operations and modes of combination, between the formal objects (or their names, the $A$ nouns) and the contensive objects of the representation.

A representation is not to be confused with an interpretation. As defined in Sec. B4, an interpretation is a correspondence between elementary statements and certain contensive statements, and it is defined for a theory whether or not that theory is a system; a representation is a correspondence between formal objects and contensive objects, and it is defined for a pure morphology without regard to the theory which is built upon it; moreover, the truth of the elementary statements is entirely unaffected by it. We shall discuss this point further in Sec. 5.

The contensive objects of the representation may be chosen in various ways. For those who are that way inclined, abstract ideas or platonistic concepts may be taken, or the contensive objects may be chosen as objects of a more concrete nature. But the choice of representation is irrelevant for the proofs of the theorems.

We shall now discuss certain particular modes of representation of a relatively concrete kind.

In the first place, the $A$ nouns themselves, since they provide a unique name for every formal object and must reflect their structure, constitute a representation by definition. This representation is called the *autonymous representation*.

In the second place, any formal system of either type has a syntactical representation. This is trivial because the autonymous representation is syntactical, but we can get further syntactical representations from the autonymous one if we replace its letters by new ones, possibly making other changes in the operators.[1] In particular we can use a Łukasiewicz presentation, standard or otherwise, as a representation even when we do not use it in the A language. Such a representation we shall call a *Łukasiewicz representation;* when it is standard we have the *standard Łukasiewicz representation.*

In connection with a syntactical representation it is worthwhile to remark that the expressions which correspond to the obs cannot constitute all the expressions of the appropriate O language in the concatenative sense, for the latter are polytectonic. If we call the expressions which actually do appear *wefs* (i.e., well-formed expressions),[2] then the obs are the wefs and the wefs form a monotectonic inductive class of expressions. Such a system might be called *eutactic,* as opposed to the *pantactic* system, in which all possible expressions may occur.[3]

By the device of enumerating the letters, and then replacing the numbers by the corresponding sams, we can reduce such a representation to one in the expressions in the alphabet $\{a, b\}$.[4]

Finally, it is possible to find a syntactical representation in which the O language has only one symbol. Since the words in a language with only one symbol are distinguishable only in the number of occurrences of that symbol, this amounts to representing the system in terms of natural numbers. Such a representation is called a *Gödel representation,* and the numerical representative of a formal object is its *Gödel number.* We can conveniently use ordinary arithmetical notation in connection with it. There are various ways of bringing it about.[5] For a syntactical system one of the simplest is to assign the prime numbers to the letters and then to assign to the sequence

$$X_1 X_2 \cdots X_n$$

the Gödel number

$$2^{g_1} \cdot 3^{g_2} \cdots \cdots p_n^{g_n}$$

where $p_k$ is the $k$th prime and $g_k$ is the Gödel number of $X_k$. For an ob system, we can assign to the ob

$$\omega_n X_1 X_2 \cdots X_{n_k}$$

the number

$$2^n \cdot 3^k \cdot 5^{g_1} \cdots \cdots p_{n+2}^{g_{n_k}}$$

This procedure can be modified in innumerable ways. For cases in which only a finite number of constituents appear, very much simpler assignments are possible (using ordered $n$-tuples or $k$-adic expansions instead of prime-factor decompositions, etc.).[6]

It is possible to present a system without having any specific representation in mind. Such a system is called *abstract.* It is clear from what has been said that an abstract system can be of either type.

---

[1] The discussion of [CFS], Sec. 4, suggests ways of getting such representations.

[2] Cf. remarks just before Example 3 in Sec. 3.

[3] For instance, the systems of Examples 1 and 2 are eutactic, the sams constituting the wefs.

[4] See below, Sec. D2.

[5] Cf. [CFS], example 6, pp. 256ff.

[6] Cf. [CFS].

**5. Interpretation of a system.** The notion of interpretation has been defined in Sec. B4 in relation to a theory. Here we shall consider specializations of this idea when we are dealing with systems.

For this purpose we first define a notion of correspondence similar to a representation except that the same contensive object may be assigned to two or more different formal objects. Let us call such a correspondence a *valuation;* this will be relative to a conceptual class $\mathfrak{B}$ consisting of the contensive objects, called *values,* which are assigned to the formal objects. For instance, we can form a valuation for Example 5 over the class $\mathfrak{B}$ consisting of 0 and 1 by assigning arbitrarily one of these values to each atom, and allowing other obs to take the values determined by the usual truth tables, with 1 in the role of truth. Again, consider Example 3. Let the values here be the words, including the empty word, in the alphabet $c_1, c_2, \ldots, c_n$. Let $a$ be assigned the empty word, and if $X$, $Y$ are assigned the words $X'$, $Y'$, respectively, let $Xb_k$ be assigned the word $X'c_k$ and $X \wedge Y$ the word $X'Y'$. Neither of these valuations can be a representation: in the first case, because infinitely many obs will be assigned the same value; in the second case, because obs giving different constructions of the same word will be assigned the same value, even though they are distinct obs.

In the examples considered in the preceding paragraph the valuation was defined by giving values to the atoms and determining the values of the other obs by contensive operations. Cases like Example 6 require slight generalization. One would begin, in this case, by assigning values to the sams, and from that point on the procedure would be the same as for Example 5. The sams constitute what are called later (Sec. 3D2) *quasi atoms.*

Given a valuation, we may define an interpretation by associating with each basic predicate as its *interpretant* a predicate defined over the values. An interpretation defined in this way will be called a *direct interpretation.* Thus we get a direct interpretation of Example 5 by taking as interpretant of ⊢ the property of being 1; and in Example 3 by taking as interpretant of equality the relation of having the same value. The interpretations in either case are valid; in the second case it is adequate also.[1]

We get another kind of interpretation by returning to Example 5. Let us say that an ob $X$ is a tautology just when it has the value 1 in every valuation of the relevant sort with standard truth tables. Then take as interpretant for the elementary statement

$$\vdash X$$

the statement that $X$ (or the associated function over the values) is a tautology. That this interpretation is valid is easily seen by an inductive argument;[2] that it is adequate, by a standard (epi)theorem of the classical propositional algebra. Yet it is not a direct interpretation, at least in a finitary sense.[3] This is typical of systems involving obs which are called

---

[1] See Exercise 3 at the end of this section.

[2] See Exercise 8.

[3] One could say that in this case the obs become certain truth functions defined over all assignments of truth values to the atoms and that we really have a valuation in which the values are truth functions. But such a value is not a finite array. Hence it is a value only in a generalized sense.

variables, for these are not assigned a single contensive value, but are allowed to vary over a range.

Interpretations of the last-mentioned sort are closely akin to what are commonly called *models*. This concept is defined for systems based on the first-order predicate calculus. As such it involves more machinery than we have here at our disposal. But the central idea is that there is a family of valuations over a certain range of values; a contensive statement function over such evaluations is assigned to each elementary statement; and the interpretant is the statement that this statement function is true for all valuations. This is a *semimodel*; a semimodel is a model just when it is valid.

**6. Comparison of syntactical and ob systems.** The interrelationships of the two types of system have been the subject of comment at various points in the foregoing. It will be expedient to bring together these more or less scattered remarks and to add to them some others, so as to have a more systematic view of the nature of these systems.[1]

In the first place, neither type of system necessarily commits one to the view that the essential subject matter of mathematics is symbols. Both types of system can be represented in terms of expressions, in fact in the expressions of any alphabet having two or more symbols and, in a generalized sense, in terms of an alphabet with only one symbol, i.e., in terms of numbers. But in neither case is one committed to a particular choice of this alphabet. The properties which one takes into account have nothing to do with the nature of the symbols themselves; one never says, for example, that '$x$' is made by two crossed lines. It would be more accurate to say that in mathematics we are concerned with structures which can form artifacts of recognizably different kinds by various modes of combination, and further, that one is interested in those properties which are not changed when one changes the elements or replaces the modes of combination by others homologous to them. These structures are such as one can objectify by chains of different kinds of links in the case of a syntactical system, or by treelike constructions in the case of an ob system.

In the second place, either of the two types of system can be reduced to the other. Thus the possibility of syntactical representation, discussed in Sec. 4, shows that an ob system can be reduced to a eutactic syntactical one. The converse reduction, for an unrestricted syntactical system, is shown in principle by Examples 3 and 4. In the case of the systems ordinarily used for logical purposes, one can go even further. For these systems are eutactic and monotectonic, and such a system, as it stands, belongs to both types simultaneously.

From a certain point of view an ob system is a more rigorous concept than a concatenative system. In the latter the associativity of the concatenation operation has to be taken for granted. Thus a proof in a concatenative system is like a proof of a geometric theorem by drawing a figure. It is perhaps true that one cannot banish such intuitive evidence entirely (since one needs some of it in checking a construction), but there is less of it in an ob system.

Again, the notion of ob system puts less emphasis on linguistic accidents.

[1] See also Secs. S3 and S4.

For example, suppose one were to take Example 6 with an autonymous representation; one would then have a concatenative system whose alphabet consisted of the letters

$$a, b, (,), \neg, \supset$$

If one were to replace these by other letters, say,

$$\alpha, \beta, [,], \sim, >$$

anyone would agree that we had merely another representation of the same concatenative system. But if one were to pass to the standard Łukasiewicz representation, the resulting concatenative system would be so different as to be a distinct system, whereas from the standpoint of ob systems, it is still only another representation of the same system. Thus an ob system is invariant of a wider class of changes in representation than is a concatenative system. Consequently it agrees with the tendency in mathematics to seek intrinsic, invariant formulations, such as vectors, projective geometries, topological spaces, etc.

Up to this point we have been discussing uninterpreted formal systems. But such systems, at least the primary ones,[1] do not come to us originally in pure form, an interpretation being added later; rather we have first some contensive discipline, from which a formal system is then created by a process of *formalization*. In the next five paragraphs we shall discuss formalization in more detail.

The first stage in formalization is the formulation of the discipline as a deductive theory, with such exactness that the correctness of its basic inferences can be judged objectively by examining the language in which they are expressed. We have then, in principle, what was called in Sec. 1C a formalized contensive theory.[2] Let $L$ be the language in which its elementary statements are expressed.

From this point on there are two distinct directions of formalization. The first direction is that in which we take $L$ as the O language of a syntactical system; we shall call the method *metasemiosis* and a system so formed a *metasystem* relative to the original discipline (or to $L$). The second direction is that in which we continue to use $L$ in the U language, but change or "abstract from" its meaning, so that it becomes the A language of a formal system; we shall call this second method *abstraction*. The new formal system may be either a syntactical system or an ob system; it may be abstract, or we may prefer to have in mind a representation or an alternative interpretation. Intermediates between these two directions are conceivable, and the distinction between them tends to break down, as we shall see, if one allows modifications of $L$ before one begins.

If one adopts the metasemiotic method, he is obliged to invent a new A language, call it $M$, for referring to $L$. For this there are several alternatives. One may, as Hilbert did, use $L$ autonymously; the discussion of

---

[1] Noneuclidean geometry, for example, did not arise by formalization from physical geometry, but by analogy from euclidean geometry. Systems so formed by analogy may well be called secondary. As this example shows, they may still be of great importance in the methodology of science.

[2] This stage was reached by Frege and Russell. One can of course distinguish a lot of earlier stages, but these do not concern us.

Sec. A2 shows that there are some dangers to this method. Again one may use for $M$ an entirely different symbolism, e.g., when we used '$a$' as name of '$\alpha$' in Sec. 2. In that case the strangeness of the new notation may increase considerably the difficulties of comprehension. (Did the reader experience no shock at the remark, toward the end of Sec. 2, that the O letters for Example 1 were the Greek letters $a$ and $b$?) The presently favored method of using nouns formed by quotation marks, although it has some advantages, nevertheless has dangers of its own, which are nearly as serious as those of the autonymous method (see Exercise A2); and even when it is used with care, as by Quine and Carnap, there is still a considerable amount of strange symbolism.

In contrast to this, if one applies the method of abstraction, one continues to use the familiar $L$ in ways which are at least analogous to its original sense. The chief danger then is that one may be misled by associations which no longer hold. However, this is a sort of difficulty to which mathematicians are accustomed; it occurs not only in logic, but whenever one generalizes. Various devices can be used to avoid the pitfalls: we can keep in mind alternative representations or interpretations, including the autonymous one; and we can make judicious changes in $L$.

Again we have included under abstraction cases in which we arrive at a formal system of either kind. If the system is syntactical, this amounts, in principle,[1] to changing $L$ to a suitable O language and then applying metasemiosis. Such cases therefore have affinities with both directions; in particular, the Hilbert type of metasystem, with autonymous representation, belongs to both. But these forms of abstraction are rather artificial. If abstraction proceeds naturally—i.e., preserving at least the essential grammatics of $L$ and ignoring the designation of the nouns of $L$ or replacing them by others—one is led, in practically all cases of logical interest, to an ob system.

So much for the process of formalization. Our discussion of it has brought out the following points bearing on our theme of comparing the two types of system. An ob system is what we arrive at if we carry out the process of abstraction in a natural manner. The concatenative structure, when it exists, only arises from excessive attention to the symbolism. It is for this reason that practically all the syntactical systems which arise by formalization are eutactic and monotectonic, and there is no need whatever, except possibly for establishing the monotectonic property of the wefs, to consider words which are not well formed. Thus an ob system is closer to actual thought. It lends itself more readily to the possibility of a representation or interpretation in terms of a contensive subject matter. Furthermore, one is naturally led to use symbolism in ways which involve less departure from familiar usage. The difficulties in such a procedure are of the same character as crop up in other branches of modern abstract mathematics, and can be met by familiar means.

On the other hand, the notion of syntactical system has some advantages of concreteness. As Hilbert remarked,[2] our thinking is only sure when it is based on operations with concrete objects of which we are immediately

---

[1] Exceptions may conceivably occur, but do not interest us.
[2] See the quotation in Sec. S3.

aware.  In order to characterize an effective process in Sec. A5, we had to presuppose certain admissible elements which can be perceived as such directly.  Now all thought is communicated by language, and therefore it is natural to take the words in a finite alphabet as the most definite such admissible elements imaginable.  The definition of a formal system requires choice of an A language whose nouns (in the fundamental cases) are such words.  Thus, in investigations of the nature of an effective process (and in other matters of a similar nature), we need syntactical considerations, especially that part of semiotics, called *tectonics*, which is concerned with the relation of linguistic expressions to constructions.

The upshot of this is that one needs both points of view, and needs to be aware of their relations to each other.  In the study of ob systems one must know that a syntactical representation is possible and that the A nouns form a monotectonic linguistic structure.  In the study of a syntactical logic, the first thing one does, whether he is aware of it or not, is to put it in the form of an ob system.[1]  Thus a system, to be useful as a logic, must (so far as present knowledge goes) belong to both.

## EXERCISES

Some of these exercises involve epitheoretical methods which are not discussed systematically until Chap. 3.  Terms defined there technically are to be taken here as self-explanatory.

**1.** Give in full the formal demonstration of the following elementary theorems of Example 3:

(a) $$ab_1b_2b_3 = ab_1 \wedge ab_2b_3$$

(b) $$(a \wedge ab_2b_1) \wedge (ab_1b_3 \wedge ab_2) = ab_2b_1b_1 \wedge ab_3b_2$$

**2.** Show that the concatenation operation of Example 3 is associative.  (Exhibit an effective process for demonstrating any instance of the theorem scheme

$$X \wedge (Y \wedge Z) = (X \wedge Y) \wedge Z$$

Cf. the proof of the associative law of addition in elementary arithmetic, e.g., in Dedekind [WSW] or Landau [GLA].)

**3.** Show that the direct interpretation of Example 3 which is described in Sec. 5 is valid and adequate.

**4.** Show that Example 4 has a direct interpretation in the nonvoid words of the alphabet $\{c_1, \ldots, c_n\}$, with equality interpreted as identity, and that this interpretation is valid and adequate.

**5.** Show that Example 4 has a representation in a subset of the obs of Example 3 and that this representation is also a valid and adequate direct interpretation.  What can you say of the converse correspondence?  Would you say that Examples 3 and 4 are equivalent, and if so, in just what sense?

**6.** Show that if $X$, $Y$, $Z$ are any obs of Example 5, the following are elementary theorems:

(a) $$\vdash Y \supset Z . \supset : X \supset Y . \supset . X \supset Z$$

(b) $$\vdash X \supset . Y \supset Z . \supset : Y \supset . X \supset Z$$

(c) $$\vdash X \supset Y \supset . X \supset Z : \supset : X \supset . Y \supset Z$$

(d) $$\vdash X \supset Y . \supset . \urcorner Y \supset \urcorner X$$

(e) $$\vdash X \supset \urcorner X \supset \urcorner X$$

---

[1] This may be true even in linguistics.  Cf. [LAG].

(Schmidt [VAL, secs. 80–82]; cf. also below, Chaps. 5 and 6.  For the dot notation see Sec. A4.)

**7.** Give an effective process for representing an arbitrary ob system in the words of an O language with an alphabet consisting of two symbols.  ([CFS], example 5.)

**8.** Check the statements made in Sec. 5 about the truth-table interpretations of Example 5.

**9.** Show that Example 5 has a representation in those obs $X$ of Example 6 for which $P(X)$ holds, and that if the $\vdash$ of Example 5 is interpreted as the $\vdash$ of Example 6, the interpretation is again valid and adequate.

**10.** Suppose that $X$, $Y$, $U$, $V$ are obs of Example 3 (or Example 4) and that

$$X \wedge U = Y \wedge V$$

Show there is an ob $Z$ such that either $X = Y \wedge Z$ and $V = Z \wedge U$ or $Y = X \wedge Z$ and $U = Z \wedge V$.

**11.** Show that if one changes the standard truth tables so that $\neg X$ has always the value 1, then the resulting tautology interpretation for Example 5 is invalid, but is valid if the third axiom scheme is omitted, or is replaced by schemes $d$ and $e$ of Exercise 6.  What can be inferred about the independence of the third axiom scheme of the example?  Show that the other two axiom schemes are independent in the same sense. (Schmidt [VAL], sec. 83.)

**12.** A group is ordinarily defined as a class $G$ of elements such that the following postulates are satisfied:

$G$1.  Corresponding to any two elements,$a,b$ of $G$,there is a unique element $a \circ b$ of $G$.

$G$2.  For all elements $a$, $b$, $c$ of $G$,

$$a \circ (b \circ c) = (a \circ b) \circ c$$

$G$3.  There exists an element $i$ of $G$ such that for all elements $a$ of $G$

$$a \circ i = a$$

$G$4.  Given an element $a$ of $G$, there exists another element $a'$ of G such that

$$a \circ a' = i$$

Let an equation involving variables for arbitrary elements of $G$ be called an elementary group identity just when it is obtained from $G$1 to $G$4 by the usual rules for equality and substitution for variables.  Formulate an ob system $G^*$ with atoms $e_1$, $e_2$, $e_3$, ... and primitive predicate $=$, such that the elementary theorems of $G^*$ are precisely those equations which become elementary group identities when $e_1$ is evaluated as $i$ and $e_2$, $e_3$, ... as unspecified elements of $G$. ([APM], p. 226; [TFD], p. 8; [LLA], pp. 33–35.)

## D. SPECIAL FORMS OF SYSTEMS

We shall consider here certain special forms to which systems can be reduced.

**1. Predicational types.**   A system in which there is a single basic predicate, and that a binary relation, is called a *(binary) relational system*.  If the theory of the system is such that the relation is reflexive and transitive, then the system will be called *quasi-ordered;* if the relation has the properties of equality, the system will be called an *equational system*.  Thus the systems of ordinary mathematics are, for the most part, equational.

A second type of system is one in which there is a single basic predicate and this one is unary.  The basic predicate picks out a class of formal objects; it is in agreement with the usual viewpoint, in which the formal objects are O sentences, to call these O theorems, or *assertions*.  Thus this type of system may be called the *assertional type;* another name, which is suitable because of the prevalence of this type in fundamental logical studies, is the *logistic*[1] *type.* For the single predicate we shall use the prefix '⊢', so that the elementary statements are of the form

$$\vdash X \tag{1}$$

where $X$ is a formal object. The sign '⊢' is called the *assertion sign.* Frequently the predicate is expressed in words, such as '—— is provable' (Hilbert) or '—— is in $T$' (Huntington).

An arbitrary system can be reduced to one of assertional type. In what sense this is true may be seen from the following account of how it may be accomplished. Referring to the standard notation at the end of Sec. C1, let us associate to each basic predicate $\phi_n^{(k)}$ a new operation $\pi_n^{(k)}$. Then let us replace every elementary statement of the form

$$\phi_n^{(k)} X_1 X_2 \cdots X_n \tag{2}$$

by the corresponding elementary statement

$$\vdash \pi_n^{(k)} X_1 X_2 \cdots X_n \tag{3}$$

We shall then have a new assertional system which is equivalent to the old, in the sense that the elementary statements of either system can be translated, preserving truth, into those of the other.

There is one reservation to be made about this reduction. In the presentation of the system, and in particular in the statement of its rules, it may be expedient to use predicates in the U language which we do not care to list explicitly among the basic predicates. In other words, we want to treat these predicates informally. Thus the property of being an ob or O expression, more generally of being a formal object, since it holds for all formal objects which enter into the discussion at all, may be left unmentioned. However, the adjunction of new operations may enlarge the domain of formal objects, and there may conceivably be an upset in regard to the rules if the U variables referring to formal objects are not restricted to the formal objects in the original sense. To be sure of avoiding difficulty in this respect it is necessary to introduce such predicates explicitly before applying the above reduction.

As an example, consider the reduction of Example 2 of Sec. C2. Let us use '□' as a binary infix for the operation which is to replace equality, and let '$\sigma$(——)' be used as a functor for the unary operation which is to replace '—— is a sam'. Then the postulates for Example 2 become

$\vdash \sigma(a)$

$\vdash a \ \square \ a$

If $\vdash \sigma(X)$, then $\vdash \sigma(Xb)$

If $\vdash \sigma(X)$, $\vdash \sigma(Y)$, $\vdash X \ \square \ Y$, then $\vdash Xb \ \square \ Yb$

In this case the introduction of $\sigma$ does not affect the equations (i.e., the elementary theorems of the form

$$\vdash X \ \square \ Y$$

---

[1] See Sec. S1.

would not be different if the first and third postulates and the premises involving $\sigma$ in the fourth postulate were omitted). But this is not true generally (see Exercises 1 and 2 at the end of this section).

In an assertional system an ambiguity arises in regard to the axioms. Up to the present an axiom has been an elementary statement; it is expressed by a sentence in the U language. Thus the statement

$$\vdash (p_1 \supset (p_2 \supset p_1)) \tag{4}$$

is an axiom of the system of Example 5. But it is natural to apply the term 'axiom' to the ob which is asserted rather than to the statement of assertion, and many persons do just this. From that point of view, instead of (4), the ob

$$(p_1 \supset (p_2 \supset p_1)) \tag{5}$$

is regarded as an axiom. Thus an axiom is something which is named rather than something which is stated. For most purposes it is permissible to use the word 'axiom' ambiguously in these two senses, the distinction being made by the context; but when the distinction is important we shall refer to (4) as an *axiomatic statement*, whereas (5) will be an *axiomatic ob*. Where, as in Example 5, it seems appropriate to call the obs "propositions," we shall speak of (5) as an "axiomatic proposition."

The assertion sign is frequently used in statements of the form

$$X_1, \ldots, X_m \vdash Y \tag{6}$$

to indicate that if $X_1, \ldots, X_m$ are adjoined to the system as new axiomatic obs, $Y$ is an assertion in the extended system. Thus (6) represents the (formal) consequence relation corresponding to (1). This usage is not inconsistent with (1)—since, for $m = 0$, the two coincide—but is an extension of it. In that sense '$\vdash$' can be read "entails" or "yields."

For some rather special purposes it is convenient to retain two or more unary predicates. Such a system might be called multiassertional. Examples 1 and 6 are examples. However, this is merely a question of expediency, for the above reduction can always be carried out in principle.

Modern logical systems are almost universally presented in assertional form. This type seems, indeed, to be intrinsically simpler and therefore to have certain advantages in questions of ultimate foundations. But relational systems are more like those used in ordinary mathematics. The earliest of the modern logical systems, the algebra of Boole,[1] was equational. Recently certain analogies between logic and algebra have led to a revival of relational logical systems[2] which promises to be fruitful. Since a relational system can be reduced to an assertional one by the procedure we have just used, and the converse reduction can be made by taking a formal object 1 and defining (1) as

$$1 = X$$

---

[1] George Boole (1815–1864), English mathematician, professor of mathematics at Queens College, Cork, 1849 to 1864. For listing of his logical works see Church [BSL]. Modern mathematical logic may be said to have begun with his principal works published in 1847 and 1854. He was also active in the theory of differential and difference equations, and he proved one of the early theorems on algebraic invariants.

[2] See, for example, Halmos [BCA] and Tarski's work on cylindrical algebras (Sec. 7S2).

it is evident that the two forms are in principle equivalent. In this work, because of its emphasis on foundational questions, the assertional type will play the principal role, but both types will occur.

This is, perhaps, the place to comment on the differences between an algebra as formulated in ordinary mathematics and an equational ob system as formulated here. In the former case one thinks of a class of "elements" as existing beforehand and the operations as establishing correspondences between them. Thus, given an $n$-tuple of elements, an operation of degree $n$ "assigns" to it some preexisting element as its value. Moreover, equality is taken for granted, and equal elements are identified, even though they may be assigned as values by different operations, or by the same operation to different $n$-tuples. In an ob system, however, only the atoms and the operations are given beforehand; an operation does not assign an element but creates a new one; the obs are generated from the atoms by the operations, and the monotectonic property requires that different constructions give different results; equality is a relation holding between these obs under circumstances specified explicitly by the conventions of the system.[1]

**2. Simplifications of the formal objects.** The preceding section dealt with transformations affecting the basic predicates. Here we shall make a few observations concerning the formal objects. The reductions considered here are rather technical, and are used only for rather special purposes. It is, however, important for foundational questions to realize that they are possible.

The first observation is that in an ob system we can get along with a single binary operation. For this purpose we assign to each operation a new atom; we can then replace a closure of the original operation by a series of steps, the $k$th step of which is the closure of a binary operation combining the result of the $(k-1)$st step (or the new atom in the case of the first step) with the $k$th argument. The new operation is called *application;* its closure is symbolized by simple juxtaposition, with parentheses omitted according to the principle of association to the left. Thus if $g$ is the atom replacing an $n$-place operation $\omega$, then we replace any ob $\omega X_1 X_2 \cdots X_n$ by $(gX_1 X_2 \cdots X_n)$, where the latter is formed by first applying $g$ to $X_1$, then the result to $X_2$, and so on. Suppose we think of such a new ob $g$ as a *function* of degree $n$; then the interpretation of application is that it combines a function of degree $n$ with an ob $X$ to form that function of degree $n-1$ which is obtained from the first function by putting $X$ in its first argument place. For example, if $A$ is the addition function, $(A1)$ would be the function which converts $x$ into $1 + x$, and $(A12)$, that is, $((A1)2)$ would be $1 + 2$, that is, $3$. Of course, this transformation introduces new obs into the system, and some of these, such as $(AA)$, would be nonsensical in the above interpretation; these difficulties can be taken care of by the same modifications of the predicational structure that were discussed in Sec. 1.

The second observation relates to the possibility, already mentioned several times, that there may be infinitely many letters or atoms. For a system of really fundamental nature, this is unsatisfactory, for the only way in which we can conceive an infinite class constructively is to think of it as an inductive

[1] Cf. [CLg], p. 17. It can be shown that the point of view of an abstract algebra can be subsumed under the other as an interpretation (see Exercise C12; also Sec. 5A4).

class of some kind. Thus such an infinite class of initial elements makes sense only when the system is founded on some other more basic one. For fundamental purposes, however, we can use the sams of Examples 1 and 2 of Sec. C. In the case of a syntactical system we can enumerate the letters and replace them, in the order of the enumeration, by the sams themselves. The result is that we shall have a new O language based on the two letters $a$ and $b$; since the $a$'s indicate the separations between letters, there will be no ambiguity about the restoration of the original letters. In the case of an ob system, we can replace the atoms by sams; the latter can be regarded as generated from $a$ by an operation indicated by postfixing $b$. If the operation of application is present, we can have just one atom and generate all obs by successive applications of that one atom to itself. Modifications of this procedure to take care of two or more separate sequences, doubly infinite sequences, etc., will not cause any difficulty of principle.[1] Naturally, some of these reductions will be artificial, and there is no point in making the most extreme reductions if the number of elements is finite and not too large.

All this discussion assumes that the number of elements concerned is enumerable. From the constructive viewpoint, however, there are no nonenumerable infinities. Considerations involving such infinities either must involve platonistic assumptions, or else they must be based on a formalized set theory which can be developed only at a much later stage.

**3. Elementary systems.** The simplification now to be considered concerns the deductive rules of the system. It may not always be possible to carry it out, but it is a special restriction defining a class of systems. Suppose that the rules are of the form

$$\mathfrak{A}_1, \ldots, \mathfrak{A}_m \to \mathfrak{B}$$

where the U sentences $\mathfrak{A}_1, \ldots, \mathfrak{A}_m, \mathfrak{B}$ are constructed from A nouns and certain U variables $x_1, \ldots, x_n$ solely by means of the operators designating the operations of the system and the verbs expressing its basic predicates— in other words, where $\mathfrak{A}_1, \ldots, \mathfrak{A}_m, \mathfrak{B}$ are elementary statements in the extension of the given system formed by adjoining $x_1, \ldots, x_n$ to it as additional atoms. The instances of the rules are then obtained by specializing the $x_1, \ldots, x_n$ to be particular expressions or obs. A rule of that character will be called an *elementary rule*, and a system containing only such rules will be called an *elementary system*.

In an assertional ob system, an elementary rule would have the form

$$X_1, \ldots, X_m \vdash Y$$

where $X_1, X_2, \ldots, X_m, Y$ are obs formed from the atoms and the $x_1, \ldots, x_n$ by the operations of the system; if the system were concatenative, the same would be true except that $X_1, \ldots, X_m, Y$ would be expressions formed from the letters and $x_1, \ldots, x_n$ by concatenation. All the systems in Examples 1 to 6 of Sec. C are elementary, provided Examples 1 and 2 are taken as concatenative.

To get an example of a nonelementary rule we can take the last rule of Example 1 if that system is taken as affixative, viz.,

If $XcY$ is a tantet, then $XbcYb$ is a tantet

---

[1] See, for example, [CLg], pp. 30–32; [DTC], p. 17, footnote 3.

Here $X$ and $Y$ are analogous to $x_1, \ldots, x_n$.  But neither $XcY$ nor $XbcYb$ can be formed by the operations of the system even if $X$ and $Y$ were admitted as letters; for affixation of $Y$ is not one of the operations of the system.  In order to state such a rule one would have to introduce an operation, viz., concatenation (or something similar) which was not used in the formation of the formal objects.  Such operations are called *auxiliary*.

Another example is the rule of substitution.  For the system of Example 5, for example, this rule would be stated in the form

$$P \vdash P^*$$

where $P$ is an ob, and $P^*$ is obtained from $P$ by substitutions.  Let us consider this in the special case where the substitution is that of an ob $M$ for $p_2$.  Then to get $P^*$ you must take a construction of $P$, replace $p_2$ at every top node, and complete the construction so altered.  An equivalent way of stating this is to use the recursive definition:

$$p_2^* \equiv M$$
$$p_i^* \equiv p_i \quad \text{for } i \neq 2$$
$$(\neg Q)^* \equiv \neg Q^*$$
$$(P \supset Q)^* \equiv P^* \supset Q^*$$

Thus substitution is an exceedingly complex auxiliary operation, and rules involving it are not elementary.[1]

If such a substitution rule were formulated as a rule of deduction, it would not be necessary to have axiom schemes in Example 5.  In fact, the axiom schemes of Example 5 could be replaced by single axioms, leading to the following example.

*Example 7.*  Like Example 5, except that there is a rule of substitution as above formulated and the axiom schemes are replaced by the following three individual axioms:

$$\vdash p_1 \supset (p_2 \supset p_1)$$
$$\vdash (p_1 \supset (p_2 \supset p_3)) \supset ((p_1 \supset p_2) \supset (p_1 \supset p_3))$$
$$\vdash (\neg p_2 \supset \neg p_1) \supset (p_1 \supset p_2)$$

This example would have exactly the same elementary theorems as Example 5 (cf. Sec. 3A3).

Although we have to admit nonelementary rules in certain cases, yet it is always a desideratum to remove them.  This demand eliminates possibilities like the affixative formulation of Example 1.

## EXERCISES

**1.** Verify that in the assertional form (Sec. 1) of Example 2 (Sec. C2), there are obs which are not sams (i.e., not obs of Example 2 itself), but that the introduction of $\sigma$ does not affect the equations.

**2.** Formulate an assertional form for Example 3 (Sec. C2) and show that in this case a predicate analogous to $\sigma$ is necessary; however, restrictions on the $X$ and $Y$ in the axiom schemes are sufficient.

**3.** Give a relational form of Example 5.

[1] For some generalizations, multiple substitution, etc., see [DSR].

4. Prove constructively (that is, by an effective process) that if one identifies similarly designated obs in the two systems, then Examples 5 and 7 have the same elementary theorems. (Sec. 3D3.)

### E. ALGORITHMS

In Sec. A5 we were concerned with the notion of effective process. We consider here ways of specifying such an effective process which can lend greater explicitness to discussions of rules, generating principles, and correspondences such as those involved in representations, valuations, interpretations, etc.

An *algorithm* is generally understood as a specification describing an effective process. Here we shall impose the additional restriction that the admissible elements for the process be formal objects of some system; however, we shall not require that all the formal objects of the system be admissible.

**1. Markov algorithms.** The algorithms described below are due to Markov. He calls them "normal algorithms," but since 'normal' has many different uses, and since these algorithms are a characteristic contribution of Markov and are the only ones he considers at any length, it is appropriate to call them *Markov algorithms*. In the future we shall often refer to them without any qualifying adjective; thus an algorithm is a Markov algorithm unless there is some indication to the contrary. There is very strong heuristic evidence that any effective process on the formal objects of a system can be specified by such an algorithm.[1]

We first suppose that we have to do with an object language with a certain alphabet $\mathfrak{A}$. We form an alphabet $\mathfrak{B}$ over $\mathfrak{A}$ by adjoining to it certain additional letters, called *auxiliary letters*. The algorithm then consists of a series of specifications of the form

$$A_i \to B_i \qquad i = 1, 2, \ldots, n \qquad (1)$$

with or without a dot following the arrow. Here $A_1, \ldots, A_n, B_1, \ldots, B_n$ are fixed expressions, possibly void, in the alphabet $\mathfrak{B}$, and the '$\to$' has a special meaning, which is not to be confused with that assigned to the same symbol in Sec. 2A4. Each of the lines of the algorithm of the form (1) will be called a *command* of the algorithm; those with a dot will be called *stop commands*, the others *nonstop commands;* and the $\mathfrak{B}$ words appearing on the left and right in (1) will be called the *antecedent* and *consequent* of the command, respectively.

Let $E$ be an expression in the alphabet $\mathfrak{B}$. The $k$th command in the algorithm (1) is said to be *applicable* to $E$ just when there is an occurrence of $A_k$ in $E$. In that case the execution of the $k$th command for $E$ will consist in the replacement of the first (i.e., left-hand-most) occurrence of $A_k$ by $B_k$.[2]

---

[1] This is Markov's thesis. It is related to Church's thesis (Kleene [IMM], sec. 62), which makes the same assertion concerning another kind of effective process defined in terms of recursive functions and Gödel representation. The two sorts of effective process have been shown to be equivalent (cf. Exercise 12); therefore Church's thesis and Markov's thesis are equivalent also.

[2] The different occurrences can be distinguished by specifying the segments of $E$ lying to the left of them. Thus the term 'occurrence' can be given an objective meaning. Cf. Sec. 3B1.

The process specified by the algorithm (1), applied to such an $E$, is now the following. We search for the first command which is applicable to $E$. If there is no such command, the process stops with $E$. Otherwise we execute the first command which is applicable to $E$, converting $E$ to $E'$. If this command is a stop command, the process stops with $E'$. If not, we start all over again with $E'$ in the place of $E$. The process continues until we reach either a stop command or an expression to which no command is applicable; in the latter case we shall say that the algorithm, or process, is *blocked*. The algorithm is said to be *applicable* to $E$ just when the process stops without being blocked.[1]

In this description there was no mention of auxiliary letters. Actually, the description would make sense if there were none, in which case $\mathfrak{B}$ would be the same as $\mathfrak{A}$ and the algorithm would be said to be *in* $\mathfrak{A}$. But in the practically interesting cases $E$ is required to be a word in $\mathfrak{A}$ and the auxiliary letters are introduced by the execution of certain commands. In that case the algorithm is said to be *over* $\mathfrak{A}$. If an algorithm $L$ is applicable to $E$, then the word which remains when the process stops will be called $L(E)$, or the result of applying $L$ to $E$.

We shall now consider some examples of algorithms. In these examples the letters '$x$', '$y$', '$z$', etc., will be used for unspecified letters of the alphabet $\mathfrak{A}$, and a command involving these variables will be understood as a whole series of commands in which all possible substitutions of $\mathfrak{A}$ letters for these variables are made in a lexicographic order. As names of auxiliary letters, lower-case Greek letters will be used.

Let us first devise an algorithm for copying an expression, i.e., for converting an expression $E$ into $EE$. Such an algorithm will be called a *duplication algorithm*.

Suppose first that we have an $\alpha$ before $E$, i.e., that we have converted $E$ into $\alpha E$; how this is done we shall see later. Then by a series of commands of the form

$$\alpha x \to x \beta x \alpha \tag{2}$$

we make a copy of the initial letter $x$ and "mark" the copy with an auxiliary $\beta$. The $\alpha$ can now form a combination with the letter next on its right, and then with the letter next to the right of that, and so on until $\alpha$ is at the end of the expression. At that point, if the original expression was, say,

$$x_1 x_2 x_3 \cdots x_n$$

we should have

$$x_1 \beta x_1 x_2 \beta x_2 x_3 \beta x_3 \cdots x_n \beta x_n \alpha$$

By a series of commands of the form

$$\beta x y \to y \beta x \tag{3}$$

the marked copies of the letters will be moved to the right of the unmarked ones without change in order, so that we should have eventually

$$x_1 x_2 \cdots x_n \beta x_1 \beta x_2 \cdots \beta x_n \alpha$$

---

[1] Occasionally one may wish to include this last possibility also. In such a case that circumstance will be explicitly mentioned.

We can now drop out the $\beta$'s and the $\alpha$ by commands:

$$\beta \rightarrow \qquad\qquad (4)$$

$$\alpha \rightarrow . \qquad\qquad (5)$$

These commands will transform $\alpha E$ into $EE$ provided that they are arranged in a proper order. The order of (2) and (3) is immaterial, but (4) must follow them in order to prevent premature dropping out of $\beta$, and (5) must follow (4) in order that the process may run to completion.

To complete the construction of the algorithm we must put the initial $\alpha$ before $E$. This can be accomplished by a command

$$\rightarrow \alpha \qquad\qquad (6)$$

called the *starting command*,[1] appearing at the end of the list of commands. For since the left sides of (2) to (5) all contain auxiliary letters and there are no auxiliary letters in $E$, none of those commands will be applicable to $E$; on the other hand, since the void word occurs at the beginning of any word, the command (6) will be applicable and will have the desired effect. After the $\alpha$ is once admitted, however, some one of the other commands will always be applicable until we reach (5), in which case the process stops. This device for starting an algorithm can be used in all cases where blocking of the rest of the algorithm is not permitted (and in some other cases in which repetition of the algorithm is allowed).

For the case where $\mathfrak{A}$ consists of the two letters $a$ and $b$, the complete algorithm is as follows:

$$\alpha a \rightarrow a\beta a\alpha$$
$$\alpha b \rightarrow b\beta b\alpha$$
$$\beta aa \rightarrow a\beta a$$
$$\beta ab \rightarrow b\beta a$$
$$\beta ba \rightarrow a\beta b$$
$$\beta bb \rightarrow b\beta b$$
$$\beta \rightarrow$$
$$\alpha \rightarrow .$$
$$\rightarrow \alpha$$

It is instructive to see how this algorithm can be modified to have the copy made in reverse order. For this it is sufficient to have the marked copy to the right of the $\alpha$ and shifted to the right immediately, and to provide that when there are no longer any adjacent pairs $\alpha x$ or $\beta x$ the $\alpha$ cleans out the $\beta$'s and disappears. The algorithm would be as follows:

$$\alpha x \rightarrow x\alpha\beta x$$
$$\beta xy \rightarrow y\beta x$$
$$\alpha\beta x \rightarrow x\alpha$$
$$\alpha \rightarrow .$$
$$\rightarrow \alpha$$

[1] More generally, any command with a void left side will be called a starting command.

Here again the order of the first two commands is immaterial but the others must follow them in the order given.

As another example, suppose that, given two algorithms $L_1$ and $L_2$, we wish to construct an algorithm $L_3$ which carries out the process of $L_1$ until it stops, and then carries out the process of $L_2$ on the result. Suppose that the algorithms $L_1$ and $L_2$ are in $\mathfrak{A}$, so that all auxiliary letters we use are introduced as extras in $L_3$. Then the construction of $L_3$ can be carried out in four stages as follows:

$1°$. Let $L_1'$ be obtained from $L_1$ by replacing commands of the forms

$$A \to B \qquad A \to. B \tag{7}$$

respectively, by

$$A \to B \qquad A \to \alpha B \tag{8}$$

(i.e., by replacing every dot by a symbol for a new auxiliary letter)[1] and then adding (6) at the end. We put $L_1'$ at the end of $L_3$. If the commands preceding $L_1'$ in $L_3$ contain auxiliary letters on the left, then $L_3$, up to the time the $\alpha$ appears, will have the same effect on any $\mathfrak{A}$ word $E$ as $L_1$ does, and if $L_1(E) = F$, the result will be an $F'$ obtained by inserting an $\alpha$ into $F$. The command (6) takes care of the possibility that $L_1$ is blocked.

$2°$. For each $\mathfrak{A}$ word, $E$, let $E^\gamma$ be obtained from $E$ by adding a $\gamma$ at the beginning and after every $\mathfrak{A}$ letter, so that if $E$ is void, $E^\gamma$ is $\gamma$, and if $E$ is, for example, $a_1 a_2 a_3$, $E^\gamma$ is $\gamma a_1 \gamma a_2 \gamma a_3 \gamma$. Let $L_2'$ be obtained from $L_2$ by replacing the commands of the form (7), respectively, by

$$A^\gamma \to B^\gamma \qquad A^\gamma \to. B^\gamma \tag{9}$$

the dot being retained just when it was originally present. Then the conditions

$$G = L_2(F) \qquad G^\gamma = L_2'(F^\gamma) \tag{10}$$

are equivalent. Let $L_2''$ be obtained from $L_2'$ by a transformation analogous to that in stage $1°$, except that $\delta$ replaces $\alpha$, and the command

$$\gamma \to \delta$$

replaces (6). We put $L_2''$ in $L_3$ immediately before $L_1'$. Then, if the commands of $L_3$ before $L_2''$ all contain on the left at least one auxiliary letter other than $\gamma$, the effect of $L_3$ (up to the introduction of $\delta$) on any $F^\gamma$ for which (10) holds is to transform it into a $G'$ formed by inserting a $\delta$ into $G^\gamma$.

$3°$. Next, we put at the head of $L_3$ commands which will transform a word $F'$, obtained as in stage $1°$, into the corresponding word $F^\gamma$. This can be done by the following commands:

$$x\alpha \to \alpha x$$
$$\alpha \to \beta$$
$$\beta x \to \gamma x \beta$$
$$\beta \to \gamma$$

---

[1] In the future I shall describe this change by saying that the dot is replaced by $\alpha$. Actually this is incorrect, for the dot is a U symbol; $\alpha$ is an auxiliary letter, hence an O symbol; '$\alpha$' is a U symbol standing for an unspecified O symbol; and what replaces the dot is neither $\alpha$ nor '$\alpha$' but the name of $\alpha$, whatever that may be. But the briefer idiom is hardly likely to be misunderstood.

Here the first command moves the $\alpha$ to the beginning; the second, which is applicable only when $\alpha$ has reached the beginning, changes $\alpha$ to $\beta$; and the third and fourth complete the transformation to $F^\gamma$.

4°. Finally, as soon as $\delta$ appears, we clean out the $\gamma$'s and close the algorithm by commands, which must precede $L_2''$ but may appear either before or after those in stage 3°, thus:

$$\gamma x \delta \to \delta \gamma x$$
$$\delta \to \epsilon$$
$$\epsilon \gamma x \to x \epsilon$$
$$\epsilon \gamma \to .$$

The complete algorithm is as follows:

$$
\left.
\begin{array}{l}
x\alpha \to \alpha x \\
\alpha \to \beta \\
\beta x \to \gamma x \beta \\
\beta \to \gamma
\end{array}
\right\} \quad \text{stage } 3°
$$

$$
\left.
\begin{array}{l}
\gamma x \delta \to \delta \gamma x \\
\delta \to \epsilon \\
\epsilon \gamma x \to x \epsilon \\
\epsilon \gamma \to .
\end{array}
\right\} \quad \text{stage } 4°
$$

$$L_2'' \qquad \text{stage } 2°$$
$$L_1' \qquad \text{stage } 1°$$

This algorithm can be seen to have all the properties required for $L_3$.

As we have seen in the foregoing examples, the order of the commands in the statement of the algorithm may be essential. But if we have two commands $\Gamma_1$ and $\Gamma_2$ such that they will never be simultaneously applicable, then it does not make any difference in what relative order $\Gamma_1$ and $\Gamma_2$ appear. This will occur in particular if there are two letters $\alpha_1$ and $\alpha_2$, such that at most one of these letters can occur in any word which the algorithm reaches, and such that $\alpha_1$ appears on the left in $\Gamma_1$ and $\alpha_2$ on the left in $\Gamma_2$.[†] On the other hand, if $\Gamma_1$ and $\Gamma_2$ are such that the left side of $\Gamma_1$ is a proper initial segment of the left side of $\Gamma_2$, then $\Gamma_2$ will never be executed if it comes after $\Gamma_1$, and therefore $\Gamma_2$ is superfluous unless it appears before $\Gamma_1$.

**2. Shuttle algorithms.** We now discuss certain specializations of a Markov algorithm which have some advantages in the technique. These specializations will be introduced in two stages. In both stages we suppose that there is a special class of auxiliary letters called *shuttles*, such that every word being operated on contains at most one of these shuttles, and these shuttles control, in a sense to be presently explained, the course of the calculation. We shall use lower-case Greek letters for shuttles; for other auxiliary letters, which will be seen to function as place markers, punctuation marks and arbitrary signs will be used.

In the first stage we impose simply the following conditions: (1) there is a unique command, appearing at the end of the algorithm, such that the left side is void and the right side is a shuttle; (2) every other command has

† This condition is fulfilled if $\Gamma_1$ is in stage 3° and $\Gamma_2$ in stage 4° of the $L_3$ of the preceding paragraph.

exactly one shuttle on the left; (3) every command which is not a stop command has exactly one shuttle on the right; (4) a stop command has no shuttle on the right. The command appearing in condition 1 will be called, as in Sec. 1, the *starting command*, and the shuttle which appears in it the *starting shuttle*. An algorithm for which the conditions 1 to 4 are satisfied will be called a *semishuttle algorithm*.

If we have an algorithm satisfying these conditions and start with a word $E$ containing no shuttles, then the only applicable command will be the starting command. Thereafter we shall have a word with exactly one shuttle unless we reach a stop command, in which case the process stops and the shuttles disappear.[1] Thus the conditions for interchangeability mentioned at the end of Sec. 1 are fulfilled for commands with different shuttles on the left. Commands with the same shuttle on the left will be said to belong to the same *phase*. The algorithm may be so stated that all commands belonging to the same phase may be brought together, and the phases themselves may be arranged in any order. It is natural to expect that the different phases correspond to significant subprocesses in the process defined by the algorithm. If, further, every phase has at its end a command with that shuttle alone on the left, it will be impossible for the algorithm to be blocked.

Before going further let us stop to examine how the above duplication algorithm can be formulated as a semishuttle algorithm. There are two main stages in the duplication process, viz., making copies of the letters and moving the copies to the right. With these two stages we associated the letters $\alpha$ and $\beta$, but although $\alpha$ functions as a shuttle, $\beta$ does not. Suppose now we replace the $\beta$ in (2) by a place marker $+$. When $\alpha$ reaches the extreme right of the expression, let it turn into a shuttle $\gamma$ which moves to the left until it meets one of these place markers, and then converts it to a $\beta$, thus:

$$x\gamma \rightarrow \gamma x$$
$$+\gamma \rightarrow \beta$$

Then the command (3) would move the $\beta x$ (i.e., the $x$ previously marked with a preceding $+$) to the right. If on its arriving there, the $\beta$ changed into $\gamma$, then the process would repeat, and it would continue until the $\gamma$ reached the beginning of the word. But in such a case the copy we should have would be reversed. To prevent the reversal there are two alternatives. The first alternative would be to let the $\beta$ restore the place marker thus:

$$\beta \rightarrow \gamma + \tag{11}$$

Then it would be necessary to have commands to clear out these markers, thus:

$$\gamma \rightarrow \delta$$
$$\delta x \rightarrow x\delta$$
$$\delta + \rightarrow \delta \tag{12}$$
$$\delta \rightarrow .$$

---

[1] Because one shuttle is brought in by the starting command, the nonstop commands do not change the number of shuttles, and the stop commands decrease it by one.

The second alternative would be to have $\alpha$ replaced by $\gamma+$ and to replace (11) by

$$\beta x+ \;\rightarrow\; \gamma+x$$

Then we could close with

$$\delta+ \;\rightarrow\;.$$

This would have the advantage that if we wished the $+$ to remain in the result as a separation—which is often convenient—we could omit the $\delta$ phase altogether.   The entire algorithm would then be

$$\alpha x \;\rightarrow\; x+x\alpha \qquad (13)$$

$$\alpha \;\rightarrow\; \gamma+$$

$$\beta xy \;\rightarrow\; y\beta x$$

$$\beta x+ \;\rightarrow\; \gamma+x \qquad (14)$$

$$x\gamma \;\rightarrow\; \gamma x$$

$$+\gamma \;\rightarrow\; \beta \qquad (15)$$

$$\gamma \;\rightarrow\;.$$

$$\rightarrow\; \alpha$$

As will be seen from this example, the new algorithm has more commands. But the semishuttle algorithm has the advantage that the composition of algorithms is simplified and systematized.   Consider, for example, the formation of $L_3$ from $L_1$ and $L_2$ (at the end of Sec. 1).   All that it is necessary to do[1] is to replace the dot in the stop commands of $L_1$ by a shuttle $\alpha$,[2] not conflicting with any already present, such that $\alpha$ moves to the beginning and then turns into the starting shuttle of $L_2$.†   Let us call this process the *substitution of $L_2$ in the outputs of $L_1$*.   More complex forms of composition can also be constructed.   Thus if $L_1$ has several outputs (i.e., stop commands), we can substitute different $L_2$'s in the different outputs; moreover, some of the $L_2$'s may be the same as $L_1$,‡ thus giving rise to an iterative process. Again, if we replace certain shuttles in an algorithm by a shuttle and a marker, and then use the new shuttle to trigger some other process, returning to the marker when that process finishes, we can interpolate processes in other processes (for example, counting the steps so as to take some action depending on the number of steps).   One can thus make a great variety of complex compositions rather simply.

Using the fact just established, we can show that, given an arbitrary Markov algorithm $L$, a semishuttle algorithm $L'$ can be found which has the same effect on every word $E$ in $\mathfrak{A}$.

Consider first the case where $L$ consists of the single command

$$A \;\rightarrow.\,B \qquad (16)$$

where

$$A \equiv a_1 a_2 \cdots a_m \qquad B \equiv b_1 b_2 \cdots b_n$$

---

[1] We must suppose, to begin with, that the shuttles in $L_1$ and $L_2$ are so chosen that there is no conflict between a command in $L_1$ and one in $L_2$.   This is always possible since shuttles are auxiliary letters and can be changed arbitrarily.

[2] Compare footnote to the corresponding construction in Sec. 1.

† The starting command of $L_2$ is, of course, omitted.

‡ There will be no conflict between the shuttles of the two $L_1$'s, since $L_1$ is to be repeated.

Let $\alpha$ be the starting shuttle for $L'$. If $m = 0$, then the commands

$$\alpha \to. B \qquad (\text{output 1}) \qquad\qquad (17)$$
$$\to \alpha$$

will do for $L'$. For $m > 0$, let $\alpha$ go through the initial word $E$ searching for $a_1$, thus:

$$\alpha x \to x\alpha \qquad \text{for } x \not\equiv a_1$$
$$\alpha a_1 \to *\beta_1$$
$$\alpha \to. \qquad (\text{output 2})$$

Let $\beta_1, \beta_2, \ldots, \beta_m$ be new shuttles not conflicting with any other letters.[1] For $i < m$, let $\beta_i$ check whether the letter next to its right is the same as $a_{i+1}$, and change into $\gamma$ if it is not, thus:

$$\beta_i a_{i+1} \to a_{i+1}\beta_{i+1} \qquad i = 1, 2, \ldots, m-1$$
$$\beta_i \to \gamma$$

If $\beta_m$ appears, it indicates that a first occurrence of $A$ has been found. Then we can erase $A$ and substitute $B$ by the commands

$$x\beta_m \to \beta_m$$
$$*\beta_m \to. B \qquad (\text{output 1}) \qquad\qquad (18)$$

If $\gamma$ appears, it indicates failure to find an occurrence of $A$, and the appropriate action is given by the commands

$$x\gamma \to \gamma x$$
$$*\gamma \to a_1\alpha$$

These commands constitute the desired $L'$. Note that we come out on output 1 if (16) is applicable and on output 2 if it is not.

Now consider the case where $L$ consists of the commands

$$A_j \to B_j \qquad j = 1, 2, \ldots, p \qquad\qquad (19)$$

Let $L'_j$ be the algorithm, as just constructed, for the $j$th command. From these we form $L'$ by the composition technique of the third preceding paragraph. From output 1 of $L'_j$ we stop if the $j$th command (19) is a stop command, and go to $L'_1$ if it is not; from output 2 of $L'_j$ we go to $L'_{j+1}$ if $j \leq p - 1$, and stop (indicating blocking) if $j = p$. Thus $L'$ is determined, and therefore a semishuttle algorithm is not less general than a Markov algorithm.

This completes the treatment of the first stage of specialization. In the second stage we have the following additional condition: (5) the left and right sides of any command contain at most one letter which is not a shuttle; on the left this letter appears always on the same side of the shuttle in all commands of the same phase. Thus a shuttle operates on a single letter

---

[1] It is necessary to have $m$ distinct shuttles (unless we use markers) even though several of the $a_i$ may be the same letter.

appearing always either on its right or on its left; in the former case the shuttle is called a right-facing shuttle, in the latter a left-facing shuttle. A semishuttle algorithm which satisfies the condition 5 will be called a *shuttle algorithm*.

The commands of the semishuttle algorithm $L'$ just shown to be equivalent to a given Markov algorithm $L$ all satisfy condition 5 except (17) and (18). But by introducing new shuttles $\delta_1, \delta_2, \ldots, \delta_n$, command (17) can be executed as follows:

$$\alpha \to \delta_1$$
$$\delta_k \to b_k \delta_{k+1} \qquad k = 1, 2, \ldots, n-1$$
$$\delta_n \to .$$

Command (18) can be executed similarly. Thus a shuttle algorithm is also not less general than a Markov algorithm. But it may require more shuttles than a semishuttle algorithm.

Let us return to the duplication algorithm. It follows from what we have just proved that this can be achieved by a shuttle algorithm. It is expedient to get this shuttle algorithm directly rather than as an instance of the general theory. Suppose for each letter $x_i$ in $\mathfrak{A}$ we have a shuttle $\xi_i$; then $\beta x_i$ can be replaced by $\xi_i$. If we do this, all commands of the semishuttle duplication algorithm satisfy condition 5 except (13), (14), and (15). To do the work of (13), we appear to need a second set of shuttles $\xi'$; then (13) could be replaced by

$$\alpha x \to \xi' x$$
$$\xi' \to x\delta$$
$$\delta \to +\epsilon$$
$$\epsilon x \to x\alpha$$

but we save one of these shuttles if we change (13) to

$$\alpha x \to xx + \alpha \tag{20}$$

which in turn can be replaced by

$$\alpha x \to x\xi'$$
$$\xi' \to x\delta$$
$$\delta \to +\alpha$$

Then for (14) we can take

$$\xi + \to \theta x$$
$$\theta \to \gamma +$$

whereas for (15) we can have (since $\beta$ has not been used)

$$+\gamma \to \beta$$

and one or the other of

$$\beta x \to \xi \qquad \text{[if we use (13)]}$$
$$x\beta \to \xi \qquad \text{[if we use (20)]}$$

The entire algorithm, if we change (13) to (20), is

$$\alpha x \to x\xi'$$
$$\alpha \to \gamma +$$
$$\xi' \to x\delta$$
$$\delta \to +\alpha$$
$$x\gamma \to \gamma x$$
$$+\gamma \to \beta$$
$$\gamma \to .$$
$$x\beta \to \xi$$
$$\xi y \to y\xi$$
$$\xi + \to \theta x$$
$$\theta \to \gamma +$$
$$\to \alpha$$

Although the transition from a Markov algorithm to a semishuttle algorithm has advantages from the standpoint of constructing an algorithm to do a given job, the insistence on a shuttle algorithm is rather a handicap from that point of view. Its chief significance is that it facilitates comparison with other forms of effective process. It is possible, for example, to show directly that a shuttle algorithm is equivalent to a certain type of Turing machine, and thus to show, by an argument of some complexity but quite elementary in principle, that the notion as a criterion of constructiveness is equivalent to other accepted ones.[1]

**3. Generalizations.** The concept of algorithm may be generalized in various ways. We shall consider some of these generalizations here, and along with them the question of how algorithms are related to other sorts of formal notions.

The notion of Markov algorithm presupposes that the specified process is to operate on an O expression in some finite O alphabet. But it is not difficult to modify it so as to apply to the obs of an ob system. In such a case we have auxiliary atoms and auxiliary operations in the place of auxiliary letters. The $A_i$, $B_i$ will be obs. The execution of the $k$th command can be explained, using the idea of labeled tree diagram, as follows. Consider that instance of $A_k$ which occurs first in the normal construction sequence; one replaces the part of the tree which gives a construction of this $A_k$ by a construction of $B_k$ and makes corresponding changes in the labels of the nodes below. We shall consider such replacements more fully later.[2] If one uses the A language proposed at the end of Sec. C3 and the autonymous representation, this replacement is a special case of that for expressions.

The notion of algorithm can evidently be modified in various other ways. For example, a simple modification is to admit explicitly commands which can be executed only once. Then a command, such as (6), so marked, can occur in its natural place at the beginning of the algorithm. Still another

---

[1] See Exercise 12 at the end of this section. For the other definitions of constructiveness see Kleene [IMM, especially sec. 62].

[2] See Secs. 3B and 3C.

modification which has been suggested is to require that after a command has been executed, the commands previous to it be excluded for later use. I do not know what the effect of some of these modifications would be.

If we were to abandon the feature by which the command to be executed at any stage, and the particular occurrence of the antecedent to be replaced, are uniquely determined, then the commands (1) for an algorithm can be regarded as the deductive rules for an assertional concatenative system. These commands are replacements; they are equivalent to the deductive rules

$$XA_iY \vdash XB_iY \qquad i = 1, 2, \ldots, n \tag{21}$$

where $X$ and $Y$ are unspecified words (in the sense that '$X$' and '$Y$' occupy places where the name of an arbitrary word may be substituted), and $A_i$ and $B_i$ are fixed words. Such a formal system is one of the kinds which Post[1] introduced; this particular kind is known as a bilateral Post system without axiom. Since the system has no axiom, it has no elementary theorems, but the rules define a consequence relation as explained in Sec. B3. The system obtained by applying this system to a word $E$ is that formed by adjoining $\vdash E$ as sole axiom.

It has been proposed[2] to call a *derivational formal system* one in which a special subclass of deductions is singled out as *derivations*. This is understood to be such that it is a definite question whether or not a deduction is a derivation. An *algorithm* may then be defined as a derivational formal system satisfying the following conditions: (1) there is no axiom; (2) the system is assertional; (3) the rules are one-place, elementary (Sec. D3) rules; (4) the rules, at least in so far as they can be applied in a derivation, are deterministic; (5) in a derivation there is not more than one rule which can be applied to a given premise; (6) there are conditions determining when a derivation terminates, and it is always a definite question whether these conditions are fulfilled. The application of such an algorithm to a formal object $E$ is the system obtained by adjoining $E$ as sole axiom. If $L$ is the algorithm, $L(E)$ is the formal object asserted in the concluding statement of a terminated derivation.

A Markov algorithm is thus an algorithm according to this definition. There are, however, other kinds. One of the most interesting is the Post algorithm, which is based on a unilateral Post system, i.e., one in which the rules, instead of being of the form (21), are of the form

$$AX \vdash XB \tag{22}$$

These have been shown to be equivalent to Markov algorithms.[3]

## EXERCISES

**1.** Construct an algorithm for transforming a finite sequence of letters into the sequence formed by writing the same letters in the reversed order.

**2.** Construct a shuttle duplication algorithm using only one set of shuttles corresponding to the letters of $\mathfrak{A}$ (in addition to shuttles independent of $\mathfrak{A}$).

[1] Post [FRG]; cf. Porte [SPA].
[2] See Porte [SPA].
[3] See Asser [NPA].

3. For each integer $n$ and each letter $a$, let $a^n$ be the word consisting of $n$ consecutive occurrences of $a$. Let $\mathfrak{A}$ be the alphabet $\{|, *\}$. Show that there are algorithms $L_1$, $L_2$, $L_3$ such that

$$L_1(|^m * |^n) = |^s \qquad \text{where } s = m + n$$
$$L_2(|^m * |^n) = |^p \qquad \text{where } p = mn$$
$$L_3(|^m * |^n) = |^q * |^r \qquad \text{where } m = qn + r \ (r \le n)$$

4. Let

$$A_{ij} \to B_{ij} \qquad i = 1, 2, \ldots, r; j = 1, 2, \ldots, n_i$$

be $r$ finite sets of commands. Devise an algorithm $L$ which has the same effect as one in which the command to be applied to any $E$ has $i$ and $j$ determined as follows: $i$ is the least number such that for some $j$ there is an occurrence of $A_{ij}$ in $E$, and $j$ is the least number such that there is an occurrence of $A_{ij}$ beginning at least as far to the left as any $A_{ik}$ (for $k \ne j$).

5. Given an ob system with a finite number of atoms and a finite number of operators. Devise an algorithm over the alphabet of the standard Łukasiewicz representation which converts a word of the alphabet into the auxiliary letter $t$ if that word is well formed and into the auxiliary letter $f$ in case it is not. ([TEA], cf. [CFS], sec. 6.)

6. Devise an algorithm, applicable to words in the alphabet $\{a, b, \neg, \supset (,)\}$ of Example 6, for testing whether a word in that alphabet is a tautology according to the ordinary binary truth tables. Let the algorithm convert the initial word into $t$ if it is a tautology, into $f$ if it is well formed and not a tautology, and into $n$ if it is not well formed. (Here $n$, $t$, $f$ are auxiliary letters.) ([TEA].)

7. Let $L_1$, $L_2$, \ldots, $L_m$ be $m$ algorithms, and let $\mathfrak{A}$ be the union of their alphabets. Let $L_i$ convert a word $X$ into $Y_i$. Let $*$ be not in $\mathfrak{A}$. Then show that there is an algorithm $L_0$ such that $L_0$ converts $X$ into

$$Y_1 * Y_2 * \cdots * Y_m$$

and is not applicable to any $X$ to which not all the $L_1, \ldots, L_m$ are applicable.

8. Let $L_1$ and $L_2$ be two algorithms. Let $L_1(X)$ be the result obtained by applying $L_1$ to $X$, and let $L_2(X)$ be defined similarly. Devise an algorithm $L_3$ such that (a) if $L_1(X)$ closes in $m$ steps, and $L_2(X)$ does not close in fewer than $m$ steps, then $L_3(X) = L_1(X)$; (b) if $L_2(X)$ closes in $m$ steps but $L_1(X)$ does not, then $L_3(X) = L_2(X)$; (c) if neither $L_1(X)$ nor $L_2(X)$ closes, $L_3(X)$ does not close.

9. (Theorem of Nagorniĭ.) If $\mathfrak{A}$ is a nonvoid alphabet, show that any algorithm $L$ over $\mathfrak{A}$ is equivalent, so far as words $E$ in $\mathfrak{A}$ are concerned, to an algorithm $L'$ in an alphabet formed by joining a single auxiliary letter to $\mathfrak{A}$. (Nagorniĭ [UTP]; cf. Chernyavskiĭ [KNA]. Let $\alpha_1, \ldots, \alpha_n$ be the auxiliary letters for $L$, $a$ a letter of $\mathfrak{A}$, $\beta$ a new auxiliary letter, $a^k$ a sequence of $k$ successive $a$'s; then replace $\alpha_k$ by $\beta\beta a\beta a^k \beta a\beta\beta$.)

10. (Universal algorithm.) If $L$ is an algorithm (1) in the fixed alphabet $\mathfrak{A}$, and $+, *, :, !$ are letters not in $\mathfrak{A}$, then the word

$$P^L \qquad A_1 + \circ B_1 : A_2 + \circ B_2 : \cdots : A_n + \circ B_n$$

where '$\circ$' is to be replaced by '$!$' if the corresponding command (1) is a stop command and is to be omitted otherwise, is called the program of $L$. Show that there is an algorithm $U$ over the alphabet $\mathfrak{A} \cup \{+, *, :, !\}$ such that for any such $L$ and any word $E$ in $\mathfrak{A}$

$$U(P^L * E) = L(E)$$

it being understood that the existence of either side entails that of the other. (Markov [TAl$_2$]. A shuttle algorithm for doing this is given explicitly in Chernyavskiĭ [KNA]; it contains 122 commands.)

**11.** (Impossibility theorem.) Let $\mathfrak{B}$ be the alphabet of Exercise 10. Show that there is no Markov algorithm $L$ over $\mathfrak{B}$ which is applicable to a word in $\mathfrak{B}$ just when it is the program of an algorithm in $\mathfrak{A}$ which is not applicable (in $\mathfrak{B}$) to its own program. (Markov [TAl$_2$, chap. 5]; the rest of the book gives many other examples of problems which are unsolvable by means of algorithms.)

**12.** (Theorem of Detlovs.) Let $\mathfrak{A}$ be the alphabet of Exercise 3. A numerical function $f(x_1, \ldots, x_n)$ is said to be algorithmic just when there is an algorithm $L$ over $\mathfrak{A}$ such that

$$y = f(x_1, \ldots, x_n) \rightleftarrows L(|^{x_1} * |^{x_2} * \cdots * |^{x_n}) = |^y$$

where '$\rightleftarrows$' is defined as in Sec. A4. The function is said to be completely algorithmic just when for arbitrary numbers $x_1, \ldots, x_n$ there is a $y$ such that the two sides of the above equivalence are both true. Show that a necessary and sufficient condition that $f$ be algorithmic is that it be partial recursive, and that it is completely algorithmic just when it is general recursive. (Detlovs [NAR]. Asser [TMM] proves directly the equivalence of being algorithmic with Turing calculability; this is done more simply, using shuttle algorithms, in Chernyavskiĭ [KNA]. Cf. below, Sec. 3C.)

## S. SUPPLEMENTARY TOPICS

**1. Historical and bibliographical comment.** For general references on Secs. A to D of this chapter see Sec. 1S5 and the references on formalism in Sec. 1S6. The following references supplement those.

For surveys of semiotics in general, see Morris [FTS] and Carnap [FLM]; for more comprehensive treatments see Morris [SLB]; Carnap [LSS], [ISm], [FLg], [MNc]; Martin [TDn], [SPr]; Kemeny [NAS]; Stegmüller [WPI]. The subject blends into the philosophy and psychology of language and linguistics. The following are examples of more philosophically or psychologically oriented works: Brown [WTh], Ogden and Richards [MMn], Quine [WOb]. On the linguistic side those to which I owe most are cited in [LAG]. The American Mathematical Society symposium [SLM], in which [LAG] appeared, contains several other papers of interest in connection with the interrelations of language and mathematics. Further references are given in [CLg], p. 37, in the bibliographies to the above-cited works, and in connection with special topics below. See also Secs. 3 and 4.

The notion of U language was first presented in an address to the Tenth International Congress of Philosophy in 1948. The abstract of this address, prepared some months before, is [LFS]; the actual text of the address is [LMF]. See also [TFD], sec. 14; [CLg], sec. 1D3.

On the use of quotation marks see Quine [MLg$_2$, pp. 23ff.]; Leblanc [IDL, pp. 2ff.]; Suppes [ILg, chap. 6]; Exner and Rosskopf [LEM, secs. 1.3 and 1.4, especially the exercises on pp. 15–16]. The first logician to use quotation marks in this way was Frege, but linguists had previously used them for the same purpose at least as far back as 1765 (see [TFD], p. 12, footnote 16).

The term 'grammatics' was proposed in [MSL] and again in [LFS]. For an account of it see [LAG] (which was prepared from [LSG]) and the papers there cited; also (for a previous edition) [TFD], sec. 15. The functionality operators $F_n$ are studied in the theory of functionality in combinatory logic ([CLg], chaps. 8–10); examples of functors belonging to various categories are given in [CLg], pp. 264 and 274.

The explanation of the dot notation is taken over from [CLg], sec. 2B5a.

For the history and collateral references the reader is referred there or to [TFD], p. 43, footnote 15.

The notion of inductive class was formulated by Kleene [IMM, pp. 20, 258ff.]. There are some minor variations in detail. He calls the conventions by which an inductive class is defined an "inductive definition"; the initial specifications he calls the "basic clauses," and the generating specifications he calls the "inductive clauses"; these together constitute the "direct clauses," in contrast to the closure specifications, which he calls the "extremal clauses."

The discussion of constructions and notions connected with them is taken with some modifications from [CLg], sec. 2B, which in turn is based on [DSR]. The idea of exhibiting a proof as a tree construction has, of course, been known for a long time; it plays a prominent role in Gentzen's [ULS] and in the work of Hertz (see his [ASB] or earlier papers cited in Church [BSL]), which preceded it, and was used explicitly in proof theory by several members of the Hilbert school (see, for example, Hilbert and Bernays [GLM.I, pp. 221, 426]); I do not know the origin.

In regard to Sec. B, the distinction here made between a theory and a system is a departure from the usage of my previous publications; for the motivation see Sec. 2. (One should observe that a system is a special kind of theory, so that the use of the term 'theory' for what is actually a system is not incorrect.) The reference for Post consistency and completeness is Post [IGT]. Tarski [FBM] derives an extensive list of theorems about formal theories (in the present sense—he calls them "sets of sentences" and uses the term 'system' for such a set which is deductively closed) in general; he uses methods which we shall consider in the next chapter. The work of Hertz (cited in the preceding paragraph) can also be regarded as having to do with theories. Otherwise there is relatively little literature regarding theories per se; one has to abstract the purely theoretical considerations from treatments of systems.

The inclusion of both syntactical and ob systems under the term 'formal system' is again an innovation of the present book (see Sec. 2). Ob systems correspond to what I previously called formal systems. For them, and particularly their relations to syntactical systems, [CFS], [DFS], and [IFI] contain some minor changes relative to the works cited in Sec. 1S5. The discussion of Sec. C contains a continuation of these changes. Ob systems have otherwise been little considered in the literature; cf., however, Herbrand [RTD, pp. 54ff.].

The systems elsewhere considered in mathematical logic are usually syntactical, and formulations of them can be found in most of the general works on mathematical logic (Secs. 1S1, 1S3, and 1S5). Especially careful formulations of them are found in Hermes [Smt]; Tarski [BBO], [WBF]; Schröter [AKB], [WIM]; Henkin [AMS], [SAT]; Rosenbloom [EML]. These show, in principle, how a syntactical system can be formulated as an ob system; on this point cf. Examples 3 and 4 and [CFS]. Precise formulations going in a slightly different direction are in Lorenzen [EOL] and Post [FRG]. The latter shows that an assertional concatenative system of a rather general sort can be reduced to one with a single axiom and rules of the form of (22) of Sec. E; this work is reported in Rosenbloom [EML, chap. 5]; also, less at length, in Porte [SPA, sec. 6], and Davis [CUn, chap. 6]. Since the algorithms of Sec. E have a syntactical basis, there is considerable detailed information on syntactical matters, mostly in the Russian language, in the literature cited for Sec. E.

The following remarks concern the examples in Secs. C to D. Example 1, complete with Hungarian names for the basic categories, was first presented in [LSG] to an audience of humanists. Example 2, here presented as a modification of Example 1, is in reality the same as Example 1 of [OFP]; Example 1 is a modification of it rather than the reverse. Example 3 is a trivial modification of Example 3 of [OFP], and is taken in principle from Hermes [Smt]. Example 4 is an obvious modification of Example 3. Example 5 is, essentially, an abbreviation obtained by Łukasiewicz (see Łukasiewicz and Tarski [UAK, p. 35, footnote 9]) of a set of axioms for the classical propositional calculus due to Frege; for a development of it at some length see Schmidt [VAL, secs. 79–84]. Examples 6 and 7 are modifications of Example 5. For other examples of ob systems see [OFP], chap. 5.

In Sec. D the term 'assertional system' is introduced here for the first time. I have used the term 'logistic system' heretofore, but that usage conflicts with other uses of that term. Otherwise the section contains little innovation over [CLg], sec. 1E. The Post investigations, just mentioned, are essentially additions to the subject matter of Sec. D.

The symbol '⊢' was introduced by Frege (see, for example, his [GGA], pp. 9ff.), who gave separate interpretations to the horizontal and vertical parts of it. Later it was adopted (but not the separate use of the horizontal and vertical parts) in Whitehead and Russell [PMt]. It is, of course, a natural modification of Frege and Whitehead and Russell [PMt] to use it as a one-place verb. Its use for stating rules, as in (6) of Sec. D, is due to Rosser ([MLV, p. 130]; see also his [LMt], pp. 56ff.). Rosser [LMt, p. 57] suggests reading '⊢' as 'yield' or 'yields' and calls '⊢' a "turnstile."

Section E on algorithms is based primarily on the work of Markov. The first 200 pages of his book [TAl$_2$] give an extremely detailed account of the general properties discussed in Sec. E1. His shorter account [TAl$_1$] is probably adequate for a reader who is willing to work out details for himself; an English translation has just appeared. At present I do not know of any other systematic account, based upon direct knowledge of the Russian, in a Western European language. Some information may be obtained from Asser's two papers [TMM] and [NPA]; the former treats the equivalence, for a numerical function, of being algorithmic (see Exercise E12) and being computable by a Turing machine; the latter treats the equivalence with "Post algorithms," mentioned at the end of Sec. E3. There is also a brief account in Porte [SPA]; but Porte admits he is unable to read the Russian, and so his treatment is based wholly on reviews.

The theory of algorithms appears to be related (as already stated) to Post [FRG]. Like Post and his followers, the Russians have been greatly interested in unsolvability theorems. This has culminated in the proof of unsolvability of the word problem of group theory in Novikov [ANP]. Naturally these considerations belong in Chap. 3.

Some of the further theorems about algorithms are included in the discussion of Secs. E2 and E3 and in the exercises at the end of Sec. E. The discussion of Sec. E2 is based on Chernyavskiĭ [KNA]. For one who knows Russian, this paper gives a readable treatment which can be understood without a detailed acquaintance with Markov's work. It includes an account of the theorem of Nagorniĭ (Exercise E9), of the universal algorithm (Exercise E10), and a proof of the equivalence of a shuttle (and hence of a Markov) algorithm with a Turing machine (cf. Exercise E12).

The references for Sec. E3 are given in the text itself.

**2. Notes on terminology.** It has already been noted that the terminology used here involves some changes from my previous publications. Such changes are necessary for progress. However, they introduce a danger of confusion, particularly when the changes occur in the work of the same author. For those readers who may have occasion to consult my previous publications, the following explanation of circumstances connected with these changes may help to avoid misunderstanding.

In [CLg] and [TFD] the term 'formal system' was restricted to what is here called an ob system. This was the central notion in a series of papers which are cited in [CLg], sec. 1S1. At first it was called an "abstract theory"; thereafter it was called indifferently a "theory" or a "system," and the adjectives 'abstract' and 'formal' were attached to it, with an increasing tendency toward the latter as time went on. In 1937 it was decided to fix on the term 'formal system'. The change affected various papers[1] (cited in [CLg], sec. 1S1) published from 1939 to 1942.

In the meantime Kleene in his [IMM], p. 62, defined the term 'formal system' so as to apply only to a syntactical system. Further on in the same book he used the term 'generalized arithmetic' for a special kind of ob system, viz., that in which there is only one operation of each degree; the general case can of course be reduced to this one by taking new obs to represent the operations (cf. Sec. D2) and then replacing the closure of an operation of degree $n$ by the closure of the Kleene operation of degree $n + 1$, using as first argument the new ob representing the original operation. Kleene's definition gives the impression (no doubt unintentionally) that his definition of 'formal system' is a true representation of Hilbert's conception. But Hilbert was interested in just one system, that for mathematics (or a significant portion of it) as a whole, and his papers contain no reference to the notion of a formal system in general; moreover, he was inclined to call his method "axiomatic" rather than "formal."

Now the resemblances between these two forms of formal system are much more important than their differences. Moreover, from the general viewpoint of Sec. 1C they are both formal. It therefore seems the best policy to use the term 'formal system' so as to apply to both of them, and to use the more specific terms 'syntactical system' and 'ob system' when it is necessary to distinguish them. The term is beginning to appear in other contexts in senses which are consistent with this.

---

[1] Of these the first to reach print was [RDN]. This was to have appeared in the *Journal of Unified Science* in the *Proceedings of the International Congress for the Unity of Science* held at Harvard University in September, 1939, but all copies except some three hundred distributed to registrants at the Congress were destroyed in the bombing of Rotterdam. The reprint which appeared mysteriously in *Dialectica* in 1954 appears to have been made from a carbon copy which was sent to Gonseth on June 22, 1939.

The reader should note that not all persons who have written about formal systems have understood the term in the sense in which it was intended. Thus Church's review of the reprint of [RDN] is based on a misunderstanding.

This crystallization of the term 'formal system' frees the terms 'abstract' and 'theory' for other uses. Hence these terms are used here in technical senses which are explained in the proper places. The usage of 'theory' is new here; that of 'abstract' extends back at least to [CLg].

So much for 'formal system'. We shall now consider some other terms concerning which changes have been made.

The formal objects of an ob system were at first called "entities." This term was translated into German as '*Etwas*'. The term 'entity' remained in use in my papers in English until the change in 1937. At that time the word 'term' was substituted for it, and with it the word 'token', which is the English cognate of German '*Zeichen*', was used for the atoms. But these words have other uses in the U language, and it became evident that it was confusing to use 'term' as a technical term. The change to 'ob' was made in 1949, too late to affect [TFD], [STC], [PBP], but in time to affect [SFL], [TCm], etc.

The replacement, in Sec. D1, of 'logistic' by assertional has already been commented on (see Sec. 1).

Another trio of terms concerning which there has been a variation in usage is 'proposition', 'statement', and 'sentence'. The term 'proposition' has a bad odor to many influential logicians because philosophers have used it in such vague ways. But since these philosophical uses do not concern us, the term is really free for us to define and use in any way we like. At first I used it in the sense of 'statement'. But with the writing of [TFD] it became desirable to indicate a distinction between what is stated in the U language and what is only named therein. The word 'statement' is here used for the former; 'proposition' for the latter. This is an important distinction regardless of what views one may have as to the ontology of the notions referred to. As for 'sentence', I use it as a grammatical term. There is at present a tendency to use it in all three of the above senses, much like the German word '*Satz*'. This tends to make the discussion less, rather than more, precise. The philosophy of this usage is discussed in [IFI]. The matter will concern us further in Secs. 3 and 5A.

In regard to functors, it was proposed in [TFD] to call the two basic kinds (with nominal and sentential values, respectively) "junctors" and "nectors." (This was a suggestion obtained from reading a work by O. Jespersen.) According to this usage, what is here called an "operator" would be called an "adjunctor" or perhaps a "conjunctor." This proposal has never been followed up. In the present context the term 'operator' seems more natural. But if it is desired to apply these considerations in contexts where it is desired to use 'operator' in some other sense, then it might still be considered.

**3. Metamathematics.** Anyone who looks at all seriously at formalistic work of modern mathematical logic can hardly avoid noticing a great variety of words beginning with the prefix 'meta-'. One meets 'metalanguage', 'metasystem', 'metatheorem', 'metalogic', 'metacalculus', 'metasemiosis' (Sec. C6), and, in German, '*Metaaussagenkalkül*'. All these terms are described as in principle due to Hilbert. Actually the only one of them which Hilbert himself used is 'metamathematics'; the rest were invented by his followers on the basis of some analogy. There is a danger that a student will lose sight of what metamathematics actually was and on what principle these analogies are based. In order to keep matters straight, I shall present here some additional discussion concerning Hilbert's ideas and their relation to the notions of this book. In doing so I shall use the terminology of the text, even though it may be quite different from Hilbert's own. Furthermore, I shall not attempt to describe the evolution of Hilbert's thought—for that see the references in Sec. 1S6—but shall describe the impression which one gets from a series of papers appearing in the early 1920s.

Hilbert took as datum ordinary mathematics expressed in a certain symbolism. The precise nature of this symbolism does not concern us, but it constituted a language $L$ in the sense of Sec. C6. Hilbert's metamathematics was then a discipline in which one reasons intuitively about the expressions of this $L$ in a syntactical system. Thus $L$ becomes a syntactical O language.

In metamathematics we make statements; however, these statements are no longer expressed by phrases of the O language, but are about them. These statements must, of course, be expressed in some language. Hilbert did not introduce any name for this language, but it is precisely what is here called the U language.

Let us now examine the activity in metamathematics. We first have to recognize the O symbols (i.e., the symbols of the original mathematics) and be able to say when a given symbol is one of them. We have to recognize O expressions as strings of O symbols, and among the O expressions a certain subclass, here called wefs (well formed expressions). Finally, we have to recognize when a sequence of such wefs, with or without auxiliary indications, constitutes a proof. A wef which is the final wef of a proof is called a provable wef, or a thesis. The principal object which Hilbert aimed at was to characterize the theses in objective terms, in particular to show that certain kinds of wefs are not theses (consistency), and if possible to find ways of deciding whether a given wef is a thesis (decision problem). All this activity is syntactical in nature; the truth or falsity of a statement of metamathematics depends solely on the structure of the O expressions being talked about. In the last analysis this depends on the equiformity (see Sec. A2) of O inscriptions.

Now the U language must contain names for the O expressions, verbs for making the above-mentioned statements about O expressions, as well as additional linguistic devices for indicating proofs, constructions, general statements for whole classes of O expressions, etc. For the names of the O expressions Hilbert used these expressions themselves. As for the second and third elements mentioned, although Hilbert recognized the need of additional symbols, called *"Mitteilungszeichen,"* to be used in the usual mathematical way (mostly as U variables in the sense of Sec. 3D4), yet he preferred to use the words of ordinary language wherever possible, and in particular expressed the verbs by words such as

$$\text{——}\textit{ist ein Zeichen}$$
$$\text{——}\textit{ist eine Formel}$$
$$\text{——}\textit{ist beweisbar}$$
$$\text{——}\textit{kommt in}\text{——}\textit{vor}$$

Metamathematics thus contains all statements to the effect that such and such an expression is provable.   Let us call these the elementary statements of metamathematics.   Further, let the part of metamathematics which is confined to the elementary statements (together with the preliminaries necessary to formulate them) be called elementary metamathematics.   Then, clearly, elementary metamathematics is a metasystem, in the sense of Sec. C6, with an autonymous representation.   Further, the part of the U language which contains the first two kinds of expressions listed in the preceding paragraph will constitute the A language of that system.   On the basis of elementary metamathematics, the rest of metamathematics can be developed in ways which we shall consider in the next chapter.   Metamathematics is thus a metasystem formed by metasemiosis from ordinary mathematics, but since it is autonymous, it can equally well be regarded as formed by abstraction (Sec. C6).   In this way the Hilbert conception can be brought under the present one.

The autonymous representation of Hilbert is—arguments to the contrary notwithstanding—quite impeccable so long as we do not associate any meaning with the O expressions (cf. Sec. 2A2).   But we do want to associate such meanings; moreover, the arguments of Gödel, Tarski, and others show that we enrich the content of our science by so doing.   Consequently, it is worthwhile to avoid the autonymous representation.   There are two essentially distinct methods of bringing this about.

The first method, which is currently the one in general use, is to change the A nouns.   This gives the various alternatives considered under metasemiosis in Sec. C6.   It has the disadvantages noted there.[1]

The second of these methods is to change the O language.   Thus Scholz [GZM.I, first edition, p. 26; second edition, p. 19] states, in effect, that his O language is to be written in heavy ("*fett*" or "*Blockschift*") type and that any such letter is to be denoted in the A language by the corresponding letter in ordinary type.   One can go further in this direction.   It is a pity that Scholz did not specify red ink for his O symbols, because then it would have been obvious on even a casual inspection that there was practically no red ink anywhere in Scholz's book except on the pages cited.   In fact, as we noticed in Sec. C3, one does not need to specify the O language at all, but can leave the reader to construct one to his fancy.   In such a case one has an abstract syntactical system.   Furthermore, since the wefs of Hilbert's O

---

[1] In support of the conclusion that the second method increases the difficulty of comprehension, the following is an extreme example.   Definition 13 of Tarski ([WBF, p. 296] or [LSM, p. 179]) contains a technical error in that the five axioms given, which are supposed to characterize a Boolean algebra, are actually satisfied by the five-element nonmodular lattice (see Chap. 4, Fig. 2d).   The error arises in that the fifth of the axioms is an incorrect transcription of the ninth axiom in one of the sets of Huntington [SIP].   This mistake appeared in the Polish edition of the work, which was published in 1933; passed unnoticed through translations, first into German and then into English; and was only discovered by accident in 1956.   As Tarski wrote on Mar. 8, 1957, "It seems strange that so far neither I nor anybody else has noticed the mistake (though the paper has probably been read by a number of people)."   Of course, Tarski's conclusions are not affected by the details of his formulation of Boolean algebra, and no one had ever attempted the somewhat arduous mechanical task of ascertaining whether the axioms actually said what they were supposed to.   If the axioms had been expressed in the usual set-theoretic language, such a situation would be hardly conceivable.

language are monotectonic, the resulting system is an ob system, and the abstraction can be pushed further until one has an abstract ob system. In this way an abstract ob system can be reached from the Hilbert standpoint. This conclusion is not affected by the fact that, at least as presented in [ALS], it was reached in a quite different way from Whitehead and Russell [PMt] as point of departure; nor by the fact (mentioned in Sec. 2) that Hilbert did not entertain, at least not explicitly, the notion of formal system in general. Hilbert has, of course, definite reasons for preferring a syntactical representation, viz., the concreteness mentioned at the end of Sec. C6. His own statement is as follows (from his [NBM], pp. 162ff., with omission of a footnote; cf. his [GLM], p. 1):

Wie wir sahen, hat sich das abstrakte Operieren mit allgemeinen Begriffsumfängen und Inhalten als unzulänglich und unsicher herausgestellt. Als Vorbedingung für die Anwendung logischer Schlüsse und die Betätigung logischer Operationen muss vielmehr schon etwas in der Vorstellung gegeben sein: gewisse ausserlogische diskrete Objekte, die anschaulich als unmittelbares Erlebnis vor allem Denken da sind. Soll das logische Schliessen sicher sein, so müssen sich diese Objekte vollkommen in allen Teilen überblicken lassen und ihre Aufweisung, ihre Unterscheidung, ihr Aufeinanderfolgen ist mit den Objekten zugleich unmittelbar anschaulich für uns da als etwas, das sich nicht noch auf etwas anderes reduzieren lässt. Indem ich diesen Standpunkt einnehme, sind mir—im genauen Gegensatz zu Frege und Dedekind—die Gegenstände der Zahlentheorie die Zeichen selbst, deren Gestalt unabhängig von Ort und Zeit und von den besonderen Bedingungen der Herstellung des Zeichens sowie von geringfügigen Unterschieden in der Ausführung sich von uns allgemein und sicher wiedererkennen lässt. Hierin liegt die feste philosophische Einstellung, die ich zur Begründung der reinen Mathematik—wie überhaupt zu allem wissenschaftlichen Denken, Verstehen und Mitteilen—für erforderlich halte: *am Anfang*—so heisst es hier—*ist das Zeichen.*

This is a point well taken. But it simply argues that one must have the possibility of a syntactical representation, not that one must actually exhibit it. If one insists on doing so, and the autonymous representation is intolerable, then it would be better for most purposes to use the device of Scholz. Furthermore, I see no reason why treelike artifacts, such as suggested in [APM] as possible representation for an ob system, would not do equally well.

In all this I have not used the term 'metalanguage'. Neither did Hilbert, although he spoke of "*Mitteilungszeichen*" and might very well have spoken of a "*Mitteilungssprache*." According to Carnap (see his [LSL], p. 9; on p. 4 he uses 'syntax language' for the language used), the general use of 'meta-' in analogy with 'metamathematics' is due to the "Warsaw logicians." The term '*Metasprache*' appears, in a generalized sense, in Tarski [BBO] and [WBF], and 'metalanguage' in this general sense appears in Carnap [ISm]. The difficulty with this term is that the process of metasemiosis can be iterated, and one forms a metametalanguage in which he talks about the metalanguage. But since metasemiosis can be applied only to a language of formalized structure, this cannot be done if the metalanguage is identified with the U language. I therefore propose to use 'metalanguage' in the sense of A language of a metasystem. Then metasemiosis can be iterated as many times as we please, and the U language will always be the top language.

This is just as close an analogy to Hilbert's idea as the prevailing ambiguous usage (cf. Sec. A2; the suggestion was made in [LMF]).

**4. Semiotic systems.** General references to semiotics were given in Sec. 1. Here I shall sketch some of the further developments of semiotics, and I shall make some critical remarks. The sources for the latter are [MSL]; [LFS]; [LMF]; [CLg], sec. 1S2.

A semiotic theory (or system) is a theory in which we talk about the symbols or expressions of some language or languages. This theory may not be a formal system, but we shall think of it as being at least partially so, so that it makes sense to talk about its A language. Such a system is necessarily an interpreted system; in it we make definite understandable statements about the language or languages concerned.

Morris and Carnap divide the field of semiotics into three parts, or "dimensions." These are called *syntactics*, *semantics*, and *pragmatics*, respectively. Suppose we have a theory $\mathfrak{S}$ concerning a language $L$. Then $\mathfrak{S}$ is said to be syntactical (relative to $L$) just when the statements of $\mathfrak{S}$ refer only to the "syntax" of $L$, that is, the structure of its expressions as strings of symbols; $\mathfrak{S}$ is semantical with respect to $L$ if the meanings—i.e., designata—of certain expressions of $L$ are also taken into account; and it is pragmatical if the relations between $L$ and its users—psychological, physiological, practical, or whatnot—are talked about as well. With pragmatics we shall not be concerned (it is the special subject of Martin [SPr]), but it will be necessary to make some remarks about the other two. In spite of the apparent objectivity of the above trichotomy, it is not always clear when a system is syntactical and when it is semantical.

The difficulty may be illustrated by reference to Example 1. Suppose that, as suggested in Sec. C3, $a$, $b$, and $c$ are, respectively, '$\alpha$', '$\beta$', and '$\gamma$'; that $L$ is the set of expressions in these three letters; and that $\mathfrak{S}$ is the system of Example 1. Then $\mathfrak{S}$ is a semiotical system relative to $L$. If the sams are the expressions formed by writing zero or more $b$'s after an $a$, the tettles are the expressions formed by writing a $c$ between two sams, and the tantets are those tettles in which the two sams flanking the $c$ are alike, then we have a valid interpretation of $\mathfrak{S}$, and $\mathfrak{S}$ is syntactical relative to $L$. On the other hand, if we know in addition that the tettles are sentences of $L$—i.e., that they designate statements when $L$ is used—then $\mathfrak{S}$ is semantical. Under these circumstances $\mathfrak{S}$ is both syntactical and semantical. This is true a fortiori if we know that the tantets are true sentences and the sams designate numbers.

A similar, but more complex, situation occurs in regard to Example 6. Let $L$ have the alphabet consisting of '$N$', '$C$', and the infinite sequence $\sigma$ consisting of '$s_0$', '$s_1$', '$s_2$', .... Let $\pi$ be the set of expressions obtained from the members of $\sigma$ by iterated use of '$N$' as unary prefix and '$C$' as binary prefix (a Łukasiewicz representation). Let $a$ be '$s_0$', and if $X$ is a member of $\sigma$, let $Xb$ be the next succeeding member of $\sigma$. Further, if $X$ and $Y$ are in $\pi$, let $\neg X$ be the result of writing '$N$' before $X$, and $X \supset Y$ the result of writing first '$C$', then $X$, then $Y$. Let $\tau$ be the subclass of $\pi$ consisting of tautologies, i.e., of those words which always have the value 1 in any valuation assigning values '0' and '1' to '$s_0$', '$s_1$', ..., the values of $\neg X$ and $X \supset Y$ being obtained from the usual truth tables as in Sec. C5. Then if $X$ is any $L$ word, there are effective processes for deciding whether $X$ is in $\sigma$, in $\pi$, or in $\tau$. Let $S(X)$, $P(X)$, $\vdash X$ be interpreted as saying, respectively, that $X$ is in $\sigma$, $\pi$, or $\tau$. Then we have a valid interpretation of Example 6, and Example 6 so interpreted is a semiotical system over $L$. Moreover, it is syntactical. If we adjoin the information that the expressions in $\pi$ become sentences if arbitrary sentences from $L$ are substituted for the variables, then it becomes semantical, and it is a fortiori semantical if we adjoin the information that any sentence obtained from those in $\tau$ by such substitution is a true sentence of $L$. But it does not cease to be syntactical on that account.

It is now clear that there is something vague about the definitions of 'syntactical' and 'semantical'. Let us attempt to be more accurate. We shall define syntactical or semantical statements relative to $L$; the extension to syntactical or semantical theories or systems can then be made in an obvious manner. Let us assume that we are dealing with statements whose truth can be determined on either syntactical grounds, i.e., considerations concerning the structure of the $L$ expressions as strings (or finite sequences) of $L$ symbols, or semantical grounds, i.e., knowledge of the $L$ designata, or both (thus excluding pragmatical considerations). Then a *syntactical statement* is one whose truth can be determined on syntactical grounds alone; a *semantical statement* is one which gives information about the designata. Then it is clear that a statement, and therefore a system, can be both syntactical and semantical at the same time. A semantical statement which is not syntactical may be called *strictly semantical;* one which has no semantical information, *purely syntactical.* We might, of course, define the terms differently, but these definitions appear to be the most useful. Statements which are both syntactical and semantical simultaneously are of great importance. We attempt to make our notions syntactical in order to have definiteness of proof; indeed, this is the purpose of deductive systems: we want them to be semantical so that they can be applied. All the examples of semantical notions in Carnap [ISm] are, from this point of view, also syntactical (cf. [LSF]). Strictly semantical notions occur in Tarski's theory of truth (his [WBF]), where they are coupled with nonconstructive notions; purely syntactical ones occur in the theory of algorithms, etc.

It should be noted, in passing, that the question of whether we have a syntactical or a semantical situation is not a question of whether or not the system is interpreted—it is interpreted, or can be, in either case—but of the information available concerning the language $L$. A syntactical system in the sense of Sec. C2 will be syntactical in the present sense—if an O language is not given it can be supplied—but it is semantical only when the O language is a language $L$ for which the information is available.

The above examples illustrate also that the semantical information may enter in several stages. With reference to the first example, we may distinguish three stages and, corresponding to them, three subdivisions of semantics, as follows:

1°. We may know what expressions of $L$ are sentences. This stage is the *grammatics* of Sec. A3. It corresponds to Carnap's "rules of formation." In Example 1 the sentences are the tettles.

2°. We may know what are the true sentences or assertions of $L$. This stage was called *aletheutics* in [LFS]. It corresponds to Carnap's "rules of truth." In Example 1 we know the aletheutics when we know that the true sentences are the tantets.

3°. We may know the designata of the remaining phrases of $L$. This is the *onomatics* of [LFS] and [CLg], sec. 1S2. It corresponds to Carnap's "rules of designation." Thus to get the onomatics of Example 1 we need to know that $a$ designates 0 (where '0' is a numeral in the U language) and that the suffix '$b$' designates the successor function among numbers.

Naturally one can, conceivably, subdivide each of these divisions. Of these subdivisions I shall mention only tectonics (Sec. C6), whose concern is with the ways in which processes of construction (Sec. A6) can be represented by expressions. This may be regarded as a subdivision of grammatics. (Suggestions made in [LFS] for division of onomatics have never been followed up.)

For each of these divisions there is a corresponding adjective which can be applied to 'statement' or 'system'. One can also apply these to 'language', as when one interprets the statement that $L$ is an aletheutical language as meaning that we know which $L$ expressions designate true sentences. Thus the A language of any formal system is aletheutical, but if the system is abstract, it is not onomatical.

There are certain subtleties to the foregoing definitions which deserve some further discussion.

Suppose we were to set up an onomatical system $\mathfrak{S}$ over an O language $L$. Let $M$ be the A language of $\mathfrak{S}$. Then $M$ must contain names for the expressions of $L$ and ways of stating the syntactical relationships between them. This part $M_1$ of $M$ will constitute the A language of a syntactical system $\mathfrak{S}_1$ over $L$; $M_1$ will hence be a metalanguage in the sense of Sec. 3. But besides $M_1$, $M$ will have to contain means for referring to the designata of $L$; thus $M$ must contain a part $M_2$, which will have to be a translation of $L$ (perhaps $L$ itself). In addition, $M$ will have to contain at least a predicate to express the designation relation. If $\mathfrak{S}$ is formalized to a formal system $\mathfrak{S}^*$, and if this formalization is by metasemiosis, then the A language of $\mathfrak{S}^*$ may be still a third language $N$; but if the formalization is by abstraction, then we may identify $N$ with $M$, and $\mathfrak{S}^*$ with $\mathfrak{S}$ (i.e., we may suppose the formalization made before we began).

The terms 'metasystem' and 'metalanguage' were defined in Secs. C3 and S3 for syntactical systems. The question is how these terms can be extended to more general semiotical theories, distinguishing the previous senses by the additional adjective 'syntactical'. Now $\mathfrak{S}_1$ is a syntactical metasystem and $M_1$ is a syntactical metalanguage (relative to $L$), but we must remember that $M_1$ itself is an aletheutical language, so that we must distinguish between the senses of our new adjectives as applied to 'language' and 'metalanguage'. By analogy it would also be natural to say that $\mathfrak{S}$ is an onomatical metasystem relative to $L$ and that $M$ is an onomatical metalanguage. But it would be confusing to call $\mathfrak{S}^*$ a metasystem of $L$, since $N$ is a metametalanguage. Thus the term 'semiotical system' is more suitable for the general situation (this is a correction of [CLg], sec. 1S2).

These considerations show the complexities of the situation in regard to semiotical systems. The terminology suggested here is not standard in the current literature, and the student will have to observe for himself how, if at all, the particular authors make these and related distinctions.

# Chapter 3
# EPITHEORY

The preceding chapter dealt with the nature of a formal system; in this chapter we study the ways in which we develop our knowledge about such a system. Such development does not consist solely in deriving elementary theorems one after the other. Once the system has been defined, we can take it as datum and formulate further statements about it in the U language. These further statements are called *epistatements* and, when true, *epitheorems;* and the prefix 'epi-' will be used generally to signalize notions involving such going beyond particular applications of the deductive rules. We shall treat in this chapter considerations about epitheoretic processes in general. Certain conventions which will be important later will be formulated, and a few epitheorems which apply to rather general sorts of systems will be stated and proved.

Henceforth we shall suppose the systems we are dealing with are ob systems. It would not be difficult to make modifications to apply to other types, but there seems to be little advantage in doing so.

The different sections of this chapter overlap with one another to such an extent that it is not possible to arrange the sections in order without anticipating notions which are not rigorously defined until later. This is particularly true in Sec. A; likewise, the terms 'U variable' and 'indeterminate', although not formally defined until Sec. D, are needed earlier. In such cases the rather vague preliminary notion which the reader has will suffice until a more adequate treatment is reached later.

## A. THE NATURE OF EPITHEORY

It is appropriate to begin the study of epitheory by listing some examples, without being concerned with how we judge their truth. After that we shall study the truth criteria for some of the simpler kinds of these epistatements. This will give an idea of the nature of epitheory. It will pave the way for the more detailed study of some involved types of epitheorems in the later sections of this chapter and, indeed, throughout the book.

**1. Examples of epistatements.** Some examples of epistatements and epitheoretic properties were considered in Chap. 2. Here certain of these examples, as well as some additional ones, are listed and classified. This classification includes several important types, but it is not intended to be

exhaustive.   Numbers in brackets refer to the Examples 1 to 6 of Chap. 2.
In dealing with sams, the ordinary notation for numbers will be used.

   *a*.  Combinations of a finite number of elementary statements by the ordinary sentential connectives; e.g.,

| | | |
|---|---|---|
| (*a*1) | $0 = 0 \,\&\, 1 = 2$ | [2] |
| (*a*2) | $2 = 4 \to 3 = 6$ | [2] |
| (*a*3) | $1 \neq 2$ (i.e., not $1 = 2$) | [2] |
| (*a*4) | $\vdash p_1 \supset p_2 \,\&\, \vdash p_2 \supset p_3 \to \vdash p_1 \supset p_3$ | [5] |

   *b*.  Statements involving extension of the system in one way or another.
Systems may be extended in any of the following ways: (1) by adding new
atoms, forming an *atomic extension;* (2) by adding new operations, forming
an *operational extension;* (3) by adding new axioms, forming an *axiomatic
extension;* and (4) by adjoining new rules, forming an *inferential extension.*
We may also have extensions in which there is a combination of these processes; if the new elements are atoms and operations only, we speak of an
*ob extension,* whereas if new axioms or rules are involved, of a *statement,* or
*theoretical extension;* we may also have combinations which involve both of
these.   Thus the idea of axiomatic extension was involved in the definition
of

$$X_1, \ldots, X_m \vdash Y$$

for an assertional system in Sec. 2D1.   Examples of epitheorems of this
type are the following.

(*b*1)   In the extension formed by adjoining the axiom

$$2 = 4$$

   we have

$$3 = 5 \qquad\qquad [2]$$

(*b*2)   If in addition to the axiom of (*b*1) we adjoin the rule

$$\text{If } X = Y \text{ and } Y = Z, \text{ then } X = Z$$

   then          $3 = 7$                     [2]

(*b*3)          $p_1 \supset p_2, p_2 \supset p_3 \vdash p_1 \supset p_3$     [5]

   *c*.  Statements involving generalization with respect to the obs and remaining true in any extension.   Examples, in which $X$, $Y$, $Z$ are arbitrary obs,
are

| | | |
|---|---|---|
| (*c*1) | $\vdash X \supset X$ | [5] |
| (*c*2) | $X = Y \to Xbb = Ybb$ | [2] |

   *d*.  Statements involving generalization with respect to the obs, but not
always unaffected by extension; e.g. (with $X$, $Y$, $Z$, as in Sec. *c*),

| | | |
|---|---|---|
| (*d*1) | $X = X$ | [2] |
| (*d*2) | $X \wedge (Y \wedge Z) = (X \wedge Y) \wedge Z$† | [3] |

   *e*.  Statements involving generalization with respect to the elementary

† Cf. Exercise 2C2.

theorems, and thus giving necessary conditions for being such a theorem; e.g.,

(e1)    Every elementary theorem is of the form

$$X = X$$

where $X$ is an ob (sam)          [2]

(e2)    (Deduction theorem.)    For any obs $X$, $Y$

If $X \vdash Y$, then $\vdash X \supset Y$          [5]

*f.* Properties of the system as a whole, such as consistency, (Post) completeness, decidability, etc.

*g.* Relations of a system to other systems, including possibly its own subsystems or extensions (thus including Sec. *b*); mappings of one system on another; etc. Thus, wherever we use numerical subscripts to indicate a sequence, we are really forming a mapping from the obs of the sequence onto sams. The relations pointed out in Sec. 2C between Examples 1 and 2, between Examples 3 and 4, and between Examples 5 and 6 are further illustrations; so also are the reductions to special form considered in Secs. 2D1 and 2D2.

*h.* Relations to extraneous considerations, infinitistic assumptions, contensive interpretations. Matters of interpretation belong here if intuitive assumptions are really involved, but interpretation of one formal system in another belongs under *g*.

Under epitheorems we must also include theorems about formal systems in general, or about systems satisfying broad general conditions or related to other such systems in stated ways. Thus the equivalences mentioned under Sec. *g* above hold for a wide variety of systems.

Epitheorems are extremely important in modern mathematical logic. The following examples are a few of the more famous ones.

Perhaps the most famous of all is the Gödel incompleteness theorem. This has already been alluded to in Sec. 1C. It says, to put it roughly, that in any formal system which is consistent and sufficiently strong to be useful for mathematical purposes, one can find constructively an elementary statement which can be neither proved nor formally disproved. For this it follows—as we saw in Sec. 1C—that it is hopeless to expect that a single formal system will serve as formalization of all of mathematics, and it shows that we must be able to prove epitheorems about a system which transcend, contensively speaking, what can be expressed by elementary theorems. Gödel went on to conclude that no such system was capable of formalizing a proof of its own consistency. The theorem was first proved for a certain formalism based on Whitehead and Russell [PMt], and as an epitheorem of that (or any other special) system, it comes under Sec. *f*, but the proof can be extended to a broad class of systems.

Other examples of famous epitheorems are the Löwenheim-Skolem theorem and the Gödel completeness theorems. These are epitheorems of the ordinary "classical" predicate calculus; they are nonconstructive in nature, so that they come under Sec. *h*. The first one states that if a "formula" (or more generally, a set of formulas) of that calculus has a model, then it has a denumerable model. The second says that a formula which is valid in every denumerable model is assertible.

**2. Fundamental truth criteria.** As explained at the beginning of Sec. A, no attempt was made in Sec. 1 to discuss the truth criteria of the epistatements there listed. Let us now turn to this problem. We shall consider in this book only a class of epitheorems which we call *constructive*. These are characterized by the fact that we accept them as true only when we have an effective process which will actually carry through to a definite decision in any particular case arising under their formulation. To grasp the significance of this restriction, let us consider the epistatement $(a3)$. Now what does it mean to say that an elementary statement is false? Well, what does it mean to say it is true? The elementary theorems form an inductive class, and an elementary statement is in the class just when there is a construction of it, i.e., a formal demonstration. In order to show effectively that an elementary statement is false, we must show that no formal demonstration of it is possible. This amounts to showing that every demonstration leads to a conclusion which is different from the given elementary statement, thus bringing the negation under Sec. 1e. In order to establish constructively the falsity of an elementary statement, we must therefore give an effective process which will show that any elementary theorem has some definite structural characteristic which the given elementary statement does not have. In the case of the epistatement $(a3)$, this is not difficult. Indeed, we shall see later that we can establish effectively the epitheorem $(e1)$; from this we have $(a3)$, since the obs on opposite sides of the equality have different constructions. But in general this cannot be done. In fact, the Gödel incompleteness theorem shows that there are systems in which we are unable to establish effectively the falsity of a single elementary statement. In such cases the negation of an elementary statement merely describes a goal which we cannot attain.

This argument, which is quite similar to that advanced by the intuitionists (but for a very different reason), shows that we cannot take the sentential connectives too naively. It is necessary to make a study of the properties which these connectives have in the present context. This study, for the complicated cases where the connectives are compounded freely, will occupy a large part of the rest of the book. The remarks made here are preliminary and, in principle, cover only the simplest case, viz., where elementary statements are being joined.

For conjunction and alternation the situation is relatively simple. In the case of conjunction, if '$A$' and '$B$' abbreviate sentences,[1] the expression

$$A \mathbin{\&} B$$

abbreviates a statement which is true just when $A$ and $B$ are both true; on the other hand, the expression

$$A \text{ or } B$$

abbreviates a statement which we regard as proved just when we have

---

[1] The letters '$A$', '$B$' (and later also '$C$') are to be understood as abbreviations of sentences; they are also used in the text as names for the same statement. This is an instance of the general convention in regard to the U language which was mentioned at the end of Sec. 2A3.

either a proof of $A$ or a proof of $B$ or both.   In particular—for reasons which
were stated in Sec. 1C1—we do not admit the possibility that we can be
sure of $A$ or $B$ without knowing which.
     The situation in the case of implication is more complex.   The ordinary
truth-table (or material) implication involves negation.   This can be applied
in cases, like $(a2)$, where the underlying system is decidable, but in general
it is open to the objections already made for negation.
     In a context where the constructive point of view is important, we shall
understand the statement
$$A \to B \tag{1}$$
where '$A$' and '$B$' are abbreviations for sentences, as being true just when
there is an effective process for obtaining a proof that $B$ holds from a proof
that $A$ holds; we understand this as including the possibility that the effective
process may show that there is no proof that $A$ holds.   In this sense $(a2)$
is true.   Another example is the following.   Let $X$ be an ob of Sec. 2C3,
Example 5, and let $Y$ be obtained from $X$ by substituting some ob $Z$ for $p_1$.
If we make this same substitution throughout a proof that $\vdash X$, we convert it
into a proof that $\vdash Y$.   Since this is an effective process, we have
$$\vdash X \to \vdash Y \tag{2}$$
     In the case where $A$ is a conjunction of elementary statements $A_1, \ldots, A_m$
and $B$ is an elementary statement, there is another way worth noting of
characterizing (1).[1]   Let us consider (1) as a rule to be adjoined to the given
system to form an inferential extension.   Let us say that (1) is an *admissible
rule* if every elementary theorem in that inferential extension is also an
elementary theorem in the original system.   Then if (1) is true as defined
in the preceding paragraph, (1) is an admissible rule.   For consider any
demonstration in the extended system whose conclusion is $C$.†   If (1) is not
used in that demonstration, then it is a demonstration in the original system.
Otherwise the first application of (1) can be eliminated, for in the tree diagram of the demonstration there must be a demonstration of each of $A_1, \ldots,$
$A_m$ above the first node where (1) is applied, and from these we can construct
a demonstration of $B$ in the original system; this demonstration can replace
the demonstration using (1).   Continuing in this way we eliminate applications of (1), one by one, until we have a demonstration of $C$ in the original.
Whether constructive admissibility implies that (1) hold as defined in the
preceding paragraph is not clear.   However, using a platonistic argument,
we can show that it does.   For if there is no proof for $A$, then (1) holds;
otherwise a proof of $A$ followed by an application of (1) to infer $B$ would be
a derivation of $B$ in the extended system; hence $B$ must be derivable in the
original system.
     Under the assumption of the last paragraph regarding $A$ and $B$, the con-
dition (1) will be fulfilled, in particular, if $B$ is a theorem in the axiomatic
extension formed by adjoining $A$ (or $A_1, \ldots, A_m$) to the axioms, for we
convert a proof that $A$ holds into a proof that $B$ holds merely by adjoining
the extra steps necessary to derive $B$ from $A$.   In such a case we say that
$B$ is *formally deducible* from $A$.   That this is a stronger relation than (1) is

---

[1] The idea due to Lorenzen.   See, for example, his [EOL].

† Note that $C$ is constructively an elementary theorem only when a formal demonstration is given.

shown by the fact that neither ($a2$) of Sec. $1^1$ nor, in general, (2) with $X$, $Y$ as there stated holds in the sense of formal deducibility.    The relation

$$X_1, \ldots, X_m \vdash Y$$

defined for an assertional system in Sec. 2D1, is the formal deducibility analogue of

$$\vdash X_1 \ \& \ \cdots \ \& \vdash X_m \to \vdash Y$$

This discussion completes the study of the simple cases of the sentential connectives which arose in the examples of Sec. 1$a$.   The discussion has also taken into account those in Sec. 1$b$ involving only axiomatic extensions [viz., ($b1$) and ($b3$)].   The generalization ideas inherent in Secs. 1$c$ to 1$e$ are of such interest that it seems expedient to devote Sec. 3 to them.

**3. Epitheoretic generalization.**    We now attack the problem, left to one side in Sec. 2, of explaining the criteria of proof for epistatements, such as those under Secs. 1$c$ to 1$e$, which involve generalization with respect to obs or elementary theorems.    Since the obs and elementary theorems are inductive classes, the problem reduces itself to that of making constructive general statements concerning the elements of an inductive class.

There are two main sources of such generalizations.    On the one hand, we may be given, as part of our data, statements involving U variables[2] or parameters which may be specialized to arbitrary elements of the class. Thus the rules and axiom schemes of Example 5 (Sec. 2C3) hold in that sense for arbitrary obs, and they will so hold by definition no matter how the system is extended.   Consequences derived from such data will hold for all determinations of the parameters; in this way, as examples to be considered presently will illustrate, the epitheorems of Sec. 1$c$ are derived.   This kind of generalization will be called *schematic generalization*.    On the other hand, it may be necessary to deduce the general statement directly from the definition of the class.    In this case we speak of *inductive generalization*, or *generalization by induction*.

To get an illustration of schematic generalization, let us work through a proof of the epitheorem ($c1$).    For this we indicate the first and second axiom schemes of Example 5 by '$PK$' and '$PS$', respectively.    Then the derivation proceeds as follows, the steps being numbered on the left and justified on the right.   Here $X$, $Y$ are arbitrary obs.

1.  $\vdash (X \supset (Y \supset X)) \supset ((X \supset Y) \supset (X \supset X))$     by $PS$
2.  $\vdash X \supset (Y \supset X)$                                                    by $PK$
3.  $\vdash (X \supset Y) \supset (X \supset X)$                                        by 1, 2, Rule
4.  $\vdash (X \supset (X \supset X)) \supset (X \supset X)$                            taking $Y$ to be $X \supset X$†
5.  $\vdash X \supset (X \supset X)$                                                    by $PK$
6.  $\vdash X \supset X$                                                                by 4, 5, Rule

---

[1] Thus if we adjoin $1 = 2$ to Example 2 as an axiom, we get $1 = 2$, $2 = 3$, $3 = 4$, etc., in the enlarged system, but without the rule of transitivity for $=$, we do not get $2 = 4$. Cf. Sec. 6A2.

[2] See Sec. D1.

† That is, we regard '$Y$' as an abbreviation for '$X \supset X$'.   We could have written '$X \supset X$' for '$Y$' in steps 1 to 3, but it is more convenient to leave '$Y$' undefined until step 4.

This gives what we shall call a *demonstration scheme* (often called a "proof scheme"); it becomes a demonstration if we determine $X$ to be any particular ob.

An alternative way of presenting such a demonstration scheme—generalized to the case of $m$ U variables in an obvious way—is the following. Let $\mathfrak{S}$ be a given system, and let $\mathfrak{S}'$ be an atomic extension of $\mathfrak{S}$ formed by adjoining new atoms $x_1, \ldots, x_m$, without any new axioms, except that it is, of course, understood that the U variables for arbitrary obs of $\mathfrak{S}$ in the formulaticn of the rules and axiom schemes of $\mathfrak{S}$ become U variables for arbitrary obs of $\mathfrak{S}'$. Such additional obs are called $(adjoined)^1$ *indeterminates*. Now any elementary theorem of $\mathfrak{S}'$ may be validly interpreted[2] as that elementary theorem scheme of $\mathfrak{S}$ which one gets by substituting distinct variables for arbitrary $\mathfrak{S}$ obs for the adjoined indeterminates; indeed, a demonstration in $\mathfrak{S}'$ will become a demonstration scheme for $\mathfrak{S}$ if such a substitution is made throughout. Moreover, the most general demonstration scheme will be obtained in just that way, for a demonstration scheme for $\mathfrak{S}$ will also be one for $\mathfrak{S}'$, and if one specializes the U variables to be the indeterminates, one gets an $\mathfrak{S}'$ demonstration from which the original demonstration scheme can be recovered by the indicated substitutions. Thus if we define an elementary theorem scheme as one which can be obtained by a demonstration scheme, then such a theorem scheme is essentially the same as an elementary theorem in a suitable atomic extension. This brings such epitheorems under Sec. 1b.

This argument applies not only to elementary theorems, but to schematic forms of certain epitheorems considered in Sec. 2. For example, the epitheorem (c2) is a schematic statement of formal deducibility. Its proof is the following demonstration in the ob extension in which $x$ and $y$ are adjoined indeterminates.

$$x = y \qquad \text{by hypothesis}$$
$$xb = yb \qquad \text{by the rule of Example 2}$$
$$xbb = ybb \qquad \text{by the same rule}$$

The argument will apply to any form of epitheorem as long as it is truly schematic.

In the case of a system like Example 5 (Sec. 2C3), it is not necessary to adjoin indeterminates, for indeterminates have already been built into the system. In that system we have an epitheorem of the type of Sec. 1e to the effect that any ob can be substituted for one of the $p_i$.

We turn now to inductive generalizations. Suppose we have an inductive class $\mathfrak{X}$ with initial elements $\mathfrak{B}$ and operations $\mathfrak{O}$. In order to show that every member $X$ of $\mathfrak{X}$ has a certain property $P$, it is sufficient to establish the two following principles: (1) Every element of $\mathfrak{B}$ has the property $P$; (2) any element formed by applying an operation $\omega$ of $\mathfrak{O}$ to elements which have the property $P$ will also have the property $P$. For knowing these two principles (constructively, of course), we can form an effective process for converting a constructive proof that $X$ is in $\mathfrak{X}$ into a constructive proof that $X$ has $P$, as follows. We can know constructively that $X$ is in $\mathfrak{X}$ only when we have actually given a construction $\mathfrak{C}$ terminating in $X$. With $\mathfrak{C}$ there is

---

[1] We shall consider the possibility of other forms of indeterminates later.
[2] See Sec. 2B4.

associated a tree diagram $\mathfrak{D}$. The elements of $\mathfrak{C}$ corresponding to the top
nodes of $\mathfrak{D}$ have property $P$ by principle 1. Then passing down the tree
from node to node—let us say, in the order of the normal construction sequence—we can show by principle 2 that the other elements of $\mathfrak{C}$ in turn
have the property $P$. Thus in due time we show that $X$ has the property $P$.

There are two forms of proof by induction corresponding to the two main
types of inductive classes, viz., the obs and the elementary theorems (or the
consequences of some basis). In the former case we shall speak of a proof by
*structural induction;* in the latter, of one by *deductive induction.* Ordinary
mathematical induction is, from the present standpoint, a structural induction on the system of sams; it occurs with such frequency, in connection
with sequences (cf. Sec. 1*g*), that it is worthwhile to name it as a third form,
*natural induction.* In any case the principles 1 and 2 are called the *basic
step* and the *inductive step,*[1] respectively, of the proof by induction.

As an example of structural induction we prove the epitheorem (*d*1). The
basic step holds since

$$a = a$$

is the axiom of Example 2. The inductive step holds since

$$X = X \rightarrow Xb = Xb$$

is an immediate application of the second rule.

The converse of the epitheorem (*d*1) is the epitheorem (*e*1), which follows
by deductive induction. The basic step holds since the sole axiom has the
indicated form. The inductive step holds since the applicable rule converts
a statement of the indicated form into another statement of the indicated
form. Other examples of epitheorems of this type are the principle of substitution for Example 5 (discussed in Sec. 2D3); this was used. in principle,
in the above discussion of demonstration schemes.

Proofs by structural or deductive induction may take several forms. Thus
in one form of proof by deductive induction we show explicitly that the
property holds for $\Gamma$ if it is an axiom; also if $\Gamma$ is obtained by a rule such that
the property holds for the premises. In another form we may suppose we
have a sequence $\Gamma_1, \Gamma_2, \ldots, \Gamma_n$ constituting a demonstration $\Delta$, and then
show that every $\Gamma_k$ has the property if all those (if any) preceding it do.
Again we may make a natural induction on $n$ (i.e., the length of $\Delta$). These
forms of proof are equivalent to one another, and the choice between them in
the sequel has often been dictated by extraneous factors.[2]

**4. Other epitheorems.**  The remaining, more complex, types of epitheorems will be discussed here more briefly. Many of them are of an advanced
character; since they thus concern the superstructure rather than the foundations of mathematical logic, they will not be treated extensively in this
book. So far as the epitheorems under Sec. 1*f* are concerned, consistency
was defined in Sec. 2B1, Post completeness in Sec. 2B2, decidability in Sec.
2B1. In view of the explanations made there and those already made here,
it seems clear what the criteria of truth for such epitheorems must be.

---

[1] These terms are due, in principle, to Kleene. In his [IMM], p. 22, he uses the terms
'basis' and 'induction step', respectively.

[2] The reader may find it a good exercise to transform certain proofs from one form to the
other.

A type of epitheoretic process coming under Sec. 1g is of sufficient importance in the sequel to be mentioned at this point.    A many-one mapping

$$a \sim a^*$$

from the obs of a system $\mathfrak{S}$ to the obs of a system $\mathfrak{S}^*$ is called a *homomorphism* of $\mathfrak{S}$ into $\mathfrak{S}^*$ just when the following conditions are fulfilled.    To each primitive operation $\omega_i$ and primitive predicate $\phi_j$ of $\mathfrak{S}$, there corresponds an operation $\omega_i^*$ and predicate $\phi_j^*$ of $\mathfrak{S}^*$, in each case of the same degree, such that $\omega_i(a_1, \ldots, a_{m_i})^*$ is the ob $\omega_i^*(a^*, \ldots, a_{m_i}^*)$, and further

$$\phi_j(a_1, \ldots, a_{n_j}) \rightarrow \phi_j^*(a_1^*, \ldots, a_{n_j}^*)$$

Evidently, a homomorphism is a valid direct interpretation of $\mathfrak{S}$ in the obs of $\mathfrak{S}^*$.    Sufficient conditions that a valuation of the obs of $\mathfrak{S}$ as obs of $\mathfrak{S}^*$ be a homomorphism are that the condition on the operations be satisfied, that the axioms of $\mathfrak{S}$ be transformed into theorems of $\mathfrak{S}^*$, and the deductive rules of $\mathfrak{S}$ be transformed into admissible rules of $\mathfrak{S}^*$.    An *endomorphism* of $\mathfrak{S}$ is a homomorphism of $\mathfrak{S}$ into itself.    These terms are most useful when $\mathfrak{S}$ and $\mathfrak{S}^*$ are interpreted systems, in which case further specializations can be made (see Sec. 5A4), but they are sometimes useful under the general circumstances here stated.

## EXERCISES

The reader should prove or disprove as many of the epitheorems listed in Sec. 1 as he can.    For other examples, see the exercises to Secs. 2C and 2D.

## B. REPLACEMENT AND MONOTONE RELATIONS

This section is concerned with definitions and theorems connected with the notion of replacement in an ob system and with properties of monotony with respect to it.    This entails some additional technical considerations related to constructions.    Some general theorems with respect to quasi-ordering and equivalence relations are proved; these theorems, although relatively simple, are applicable to many existing systems.

**1. Preliminary explanations.**    The notion of an inductive class of elements $\mathfrak{X}$ formed from initial elements $\mathfrak{A}$ by generating principles $\mathfrak{O}$ is familiar to us from Sec. 2A5.    Here it is supposed that $\mathfrak{X}$ is a class of obs; that $\mathfrak{A}$ is also a class of obs, usually but not necessarily atoms or quasi atoms; and that $\mathfrak{O}$ is a class of operations for forming obs from obs.    Under these circumstances we call $\mathfrak{X}$ the *combinations* of $\mathfrak{A}$ by $\mathfrak{O}$.    A *proper combination* of $\mathfrak{A}$ by $\mathfrak{O}$ is one which is not a combination of any proper subset of $\mathfrak{A}$.

When $\mathfrak{A}$ is the class of atoms and $\mathfrak{O}$ is the class of primitive operations of a system $\mathfrak{S}$, then the combinations of $\mathfrak{A}$ by $\mathfrak{O}$ are precisely the obs of $\mathfrak{S}$.    Other possibilities are, however, admissible.    Thus $\mathfrak{A}$ need not include all the atoms, and it may include obs which are not atoms[1]; likewise $\mathfrak{O}$ need not include all the primitive operations, and it may include operations which are defined in some way.

---

[1] They need not even be quasi atoms in the sense of Sec. D2.

In the discussion of Sec. 2A6, certain technical terminology relating to constructions and tree diagrams was omitted so as not to clutter up the treatment with unnecessary technicalities. Since this terminology will be useful later, it will now be explained. Note that, under the general circumstances admitted here, a construction is not necessarily unique. Indeed, in the system of sams, *abbb*, as a combination of *a* by the operations of affixing *b* and of affixing *bb*, has three different constructions.

Let $\mathfrak{C}$ be a construction of an ob $X$ of $\mathfrak{X}$, and let $\mathfrak{D}$ be its associated tree diagram. Then we shall call $X$ the *terminus* of $\mathfrak{C}$. Further the *data* of $\mathfrak{C}$ will be the obs in $\mathfrak{A}$ which correspond to the top nodes of $\mathfrak{D}$. An ob $Y$ will be called a *component* of $X$ just when it corresponds to some node in $\mathfrak{D}$; a *proper component* is one which corresponds to a node other than the bottom node. A *branch* of $\mathfrak{D}$ is a sequence of nodes such that the successor of any node is joined to its predecessor from below; a branch which begins with a top node and ends with the bottom node will be called a *maximal branch*. A node $Y$ is *over* a node $Z$ (and $Z$ is *under* $Y$) if $Y$ and $Z$ are on a common branch and $Z$ comes after $Y$ in the sequence. These terms relating to $\mathfrak{D}$ may be extended by metonymy to $\mathfrak{C}$; in fact, it is permissible to identify nodes with the corresponding components. Moreover, if $Y$ is a component, those components which are above $Y$, together with $Y$ itself, constitute a construction terminating in $Y$; this construction is called the *construction of Y determined by* $\mathfrak{C}$, and the associated tree diagram, the *subtree determined by Y in* $\mathfrak{D}$. Finally, if $Y$ is a component of $X$ in $\mathfrak{C}$, there will be a branch which begins with the node associated with $Y$ and ends with the bottom; the corresponding sequence of components beginning with $Y$ and ending with $X$ will be called the *composition from Y to X in* $\mathfrak{C}$.

Let $\mathfrak{C}$ be a construction of $X$, and $Y$ be a component in $\mathfrak{C}$. Then there will exist a composition from $Y$ to $X$ in $\mathfrak{C}$. If $Y$ occurs more than once as a component, then there will be a separate composition for each occurrence. Accordingly, we define an *occurrence* of $Y$ in $X$ (relative to $\mathfrak{C}$) as a composition from $Y$ to $X$ (in $\mathfrak{C}$). This corresponds to current practice with syntactical systems, where an occurrence is identified with the initial segment which ends at the last letter of the occurrence. If the syntactical system is taken in the affixative sense, that notion of occurrence can be considered a special case of that considered here.

Such a composition can be characterized as a sequence $U_1, U_2, \ldots, U_n$, such that $U_1 \equiv Y$, $U_n \equiv X$, and

$$U_{k+1} \equiv \phi_k(U_k) \qquad k = 1, 2, \ldots, n-1$$

where $\phi_k$ is a unary operation obtained by fixing all but one of the arguments of some operation $\omega$ of $\mathfrak{D}$. We shall sometimes call the $\phi_k$ the *component operations* of the composition.

Now let there be given an occurrence of $Y$ in $X$. The ob $X'$ arising from $X$ by *replacement* of this occurrence of $Y$ by $Y'$ is defined thus: Let the $U_k$, $\phi_k$ be as above, and let

$$U'_1 \equiv Y'$$
$$U'_{k+1} \equiv \phi_k(U'_k) \qquad k = 1, 2, \ldots, n-1$$

Then          $$X' \equiv U'_n$$

We can extend this definition to replacement in an elementary theorem simply by treating the predicate as if it were an operation.

**2. The replacement theorem.** Let an infixed '*R*' indicate a relation (i.e., a binary predicate between obs), and let $\phi$ be a unary operation converting an ob to an ob. Then $\phi$ will be said to be *directly monotone* with respect to *R* just when for all obs *X*, *Y*

$$X \ R \ Y \to \phi X \ R \ \phi Y$$

It will be said to be *inversely monotone* with respect to *R* just when for all obs *X*, *Y*

$$X \ R \ Y \to \phi Y \ R \ \phi X$$

Now let $\psi(X)$ be derived from *X* by a sequence of operations $\phi_1, \ldots, \phi_n$, and let every $\phi_k$ be either directly or inversely monotone with respect to *R*. If the number of inversely monotone $\phi$'s is even, then $\psi$ is directly monotone with respect to *R*; if that number is odd, it is inversely monotone. This is easily shown by natural induction on *n*. Any inversely monotone $\phi$ changes *R*, so to speak, into its converse; an even number of such reversals leaves *R* unchanged, whereas an odd number is equivalent to a single reversal.

The replacement theorem, hereafter referred to as Rp, is as follows.

**Theorem 1.** *Let U be a component of X, and let* $\phi_1, \ldots, \phi_n$ *be the component operations of an occurrence of U in X. Let each* $\phi_k$ *be directly or inversely monotone with respect to R. Let replacement of this occurrence of U by V convert X into Y. Then, if the number of inversely monotone* $\phi_k$ *is even, we have*

$$U \ R \ V \to X \ R \ Y$$

*whereas if that number is odd,*

$$U \ R \ V \to Y \ R \ X$$

*Proof.* This follows immediately from the preceding discussion.
*Examples.* In propositional algebra, let

$$\phi_1(X) \equiv Z \supset X$$
$$\phi_2(X) \equiv X \supset Z$$
$$\phi_3(X) \equiv \neg X$$

where *Z* is a fixed ob, and let *R* be such that

$$X \ R \ Y \rightleftarrows \vdash X \supset Y$$

Then $\phi_1$ is directly monotone with respect to *R* (for each fixed *Z*), whereas $\phi_2$ and $\phi_3$ are inversely monotone. (These facts are shown in elementary propositional algebra.[1]) Thus, if for particular obs *X*, *Y*, *Z*, we have

$$\vdash X \supset (Y \supset Z)$$

and if $\qquad\qquad \vdash X' \supset X, \qquad \vdash Y' \supset Y, \qquad \vdash Z \supset Z'$

then $\qquad\qquad\qquad \vdash X' \supset (Y' \supset Z')$

---

[1] They will be established in Chap. 5.

In case $R$ is symmetric, there is no distinction between direct and inverse monotony. Thus the rule for replacement of equivalents in propositional algebra is a special case of Rp.

In certain later connections it will be convenient to interchange the roles of $R$ and $\phi$, that is, to say that $R$ is directly (inversely) monotone with respect to $\phi$, rather than that $\phi$ is directly (inversely) monotone with respect to $R$.

**3. Monotone relations.**  A *monotone relation* is a relation $R$ such that

$$(\pi) \qquad\qquad U\ R\ V \to X\ R\ Y$$

whenever $Y$ is obtained from $X$ through replacement of an occurrence of a component $U$ of $X$ by $V$. Then *a necessary and sufficient condition that $R$ be monotone is that every primitive operation be directly monotone with respect to each of its arguments* (i.e., with all arguments but the one in question fixed). The sufficiency is shown by Rp; the necessity by specializing $X$ to $\phi(U)$, with $\phi$ any operation with all arguments but one held fixed.  It will be convenient to call the property of monotony $(\pi)$, as indicated.

A *monotone quasi ordering* is a monotone relation which is also reflexive and transitive; a *monotone equivalence* is a monotone quasi ordering which is also symmetric.  In an applicative system the characteristic properties of a monotone equivalence are

$$(\rho) \qquad\qquad\qquad X\ R\ X$$

$$(\sigma) \qquad\qquad X\ R\ Y \to Y\ R\ X$$

$$(\tau) \qquad X\ R\ Y\ \&\ Y\ R\ Z \to X\ R\ Z$$

$$(\mu) \qquad\qquad X\ R\ Y \to ZX\ R\ ZY$$

$$(\nu) \qquad\qquad X\ R\ Y \to XZ\ R\ YZ$$

Here $(\pi)$ breaks into the two properties $(\mu)$ and $(\nu)$.

**4. Monotone quasi ordering generated by a given relation.**  Given a relation $R_0$, the *monotone quasi ordering generated* by $R_0$ is the relation $R$ defined by the postulates $(\rho)$, $(\tau)$, $(\pi)$, and

$$(\epsilon) \qquad\qquad X\ R_0\ Y \to X\ R\ Y$$

The *monotone equivalence generated* by $R_0$ is that defined by these same postulates together with $(\sigma)$.  Here the word 'defined' is intended in the sense that the true statements of the form

$$X\ R\ Y \qquad\qquad\qquad\qquad (1)$$

form an inductive class having those which follow by $(\rho)$ or $(\epsilon)$ as initial elements, and $(\tau)$, $(\pi)$, and $(\sigma)$, if relevant, as generating rules; the rule $(\pi)$ can be replaced, by virtue of Theorem 1, by rules stating the directness of each of the operations formed from those of $\mathfrak{D}$ by fixing all but one of the arguments—in the case of an applicative system, this gives the rules $(\mu)$ and $(\nu)$.

**Theorem 2.**  *Let $R$ be the monotone quasi ordering generated by $R_0$.  Then a necessary and sufficient condition that* (1) *hold is that there exist a sequence $X_0, X_1, \ldots, X_n$, $(n \geq 0)$, such that $X_0 \equiv X$, $X_n \equiv Y$, and, if $n \geq 1$, for every $k = 0, 1, \ldots, n-1$, $X_{k+1}$ is obtained from $X_k$ by replacement of an occurrence of a component $U_k$ by an ob $V_k$ such that*

$$U_k\ R_0\ V_k \qquad\qquad\qquad\qquad (2)$$

*Proof of Sufficiency.* If $n = 0$, then (1) holds by $(\rho)$. If not, then we have, for all $k < n$,

$$U_k \; R \; V_k \qquad \text{by (2) and } (\epsilon)$$
$$X_k \; R \; X_{k+1} \qquad \text{by Rp}$$
$$X \; R \; Y \qquad \text{by } (\tau)$$

*Proof of Necessity.* Let $S$ be the relation described in the theorem. Then what we have to show is that

$$X \; R \; Y \to X \; S \; Y \tag{3}$$

To do this we show that $S$ satisfies the postulates for $R$; then (3) follows by deductive induction on the proof of the premise.

If $n = 0$ in the definition of $S$, then $Y$ is the same as $X$, and so $S$ satisfies $(\rho)$. If $X \; R_0 \; Y$, then the case of the definition of $S$ where $n = 1$, $U_1 \equiv X$, $V_1 \equiv Y$ applies; so $S$ satisfies $(\epsilon)$. Further, $S$ satisfies $(\tau)$, since a series of replacements carrying $X$ into $Y$, followed by a series carrying $Y$ into $Z$, will give a series carrying $X$ into $Z$. Finally, $S$ satisfies $(\pi)$. For let $Z \; S \; W$, where $Z$ is a component of $X$ and replacement of an occurrence of $Z$ by $W$ carries $X$ into $Y$. Let $Z_0, Z_1, \ldots, Z_n$ ($Z_0 \equiv Z$, $Z_n \equiv W$) be the sequence formed as in the theorem by replacement of the $U_k$ by the $V_k$. Let replacement of $Z$ by $Z_k$ convert $X$ into $X_k$. Then $U_k$ is a component of $X_k$, and the replacement of the appropriate occurrence of $U_k$ by $V_k$ will convert $X_k$ into $X_{k+1}$. Then $X \; S \; Y$, showing that $S$ has $(\pi)$, and hence that $S$ satisfies all the postulates for $R$.

This completes the proof of Theorem 2.

Examples of such generated relations are: (*a*) if $R_0$ is the relation of parent to child, $R$ is that of ancestor to descendant (on this account $R$ is often called an ancestral relation, or the ancestral of $R_0$); (*b*) in the system of natural numbers, if $R_0$ is the pair $(0,1)$, $R$ is the relation "less than or equal to," and if $R_0$ is the pair $(1,3)$, $R$ is that same relation among the odd numbers. Other examples will occur shortly. Thus, in Sec. 3C3, $\equiv$ is the monotone equivalence generated by $D$, whereas in Sec. 3D4, the relation of $\lambda$ convertibility is the monotone equivalence generated by $(\alpha)$ and $(\beta)$. Various reducibility relations are examples of monotone quasi orderings. Other examples are in the diagrams of Sec. 4A2, item 8°.

## EXERCISES

**1.** Among the natural numbers with '$\leq$' understood as 'less than or equal to', what sort of monotony characterizes the following operations: addition, subtraction, multiplication, division, greatest common divisor?

**2.** In the system of natural numbers, let $R_0$ be the two pairs $(2,4)$ and $(5,8)$. Characterize exhaustively the pairs for which $R$ holds as well as those for which $R$ does not hold.

**3.** Show that if $R_0$ is symmetric, the monotone quasi ordering generated by $R_0$ is an equivalence. ([CLg], corollary 2D1.1.)

**4.** Show that the monotone equivalence generated by $R_0$ is the same as the monotone quasi ordering generated by $S_0$, where

$$X \; S_0 \; Y \rightleftarrows X \; R_0 \; Y \text{ or } Y \; R_0 \; X$$

([CLg], theorem 2D2.)

## C. THE THEORY OF DEFINITION

One of the methods of developing our knowledge of a system is the introduction of new terms by definitions. We shall consider this process here in some detail, bringing it into relation with the algorithms of Sec. 2E. The definitions considered are those which are often called nominal definitions. Other types of definitions, such as semantic definitions and real definitions, are not considered.

**1. Preliminaries.** A definition is traditionally conceived as a convention in regard to the use of language. By such a convention we introduce a new symbol or symbol combination called a *definiendum*, with the stipulation that it is to stand for some other symbol combination, called the *definiens*, whose meaning is already known on the basis of the data and previous definitions. It is then expected that, by successive replacement of definienda by their respective definientia, we can reduce any properly formed expression containing primitive and defined symbols to one which is understood in terms of the primitive symbols alone; this latter expression, here called the *ultimate definiens*, is supposed to always exist and be unique. Thus definitions are regarded essentially as abbreviative linguistic devices, which are theoretically eliminable, but practically necessary in order to cut our discussions down to manageable size

Now we frequently want to apply the term 'definition' to conventions of a more complex nature than this. We define not only single combinations, but whole schemes or families of them, and we do this by stipulations in which the new symbol may appear in some of the definientia. Thus if we use '$D$' as an infix separating the definiendum from the definiens (we can read '$D$' as 'is defined to be'), the statement schemes

$$a + Y \, D \, Y$$

$$Xb + Y \, D \, X + Yb$$

relative to the system of sams (Sec. 2C2, Example 2), '$X$' and '$Y$' being U variables for arbitrary sams, would be considered as constituting a definition of the operation of addition among sams. That they can indeed so function is shown by the fact that '$+$' can be eliminated by successive replacements of definienda by definientia. For example, starting with $abb + ab$, we have, successively,

$$ab + abb$$

$$a \ + abbb$$

$$abbb$$

It is desirable to extend the notion of definition so as to include such conventions involving recurrence.

In the case where the definiendum is a noun denoting a formal object, the definition process amounts to extending the A language to include some new nouns. These new nouns are then alternative names for the same obs as before. But it is evidently permissible, in view of the arbitrariness associated with the word 'ob', to conceive the process of definition as one which forms new obs of which the new A nouns are representatives.  The introduction of these defined obs is thus a form of extension, here called a definitional extension.  The association of these new obs to the original basic obs is then a valuation of this extension in the original system.

This notion of definitional extension will be adopted as a basis in what follows.

**2. Definitional reductions.**  Let us now describe the notion of definitional extension more formally.  Let the original system be $\mathfrak{S}_0$; we shall use the adjective *'basic'* in reference to $\mathfrak{S}_0$, its obs, operations, axioms, etc. The extended system we call $\mathfrak{S}_1$, and we use the adjective *'new'* to indicate constituents which appear in it but not in $\mathfrak{S}_0$.  We use the letters *'A'*, *'B'*, *'C'*, with or without affixes, for basic obs, and *'X'*, *'Y'*, *'Z'*, for arbitrary obs, new or basic, of $\mathfrak{S}_1$.  Then $\mathfrak{S}_1$ is a *definitional extension* of $\mathfrak{S}_0$ just when the following conditions are satisfied.

*a.*  The obs of $\mathfrak{S}_1$ are formed by adjoining to $\mathfrak{S}_0$ certain new operations and new atoms.  It will be convenient to consider the new atoms as operations of degree zero, so that the new primitive constituents of $\mathfrak{S}_1$ are operations in this extended sense.

*b.*  There is a new binary predicate, expressed by *'D'* used as an infix.  The new elementary statements are thus of the form

$$X \, D \, Y \qquad (1)$$

We call $X$ the *definiendum* and $Y$ the *definiens* of (1).

*c.*  The new axioms of $\mathfrak{S}_1$ consist of all statements

$$X \, D \, X \qquad (2)$$

together with a certain set $\mathfrak{E}$ of *defining axioms*, each of which is of the form

$$\phi(A_1, \ldots, A_m) \, D \, Z \qquad (3)$$

where $\phi$ is a new operation of degree $m$.  This $\phi$ will be called the *principal operation* of (3).  Note that the arguments of the principal operation must be basic obs.

*d.*  There is a *rule of definitional reduction*, called Rd, allowing inferences of the form

$$X \, D \, Y \rightarrow X \, D \, Y'$$

where $Y'$ is obtained from $Y$ by replacing an occurrence of the definiendum of a defining axiom by its definiens.  An application of this rule will be called a *contraction* of the component replaced.

Let us call a demonstration from the new axioms by Rd a *definitional reduction*.  It must evidently start with an instance of (2) or a defining axiom,[1] the replacements being made in the definiens.  Consequently, we can represent such a reduction by simply giving the sequence of definientia.  Just when we eventually reach a definiens which is basic, this definiens will be called an *ultimate definiens* of $X$.

It follows, since the arguments of the principal operations of the defining axioms are basic obs, that the different components which can be replaced

---

[1] This case can be reduced to the case where we start with (2).  For we can take $X$ to be the definiendum of the axiom.

in a given $Y$ are nonoverlapping. They can therefore be contracted in any order without change in the final effect. If we fix on an order, say, from left to right, or according to a fixed order of the principal operations and then from left to right for the operation whose turn it is, we call a *standard reduction* one conforming to that order. Such a reduction is an essentially unique process. Indeed, the only possible freedom left lies in the fact that if there are two or more defining axioms with the same definiendum, we can choose which one to apply. Furthermore, any reduction can be transformed into a standard reduction without affecting the result.

Let us call a set of defining axioms, and also a definitional extension based on it, *proper* just when this last possibility does not arise, i.e., when there is at most one defining axiom for any possible definiendum. In such a case the reduction will proceed in a unique manner. It may go through to an ultimate definiens, or it may terminate in a new ob containing a possible definiendum for which there is no defining axiom (in which case we say it is *blocked*), or it may continue indefinitely.

Even in case the defining axioms are an improper set, we may remove the ambiguity by arranging the defining axioms in an order, and then agreeing that in case of doubt the first axiom in the ordering shall be applied. Even though the number of defining axioms be infinite, which it generally is, they can be enumerated, and this gives an order which can be used for the purpose mentioned. We shall suppose that such an order is fixed as part of the definition of a standard reduction. We shall call a definitional extension with such a definition of standard reduction a *standardized definitional extension*.

A standard reduction, then, has many of the characteristics of a process specified by an algorithm in which the defining axioms (with '$D$' replaced by '$\rightarrow$') are the commands. It differs from such a process chiefly in that the number of defining axioms is generally infinite. This possibility of an infinite number of commands introduces an indefiniteness in that it may, conceivably, not be a definite question whether the process has terminated at a given stage. When this eventuality is taken account of, we have an effective process. However, instead of investigating this question, we use a different approach which applies to all cases of practical interest.

Let us call a definitional extension *schematic* just when the defining axioms are given by means of a finite number of axiom schemes containing U variables for arbitrary basic obs. The definitional extensions arising in the formalization of ordinary definitions as considered in Sec. 1 are schematic. A schematic extension can be standardized by specifying an order of the axiom schemes and requiring that they be applied in that order. As in Sec. A3, we can consider such an axiom scheme as an axiom in an atomic extension $\mathfrak{S}_1^*$ of $\mathfrak{S}_1$ in which these U variables are taken as indeterminates, and thus as commands in an algorithm in which the U variables are auxiliary letters. Now the application of such an axiom scheme consists in finding the first occurrence of its principal operation such that the arguments are basic and are obtained by substitution of basic obs for the variables in the axiom scheme, ascertaining the obs to be substituted, putting the appropriate $Z$ in the place of the definiendum formed by that occurrence and its arguments, and effecting the substitution. All this can be done by a Markov algorithm over $\mathfrak{S}_1^*$

formed by adding certain commands in an appropriate order to those obtained from the axiom schemes.[1]

It is expedient to define here certain terms for describing special kinds of definitional extensions (d.e.). A *complete* d.e. is one such that for every new ob $X$, there is at least one definitional reduction to an ultimate definiens. A *univalent* d.e. is one in which there is at most one such ultimate definiens. Then we have seen that a proper d.e. is necessarily univalent, but the example given later (in Sec. 3) shows that the converse is not true. Also, any standardized d.e. is univalent. A *partial recursive* d.e. is a univalent schematic one. A *recursive* d.e. is a partial recursive one which is also complete. These terms agree with the usual ones[2] in case $\mathfrak{S}_0$ is the system of sams (i.e., numbers). Also, an *explicit* definition of an operation $\phi$ in terms of $\psi_1, \ldots,$ $\psi_n$ is one in which $\phi$ occurs only as principal operation in certain axioms, the remaining axioms constituting a definition of $\psi_1, \ldots, \psi_n$; if the $\psi_i$ are not mentioned, it is understood that $\phi$ is the only new operation.

**3. Definitional identity.** We shall use the infix '$\equiv$' to designate the relation, called definitional identity, which is the monotone equivalence generated by the defining axioms. The following example shows that this relation may have some strange properties if the definitional extension is not proper.

*Example* 1. Let $\mathfrak{S}_0$ be the system of natural numbers (cf. Example 2 of Sec. 2C). Let $\mathfrak{E}$ consist of the axiom and axiom scheme as follows:

$$\phi(0) \ \boldsymbol{D} \ 0$$

$$\phi(A) \ \boldsymbol{D} \ \phi(A')$$

These generate a partial recursive definitional extension of $\mathfrak{S}_0$; in fact, $\phi(0)$ has the ultimate definiens 0, while for any other $A$, the reduction of $\phi(A)$ continues indefinitely. If we standardize the extension by taking the axiom and axiom schemes in the order given, then we have exactly the same situation; but if we reverse the order, then $\phi(A)$ is undefined for every $A$. Nevertheless we have, in either case,

$$\phi(A) \equiv 0$$

for every $A$.

On account of this example it is necessary to distinguish between the relations $\boldsymbol{D}$ and $\equiv$, and some authors are careful, in making definitions, to use a notation which suggests the asymmetry of $\boldsymbol{D}$.[3] However, in the important case of proper definitions, it is not difficult to show by a deductive induction on the proof of

$$X \equiv Y$$

that if either $X$ or $Y$ has the ultimate definiens $A$, then they both do.[4]

In view of this result, it is not necessary to insist on the distinction between $\boldsymbol{D}$ and $\equiv$, provided that we see to it that our definitional extensions are

---

[1] See Exercise 1 at the end of this section. It follows from the heuristic principle mentioned in Sec. 2C, together with the fact that the process is effective, that such an algorithm exists.

[2] See Exercise 2.

[3] Notably Church; see, for example, his [Dfn]. However, one should not hastily infer that these authors were influenced by reasons similar to those advanced here.

[4] See Exercise 3.

proper. Accordingly, we shall use the symbol '≡' in making definitions throughout this book.

These definitions are not always permanent. Even when we make definitions which are to hold only in the immediate context, the sign '≡' will be used without thereby implying any commitment that the defining axiom so made is to hold several lines farther on. Such temporary definitions occur, for example, when we specify a temporary value for some U variable.

**4. Sentential concepts.** So far we have been concerned solely with the nouns in the A language. When defined predicates or other sentential concepts occur, we shall use the sign of equivalence '⇌' in making definitions. Such definitions are much less formal—if we wished to be formal with them, we should shift them into the ob structure by the process of reduction to assertional form used in Sec. 2D1; in connection with them we hardly pass beyond the point of view of Sec. 1. With this reservation what we have said in regard to ob definitions applies to them also.

In the case of a definitional extension we shall understand that the axioms and rules of $\mathfrak{S}_0$ hold for new obs in $\mathfrak{S}_1$, just as they hold when those obs are replaced by their ultimate definientia.

### EXERCISES

**1.** Exhibit in detail the Markov algorithm for determining the ultimate definiens of a schematic definitional extension as described in Sec. 2. ([TEA].)

**2.** Show that for numerical functions (which have operations, in the present sense, over the system of Example 2 of Sec. 2C2), the definitions of recursive and partial recursive functions given in Sec. 2 coincide with the usual ones (e.g., in Kleene [IMM, pp. 266, 326]). Hence derive from Exercise 1 that part of the theorem of Detlovs (Exercise 2E12) which says that every partial recursive numerical function is algorithmic.

**3.** Suppose that we have a proper definitional extension and that ≡ is defined as in Sec. 3. Let

$$X \equiv Y$$

Show that if either $X$ or $Y$ has an ultimate definiens, then they both do, and the two ultimate definientia are the same. ([CLg], sec. 2E3; cf. *ibid.*, sec. 4B4.)

**4.** Let $\mathfrak{S}_1$ be a definitional extension of $\mathfrak{S}_0$, and let $\mathfrak{P}$ be a set of new operations of $\mathfrak{S}_1$. Let $\mathfrak{E}_2'$ consist of all statements of the form

$$\psi(C_1, C_2, \ldots, C_p) \, D \, C_{p+1}$$

where $\psi$ is any operation in $\mathfrak{P}$, which are demonstrable in $\mathfrak{S}_1$. Let $\mathfrak{E}_2''$ be a set of defining axioms such that the principal operations are not in $\mathfrak{S}_1$ and all other new operations which are in $\mathfrak{S}_1$ are also in $\mathfrak{P}$. Let $\mathfrak{S}_2$ be the definitional extension whose defining axioms consist of $\mathfrak{E}_2'$ and $\mathfrak{E}_2''$, and let $\mathfrak{S}_3$ be that definitional extension formed by adjoining $\mathfrak{E}_2''$ to $\mathfrak{S}_1$. Show that (1) holds in $\mathfrak{S}_2$ if and only if it holds in $\mathfrak{S}_3$. ([CLg], sec. 2E4, corollary 3.1. The extension $\mathfrak{S}_2$ defines the operations of $\mathfrak{S}_2$ "relative to" $\mathfrak{P}$. Some generalizations are also considered in [CLg], sec. 2E4.)

**5.** Let $\mathfrak{S}_0$ have the standard Łukasiewicz representation. A set of defining axiom schemes (with '$x_1$', '$x_2$', ... as U variables for basic obs) of the form

$$\phi(a_i) \, D \, \psi(a_i) \qquad i = 1, 2, \ldots$$

$$\phi(\omega_n^k x_1 x_2 \cdots x_n) \, D \, \chi[x_1, \ldots, x_n, \phi(x_1), \ldots, \phi(x_n)]$$

will be said to constitute a primitive recursion scheme for $\phi$ in terms of $\psi$ and $\chi$ (the latter being not necessarily new); this convention is understood to hold with the obvious modifications in case $\phi$ depends on additional parameters. Let $\mathfrak{S}_1$ be a definitional extension whose new operations, in a certain fixed order, are $\phi_1, \phi_2, \ldots,$ $\phi_m$. Let the defining axioms for $\mathfrak{S}_1$ be such that for every $k = 1, 2, \ldots, m$, $\phi_k$ is either (a) defined explicitly in terms of $\phi_1, \phi_2, \ldots, \phi_{k-1}$ or (b) defined by a primitive recursion scheme in terms of $\phi_i$ and $\phi_j$, where $i < k, j < k$. Show that $\mathfrak{S}_1$ is partial recursive; further, if the explicit definitions under (a) are complete, in particular if they are of the form

$$\phi(x_1, \ldots, x_n) \, D \, \mathfrak{X}$$

then $\mathfrak{S}_1$ is recursive. (Cf. [CLg], theorem 2E6.)

   **6.** State and prove a theorem giving a sufficient condition that an operation $\phi$ defined by primitive recursion be monotone with respect to a relation $R$, which, in the basic theory, is the monotone quasi ordering generated by $R_0$. ([CLg], theorem 2E7.)

   **7\*.** Can an improper definition always be transformed constructively into an equivalent proper one by omitting axioms? (Cf. discussion in [CLg], sec. 2E2, footnote 52.)

   **8\*.** Develop an analogue of the $\mu$ function, normal form theorem, etc., for (partial) recursive definitions over an arbitrary $\mathfrak{S}_0$.

## D. VARIABLES

   This section deals with various matters concerning the uses of the word 'variable'. Two different senses of the word are contrasted in Sec. 1. Then various matters necessary to make the notions precise are taken up. The section concludes with a brief sketch of combinatory logic, which eliminates certain kinds of variables.

   **1. Classification of variables.** We begin by distinguishing two different senses of the word 'variable'.

   On the one hand, the term is applied to certain phrases of the U language whose meaning is not fixed. We shall call these phrases *U variables* in contradistinction to the *U constants*, which have a fixed meaning. Thus the letters '$X$', '$Y$', '$Z$', etc., have been used systematically throughout the foregoing for unspecified obs, and we needed these symbols in order to state rules, axiom schemes, and general epitheorems. Evidently U variables of some sort are necessary in order to make any general statements whatever; without them we should not even be able to formulate our systems.

   On the other hand, certain systems contain obs which are called "variables," usually because certain substitutions can be made for them. For instance, the $p_1, p_2, \ldots$ of Sec. 2C3, Example 5, are often called "propositional variables." Such variables we shall call *formal* variables. These variables are not, except for the possibility of an autonymous representation, U expressions; they may, of course, be expressions of an O language, but from the point of view of the U language, they are objects, not symbols.

   Since the formal variables are obs, they have names in the A language, for example, '$p_1$', '$p_2$', etc., in Sec. 2C3. These names are proper nouns in the U language, hence U constants. Since we rarely have occasion to use these proper nouns, it will be convenient to suppose—unless otherwise stated—that the formal variables are

$$e_1, e_2, e_3, \ldots$$

constituting a sequence e which is the set of all formal variables. This leaves the letters '$x$', '$y$', '$z$', etc., free for use as U variables referring to unspecified members of e. Thus $e_1$, $x$, $y$, $z$ will be formal variables because they are members of e; '$e_1$' is a U constant; '$x$', '$y$', '$z$' are U variables because it is not specified which member of e they refer to.

There are three main types of formal variables; these are here called (a) indeterminates, (b) substitutive variables, and (c) bound variables. In contrast to bound variables, indeterminates and substitutive variables together are called *free variables*. We shall discuss these later in more detail; in the meantime, brief definitions of them are as follows.

a. An *indeterminate* is an atom concerning which the primitive frame of the system makes no specific statement beyond the fact that it is an atom; the only restrictions on it are that obs involving it may be values of the U variables for obs in general, which appear in the rules and axiom schemes. Thus indeterminates may appear in axioms derived by substitution from an axiom scheme, but not otherwise. The atoms of Sec. 2C3, Example 5, are examples. Modifications of this definition will be considered in Sec. 2.

b. *Substitutive variables* are those obs for which substitutions are permitted in a rule of substitution explicitly formulated as a rule of deduction—as in Sec. 2D3, Example 7—or in some rule or axiom (scheme) in which substitution is essential.[1]

c. *Bound variables* occur in a system with formal variables when there is at least one operation one or more of whose arguments are restricted to formal variables, which variables are said to be bound by the operation, such that substitutions involving the variables so bound are restricted. Bound variables occur in ordinary mathematics. Thus in the statement

$$\int_0^3 x^2\, dx = 9 \tag{1}$$

we say that the left side of this equation is an operation on four arguments, namely, $x$, $x^2$, 0, and 3; the variable $x$† is bound, and it is not possible to substitute anything for it except to change it to another variable. Again, in the equation

$$\int_0^3 (xy)^2\, dx = 9y^2$$

we are restricted in the substitutions we can make for $y$; if we substitute anything which involves $x$, the resulting equation is false. Bound variables evidently bring in all the complications associated with substitutive variables, and some more besides.

**2. Indeterminates.** An indeterminate was defined in Sec. 1 as an atom on which the primitive frame of the system imposes no restriction except that it be an ob. Such a notion is significant only when there are axiom schemes stated with U variables for all obs, so that obs constructed with indeterminate components can be substituted for these U variables and thus enter into the

---

[1] An example is the axiom scheme ($\beta$) in the theory of $\lambda$ conversion (see Sec. 4).

† Some persons would say that '$x$' is bound. This would be appropriate if we thought of statement (1) as an expression in some O language rather than, as here, in the U language. It is perhaps worth noting that the statement (1), as usually interpreted, does not say anything about either $x$ or '$x$'.

axioms.   Otherwise, this notion of indeterminate would be vacuous, because the premises of a rule would never be fulfilled for an ob with an indeterminate component.   But there are systems, e.g., Sec. 2C3, Example 6, which do not have axiom schemes of this sort.   It is desirable to modify the notion of indeterminate so as to be applicable to such a system.

One type of case where this situation arises is where the property of being an ob is expressed by a basic predicate of the system.   Let us call such a predicate, if it exists, a *universal predicate*, or since it is a unary predicate, a *universal category*.   In such a case we say that an indeterminate is an ob concerning which nothing is postulated except that it is an ob to which the universal predicate applies.   For example, in one type of combinatory logic there is an ob E such that

$$\vdash \mathsf{E}X \tag{2}$$

is postulated as an axiom for all cases where $X$ is an atom; the system, furthermore, is applicative, and there is a rule.

$$\mathsf{E}X, \mathsf{E}Y \vdash \mathsf{E}(XY) \tag{3}$$

It then follows, by structural induction, that (2) holds wherever $X$ is an ob. The system does not contain any indeterminates, but the point is that one cannot form an atomic extension of the system without postulating (2) for the new atoms.   When nothing else is postulated, they are indeterminates according to the modified definition.

Again, it may happen that, although there is no basic predicate applying to all obs, there is one which holds for all obs which we regard, in one sense or another, as significant.   In such a case it is still appropriate to call the predicate universal, and to call the obs to which it applies *proper obs*.   Thus, in another form of combinatory logic, the scheme (2) does not hold for all obs, but does for all those which enter into the formal deduction; hence it is appropriate to call any atom, for which (2) and nothing else is postulated, an indeterminate also.   Although they are not ob systems, the concatenative systems of ordinary mathematical logic are such that the expressions which are not wefs do not play any role, and it would therefore be appropriate to call any O symbol an indeterminate for which it was postulated only that it was a wef.

A different, but analogous, situation occurs in the case of systems, like Sec. 2C3, Example 6, which have been derived from other systems by the process, described in Sec. 2D2, of reduction in the number of atoms.   In such a case let us take as proper obs those obs which represent the original obs and call *quasi atoms* those representing the original atoms.   If the original system contained indeterminates (or other formal variables), they will not in general be atoms in the new system, but quasi atoms; our previous discussion still applies with suitable changes.

In all these cases the notion of indeterminate involves arbitrary conventions which have to be made separately in each case.   But in principle, an indeterminate is an atom or quasi atom concerning which nothing is postulated except that it be admissible into the system as a (perhaps proper) ob.

Indeterminates are involved whenever we form an atomic extension. The additional atoms adjoined in such an extension are indeterminates by definition. But evidently we cannot derive any theorems containing them unless there are axiom schemes into which they can be substituted. Thus we have the same situation with respect to atomic extensions that we had with respect to indeterminates. We therefore extend the definition of atomic extension so as to include all ob extensions in which the new atoms are indeterminates. Indeterminates adjoined in this way we call *adjoined indeterminates*, to distinguish them from indeterminates already present.[1]

**3. Substitutive variables.** Substitutive variables were defined in Sec. 1 as those obs for which substitutions were permitted by a rule of substitution. Thus they are a class of obs which are formulated explicitly in the primitive frame of the system[2] (although they may not be called variables). For example, in the modification of propositional algebra considered in Sec. 2D3, Example 7, the substitutive variables are precisely the atoms.

The substitution rule referred to in the definition was formulated in Sec. 2D3 for a special case. In general, if there are no bound variables to restrict the substitution, we define the result of substituting an ob $M$ for $x$, symbolized as

$$[M/x]X \tag{4}$$

as that ob $X^*$ whose construction is obtained from a construction of $X$ by replacing subconstructions leading to $x$ by constructions of $M$. This definition could be so phrased as to make sense even if $x$ is not an atom, but there is little point in this since, in the substitution rules we are actually interested in, the substitutive variables are atoms (or at any rate quasi atoms, with $X$ proper). In that case, supposing that we have a standard presentation, the definition can be given recursively as follows:

$$x^* \equiv M$$
$$y^* \equiv y \qquad \text{for } y \text{ any other (quasi) atom} \tag{5}$$
$$(\omega_n^k X_1 \cdots X_n)^* \equiv \omega_n^k X_1^* \cdots X_n^*$$

We noted in Sec. 2D3 that Example 7 introduced there had the same elementary theorems as Example 5 of Sec. 2C3. This is a general situation, provided that the rules of deduction are invariant of substitution, i.e., that if we make corresponding substitutions in premises and conclusion of an instance of a rule, we get another instance of the same rule. Let us call this condition the *invariance condition*. We then have the following theorem:

**Theorem 1.** *Let $\mathfrak{S}_1$ be a system with axiom schemes and indeterminates. Let $\mathfrak{S}_2$ be a system having the same morphology as $\mathfrak{S}_1$. Let the axioms of $\mathfrak{S}_2$ be obtained by having the distinct U variables of the axiom schemes of $\mathfrak{S}_1$ denote distinct indeterminates. Let the rules of $\mathfrak{S}_2$ be the rules of $\mathfrak{S}_1$ and in addition a rule of substitution in which the indeterminates of $\mathfrak{S}_1$ are taken as substitutive variables. Let the invariance condition hold. Then $\mathfrak{S}_1$ and $\mathfrak{S}_2$ have the same elementary theorems.*

*Proof.* Every axiom of $\mathfrak{S}_1$ is derived by substitution from an axiom of $\mathfrak{S}_2$ and hence is true in $\mathfrak{S}_2$. Since the rules of $\mathfrak{S}_1$ are valid in $\mathfrak{S}_2$, it follows by deductive induction that every theorem of $\mathfrak{S}_1$ is true in $\mathfrak{S}_2$.

---

[1] Cf. Sec. A3.

[2] In this they differ from indeterminates which may have no special role assigned in the primitive frame.

Next, we shall see that if we take the indeterminates of $\mathfrak{S}_1$ as substitutive variables, the rule of substitution is admissible for $\mathfrak{S}_1$. In fact, we show by deductive induction that every substitution instance of an (elementary) theorem of $\mathfrak{S}_1$ is again a theorem of $\mathfrak{S}_1$. The basic step in this induction holds since every substitution instance of an axiom is an axiom. (Note that it is essential in this that the indeterminates do not occur except by giving values to the U variables.) The inductive step holds by the invariance condition.

From this it follows, by deductive induction, that every theorem of $\mathfrak{S}_2$ is a theorem of $\mathfrak{S}_1$. In fact, every axiom of $\mathfrak{S}_2$ is an axiom of $\mathfrak{S}_1$; the inductive step follows by the preceding paragraph.

Another way of proving the second half of this theorem is to show that in a demonstration in $\mathfrak{S}_2$ the substitutions can be pushed back to the axioms. The axioms then become essentially axiom schemes. This method is called by some German writers "*Rückverlegung der Einsetzungen*."

The invariance condition certainly holds if the rules are elementary (Sec. 2D3). Whether there are other significant cases of it is not clear.

According to this theorem, indeterminates and substitutive variables have much in common. The term '*free variable*' will be used for any formal variable which is not bound, hence for indeterminates and substitutive variables together.

It will be convenient to extend the term 'substitutive variable' to cases where the operation of substitution is necessary in the statement of some rule or axiom scheme, but not necessarily of a rule of substitution in the above sense. We shall meet examples in connection with the axiom scheme ($\beta$) of Sec. 4.

**4. Bound variables.** The instances of bound variables which were mentioned in the definition in Sec. 1 were all devices for making statements, not about the variables themselves, but about certain functions. Thus statement (1) would ordinarily be considered not as a statement concerning four objects, namely, $x$, $x^2$, 0, 3, but about three objects, namely, the square function, 0, 3. Since a function is a law of correspondence assigning a "functional value" to every admissible argument, one can only indicate a function by a device which gives the functional value for an unspecified argument; bound variables are such a device. The reader will speedily convince himself that this is the case in all instances of bound variables with which he is familiar. There is indeed reason to believe that all instances of bound variables which arise under the definition in Sec. 1 are of similar character. However, it is not necessary to demonstrate this; we can simply take it as a revision of the definition. One can then conclude that if we had a means of representing functions, all operations involving binding of variables could be replaced by ordinary operations.

One such device for indicating functions is that used by Alonzo Church in his calculuses[1] of $\lambda$ conversion. Let $M$ be an ob formed from a free variable $x$ and other atoms; then

$$\lambda x(M) \tag{6}$$

---

[1] I take the liberty of changing Church's 'calculi' to 'calculuses' for the reason stated in Fowler [DME, under 'calculus']. Cf. [rev Church].

will be the function whose value for any argument is obtained by substituting that argument for $x$ in $M$.   The operation of forming (6) from $x$ and $M$ will be called the $\lambda$ *operation*, or *functional abstraction*.   It is not required here that $M$ actually contain $x$.†   We call the ob $M$ the *base* of (6) and $x$ the *bound variable*.   A dot before the name of the base will avoid the necessity of enclosing the latter in parentheses.   Thus $\lambda x.x^2$, that is, $\lambda x(x^2)$, will be the square function; moreover, if we define

$$\int_a^b M \, dx \equiv J(\lambda x.M,a,b)$$

then (1) can be written

$$J(\lambda x.x^2,0,3) = 9$$

The natural extension of (6) for functions of several variables is

$$\lambda^n x_1 \cdots x_n . M \tag{7}$$

Using the idea back of the reduction of operations to application in Sec. 2D2, this can be defined recursively in terms of (6) thus:

$$\lambda^1 x.M \equiv \lambda x.M$$
$$\lambda^{n+1}x y_1 \cdots y_n.M \equiv \lambda x.(\lambda^n y_1 \cdots y_n.M) \tag{8}$$

Various operations involving bound variables which are of some importance in modern logic can be defined in terms of functional abstraction and ordinary operations; e.g.,

$$(\forall x)X \equiv \Pi(\lambda x.X) \qquad\qquad X \text{ for all } x$$
$$(\exists x)X \equiv \Sigma(\lambda x.X) \qquad\qquad X \text{ for some } x \tag{9}$$
$$X \supset_x Y \equiv \Xi(\lambda x.X, \lambda x.Y) \qquad Y \text{ for all } x \text{ such that } X$$

Here the interpretations, written briefly at the right, are, strictly speaking, the interpretations of the sentences which occur when the corresponding obs are asserted.   Here $\Pi$, $\Sigma$, $\Xi$ are ordinary operations; in an applicative system they can be taken as obs.   For the rest of this section it will be supposed that all operations involving binding of variables are similarly defined in terms of the $\lambda$ operation and ordinary operations.   Then the functional abstraction is the only operation which binds variables.

The formulation of substitution in a system with bound variables involves considerable complexity.   We have seen in Sec. 1 that if we substitute for a free variable an ob in which the bound variable occurs, we get something intuitively absurd.   This phenomenon is known as *confusion of bound variables*.   To avoid it we must regard $[M/y]\,(\lambda x.X)$ as undefined whenever $x$ is free in $M$.   But the relation between $x$ and $M$ expressed here by '$x$ is free in $M$' has itself to be defined by recursion, viz.: $(a)$ $x$ is free in itself but not in any other formal variable; $(b)$ $x$ is free in the closure of an ordinary operation just when it is free in one or more of the arguments; $(c)$ $x$ is free in $\lambda z.N$ just when it is distinct from $z$ and free in $N$.   With this understanding, we can add to (5)

$$(\lambda x.X)^* \equiv \lambda x.X$$
$$(\lambda y.X)^* \equiv \lambda y.X^* \qquad \text{if } y \text{ is not free in } M \text{ and is distinct from } x$$

† In this respect the notation differs from that of Church.

This gives a partial recursive definition of (4). To make the definition recursive, we should need to change the $y$, let us say to the first $z$ in $e$ which is distinct from $x$ and not free in $M$ or $X$, replacing the last clause by

$$(\lambda y.X)^* \equiv \lambda z([z/y]X^*)$$

This definition is complex and difficult to deal with.[1]

If the only operations in the system are application and the $\lambda$ operation, the latter will have the following properties:

($\alpha$)             $\lambda x.X = \lambda y.[y/x]X$      if $y$ is not free in $X$

($\beta$)             $(\lambda x.X)M = [M/x]X$

provided the right sides exist. Here '$=$' is taken intuitively as designating identity in meaning. But ($\alpha$) and ($\beta$) can be taken as axiom schemes in a system in which '$=$' is taken as designating a basic predicate with the properties of equality, i.e., as the monotone equivalence generated by ($\alpha$) and ($\beta$). That would give us a formalization of a calculus of $\lambda$ conversion.

**5. Combinatory logic.** Systems involving formal variables have the peculiarity that their statements appear to be about certain objects called variables—and this, from the formal point of view, is what they are—but when the system is interpreted in a natural manner, there are no contensive objects corresponding to these variables. In other words, such a system does not have a natural[2] direct interpretation. Statements involving formal variables are interpreted as statements about the functions obtained from those statements by some contensive analogue of functional abstraction. If we could find a way of defining functional abstraction in terms of ordinary operations alone, then formal variables would not be needed in the formulation of systems. They would be useful only for epitheoretic purposes.

Such a definition has been given in combinatory logic. One form of this is an applicative system containing no bound variables, such, however, that in an appropriate atomic extension a functional abstraction

$$[x]\mathfrak{X} \tag{10}$$

can be defined having the formal properties of the $\lambda$ operation. Without going into details we shall see how, in principle, such a definition is possible.

Suppose, then, we have such an applicative system $\mathfrak{H}$ and that there is an equality relation in $\mathfrak{H}$ with the usual properties. If $\mathfrak{X}$ is an ob in an atomic extension of $\mathfrak{H}$ in which $x$ is one of the adjoined indeterminates, then (10) will have to be a combination, formed by application, of those atoms of that extension which are distinct from $x$. Suppose these atoms belong to an extension $\mathfrak{H}'$ of $\mathfrak{H}$ which does not include $x$, and let $\mathfrak{H}'(x)$ be the extension formed by adjoining $x$ to $\mathfrak{H}'$; then the prefix $[x]$ will associate to each ob $\mathfrak{X}$ of $\mathfrak{H}'(x)$ an ob $X$ of $\mathfrak{H}'$ such that—this is a special case of ($\beta$) where $M \equiv x$—

$$Xx = \mathfrak{X} \tag{11}$$

Let us use German letters for obs of $\mathfrak{H}'(x)$ and italic letters for obs of $\mathfrak{H}'$. Then in order to define (10) recursively for every $\mathfrak{X}$ in $\mathfrak{H}'(x)$, it is sufficient to

---

[1] See [CLg], sec. 3E.

[2] They may, conceivably, have an artificial one. Cf. Sec. 2C5.

define it $(a)$ when $\mathfrak{X} \equiv x$, $(b)$ when $\mathfrak{X} \equiv U$, where $U$ is an ob of $\mathfrak{H}'$, and $(c)$ when $\mathfrak{X} \equiv \mathfrak{Y}\mathfrak{Z}$, where the functional abstracts $Y$, $Z$ of $\mathfrak{Y}$, $\mathfrak{Z}$, respectively, are known.   Let I, K, S be three fixed obs of $\mathfrak{H}$ and define

$$[x].x \equiv \mathsf{I}$$
$$[x].U \equiv \mathsf{K}\,U \tag{12}$$
$$[x].\mathfrak{Y}\mathfrak{Z} \equiv \mathsf{S}YZ$$

where $\qquad\qquad Y \equiv [x]\mathfrak{Y} \qquad Z \equiv [x]\mathfrak{Z}$

Then (11) will hold provided we have[1]

$$\mathsf{I}x = x$$
$$\mathsf{K}xy = x \tag{13}$$
$$\mathsf{S}xyz = xz(yz)$$

Here the '$x$', '$y$', '$z$' indicate that these equations are to hold schematically (i.e., as formal consequences of the assumption that $x$, $y$, $z$ are obs), and thus when $x$, $y$, $z$ are any adjoined indeterminates, the $x$ not necessarily the same as the variable in (10).

The functional abstract (10) so defined will be a combination of I, K, S and the atoms other than $x$ in $\mathfrak{X}$.   It can be shown to have the properties analogous to the $(\alpha)$ and $(\beta)$ of Sec. 4 [the property $(\beta)$ follows from (11) by the epitheorem which allows substitution for an indeterminate].   The combinations will be very long and complex.

The obs I, K, S, as well as combinations formed from them alone, are called *combinators*.   Among these combinators certain ones, called B, B', C, I', K', $\mathsf{K}_{(1)}$, W, will be mentioned from time to time in this book because they form simple combinations.   They have "reduction rules," analogous to (13), as follows:

$$\mathsf{B}xyz = x(yz)$$
$$\mathsf{B}'xyz = y(xz)$$
$$\mathsf{C}xyz = xzy$$
$$\mathsf{I}'xy = yx \tag{14}$$
$$\mathsf{K}'xy = y$$
$$\mathsf{K}_{(1)}xyz = xy$$
$$\mathsf{W}xy = xyy$$

They may be defined in terms of I, K, S thus:

$$\mathsf{B} \equiv \mathsf{S}(\mathsf{KS})\mathsf{K}$$
$$\mathsf{B}' \equiv \mathsf{CB}$$
$$\mathsf{C} \equiv \mathsf{S}(\mathsf{BBS})(\mathsf{KK})$$
$$\mathsf{I}' \equiv \mathsf{CI}\ (= \mathsf{B}(\mathsf{SI})\mathsf{K})$$
$$\mathsf{K}' \equiv \mathsf{CK} \text{ or } \mathsf{KI}$$
$$\mathsf{K}_{(1)} \equiv \mathsf{BK}$$
$$\mathsf{W} \equiv \mathsf{SSK}'$$

---

[1] Recall that, since $\mathfrak{H}$ is applicative, we are using the principle of association to the left (Sec. 2D2).

*Combinatory logic* is the branch of mathematical logic which deals with combinators and their properties. One form of it has been sketched here; there are various other forms and modifications. Combinators, or operators analogous to them, can be defined in terms of $\lambda$ conversion, and therefore the various calculuses of $\lambda$ conversion are considered as belonging to combinatory logic.

The obs $X$ which satisfy (11) for given $\mathfrak{X}$ are not unique. In fact, there exist combinators $X$ and $Y$ such that

$$Xx = Yx \qquad (15)$$

but not

$$X = Y \qquad (16)$$

Examples are SKS and KIK. However, if about six particular axioms, for example,

$$SK = KI$$

are adjoined to the system, then (16) will hold whenever (15) does. The new axioms are called the "combinatory axioms."

The indicated form of combinatory logic is a system of an exceedingly fundamental sort. It is assertional, applicative, elementary, and completely finite in structure; i.e., there is a relatively small finite number of atoms, axioms, and elementary rules. The system contains no formal variables, yet it is adequate to form a foundation for doing anything which can be done with variables in the more usual systems. Moreover, the reasoning is so fine that an elementary demonstration of even a comparatively simple theorem would contain a very large number of steps; the epitheoretic method is necessary in order to develop it profitably.

In this book we shall not be concerned further with combinatory logic, but will pass on to the development of systems more nearly like those ordinarily considered.

### EXERCISES

**1.** Show that the substitution prefix $[M/x]X$ has the following properties:

(a)
$$[x/x]X \equiv X$$

(b) If $x$ does not occur in $X$, then $[M/x]X \equiv X$.

(c) If either $y$ does not occur in $M$ or $x$ does not occur in $X$,

$$[M/x][N/y]X \equiv [N^*/y][M/x]X$$

where
$$N^* \equiv [M/x]N$$

([CLg], secs. 6D, 3E; [DSR].)

**2.** As definition of simultaneous substitution, let it be specified that

$$[M_1/x_1, M_2/x_2, \ldots, M_n/x_n]X$$

is the $X^*$ defined recursively by

$$x_i^* \equiv M_i \qquad i = 1, 2, \ldots, n$$
$$y^* \equiv y \qquad \text{for } y \text{ any other (quasi) atom}$$
$$(\omega_m^k X_1 \cdots X_m)^* \equiv \omega_m^k X_1^* \cdots X_m^*$$

Show that
$$[M_1/x_1, M_2/x_2, \ldots, M_n/x_n]X \equiv [M_1/z_1][M_2/z_2]\cdots[M_n/z_n][z_1/x_1]\cdots[z_n/x_n]X$$

where $z_1, \ldots, z_n$ are distinct variables which do not occur in $M_1, \ldots, M_n$ or $X$. What is the distinction between this and

$$[M_1/x_1][M_2/x_2] \cdots [M_n/x_n]X$$

3. Express the following by means of functional abstraction in combination with suitable ordinary operations:

(a) The ob $X$ is assertible if some (unspecified) ob is substituted for $x$ in $X$.

(b) The derivative of $x^2$ is $2x$.

(c) The ob $E$, applied to an $X$ which is a function of $x$, gives the function which is obtained from $X$ by substituting $x + 1$ for $x$.

4. If $P$ is an operation on functions, show that there are two possible interpretations to

$$Pf(x + 1)$$

Distinguish these by the use of the $\lambda$ notation. Show that the two are distinct if $f(x) = x^2$ and

$$Pf(x) = \begin{cases} f'(0) & \text{if } x = 0 \\ \dfrac{f(x) - f(0)}{x} & \text{if } x \neq 0 \end{cases}$$

([CLg], p. 81.)

5. Under what circumstances is '$D$' a suitable notation for the differentiation operator in the differential calculus?

6. By means of the combinators S, K, I, construct combinators $S_2$, $\Phi$, $Z_2$ such that

$$S_2 xyzu = xu(yu)(zu)$$
$$\Phi xyzu = x(yu)(zu)$$
$$Z_2 xy = x(xy)$$

([CLg], secs. 6A3, 5E1, 5B1, 5E5; for the general technique see [CLg], secs. 5B, 6A3.)

## S. SUPPLEMENTARY TOPICS

**1. Historical and bibliographical comment.** This chapter is a revision of [CLg], chap. 2. General references relating to the chapter as a whole may be found there and in Sec. 1S5. As in Chap. 2, the references given here supplement the general ones.

The notion of epitheory is an outgrowth of Hilbert's metamathematics. As remarked in Sec. 2S3, the elementary statements of metamathematics are only statements to the effect that particular wefs are demonstrable; these would be of no interest if it were not for the fact that the subject can be developed by epitheoretical methods. The whole purpose of metamathematics was to provide a basis whereby the proofs of certain sorts of epitheorems might be made objectively intelligible.

On this ground it would be natural to use the term 'metatheory' rather than 'epitheory'. Before [TFD], e.g., in [APM], [OFP], I did just that. The new term was introduced in [TFD] on account of the pressure of certain criticisms. I shall return to this point in Sec. 2.

A list of further examples of epitheorems, together with some discussion of philosophy, may be found in [OFP], chap. 9.

The incompleteness theorem of Gödel appeared for the first time in his [FUS]; a little later he revised and generalized it in his [UPF]. An important extension occurred in Rosser [ETG]. The literature, technical and nontechnical, growing out of this theorem is huge; I can only mention a few items which I have found particularly interesting. For an excellent popular exposition see Nagel and Newman [GPr]; somewhat more technical is Mostowski [SUF] (contains a full, but concise treatment with some innovations). Other treatments of some interest are Smullyan [TFS] (contains very modern and essential simplifications); Fraenkel and Bar-Hillel [FST] (history and comment, with numerous references); Myhill [PIM] (discusses philosophical implications); Findlay [GSN] (favorably cited by Myhill); Ladrière [LIF] (an encyclopedic report on limitative—i.e., incompleteness—theorems and everything connected with them); Rosser [IEP]. Other undecidability theorems, which one may regard as inspired by it, are Church [UPE] (undecidability of $\lambda$ conversion) and [NEP] (unsolvability of decision problem); Tarski et al. [UDT] (three short papers on undecidability of systems based on predicate calculus). See also Sec. 3.

For a discussion of the Löwenheim-Skolem theorem, with references to sources and other expositions, see Fraenkel and Bar-Hillel [FST, pp. 105ff.], Skolem [PTL], Church [IML$_2$, secs. 45 and 49]. A philosophical discussion appears in Berry and Myhill [OSL]. For some recent work see Quine [ISC] (cf. also his [CQT]); Beth [CTL], [TPT]; Rasiowa and Sikorski [PSL]; Vaught [ALS]. See also the next paragraph.

The Gödel completeness theorem appeared in his [VAL]. There have been many other proofs and an extension by Malcev. For references to these see Fraenkel and Bar-Hillel [FST, pp. 105 and 289], Beth [FMt, sec. 186], Mostowski [PSI], Robinson [OIM.II], Rasiowa and Sikorski [GTh]. The proof given later (in Chap. 7) is similar to the last-named one of Rasiowa and Sikorski. The theorem as stated here includes the Löwenheim-Skolem theorem.

For the higher types of epitheorems in general see Sec. 3.

The discussion of Sec. B is a revision of that in [CLg], secs. 2B and 2C. The replacement theorem appeared, essentially as here formulated, in MacLane [ABL]; a similar theorem, under somewhat more special assumptions, is in Herbrand [RTD, p. 21]. The replacement theorem for an equivalence relation appeared already in Post [IGT].

The present treatment of definitions is a revision of that which appeared in [CLg], sec. 2E; a previous revision appeared in [DFS]. This was based not so much on previous work on the philosophy of definitions as on the observation that Kleene's theory of recursive definitions, as presented in his [IMM] (and originally in his [GRF]), lends itself naturally to an extension to definitions over an arbitrary inductive class. The following is a sample of the extensive literature on definitions in general: Church [Dfn], [IML$_2$]; Dopp [VSD]; Dubislav [Dfn]; Ajdukiewicz [TCD]; Suppes [ILg, chap. 8]. Definitions should not be confused with definability theorems; the latter have significance only in systems with a formal equality relation.

The account of variables in Sec. D is based on [CLg], sec. 2D. Additional information about $\lambda$ conversion in Sec. D4 is from [CLg], chap. 3. The authoritative treatment of $\lambda$ conversion is Church [CLC]. The sketch of combinatory logic in Sec. D5 is, of course, an epitome of [CLg], especially chaps. 5 and 6. For a shorter treatment of the subject see Cogan [FTS]. An extremely condensed summary is also given in [DTC], sec. 2. There is an independent analysis of variables due to Menger. His papers to 1956 are cited in [CLg]; his more recent ones are listed in the present bibliography.

**2. Terminological note.** In Sec. 1 we noticed that it would be appropriate to call the subject of this chapter metatheory and that this was my own usage before 1947. However, there was some criticism of this usage, e.g., in Kleene [rev. C]. The upshot of this criticism was that the prefix 'meta-' connoted that abstraction was made by metasemiosis, and that it was confusing to use this term in contexts where there was no explicit commitment to metasemiosis. Accordingly, the new term 'epitheory' was introduced in [TFD]. In this book the prefix 'meta-' is reserved for situations where some sort of metasemiosis is explicitly postulated. The prefix 'epi-' is noncommittal; it neither excludes the metasemiotic point of view nor implies any commitment to it. It may well be argued that this is the original Hilbert intention regarding 'meta-'; we have seen (Sec. 2S3) that metamathematics is just as compatible with formalization by abstraction as with metasemiosis. Be that as it may, influential persons since that time—notably Tarski and Carnap—have given it a distinctly semiotic flavor. I respect this influence by using a new term.

Since we have a new term, some comment on other aspects of it is appropriate.

'Epitheory' does connote that we have considerations going beyond the elementary stage. We have seen that there are elementary statements in metamathematics (although they are scarcely interesting), but an elementary epitheorem, at least in the technical sense of 'elementary', would be a contradiction in terms. If we want a term to include elementary theorems and epitheorems under one head, we can say simply "theorem." Again, the prefix 'epi-' comes from a Greek preposition '$\epsilon\pi\iota$-', meaning "upon," or "on top of," whereas 'meta-', from '$\mu\epsilon\tau\alpha$', means "beyond," or "after." One therefore thinks of a metatheorem as something which lies on the far side of a boundary; an epitheorem as something which rests upon something else as a support. This difference in etymology comes into play when one uses figurative speech; the figures which naturally come to mind are different in the two cases.

Finally, it conforms to the usage of Secs. 2S3 and 2S4 to use 'metatheory' as a common noun; one can form a metatheory over a language L. There has, as yet, been little occasion to use 'epitheory' in an analogous sense. Confusion between this individual one and the collective one could occur for either term, but it has not done so (cf. the situation in regard to 'logic' in Sec. 1A). If the need should arise, we could use 'metatheoretics' and 'epitheoretics' in the collective senses.

**3. Higher epitheory.** The discussion has now reached the proper point for completing the program of Sec. 1S4. As promised there, I shall give here a sketch of the epitheoretical methods which go beyond the scope of this book. These methods may be, and often are, nonconstructive; but there are constructive topics, like the Gödel theorem, which involve considerations going beyond our domain in other respects. Since this is a continuation of Sec. 1S4, the general remarks made there will apply here.

The first step in this direction may be said to be the theory of recursive numerical functions. This may be regarded as the epitheory, along the lines

of Sec. 3C, of the system of natural numbers, which in turn can be taken as
Example 2 of Sec. 2C3, regarded as an affixative system. The theory has
been traced back to Dedekind [WSW] and even further, but in one sense
may be said to have begun with Skolem [BEA] (cf. Sec. 1S4$d$). It received
a great impetus from the work of Hilbert, and several sections of Hilbert
and Bernays [GLM] are devoted to it. Gödel in his [FUS] used it essentially. So far only what have since been called "primitive recursions" were
taken into account. In his [UPF], Gödel proposed the definition which is
the basis of the modern theory of general recursive functions (he credits the
basic idea to a verbal suggestion of J. Herbrand). The latter theory was
developed principally by Kleene, starting with his [GRF]; his [IMM] is thus
the authoritative treatise on the subject. R. Péter, in Budapest, studied
the subject independently; her book [RFn] is easier to read than Kleene's,
but treats a lot of special problems which have not attracted widespread
interest. She is the principal authority on recursions intermediate between
primitive and general recursion. Goodstein in his [RNT] studied primitive
recursion from a somewhat different viewpoint. The notion of general
recursiveness has been shown to be equivalent, in principle, to several other
notions of effectiveness, e.g., computability by an idealized machine (Turing
[CNA]), definability in combinatory logic, Markov algorithms (Sec. 2E,
especially Detlovs' theorem in Exercises C2 and 2E12). This thesis that
general recursiveness and effective calculability are to be identified is known
as Church's thesis; see Kleene [IMM, sec. 62]. By the theorem of Detlovs
this is equivalent to Markov's thesis (cf. Sec. 2E1). The thesis has important
consequences. For one thing, the Gödel numbers of the assertions of an
assertional formal system can then be enumerated by a recursive function,
and consequently a formal system can be identified with a "recursively
enumerable set" (r.e. set). On the other hand, one would expect close
contact with intuitionistic mathematics, and this contact has been (and, I
think, still is being) exploited. There are also contacts with more practical
problems such as the design and operation of computing automata. These
investigations have usually been made quite constructively. But the higher
parts of the theory, leading to "recursive hierarchies" of various sorts, de-
part considerably from the constructive viewpoint. The whole subject has
become a major specialty in itself, and one which is presently very active.
For works of introductory character see, besides those already cited, the
following: Post [RES] (the pathmaking work on r.e. sets); Davis [CUn];
Rogers [TRF]; Myhill-Dekker [RET] (types of r.e. sets invariant with respect
to recursive transformations).

Another direction of inquiry has arisen from the attention to symbolic
structures. On this see Sec. 2S3. The development shows plainly the
influence of Hilbert's metamathematics, but there is also evidence of influence
from several other sources (Frege, C. S. Peirce, E. Husserl, etc.).

The incompleteness theorem of Gödel [FUS] may be regarded as a combination of the last two directions. Gödel showed that the formulas of a
system could be represented as numbers (see the "Gödel representation"
of Sec. 2C4) and that in terms of this translation a formula could be constructed which asserted its own nonprovability. Thus the liar paradox could
be set up in the system. On the basis of reasonable assumptions, one can

show that the formula so constructed is neither provable nor disprovable. The catastrophic effect of this theorem on mathematical logic has already been discussed in Sec. 1, and also in Secs. 1C and 3A1.

The Gödel incompleteness theorem, as well as many other such theorems, uses methods which are strictly constructive. But, as Tarski early pointed out, one gets results of great significance by using nonconstructive semantical methods. One develops a metatheory about a certain object language using a certain metalanguage; one is then free to reason more or less platonistically in the metalanguage. The Löwenheim-Skolem and Gödel completeness theorems (Secs. 1 and 3A1) are examples of theorems which can be so regarded. Tarski has been the leader in a series of investigations of this sort. This is at present an extremely important development, with a large and rapidly growing literature. I can list only a few suggestive papers. For summaries see Mostowski [PSI], Beth [FMt, especially parts IV and VII]. Tarski's fundamental paper is his [WBF]; this and some of his other early papers have been collected and translated into English in Tarski and Woodger [LSM]. Among his later papers, [NMB] and [GTP] indicate the general direction. See also Tarski *et al.* [UTh]. The algebraic applications which Abraham Robinson developed in his [MMA], [CTh] are closely related. The field is at present being actively developed by Tarski and his group on the Pacific coast of the United States, as well as by some of his former colleagues in Poland.

Another direction of growth in epitheory is the search for methods which may be called constructive in an extended sense, but not in the strict sense in which that term is intended here. The most famous of these is Gentzen's proof of the consistency of arithmetic (Gentzen [NFW], originally in his [WFR]; discussed, for example, in Bernays [QMA]), which admits as nonconstructive principle that a descending sequence of ordinal numbers beginning with any ordinal up to $\epsilon_0$ (the first Cantor $\epsilon$ number) can have only a finite number of terms. For other extensions of the notion of constructivity, see Bernays [BSB], Gödel [BNB], Kreisel [INF].

Further remarks on the significance of constructive and nonconstructive methods may be found in Goodstein [NMS], Kreisel [HPr], Lorenzen [LRF], Mostowski [OUM], Myhill [PIM], Shanin [LPA], Skolem [CRF]. See also Heyting [CMt], which contains a series of papers on various aspects of constructiveness presented at a conference in Amsterdam in 1957.

# Chapter 4
# RELATIONAL LOGICAL ALGEBRA

In this chapter we shall consider certain systems of algebraic character which have some importance for modern mathematical logic. The algebraic character of these systems consists in the fact that they contain no bound variables; furthermore, they are relational (Sec. 2D1), the basic relation being either a quasi ordering or an equality (Sec. 3B), and their operations have certain analogies with the operations of ordinary algebra. These systems are considered here, not because one is obligated to begin with them by the fact that they are inherently fundamental (in the sense that one has to introduce them before one can proceed with the development of the elementary theorems of a system), but because they are rather simple systems which have interpretations in other systems, so that they are important in the comparative epitheoretic study of systems in general. Attention will be confined to those properties, generally rather simple, which are of logical interest.

The treatment will begin in Sec. A with a preliminary discussion of logical algebras in general; this will include a statement of the notational conventions which apply to the chapter as a whole. Then in Sec. B we shall proceed to the study of lattices, in which there are at most two operations, called join and meet, which have properties analogous to the algebraic sum and product. The last two sections (Secs. C and D) will be devoted to systems in which operations of subtraction and implication, which are analogous in some ways to the inverse operations of ordinary algebra, are adjoined to the lattice operations. The discussion will terminate, at the end of Sec. D, with a discussion of Boolean rings and their duals.

The following topics, which also have an algebraic character, are not treated in this chapter. All matters relating to negation, including Boolean algebras, are postponed to Chap. 6. Although there are important algebraic theories related to quantification, they are beyond the scope of this book and are barely mentioned in Chap. 7. Finally, everything concerning modal operations is postponed to Chap. 8.

## A. LOGICAL ALGEBRAS IN GENERAL

This section will include, in Sec. 1, a general characterization of logical algebras as quasi-ordered or equational algebras with certain idempotent operations, together with a statement of notational conventions for the chapter as a whole. Then, in Sec. 2, there will be described a number of different interpretations of such algebras; these will be referred to from time to time in what follows.

**1. Preliminary conventions.** The term *'algebra'* is used in this book as a name for a system with free variables but no bound variables. Thus the system of Example 5 (Sec. 2C3) is an algebra and is aptly called propositional algebra. In contradistinction the term *'calculus'* will, as a rule, be used to describe a system with bound variables, so that it is suitable to speak of a calculus of $\lambda$ conversion, a predicate calculus, etc. These terms agree with ordinary mathematical usage, where the distinguishing characteristic of the infinitesimal calculus, as opposed to elementary algebra, is the presence of bound variables in the former.[1]

The algebras considered in this chapter are relational (Sec. 2D1). There is one basic predicate, a binary relation which is either a quasi ordering or an equality. This relation will be designated by the infix '$\leq$' in the quasi-ordered case and by the infix '$=$' in the case of equality. Even in the quasi-ordered case an equality relation is present, defined thus:

$$x = y \rightleftarrows x \leq y \ \& \ y \leq x \tag{1}$$

where '$x$' and '$y$' are U variables for arbitrary obs. Such a quasi-ordered system, when interpreted so as to have a contensive equality, is called *partially ordered* when, in the interpretation, the relation of equality defined by (1) coincides with that contensive equality; but the distinction between partial order and quasi order is not relevant to an uninterpreted system. However, it will be expedient to describe a quasi-ordered system as partially ordered when only partially ordered interpretations are intended.

The principal operations considered in this chapter—and the only ones admitted until we come to Sec. C—are two binary ones called *meet* and *join* and indicated, respectively, by the infixes '$\wedge$' and '$\vee$'. The obs are then constructions from the atoms by these two operations. The operations will turn out to be commutative and associative, and they may or may not have certain properties analogous to the distributive law of ordinary algebra. What is peculiar about these algebras, however, is that the operations will be idempotent; i.e., we shall have

$$a \wedge a = a \qquad a \vee a = a \tag{2}$$

for all obs $a$. These laws hold for most algebras which have logical interest, and therefore they may be regarded as characterizing logical algebras in the sense of Sec. 1A.

As already exemplified in the discussion of (1) and (2), it will be expedient to abandon the convention whereby capital italic letters are used as U variables and lower-case italics are reserved for U constants. This convention served its purpose throughout Chaps. 2 and 3. In this chapter we use '0' and '1' as U constants, and we also, in agreement with Sec. 3D1, reserve the letters '$e_1$', '$e_2$', ... for that purpose (although we scarcely need them); all other lower-case italic letters are U variables for unspecified obs. This is the usage of ordinary algebra, and it is expedient to follow it. We shall revert to a notation resembling the earlier one in Chap. 5.

---

[1] Systems like combinatory logic (Sec. 3D5) which contain no variables do not come under either term.

In what follows, various species of logical algebras will be defined. In these definitions it is understood that a system which satisfies the postulates of a species belongs to that species regardless of whether or not it satisfies additional postulates, i.e., whether it also belongs to a more restricted species. On the other hand, when we speak of a *general system* of such and such a species, we mean that nothing else is postulated beyond the postulates defining the species. The general system of a species is then a particular formal system whose elementary theorems are those statements which are derivable from the postulates of the species. This is closely related to what is often called a *free* system, with the elements of e as generators.

Since the basic relation is of the sort considered in Sec. 3B, we may use the conventions explained and the results established there. Thus ($\rho$) and ($\tau$), which are, respectively, the reflexive and transitive properties of the basic relation, apply by definition to both $\leq$ and $=$; and ($\sigma$), the symmetric property, applies to $=$. The monotonic property ($\pi$) will be established as an epitheorem for lattices, and the replacement theorem, Rp, will be shown to hold in all the algebras here considered. In connection with proofs, a notation such as

$$a \leq b \qquad \text{by } X$$
$$\leq c \qquad \text{by } Y$$

where '$X$' and '$Y$' are replaced by citations of sources, is to be understood as meaning

$$a \leq b \qquad \text{by } X$$
$$b \leq c \qquad \text{by } Y$$

Therefore $\qquad\quad a \leq c \qquad$ by ($\tau$)

## 2. Interpretations of logical algebras.

Before we enter on the formal developments, let us look at some examples of interpretations of these systems.[1] (For further examples and discussion, see Sec. C5.)

1°. *Class Interpretation.* This is the oldest of the interpretations. According to this interpretation, the (interpreted) obs are classes and the basic relation $\leq$ is inclusion. The meet of two classes is their intersection, i.e., the class of those elements which belong to both. The join, or union, is the class of elements which belong to one or the other or both.

2°. *Relation Interpretation.* The obs are relations. Let $R$, $S$ be such relations. Then we interpret $R \leq S$ as saying that whenever $x$ stands in the relation $R$ to $y$, $x$ stands also in the relation $S$ to $y$; $R \wedge S$ is the relation which holds between $x$ and $y$ just when both of the relations $R$, $S$ so hold; and $R \vee S$ is the relation which holds between $x$ and $y$ just when either one or both of $R$, $S$ so hold. Since relations can be construed as classes of ordered pairs, this is a special case of the preceding.

3°. *Propositional Interpretation.* According to this the obs are propositions. Views as to the nature of propositions vary, and we shall have to postpone discussion of this question until later. Suffice it to say that propositions are objects which we talk about but are interested in interpreting

---

[1] The examples 1° to 7° are presented by describing a mode of valuation, which can be made more precise in various ways, together with an interpretant for the basic relation; it is then understood that the interpretant of an elementary statement affirms that the indicated relation holds for all valuations admitted by the mode.

as statements; we might, for example, regard them as sentences in an O language, which we mention but do not use in the A language. Then $\leq$ is the relation of implication which holds between two propositions just when the conditional connection (i.e., the 'if ——$_1$, then ——$_2$' connection, symbolized by '→' in Sec. 3A2, but not necessarily with a constructive connotation) between the corresponding interpreted statements is true; $\wedge$ is the conjunction operation which forms from the propositions a third one whose corresponding statement is true just when those corresponding to the two operands are both true; and $\vee$ is similarly related to the 'or' connection between statements.

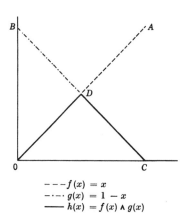

$$----f(x) = x$$
$$-\cdot-\cdot\ g(x) = 1 - x$$
$$———\ h(x) = f(x) \wedge g(x)$$

FIGURE 1

4°. *Order Interpretation.* The obs are elements of some ordered or partially ordered set, say, real numbers; $\leq$ is the relation of either preceding in the order (which we interpret as the relation of less to greater) or being the same; $a \wedge b$ is the greatest lower bound, and $a \vee b$ the least upper bound, of $a$, $b$. We may also use this interpretation when the obs are functions from some range to the ordered set.[1] Thus if $f$ and $g$ are two real functions defined for real numbers $x$, $0 \leq x \leq 1$, $f \wedge g$ would be the function whose value for any $x$ is the smaller of $f(x)$ and $g(x)$; thus if $f(x)$ is $x$ and $g(x)$ is $1 - x$, $f \wedge g$ is the function $h(x)$, namely,

$$h(x) = \tfrac{1}{2} - |x - \tfrac{1}{2}|$$

In Fig. 1 the graph of $f(x)$ is the straight line $OA$, that of $g(x)$ is the line $BC$, that of $h(x)$ is the broken line $ODC$.

5°. *Divisibility Interpretation.* The obs are natural numbers; $\leq$ is the relation of divisibility; $a \wedge b$ is the greatest common divisor of $a$ and $b$, and $a \vee b$ is their least common multiple. (This is a special case of interpretation 4°.)

6°. *Closure Interpretations.* Let $K$ be a class of sets, and let $\leq$ be the relation of inclusion between the elements of $K$. Suppose there is a "closure operation" which assigns to every element $x$ of $K$ an element $x^*$ called its closure, such that for all $x$ in $K$,

(c1)                                        $x \leq x^*$
(c2)                                    $(x^*)^* \leq x^*$
(c3)                              $x \leq y \to x^* \leq y^*$

We call an element $a$ of $K$ *closed* just when $a^* \leq a$. Then $a^*$ is the least closed set which contains $a$. We now can form an interpretation in which the obs are the closed sets of $K$, $\leq$ is set inclusion as before, $\wedge$ is set intersection, and

---

[1] This sort of generalization can be applied also to the other interpretations (cf. Sec. C5).

∨ is the operation which assigns to a pair of closed sets $a$, $b$ the least closed set which contains them both. The following are examples of this sort of interpretation (it suffices to specify the closed sets in each case):

*a.* Closed sets in the sense of ordinary point-set topology.

*b.* Linear spaces (points, lines, planes, etc.) in any number of dimensions. Such a linear space can be characterized as one containing all points which are on the same straight line with any two of its points. This example leads to projective geometry and some generalizations of it.

*c.* Convex sets, i.e., sets which contain, with any two points, all points on the line segment connecting them.

*d.* The subgroups of a mathematical group and, more generally, the sub-algebras of an algebra.

*e.* Deductive theories (cf. Sec. 2B3).

7°. *Open-set Interpretations.* These are obtained from those considered in interpretation 6° by interchanging ∧ and ∨, ≤ and its converse. The name is derived from the fact that in ordinary point-set topology, corresponding to interpretation 6°*a*, this gives rise to open sets.

8°. *Artificial Interpretations.* This includes interpretations set up by means of tables, diagrams, and the like. These are often deliberately designed to be invalid for some axiom scheme, and thus to furnish a proof that that axiom scheme is not deducible from certain others.

Some samples of diagrams furnishing interpretations are given in Fig. 2. Here the obs are the points indicated by small circles; a·line segment with an arrow from a first point to a second indicates that the two points in that order stand in a relation $R_0$, and ≤ is the quasi ordering generated by this $R_0$. Here in Fig. 2g and $h$ the equality relation defined by (1) holds between obs which are not identical, so that these cases are not partially ordered. Of course, the interpretations of ∧, ∨ have to be made by supplementary conventions.

Another way of making artificial interpretations is by tables. This is most natural for assertional systems, but it can be used also for relational ones. Thus one may be given a certain set $V$ of values and regard the operations as assigning to each pair of values from $V$ for the arguments a value from $V$ as assigned value, these assignments being exhibited by a table like the following pair, where $V$ is 0, 1, 2, and the left-hand table gives the values of $x \wedge y$, the right hand those for $x \vee y$:

| $x$ \ $y$ | 0 | 1 | 2 |
|---|---|---|---|
| 0 | 0 | 1 | 2 |
| 1 | 1 | 1 | 2 |
| 2 | 2 | 2 | 2 |

| $x$ \ $y$ | 0 | 1 | 2 |
|---|---|---|---|
| 0 | 0 | 0 | 0 |
| 1 | 0 | 1 | 1 |
| 2 | 0 | 1 | 2 |

One can have a similar table associated with ≤ and agree that $x \leq y$ holds just when the value obtained from that table belongs to a subclass of "designated" values of $V$. One can generalize this, as in the case of the tautologies considered in Sec. 2C5, to the case where such a designated value is taken not for just one assignment of values to the $e_i$, but for all such assignments.

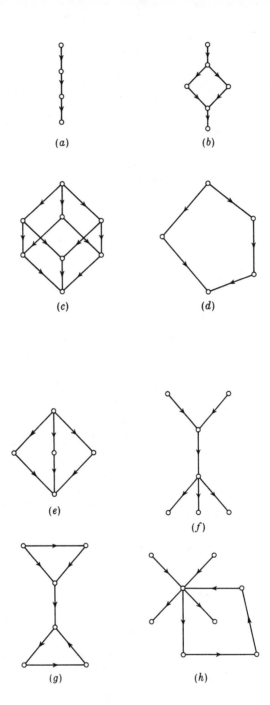

FIGURE 2
130

These artificial interpretations do not always lead to logical algebras as
defined in Sec. 1. But when they have some analogy with such algebras,
they may be considered as logical algebras in an extended sense (cf. Sec. 1A).

## B. LATTICES

In this section we shall consider logical algebras which have no operations
other than the meet and join described in Sec. A. A species of algebra,
called a lattice, having both these operations, will be defined and studied in
Sec. 2. As a preliminary to this we shall study in Sec. 1 a species, called a
semilattice, in which only one operation is present. In Sec. 3 the notion of
lattice will be specialized by adjoining a postulate of distributivity. Distributive lattices are the most important kind of lattices for logic, and hence
it is worthwhile to devote some space to them; in connection with them we
shall note, but not develop, an intermediate kind of lattice, called a modular
lattice, which has important applications in other branches of mathematics.
Finally, in Sec. 4, some relatively trivial matters connected with special
elements 0 and 1 will be listed for the sake of completeness.

In accordance with the general policy of this chapter, only very simple
and elementary properties of lattices are considered here. When combined,
for example, with set-theoretic methods, the theory of lattices becomes a
rather extensive subject. These extensions go beyond the scope of this
book.

**1. Semilattices.** We begin this discussion by consideration of systems
in which only one of the operations is postulated. We take this to be the
operation $\wedge$. The important species of this type is the semilattice. This is
defined in the next paragraph; the definition will be followed by some theorems about it.

A *semilattice* is a partially ordered algebra with a single binary operation,
which we indicate by '$\wedge$' used as infix, such that the following postulates
hold:

$$\Lambda\text{K} \qquad\qquad a \wedge b \leq a$$

$$\Lambda\text{K}' \qquad\qquad a \wedge b \leq b$$

$$\Lambda\text{S} \qquad\qquad c \leq a \ \& \ c \leq b \rightarrow c \leq a \wedge b$$

Here $a$, $b$, $c$ are arbitrary obs. The expressions '$\Lambda$K', etc., written at the
left, are used as names for the corresponding postulates.[1] Note that $\Lambda$K
and $\Lambda$K' are axiom schemes, whereas $\Lambda$S is a rule. Since the system is
quasi-ordered, the rule $(\tau)$ and the axiom scheme $(\rho)$ are also postulated.
The postulates may be summed up by saying that $a \wedge b$ is the greatest lower
bound of $a$, $b$ with respect to $\leq$ as quasi ordering; it is a lower bound by $\Lambda$K
and $\Lambda$K', and it is the greatest lower bound by $\Lambda$S.

---

[1] These names are suggested by analogies with the combinators denoted by corresponding letters in Sec. 3D5. However, these analogies are not relevant here. The
names can be taken as arbitrary; as such, they are no worse than others which might be
proposed.

**Theorem 1.**    *In any semilattice, the following hold for any obs $a$, $b$:*

$\Lambda$W $\qquad\qquad\qquad a \leq a \wedge a$

$\Lambda$C $\qquad\qquad\qquad a \wedge b \leq b \wedge a$

$\Lambda$B $\qquad\qquad\qquad a \leq b \rightarrow c \wedge a \leq c \wedge b$

$\Lambda$B$' \qquad\qquad\qquad a \leq b \rightarrow a \wedge c \leq b \wedge c$

*Conversely, if $\mathfrak{S}$ is a relational system with relation $\leq$ and a binary operation $\wedge$ such that $(\tau)$, $\Lambda$K, $\Lambda$C, $\Lambda$W, and either $\Lambda$B or $\Lambda$B$'$ are satisfied, then $\mathfrak{S}$ is a semilattice.*

*Proof.*   Suppose, first, that $\mathfrak{S}$ is a semilattice with basic relation $\leq$ and operation $\wedge$. Then, by $(\rho)$,

$$a \leq a$$

Hence, by $\Lambda$S, taking $b$ and $c$ both identical to $a$, we have

$$a \leq a \wedge a$$

which is $\Lambda$W.   Further, we have

$\qquad\qquad\qquad a \wedge b \leq b \qquad\qquad$ by $\Lambda$K$'$

$\qquad\qquad\qquad a \wedge b \leq a \qquad\qquad$ by $\Lambda$K

$\qquad\qquad\qquad a \wedge b \leq b \wedge a \qquad\quad$ by $\Lambda$S

so that we have $\Lambda$C.   To prove $\Lambda$B, assume the premise, viz.,

$$a \leq b$$

The desired conclusion follows by $\Lambda$S from

$$c \wedge a \leq c \qquad\quad c \wedge a \leq b$$

Of these, the first follows by $\Lambda$K; the second, since

$\qquad\qquad\qquad c \wedge a \leq a \qquad\qquad$ by $\Lambda$K$'$

$\qquad\qquad\qquad\qquad\;\; \leq b \qquad\qquad$ by hypothesis

Thus $\Lambda$B holds.   There is a similar proof of $\Lambda$B$'$, or we may note that, in the presence of $(\tau)$ and $\Lambda$C, either one of $\Lambda$B, $\Lambda$B$'$ entails the other.

Conversely, suppose $(\tau)$, $\Lambda$K, $\Lambda$C, $\Lambda$W, and either $\Lambda$B or $\Lambda$B$'$ hold for $\mathfrak{S}$.   Then

$\qquad\qquad\qquad a \leq a \wedge a \qquad\qquad$ by $\Lambda$W

$\qquad\qquad\qquad\quad\; \leq a \qquad\qquad\;\;$ by $\Lambda$K

so that $(\rho)$ holds.   The postulate $\Lambda$K holds by hypothesis.   Further, we have

$\qquad\qquad\qquad a \wedge b \leq b \wedge a \qquad\qquad$ by $\Lambda C$

$\qquad\qquad\qquad\qquad\;\; \leq b \qquad\qquad\;\;$ by $\Lambda K$

so that $\Lambda$K$'$ holds.   It remains to prove $\Lambda$S.   To do this we note first that, according to a remark made in the previous paragraph, we have both $\Lambda$B and $\Lambda$B$'$.   Then, assuming the premises of $\Lambda$S, viz.,

$$c \leq a \qquad\quad c \leq b$$

we have

$\qquad\qquad\quad c \leq c \wedge c \qquad\quad$ by $\Lambda$W

$\qquad\qquad\quad\;\; \leq a \wedge c \qquad\quad$ by first premise and $\Lambda$B$'$

$\qquad\qquad\quad\;\; \leq a \wedge b \qquad\quad$ by second premise and $\Lambda$B

This completes the proof.

Note that $\Lambda$B and $\Lambda$B' together give the property $(\pi)$ of Sec. 3B3. Thus we can use the replacement theorem Rp in connection with a semilattice.

**Theorem 2.**  *A necessary and sufficient condition that*

$$a \leq b \tag{1}$$

*hold in a general semilattice is that every atom which occurs as component in $b$ occur also at least once as a component of $a$.*

*Proof.* The necessity of the condition follows by deductive induction. For every axiom (viz., every instance of $(\rho)$, $\Lambda$K, $\Lambda$K') has the property, and each of the rules $(\tau)$ and $\Lambda$S leads from premises having the property to a conclusion having the property.

It remains, therefore, only to prove the sufficiency. We suppose that the condition is fulfilled and show that (1) holds.

Suppose first that $b$ is an atom. Then it is a component of $a$, and hence there is a composition (Sec. 3B1) from $b$ to $a$. Let the steps of this composition be $a_0, a_1, \ldots, a_n$, where $a_0 \equiv b$, $a_n \equiv a$. By $(\rho)$, $a_0 \leq b$, and if $a_k \leq b$, then $a_{k+1} \leq b$ by $\Lambda$K or $\Lambda$K'. Hence, by induction on $k$, it holds for all $k$ and hence for $k = n$.

This argument shows that

$$a \leq c$$

holds for every atomic component $c$ of $b$. By structural induction and $\Lambda$S, it holds for every component and hence for $b$ itself. This completes the proof.

COROLLARY 2.1.  *The operation of a semilattice is associative; i.e.,*

$$a \wedge (b \wedge c) = (a \wedge b) \wedge c$$

*holds for all obs $a$, $b$, $c$.*

For some purposes it is advantageous to conceive a semilattice as an equational system. The following theorem is concerned with this.

**Theorem 3.**  *If equality is defined in a semilattice by*

$$a = b \rightleftarrows a \leq b \,\&\, b \leq a, \tag{2}$$

*then equality is a monotone equivalence such that for all obs $a$, $b$, $c$,*

(i) $\qquad\qquad\qquad a \wedge a = a$

(ii) $\qquad\qquad\qquad a \wedge b = b \wedge a$

(iii) $\qquad\qquad a \wedge (b \wedge c) = (a \wedge b) \wedge c$

*and*

(iv) $\qquad\qquad\qquad a \leq b \rightleftarrows a = a \wedge b$

*Conversely, if $\mathfrak{S}$ is an algebra with an operation $\wedge$ and basic relation $=$ satisfying $(\sigma)$, $(\tau)$, $(\pi)$ and (i), (ii), (iii), then $\mathfrak{S}$ is a semilattice with the relation defined by (iv) as basic relation; moreover, (2) holds.*

*Proof.* Let $\mathfrak{S}$ be a semilattice and let $=$ be defined by (2). Then (i), (ii), (iii) hold by Theorem 2. As for (iv), we have

$$\begin{aligned}
a \leq b &\rightarrow a \leq a \,\&\, a \leq b &&\text{by } (\rho)\\
&\rightarrow a \leq a \wedge b &&\text{by } \Lambda\text{S}\\
&\rightarrow a = a \wedge b &&\text{by } \Lambda\text{K}\\
&\rightarrow a \leq a \wedge b &&\text{by } (2)\\
&\rightarrow a \leq b &&\text{by } \Lambda\text{K}'
\end{aligned}$$

The monotonic character of $=$ follows from that of $\leq$. Thus $\mathfrak{S}$ has all the required properties.

Conversely, let $\mathfrak{S}$ be an equational system as stated in the theorem. Then the new relation defined by (iv) is reflexive by (i); it is transitive, since

$$a = a \wedge b \ \& \ b = b \wedge c$$
$$\rightarrow a = a \wedge (b \wedge c) = (a \wedge b) \wedge c = a \wedge c$$

It has $\Lambda$K, $\Lambda$K', since

$$a \wedge b = (a \wedge a) \wedge b = a \wedge (a \wedge b) = (a \wedge b) \wedge a$$
$$a \wedge b = a \wedge (b \wedge b) = (a \wedge b) \wedge b$$

It satisfies $\Lambda$S, since

$$c = c \wedge a \ \& \ c = c \wedge b$$
$$\rightarrow c = c \wedge b = (c \wedge a) \wedge b = c \wedge (a \wedge b)$$

Finally, we have (2) since

$$a = a \wedge b \ \& \ b = b \wedge a \rightarrow a = a \wedge b = b \wedge a = b$$

and

$$a = b \rightarrow a = a \wedge a = a \wedge b \ \& \ b = b \wedge b = b \wedge a$$

This completes the proof.

**2. Lattices in general.** A *lattice* is a partially ordered algebra with two operations $\wedge$ and $\vee$, such that $\Lambda$K, $\Lambda$K', and $\Lambda$S hold, and also the postulates

| VK | $a \leq a \vee b$ |
| VK' | $b \leq a \vee b$ |
| VS | $a \leq c \ \& \ b \leq c \rightarrow a \vee b \leq c$ |

In other words, a lattice is an algebra which is a semilattice both with respect to $\leq$ and $\wedge$ and with respect to $\geq$ (converse of $\leq$) and $\vee$. Thus for every pair $a$, $b$ there is a greatest lower bound $a \wedge b$ and a least upper bound $a \vee b$.

All the examples 1° to 7° in Sec. A2 are lattices; those in Fig. 2a to e are also lattices, but those of Fig. 2g and h are not lattices because they are not partially ordered.

The following theorem, called the *principle of duality*, follows by deductive induction and examination of the postulates.

**Theorem 4.** *Given any theorem concerning a lattice, one obtains another by interchanging $\leq$ and $\geq$, $\wedge$ and $\vee$.*

The duals of the theorems of Sec. 1 thus hold in a lattice. The duals of $\Lambda$B, $\Lambda$B', $\Lambda$C, $\Lambda$W will be called VB, VB', VC, VW, respectively.

Special interest attaches to the equational form of a lattice. The equational postulates of the following theorem are those given by Birkhoff [LTh].

**Theorem 5.** *In a lattice with equality defined by (2), the relations*

| L1 | $a \wedge a = a$ | $a \vee a = a$ |
| L2 | $a \wedge b = b \wedge a$ | $a \vee b = b \vee a$ |
| L3 | $a \wedge (b \wedge c) = (a \wedge b) \wedge c$ | $a \vee (b \vee c) = (a \vee b) \vee c$ |
| L4 | $a \wedge (a \vee b) = a$ | $a \vee (a \wedge b) = a$ |
| | $a = a \wedge b \quad \rightleftarrows \quad b = a \vee b$ | (3) |

*hold for all obs; moreover, either side of* (3) *is equivalent to* $a \leq b$. *Conversely, if* L2 *and* L3, *and either* L4 *or both* L1 *and* (3) *hold in an equational algebra with operations* $\wedge$, $\vee$ *and monotone equality, and if either side of* (3) *is taken as definition of* $a \leq b$, *then the algebra is a lattice with respect to* $\leq$, $\wedge$, *and* $\vee$, *and* (2) *holds as an equivalence.*

*Proof.* L1 to L3 are all true in a lattice by Theorem 3 and its dual, and by Theorem 3, $a \leq b$ is equivalent to the left side of (3). The following argument shows that the left side of L4 also holds:

$$a \wedge (a \vee b) \leq a \qquad \text{by } \Lambda\text{K} \qquad (4)$$
$$a \leq a \qquad \text{by } (\rho)$$
$$a \leq a \vee b \qquad \text{by } \text{VK}$$
$$a \leq a \wedge (a \vee b) \qquad \text{by } \Lambda\text{S} \qquad (5)$$

From (4) and (5) we have the left side of L4 by (2). The rest of L4 and (3) follows by duality.

Next, we show that in an equational system, with operations $\wedge$, $\vee$ and monotone equality, L1 and (3) are consequences of L2, L3, and L4. By duality (in combination with L2), it suffices to show the left half of L1 and the implication from right to left in (3). This can be done as follows:

$$a \wedge a = a \wedge (a \vee (a \wedge b)) \qquad \text{by L4, Rp}$$
$$= a \qquad \text{by L4}$$

which takes care of L1. Also assuming $a \vee b = b$,

$$a = a \wedge (a \vee b) \qquad \text{by L4}$$
$$= a \wedge b \qquad \text{by hypothesis}$$

which establishes (3).

Finally, we show that an equational system satisfying L1 to L3 and (3) is a lattice for which (2) is a valid equivalence. In fact, by Theorem 3, $\leq$ being defined by the left half of (3), we have a semilattice with respect to $\leq$, $\wedge$; likewise, if $\geq$ is defined by

$$a \geq b \rightleftarrows a = a \vee b$$

we have a semilattice with respect to $\geq$, $\vee$. Since these relations $\leq$, $\geq$ are converses of one another by virtue of (3), we have a lattice. This completes the proof.

The laws L1 to L4 are known, respectively, as the *idempotent laws*, the *commutative laws*, the *associative laws*, and the *laws of absorption*.

The following theorem concerns relations of distributivity which are valid in every lattice.

**Theorem 6.** *For all obs* $a$, $b$, $c$, *in any lattice,*

$$(a \wedge b) \vee (a \wedge c) \leq a \wedge (b \vee c) \qquad (6)$$
$$a \vee (b \wedge c) \leq (a \vee b) \wedge (a \vee c) \qquad (7)$$

*Proof.* Since (6) and (7) are dual to one another, it will suffice to prove (6). This is done as follows:

$$a \wedge b \leq a \wedge (b \vee c) \qquad \text{by VK, Rp}$$
$$a \wedge c \leq a \wedge (b \vee c) \qquad \text{by VK', Rp}$$

From these, (6) follows by VS. This completes the proof.

**3. Distributive lattices.** The converses of the relations (6) and (7) do not hold in a general lattice. For example, in the interpretation $6°b$ of Sec. A2, if $b$ and $c$ are points and $a$ is a point on $b \lor c$ (i.e., the line $bc$) distinct from $b$ and $c$, then $a \land (b \lor c)$ is the point $a$, whereas $(a \land b) \lor (a \land c)$ is null. These converses also fail in the interpretation $6°c$ of Sec. A1 and in Fig. $2d$ and $e$.

DEFINITION.    A *distributive lattice* is a lattice in which the converses of (6) and (7), and hence

$$a \land (b \lor c) = (a \land b) \lor (a \land c) \tag{8}$$
$$(a \lor b) \land (a \lor c) = a \lor (b \land c) \tag{9}$$

hold.

**Theorem 7.**    *A necessary and sufficient condition that a lattice L be distributive is that*

$$a \land (b \lor c) \leq (a \land b) \lor c \tag{10}$$

*hold for all obs a, b, c of L.*

*Proof.*    Suppose first that (8) holds in a lattice $L$.    Then we have

$$\begin{aligned} a \land (b \lor c) &\leq (a \land b) \lor (a \land c) &&\text{by (8)} \\ &\leq (a \land b) \lor c &&\text{by } \Lambda K', \text{ Rp} \end{aligned}$$

Conversely, if (10) holds, we have

$$\begin{aligned} a \land (b \lor c) &\leq a &&\text{by } \Lambda K \\ a \land (b \lor c) &\leq a \land ((a \land b) \lor c) &&\text{by (10) and } \Lambda S \\ &\leq a \land (c \lor (a \land b)) &&\text{by VC, Rp} \\ &\leq (a \land c) \lor (a \land b) &&\text{by (10)} \\ &\leq (a \land b) \lor (a \land c) &&\text{by VC} \end{aligned}$$

Thus (8) and (10) are deducible from each other.

Next we note that (8) and (9) are dual to one another.    On the other hand, if we dualize (10), we have

$$(a \lor b) \land c \leq a \lor (b \land c)$$

By $\Lambda$C, VC (and of course Rp), this is equivalent to

$$c \land (b \lor a) \leq (c \land b) \lor a$$

This is obtained from (10) by interchanging $a$ and $c$.    Since '$a$' and '$c$' are U variables for arbitrary obs, this shows that (10) is self-dual.    Then the dual of the argument of the preceding paragraph shows that (10) is also equivalent to (9).

Thus if (10) holds, so do (8) and (9), and if either (8) or (9) holds, and a fortiori if they both hold, so does (10), Q.E.D.

COROLLARY 7.1.    *Either* (8) *or* (9) *implies the other.*

In a distributive lattice we can calculate with the operations $\land$, $\lor$ just as we do with the sum and product in ordinary algebra; the only differences are $(a)$ that we can take either of the operations as the sum, the other as the product, and $(b)$ that, on account of the laws of idempotency, we have no need of exponents or numerical coefficients.    Hence every ob can be proved equal to one which is a meet of joins of the atoms, also to one which is a join of meets, and either of these "normal forms" is essentially unique.

From this remark we can deduce a decision procedure for a general distributive lattice. Let

$$a \leq b \tag{11}$$

be a given elementary statement. Let

$$a = a_1 \vee a_2 \cdots \vee a_m$$
$$b = b_1 \wedge b_2 \cdots \wedge b_n$$

where each $a_i$ is a meet of atoms, each $b_j$ is a join of atoms, and $a$ and $b$ are obtained by multiplying out as described in the preceding paragraph. Then a necessary and sufficient condition that (11) hold is that

$$a_i \leq b_j \tag{12}$$

hold for all $i = 1, 2, \ldots, m$, and $j = 1, 2, \ldots, n$. Now each of the statements (12) is of the form

$$u_1 \wedge u_2 \cdots \wedge u_p \leq v_1 \vee v_2 \cdots \vee v_q \tag{13}$$

where each of $u_1, \ldots, u_p, v_1, \ldots, v_q$ is an atom. A sufficient condition for (13) is that some atom occur on both sides.[1] That this condition is also necessary may be shown by considering valuations over the ring consisting of 0 and 1. By this we mean, in agreement with Sec 2C5, valuations formed by assigning values 0 and 1 to the formal variables $(e_1, e_2, \ldots)$ in an arbitrary manner and interpreting $\wedge$ as the product and $\vee$ as the maximum.[2] Then $\leq$ can be interpreted as the predicate such that (11) is true for a given valuation whenever $a$ is 0 or $b$ is 1 (or both) and is false when $a$ is 1 and $b$ is 0. Then it follows by deductive induction that the interpretant of every demonstrable statement (11) is true for every valuation, whereas if no variable occurs on both sides of (13), the latter can be made false if we give all variables on the left the value 1 and all those on the right the value 0. This argument proves the following:

**Theorem 8.**    *The general distributive lattice is decidable.*

Although distributive lattices are a very specialized kind of lattice, and modern algebra, and mathematics generally, is full of examples of lattices which are not distributive, most of the lattices which come up in logic are distributive for reasons which will appear later. It is therefore expedient to consider a few other properties. The following theorem and corollary represent one of the most interesting of the simpler properties.

**Theorem 9.**    *A necessary and sufficient condition that a lattice L be distributive is that for all obs a, b, c of L,*

$$a \wedge b \leq c \,\&\, a \leq b \vee c \rightarrow a \leq c \tag{14}$$

---

[1] This can be shown by $(\rho)$, $\Lambda K$, $\Lambda K'$, $VK$, $VK'$, $(\pi)$.

[2] The tables for $\wedge$, $\vee$, and $\leq$ are, explicitly,

|   | 0 | 1 |
|---|---|---|
| 0 | 0 | 0 |
| 1 | 0 | 1 |

|   | 0 | 1 |
|---|---|---|
| 0 | 0 | 1 |
| 1 | 1 | 1 |

|   | 0 | 1 |
|---|---|---|
| 0 | T | T |
| 1 | F | T |

*Proof of Necessity.*  Suppose that (10) holds schematically and that the premises of (14) hold for particular $a, b, c$.  Then

$$a \leq a \wedge (b \vee c) \qquad \text{by } \Lambda\text{W, Hp, Rp}$$
$$\leq (a \wedge b) \vee c \qquad \text{by (10)}$$
$$\leq c \qquad \text{by Hp, Rp, VW}$$

*Proof of Sufficiency.*  Suppose that (14) holds schematically.  Let

$$p \equiv a \wedge (b \vee c) \qquad q \equiv (a \wedge b) \vee c$$

Then

$$p \wedge b = a \wedge (b \vee c) \wedge b = a \wedge b \leq q$$
$$b \vee q = b \vee (a \wedge b) \vee c = b \vee c \geq p$$

Thus

$$p \wedge b \leq q \ \& \ p \leq b \vee q \qquad (15)$$

Hence by (14)

$$p \leq q$$

which is (10), Q.E.D.

COROLLARY 9.1.  *In order that $L$ be distributive it is necessary and sufficient that*

$$a \wedge c \leq b \wedge c \ \& \ a \vee c \leq b \vee c \rightarrow a \leq b \qquad (16)$$

*Proof.*  The premises of (16) imply those of (14), with $b$ and $c$ interchanged, by $\Lambda$K, VK, VC; the converse implication holds by $\Lambda$S, VS, $\Lambda$K', VK', VC.

COROLLARY 9.2.  *In any distributive lattice*

$$a \wedge c = b \wedge c \ \& \ a \vee c = b \vee c \rightarrow a = b \qquad (17)$$

*Proof.*  One has only to apply Corollary 9.1 twice, once with the predicate $\leq$ and once with $\geq$.

Some other properties of distributive lattices are contained in the exercises.

In mathematical applications the class of lattices in which (10) holds under the restriction that $c \leq a$ is of great importance.  Such lattices are called *modular lattices*.  Thus the lattice of linear spaces (interpretation 6° of Sec. 2) is modular, so also is the lattice of normal subgroups of a group; on the other hand, the lattice of Fig. 2d is nonmodular.  A closely related group of lattices are of some interest in certain investigations of abstract linear dependence.  Various suggestions have been made for using such lattices in logic, particularly in connection with quantum mechanics.  However, such suggestions have not yet borne fruit.  From the standpoint of logic, modular and other nondistributive lattices do not yet have the importance that they have in mathematics as a whole.

**4. Special elements.**  We are now concerned with obs 0 and 1 such that

$$0 \leq a \qquad a \leq 1 \qquad (18)$$

hold, respectively, for all obs $a$.  An ob 0 satisfying the left-hand half of (18) will be called a *zero element*; an ob 1 satisfying the right-hand half, a *unit element*.

From (18) it follows that the notions of zero and unit element are dual to one another.  In the following theorem, items in the separate columns apply if the corresponding half of (18) is postulated.

**Theorem 10.** *Any two*

   *zero elements*      *unit elements*

*are equal.   Further, if* (18) *holds in a semilattice, then*

$$a \wedge 0 = 0 \qquad a \wedge 1 = a \qquad\qquad (19)$$

*holds for all obs a; if* (18) *holds in a lattice, then we have in addition*

$$a \vee 0 = a \qquad a \vee 1 = 1 \qquad\qquad (20)$$

*Proof.* The proof of uniqueness is immediate by (18); that of (19) follows by (18) and part (iv) of Theorem 3; whereas (20) follows from (19) by duality.

#### EXERCISES

**1.** Give the complete formal demonstration of the following elementary theorems of a semilattice:

(a)                     $e_1 \wedge (e_2 \wedge e_3) \leq e_2 \wedge (e_3 \wedge e_1)$

(b)                     $e_1 \wedge e_2 \leq (e_2 \wedge e_1) \wedge e_2$

(c)                     $e_1 \wedge (e_2 \wedge e_3) \leq e_2 \wedge e_1$

**2.** Show that the schemes (ii) and (iii) of Theorem 3 are equivalent to the single scheme

$$(a \wedge b) \wedge c = (b \wedge c) \wedge a$$

(Byrne [TBF]; Dilworth [ARL].)

**3.** A partially ordered set is called a chain if, for any $a$, $b$, either $a \leq b$ or $b \leq a$. Show that a chain is always a distributive lattice.

**4.** Show that example $6°a$ of Sec. A2 is a distributive lattice.   (Sec. C5, below.)

**5.** Show that example $6°b$ of Sec. A2 is a modular lattice.

**6.** Show that example $6°c$ of Sec. A2 is a nonmodular lattice.

**7.** Show that the following are elementary theorem schemes of a general lattice:

(a)                 $(a \wedge b) \vee (c \wedge d) \leq (a \vee c) \wedge (b \vee d)$

(b)             $(a \wedge b) \vee (b \wedge c) \vee (c \wedge a) \leq (a \vee b) \wedge (b \vee c) \wedge (c \vee a)$

**8.** Show that a necessary and sufficient condition that a lattice be distributive is that

$$(a \vee b) \wedge (b \vee c) \wedge (c \vee a) \leq (a \wedge b) \vee (b \wedge c) \vee (c \wedge a)$$

**9.** Construct a Markov algorithm for reducing an ob of a general distributive lattice to a join of meets of atoms.

**10.** Show that every lattice which is nonmodular contains a five-element nonmodular sublattice with the same structure as Fig. 2d.

**11.** Construct the lattice diagram for the free modular lattice generated by $e_1$, $e_2$, $e_3$. Show that it contains 28 elements and that 5 of these form a sublattice with the structure of Fig. 2e.

**12.** Show the converse of Corollary 9.2 (use Exercise 11).

### C. SKOLEM LATTICES

We shall now consider lattices in which, in addition to the operations $\wedge$ and $\vee$, there is a third operation, symbolized by the infix ' $\supset$ ', called the *ply operation*. This will have properties such that in the propositional interpretation, where $a$ and $b$ are propositions—i.e., obs with which there are associated, in some manner, certain statements $A$ and $B$—$a \supset b$ is an ob to which there is associated in the same interpretation the statement formed from $A$ and $B$ by the conditional connective 'if ——$_1$, then ——$_2$'. This interpretation motivates the postulates $P_1$ and $P_2$ chosen in Sec. 1. From then on we shall study the consequences of those postulates and properties of the lattice, called an *implicative lattice*, formed by adjoining them. We shall find, among other things, that any such lattice is necessarily distributive; that conversely, any finite distributive lattice (and any distributive complete lattice) is implicative; that such a lattice has certain topological interpretations; and that certain properties ordinarily associated with classical implication fail to hold in it.

We shall also study the dual situation. In this case the lattice will be called a subtractive lattice. The term *'Skolem lattice'* seems to be the best available term for a lattice which is either implicative or subtractive.[1]  Skolem lattices which satisfy no other postulates than those considered in this section will be called *absolute*, to distinguish them from the "classical" ones considered in Sec. D (the reason for these terms will become apparent in Chap. 5).

The postulates $P_1$ and $P_2$ and their motivation will concern us in Sec. 1; their adjunction to a semilattice, in Sec. 2; and their adjunction to a lattice, to form an implicative lattice, in Sec. 3. The dual situation will be studied in Sec. 4. Finally, matters connected with interpretations will be discussed for both kinds of Skolem lattices simultaneously in Sec. 5.

From now on it will be convenient to drop the infix '∧' and to express the meet operation by simple juxtaposition. In discussions concerned with interpretation, we shall suppose that ∧ is propositional conjunction, whereas the join operation ∨ is propositional alternation.

**1. Postulates for the ply operation.** If the ply operation is to be a formalization, in the indicated sense, of the conditional connective, we must postulate for it an analogue of the characteristic property of the connective, viz., the rule of modus ponens. Such an analogue is given by the postulate

$P_1$ $$a(a \supset b) \leq b$$

Accordingly, we adopt $P_1$ as one of the postulates for an implicative lattice.

Now $P_1$ says that $a \supset b$ is a solution for $x$ of the relation

$$ax \leq b \tag{1}$$

There may be many solutions of (1); one of these is $b$ itself, and any $x$ such that $x \leq b$ is also a solution. The totality of all solutions is a conceptual class $X$ such that

$$y \text{ is in } X \,\&\, x \leq y \to x \text{ is in } X \tag{2}$$

Such a class we call an *ideal*.[2]  Now we have met such ideals before. In fact, the set of all $x$ such that

$$x \leq a \,\&\, x \leq b$$

---

[1] On the reasons for this see Sec. S2.

[2] This is the usual definition of ideal in a partially ordered set.

is an ideal, and the postulates $\Lambda K$, $\Lambda K'$ assert that $a \wedge b$ is a member of it. In that case we completed our postulates by $\Lambda S$, which said, in effect, that $a \wedge b$ is maximal in the ideal. This suggests what to do here. We adopt as second postulate for the ply operation that $a \supset b$ shall be maximal among the solutions of (1); viz.,

$P_2$ $\qquad\qquad\qquad\qquad ac \leq b \to c \leq a \supset b$

This merely has the effect of making $a \supset b$ unique. Note that $P_1$ and $P_2$ together give

$$ac \leq b \rightleftarrows c \leq a \supset b \tag{3}$$

The ply operation is thus a kind of inverse to the meet operation. If we think of the latter as multiplication, the former is analogous to division. However, the logical interpretation has suggested an interchange of the arguments compared with that which is customary in algebra; thus $a \supset b$ is a quotient of $b$ by $a$, not of $a$ by $b$. Where the algebraic aspect is dominant, notations more in agreement with tradition have been used. Thus Skolem [UAK] used $b/a$; Dilworth [ARL], $b:a$.

**2. Implicative semilattices.** The postulates $P_1$ and $P_2$ do not require the presence of any operation other than meet. Accordingly, it makes sense to define an *implicative semilattice* as a system formed by adjoining a ply operation satisfying $P_1$ and $P_2$ to a semilattice with relation $\leq$ and operation $\wedge$. Some properties of such a system will now be derived.

**Theorem 1.** *In an implicative semilattice, the operation $\supset$ is inversely monotone with respect to its left argument and directly monotone with respect to its right argument. The semilattice has a unit 1 such that*

$$a \leq 1$$

*Further, the following conditions hold for all obs $a$, $b$, $c$:*

(i) $\qquad\qquad\qquad\qquad b \leq a \supset b$

(ii) $\qquad\qquad a \supset (b \supset c) = ab \supset c = b \supset (a \supset c)$

(iii) $\qquad\qquad a \supset (b \supset c) \leq (a \supset b) \supset (a \supset c)$

(iv) $\qquad\qquad a \supset bc = (a \supset b)(a \supset c)$

(v) $\qquad\qquad\qquad\qquad a = 1 \supset a$

(vi) $\qquad\qquad a \leq b \rightleftarrows 1 \leq a \supset b$

*Proof of Left Monotony.* Assume

$$a \leq b \tag{4}$$

Then $\qquad\quad a(b \supset c) \leq b(b \supset c) \leq c \qquad$ by $\Lambda B'$, $P_1$

Hence $\qquad\quad b \supset c \leq a \supset c \qquad\qquad$ by $P_2$, Q.E.D.

*Proof of Right Monotony.* Again assuming (4), we have

$$c(c \supset a) \leq a \leq b \qquad\qquad \text{by } P_1$$

Then $\qquad\quad c \supset a \leq c \supset b \qquad\qquad$ by $P_2$, Q.E.D.

In view of these two properties, we can use Rp in an implicative semilattice.

*Proof of Unit Property.*   Define

$$1 \equiv e_1 \supset e_1 \tag{5}$$

Then since                     $e_1 a \le e_1$     by $\Lambda$K

we have                        $a \le 1$     by $P_2$, Q.E.D..

*Proof of* (i)

                               $ab \le b$              by $\Lambda$K$'$

Hence                          $b \le a \supset b$      by $P_2$, Q.E.D.

*Proof of* (ii)

                               $a(a \supset (b \supset c)) \le b \supset c$              by $P_1$

Hence        $ab(a \supset (b \supset c)) \le b(b \supset c) \le c$       by Rp, $P_1$

Then              $a \supset (b \supset c) \le ab \supset c$              by $P_2$

This proves the first half of the left equation.   Conversely,

                               $ab(ab \supset c) \le c$              by $P_1$

                               $a(ab \supset c) \le b \supset c$              by $P_2$

                               $ab \supset c \le a \supset (b \supset c)$              by $P_2$

This proves the left equation.   The right equation follows by interchanging
$a$ and $b$ and using $\Lambda$C and Rp.

*Proof of* (iii).   We have

$a(a \supset b)\,(a \supset (b \supset c))$

               $\le a(a \supset b)a(a \supset (b \supset c))$              by $\Lambda$W, $\Lambda$C

               $\le b(b \supset c)$              by $P_1$, Rp

               $\le c$              by $P_1$

Hence

        $(a \supset b)(a \supset (b \supset c)) \le a \supset c$              by $P_2$

               $a \supset (b \supset c) \le (a \supset b) \supset (a \supset c)$              by $P_2$, Q.E.D.

*Proof of* (iv)

                               $a \supset bc \le a \supset b$              by $\Lambda$K, Rp

                               $a \supset bc \le a \supset c$              by $\Lambda$K$'$, Rp

Hence

                               $a \supset bc \le (a \supset b)(a \supset c)$              by $\Lambda$S

Conversely

                               $a(a \supset b)(a \supset c) \le bc$              by $P_1$†

Hence

                               $(a \supset b)(a \supset c) \le a \supset bc$              by $P_2$

*Proof of* (v)

                               $a \le 1 \supset a$              by (i)

Conversely                    $1 \supset a = 1(1 \supset a)$              by Theorem B10

                               $\le a$              by $P_1$

† With $\Lambda$W as in the proof of (iii).

*Proof of* (vi).    Suppose

$$a \leq b$$

Then                $$1a \leq b$$

Hence               $$1 \leq a \supset b \qquad \text{by } P_2$$

Conversely, assume  $$1 \leq a \supset b$$

Then                $$a1 \leq b \qquad \text{by } P_1$$

Hence               $$a \leq b \qquad \text{by Theorem B10}$$

This completes the proof of Theorem 1.

The following theorem shows that an implicative semilattice can be characterized in a certain sense by axiom schemes.

**Theorem 2.** *Let $L$ be a semilattice with operation $\wedge$, and let $\supset$ be a binary operation in $L$ which is monotone with respect to equality. Then a necessary and sufficient condition for $L$ to be an implicative semilattice with $\supset$ as ply operation is that $P_1$ and the following properties hold:*

(i) $$b \leq a \supset ab$$

(ii) $$a \supset bc \leq a \supset b$$

*Proof of Necessity*

$$a(a \supset b) \leq b \qquad \text{by } P_1$$
$$a(a \supset b) \leq a \qquad \text{by } \Lambda K$$
$$a(a \supset b) \leq ab \qquad \text{by } \Lambda S$$
$$a \supset b \leq a \supset ab \qquad \text{by } P_2$$
$$b \leq a \supset ab \qquad \text{by (i) of Theorem 1, } (\tau)$$

This proves (i). As for (ii), it follows immediately from (iv) of Theorem 1 and $\Lambda K$.

*Proof of Sufficiency.* It suffices to derive $P_2$ from (i) and (ii). Suppose then that

$$ac \leq b$$

Then by Theorem B3,

$$ac = acb = b(ac)$$

Hence              $$c \leq a \supset ac \qquad \text{by (i)}$$
$$\leq a \supset b(ac) \qquad \text{by Rp}$$
$$\leq a \supset b \qquad \text{by (ii), Q.E.D.}$$

**3. Implicative lattices.** An *implicative lattice* is a lattice with relation $\leq$ and operations $\wedge$, $\vee$, to which is adjoined an operation $\supset$ satisfying the postulates $P_1$ and $P_2$.

Naturally, all the properties derived in Sec. 2 apply to an implicative lattice; in addition, we have the following:

**Theorem 3.** *An implicative lattice is distributive. Furthermore, the following holds for all obs $a$, $b$, $c$.*

$$(a \vee b) \supset c = (a \supset c)(b \supset c) \qquad (6)$$

*Proof of the Distributive Law* (Skolem).   We have

$$ab \le ab \vee ac \qquad\qquad \text{by VK}$$
$$b \le a \supset (ab \vee ac) \qquad \text{by } P_2$$
$$c \le a \supset (ab \vee ac) \qquad \text{similarly, using VK}'$$
$$b \vee c \le a \supset (ab \vee ac) \qquad \text{by VS}$$
$$a(b \vee c) \le ab \vee ac \qquad\quad \text{by } P_1, \text{ Q.E.D.}$$

*Proof of* (6).   By Rp (Theorem 1) and $\Lambda$S,

$$(a \vee b) \supset c \le (a \supset c)(b \supset c) \tag{7}$$

On the other hand, using the distributive law, we have

$(a \vee b)(a \supset c)(b \supset c)$

$$\le a(a \supset c)(b \supset c) \vee b(a \supset c)(b \supset c)$$
$$\le c(b \supset c) \vee c(a \supset c) \qquad\qquad \text{by } P_1$$
$$\le c \qquad\qquad\qquad\qquad\qquad\qquad\quad \text{by } \Lambda K, \text{ VW}$$

Hence

$$(a \supset c)(b \supset c) \le (a \vee b) \supset c \qquad \text{by } P_2$$

From this and (7) we have (6), Q.E.D.

Thus every lattice in which the class of $x$ satisfying (1) has a maximum is distributive.  The great importance in logic of distributive lattices, as opposed to other kinds of lattices, stems from the fact that every reasonable implication operation yet considered in logic satisfies $P_1$ and $P_2$.

Conversely, given any distributive lattice $L$, we can define $a \supset b$ as the l.u.b. (least upper bound) of the class of all $x$ satisfying (1).  Then $P_2$ is equivalent to the existence of $a \supset b$.  If in addition we have

$$a\{\mathfrak{U}_x\{x \mid B(x)\} \le \mathfrak{U}_x\{ax \mid B(x)\} \tag{8}$$

where $\mathfrak{U}_x\{A(x) \mid B(x)\}$ is, for any ob function $A(x)$ and condition $B(x)$, the l.u.b. of all obs of the form $A(x)$ for $x$ such that $B(x)$, then $P_1$ will also hold. In fact, if

$$c \equiv \mathfrak{U}_x\{x \mid ax \le b\}$$

then

$$ac \le \mathfrak{U}_x\{ax \mid ax \le b\} \qquad \text{by (8)}$$
$$\le b$$

The condition (8) is a generalized distributive law.  It reduces to the ordinary distributive law in case the lattice is finite.  Hence we have:

**Theorem 4.**   *Every finite distributive lattice is implicative, $a \supset b$ being defined as the join of all $x$ satisfying* (1).

**4. Subtractive lattices.**   Let us now consider the situation dual to that treated in Secs. 1 to 3.

The operation dual to ply will be a kind of subtraction.   We use the infix '$-$' for it, the argument being written, however, in the order which is customary in algebra, so that $a - b$ is the dual of $b \supset a$.  The postulates for it will be

$(-)_1$ $\qquad\qquad\qquad\qquad a \le b \vee (a - b)$
$(-)_2$ $\qquad\qquad\qquad\qquad a \le b \vee c \to a - c \le b$

A lattice with such an operation will be called *subtractive*.

**Theorem 5.** *A subtractive lattice is distributive and has* $e_1 - e_1$ *as a zero element. The subtractive operation is directly monotone with respect to its left argument, inversely monotone with respect to its right argument. Furthermore, the following hold:*

| | |
|---|---|
| (i) | $a - b \leq a$ |
| (ii) | $a - (b \vee c) = (a - b) - c = (a - c) - b$ |
| (iii) | $(a - c) - (b - c) \leq (a - b) - c$ |
| (iv) | $(a - c) \vee (b - c) = (a \vee b) - c$ |
| (v) | $a = a - 0$ |
| (vi) | $a \leq b \rightleftarrows a - b = 0$ |
| (vii) | $a - bc = (a - b) \vee (a - c)$ |
| (viii) | $a = ab \vee a - b$ |

*Proof.* All except (viii) follow by duality from Theorems 1 and 3. The proof of (viii) is as follows:

$$a \leq a(b \vee (a - b)) \qquad \text{by } (\rho), (-)_1, \Lambda S$$
$$\leq ab \vee (a - b) \qquad \text{by (10) of Sec. B}$$
$$\leq a \qquad \text{by } \Lambda K, \text{(i), VS}$$

This completes the proof.

*Remark.* The dual of (viii) of Theorem 5 is

$$(a \vee b)(a \supset b) = b \tag{9}$$

This, although true for an implicative lattice, is not of much interest, whereas (viii) is constantly used in Sec. D.

**5. Examples.** Let us now examine some of the interpreted lattices exhibited in Sec. A2 with reference to their character as implicative or subtractive lattices or both. We shall also consider some other illustrations similar to those there given.

Interpretations 1° to 3° are illustrations of stronger systems considered later. However, one can see intuitively (with formal verification, if desired, postponed until later) that if $a'$ is the complement or negative of $a$, then $a' \vee b$ has the properties of $a \supset b$, and $ab'$ those of $a - b$. These lattices are thus both implicative and subtractive.

With reference to interpretation 4°, we may observe that it may be generalized to functions from an arbitrary set or space $X$ to a partially ordered set $Y$. If for two such functions $f, g$ we define

$$f \leq g$$

as equivalent to

$$f(x) \leq g(x) \qquad \text{for all } x \text{ in } X$$

then we have the partially ordered set which is commonly written $Y^X$. This will be a lattice provided $Y$ is a lattice. Suppose, then, that $Y$ is a linearly ordered set.[1] If $Y$ has a unit 1, then given $f, g$ in $Y^X$, the function $h$ such that

$$h(x) = \begin{cases} 1 & \text{if } f(x) \leq g(x) \\ g(x) & \text{otherwise} \end{cases}$$

[1] That is, such that for any $a, b$, at least one of the statements $a \leq b$ or $b \leq a$ holds.

will have the properties of $f \supset g$.   On the other hand, if $Y$ has a zero, namely, 0,

$$h(x) = \begin{cases} 0 & \text{if } f(x) \leq g(x) \\ f(x) & \text{otherwise} \end{cases}$$

will have the properties of $f - g$.   But if $Y$ has no unit, then there will not in general be any $f \supset g$, for the value of $h(x)$ such that

$$f \wedge h \leq g$$

is not bounded above if $f(x) \leq g(x)$; similarly, there will in general be no $f - g$ if $Y$ has no zero.

This example can, in a certain sense, be subsumed under a generalization of interpretation 1°.   For we can interpret $f$ as the set of all pairs $(x,y)$ in which $x$ is in $X$, $y$ is in $Y$, and $y \leq f(x)$.   This is a generalization of interpretation 1° in that the obs are not all possible subsets of a certain universal set, but only certain special ones, forming a "ring of sets."   According to a theorem of Stone and Birkhoff, every distributive lattice can be so represented.

Interpretation 5° can be subsumed under the generalized interpretation 4°.   In fact, let

$$a = \prod p_i{}^{m_i} \qquad b = \prod p_i{}^{n_i}$$

be the expressions of $a$ and $b$, respectively, as products of powers of the prime numbers $p_1, p_2, \ldots$ .   Then $a$ and $b$ are biuniquely correlated with certain functions $Y^X$ in which $X$ are the positive rational integers and $Y$ are the natural numbers (i.e., rational integers $\geq 0$); further, $a \leq b$ if and only if $m_i \leq n_i$ for all $i$ (the first '$\leq$' denotes the divisibility relation, the second the natural ordering of the natural numbers).   Hence we have, by the argument in the case of interpretation 4°, a subtractive lattice, while the divisors of a fixed integer form a finite lattice which is both subtractive and implicative.

The illustration 6°$a$ gives rise to a subtractive lattice, as follows.   We define a topological space as a system of sets with a closure operation satisfying the schemes listed under interpretation 6° in Sec. A2 and in addition the following:

$$(a \vee b)^* \leq a^* \vee b^* \tag{10}$$

In such a system the lattice union among closed sets is the same as the set union.   If now we define $a - b$ as $(a \dotminus b)^*$, where $a \dotminus b$ is the ordinary set-theoretic difference (i.e., the set of points in $a$ but not in $b$), then we have for all sets $a$, $b$, $c$, closed or not,

$$a \leq b \vee (a \dotminus b) \leq b \vee (a - b)$$
$$a \leq b \vee c \rightarrow a \dotminus b \leq (b \vee c) \dotminus b \leq c$$
$$\rightarrow a - b \leq c^*$$

Hence if $a$, $b$, $c$ are closed sets, we have $(-)_1$ and $(-)_2$, so that the lattice of closed sets in a topological space is a subtractive lattice.

This gives a simple intuitive example of a subtractive lattice, viz., in terms of the closed point sets of the ordinary euclidean plane.   For example, if $a$ and $b$ are the closures of the interiors of two intersecting circles in the plane, $a - b$ is the closure of that part of the interior of $a$ which is exterior to $b$ (thus the crosshatched area in Fig. 3).

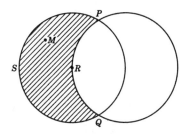

FIGURE 3

By duality the open sets in a topological space[1] form an implicative lattice. Thus, in Fig. 3, if $a$ and $b$ are the interiors of the two circles, $a \supset b$ is the interior of the open set formed by adjoining to $b$ the entire exterior of $a$.

Finally, by Theorem 4 and its dual, every finite distributive lattice is an example of a lattice which is both subtractive and implicative. Especially useful in this connection is the "free" distributive lattice generated by three indeterminates (called *generators*) $e_1$, $e_2$, $e_3$. The lattice diagram for this lattice is given in Fig. 4.

These examples enable us to show that certain elementary statement schemes are not universally true. Consider, for example, the schemes

$$b(a - b) = 0 \tag{11}$$

$$b \le b - (a - b) \tag{12}$$

and their duals

$$a \vee (a \supset b) = 1 \tag{13}$$

$$(a \supset b) \supset a \le a \tag{14}$$

In Fig. 3, $b(a - b)$ is not void, but consists of the arc $PRQ$ of the boundary of $b$; hence (11) is false in that interpretation. Again, if the point $M$ in $a$ were an isolated point of $b$, it would not be a point of $b - (a - b)$, so that (12) also does not hold. Thus (11) and (12) are not theorem schemes of an absolute subtractive lattice; and by duality, (13) and (14) are not theorem schemes of an absolute implicative lattice.

**EXERCISES**

**1.** Show that the following hold in an implicative semilattice:

(i)         $a \le b \to c \le a \supset b$

(ii)        $a \le a \supset b \rightleftarrows a \le b$

(iii)       $a(a \supset b) = ab$

(iv)        $a \supset b \le ac \supset bc$

(v)         $(a \supset b)(b \supset c) \le (a \supset c)$

(vi)        $a \supset ab = a \supset b = a \supset (a \supset b)$

---

[1] In topological works it is shown that a topological space is a self-dual concept; i.e., if interiors are suitably defined, the duals of the closure properties follow for the interior properties and vice versa.

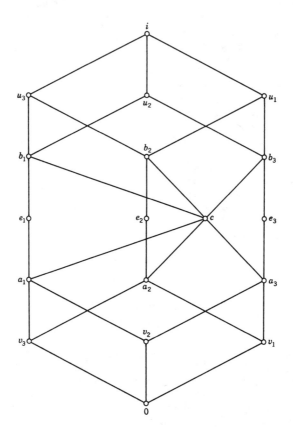

FIGURE 4

**2.** Show that the following hold in an implicative lattice:

(i) $$(a \supset b) \vee (a \supset c) \leq a \supset (b \vee c)$$

(ii) $$(a \supset c) \vee (b \supset c) \leq ab \supset c$$

(iii) $$c \vee (a \supset b) \leq (a \vee c) \supset (b \vee c)$$

(iv) $$(a \vee b)(a \supset b) = b$$

(v) $$a \vee b \leq (a \supset b) \supset b$$

(vi) $$(a \supset b)(b \supset a) = (a \vee b) \supset ab$$

**3.** Dualize the statements of Exercises 1 and 2 so as to get theorems for a subtractive lattice.

**4.** Show that a necessary and sufficient condition that a semilattice with a unit and an additional operation $\supset$ be an implicative semilattice with $\supset$ as ply operation is that the two following conditions hold:

$$a \leq b \rightleftarrows a \supset b = 1$$

$$(a \wedge b) \supset c = a \supset (b \supset c)$$

for all $a, b, c$. What is the dual situation for a subtractive lattice? (Birkhoff [LTh$_1$, p. 128].)

5. Show directly that in a subtractive lattice the postulate $(-)_2$ can be replaced (assuming monotony) by the two postulates

$$a \leq b \rightarrow a - c \leq b - c$$

$$(a \vee b) - b \leq a$$

Discuss the relation of this to Theorem 2.

6. Find counterexamples for (13) and (14) in the lattice of interpretation 4° of Sec. A2 ([LLA], p. 78).

7. Show that the following are not universally true in a general implicative lattice:

$$a \supset (b \vee c) \leq (a \supset b) \vee (a \supset c) \qquad \text{[converse of (ii) of Exercise 2]}$$

$$(a \vee c) \supset (b \vee c) \leq c \vee (a \supset b) \qquad \text{[converse of (iii)]}$$

Investigate converses of other theorem schemes stated in terms of $\leq$ but not $=$.

8. Show that an equational algebra with monotone operations $\wedge$, $\vee$, $\supset$ is an implicative lattice if and only if the following axiom schemes are satisfied:

$A1$                               $a \supset a = b \supset b$

$A2$                          $(a \supset b) \wedge b = b$

$A3$                        $a \wedge (a \supset b) = a \wedge b$

$A4$                 $a \supset (b \wedge c) = (a \supset b) \wedge (a \supset c)$

$A5$                 $(a \vee b) \supset c = (a \supset c) \wedge (b \supset c)$

Show further that these schemes are independent and that $A1$ to $A4$ characterize in the same way an implicative semilattice. (Monteiro [AIA]. The complications he gets into with the associative law can be avoided by use of Theorem B2.)

## D. CLASSICAL SKOLEM LATTICES

In the preceding section we have seen that the schemes

$$a \vee (a \supset b) = 1 \tag{1}$$

$$(a \supset b) \supset a \leq a \tag{2}$$

are not elementary theorem schemes of an absolute implicative lattice, and their duals,

$$b(a - b) = 0 \tag{3}$$

$$b \leq b - (a - b) \tag{4}$$

are not theorem schemes of an absolute subtractive lattice. A lattice formed by strengthening the postulates for an implicative lattice so that (1) and (2) hold will be called a *classical implicative lattice*; dually, a *classical subtractive lattice* is a subtractive lattice such that (3) and (4) hold.

In this section we shall study such lattices. We shall put here the primary emphasis on the subtractive rather than the implicative form, because that seems, at the moment, to have the more interesting applications. Various alternative postulate systems will be considered in Sec. 1. In Sec. 2 it will be shown that a classical subtractive lattice is equivalent to a Boolean ring, i.e., to a ring, in the sense of modern algebra, in which multiplication is idempotent. Finally, in Sec. 3, there will be some remarks about the duals of these properties.

**1. Classical subtractive lattices.**   By definition, a classical subtractive lattice is a subtractive lattice in which (3) and (4) are elementary theorems. These properties are valid in the set interpretation ($1°$ of Sec. A2), if $a - b$ is taken as the set of elements which are in $a$ but not in $b$.   We call this interpreted operation the *set difference*.   A family of sets which is closed with respect to union, intersection, and set difference is called a "field of sets"; any such field gives a valid interpetation of a classical subtractive lattice.

We begin with a proof that the schemes (3) and (4) are equivalent to one another and to

$$ab \leq a - (a - b) \tag{5}$$

In fact, if (4) holds, then

$$
\begin{aligned}
ab &\leq ab - (a - ab) && \text{by (4)} \\
&\leq a - (a - b) && \text{by } \Lambda\text{K}, \Lambda\text{K}', \text{Rp}
\end{aligned}
$$

where Rp holds by Theorem C5.   Thus (5) is a consequence of (4).   On the other hand, if (5) holds, then

$$
\begin{aligned}
b(a - b) &\leq (a - b) - ((a - b) - b) && \text{by (5)} \\
&= (a - b) - (a - b) && \text{by (ii) of Theorem C5, VW, Rp} \\
&= 0 && \text{by (vi) of Theorem C5}
\end{aligned}
$$

so that (3) is a consequence of (5).   To show that (4) is a consequence of (3), we have from (viii) of Theorem C5, putting $b$ for $a$ and $a - b$ for $b$,

$$
\begin{aligned}
b &\leq b(a - b) \vee (b - (a - b)) \\
&\leq 0 \vee b - (a - b) && \text{by (3)} \\
&= b - (a - b), \text{ Q.E.D.}
\end{aligned}
$$

From this discussion we conclude the following:

**Theorem 1.**   *In order that a subtractive lattice be classical, it is necessary and sufficient that any one of the schemes* (3) *to* (5) *be a valid theorem scheme.*

The characterization given by Theorem 1 has the disadvantage that it contains the rather complex rule $(-)_2$.   There is a certain interest in seeking a formulation based on axiom schemes alone.   We proceed with such a search.

In the first place the scheme

$$a - b \leq a \tag{6}$$

is valid in any subtractive lattice by (i) of Theorem C5.   Further, the rule

$$a \leq b \ \& \ ac = 0 \rightarrow a \leq b - c \tag{7}$$

is valid in the interpretation as set difference.   We shall see that $(-)_2$ is derivable from (3), (6), (7), and $(-)_1$.   In fact, assume the premise of $(-)_2$, namely,

$$a \leq b \vee c$$

and let
$$p \equiv (a - b) - c$$

Then
$$pc \leq 0 \qquad \text{by (3)}$$
$$pb \leq (a - b)b \leq 0 \qquad \text{by (6), (3)}$$

From these we conclude, by (7), that

$$b \leq (b \vee c) - p$$
$$c \leq (b \vee c) - p$$

Then
$$b \vee c \leq (b \vee c) - p \qquad \text{by VS}$$

Hence
$$p(b \vee c) \leq 0 \qquad \text{by (3)}$$

On the other hand, by (6) and the premise of $(-)_2$,

$$p \leq (a - b) \leq a \leq b \vee c$$

Hence
$$p = p(b \vee c) = 0$$

Therefore, by $(-)_1$, we have

$$a - b \leq c \vee p = c$$

This is a consequence of the premise of $(-)_2$, so that $(-)_2$ is established.

Next we note that (7) and $(-)_1$ are both consequences of

$$ab = abc \vee a(b - c) \tag{8}$$

In fact, taking $b \equiv a$ in (8), we have

$$a = ac \vee a(a - c)$$
$$\leq c \vee (a - c) \qquad \text{by } \Lambda\text{K}', \text{Rp}$$

which gives us $(-)_1$. Again, assuming the premises of (7), we have from (8)

$$a = ab = a(b - c)$$

Hence
$$a \leq b - c$$

so that (7) holds.

On the other hand, we see that (8) is true in an arbitrary subtractive lattice as follows:

$$ab = a(bc \vee (b - c)) \qquad \text{by (viii) of Theorem C5}$$
$$= abc \vee a(b - c) \qquad \text{by the distributive law}$$

Summing this up, we have the following:

**Theorem 2.** *Let L be a lattice with a zero element and an extra operation of subtraction indicated by the infix '—'. Then a necessary and sufficient condition that L be a classical subtractive lattice is that (3), (6), and (8) be theorem schemes; another is that $(-)_1$, (3), (6), and (7) hold.*

For use as lemmas in Sec. 2, we now show that the following are theorem schemes of a classical subtractive lattice:

$$a(b - c) = ab - ac \tag{9}$$
$$a - (b \vee c) = (a - b)(a - c) \tag{10}$$
$$a - (b - c) = ac \vee (a - b) \tag{11}$$

*Proof of* (9).   From right to left we argue thus:

$$ab \ \leq a(c \vee (b - c)) \qquad \text{by } (-)_1$$
$$\leq ac \vee a(b - c) \qquad \text{by (8) of Sec. B}$$

Hence          $ab - ac \leq a(b - c) \qquad \text{by } (-)_2$

To prove the converse, let

$$p \equiv a(b - c)$$

Then          $p \leq ab \qquad\qquad \text{by (6), Rp}$
$$acp \leq c(b - c) \qquad \text{by Sec. B}$$
$$\leq 0 \qquad\qquad\ \text{by (3)}$$

Hence          $p \leq ab - ac \qquad \text{by (7), Q.E.D.}$

*Proof of* (10).   From left to right we have

$$a - (b \vee c) \leq a - b \qquad\qquad\quad \text{by VK, Rp}$$
$$a - (b \vee c) \leq a - c \qquad\qquad\quad \text{by VK', Rp}$$
$$a - (b \vee c) \leq (a - b)(a - c) \qquad \text{by } \Lambda S$$

To prove (10) from right to left, let

$$p \equiv (a - b)(a - c)$$

Then          $p \leq a \qquad\qquad\qquad\qquad\quad \text{by (6)}$
$$(b \vee c)p \leq bp \vee cp \qquad\qquad \text{by (8) of Sec. B}$$
$$\leq b(a - b) \vee c(a - c)$$
$$\leq 0 \qquad\qquad\qquad \text{by (3)}$$

Then (10) follows by (7).

*Proof of* (11).   From left to right we have

$$a \ \leq b \vee (a - b) \qquad\qquad\quad \text{by } (-)_1$$
$$\leq c \vee (b - c) \vee (a - b) \qquad \text{by } (-)_1, \text{Rp}$$
$$a - (b - c) \leq c \vee (a - b) \qquad\qquad\qquad \text{by } (-)_2$$
$$a - (b - c) \leq a(c \vee (a - b)) \qquad\qquad\ \text{by (6), } \Lambda S$$
$$\leq ac \vee (a - b) \qquad\qquad\quad \text{by (10) of Sec. B}$$

Conversely, let $p \equiv ac \vee (a - b)$.

Then          $p \leq a \qquad\qquad\qquad\qquad\qquad\ \text{by } \Lambda K, (6), VS$
$$p(b - c) \leq ac(b - c) \vee (a - b)(b - c) \qquad \text{by (8) of Sec. B}$$
$$\leq c(b - c) \vee (a - b)b \qquad\qquad \text{by } \Lambda K', (6)$$
$$\leq 0 \qquad\qquad\qquad\qquad\quad \text{by (3)}$$

Hence (11) follows by (7).

This discussion is summed up in the following theorem:

**Theorem 3.**   *In every classical subtractive lattice the schemes* (3) *to* (6) *and* (8) *to* (11) *are elementary theorem schemes and* (7) *holds for all a, b, c.   Conversely, every subtractive lattice in which any one of the schemes* (3) *to* (5) *holds is a classical subtractive lattice; so also is every distributive lattice with*

*subtraction operation and zero element which satisfies* $(-)_1$ *and both* (3)
*and* (6).[1]

**2. Boolean rings.** We now study the relation between a classical subtractive lattice and a ring in the sense of modern algebra.

DEFINITION 1. A ring is an equational algebra with a special element 0; two binary operations, called the sum and product and indicated, respectively, by the infix '+' and simple juxtaposition; and a unary operation indicated by the suffix '*', all these operations being monotone with respect to equality, so that the following axiom schemes are satisfied:

| | |
|---|---|
| R1 | $a + (b + c) = (a + b) + c$ |
| R2 | $a + 0 = a$ |
| R3 | $a + a^* = 0$ |
| R4 | $a + b = b + a$ |
| R5 | $a(bc) = (ab)c$ |
| R6 | $a(b + c) = ab + ac$ |
| R7 | $(a + b)c = ac + bc$ |

DEFINITION 2. A Boolean ring is a ring in which multiplication is idempotent; i.e., for all obs $a$,

| | |
|---|---|
| R8 | $aa = a$ |

We proceed to show that the notions of Boolean ring and of classical subtractive lattice are equivalent. In this discussion we shall revert to the notation of Sec. B and use the infix '∧' for the lattice meet operation in stating the definitions; but since the lattice meet and the ring multiplication are the same, it is not necessary to make this distinction throughout.

**Theorem 4.** *Let $L$ be a classical subtractive lattice with relation $\leq$, meet operation $\wedge$, join operation $\vee$, and subtraction $-$. Let the ring operations be defined thus:*

| | |
|---|---|
| (i) | $a + b \equiv (a - b) \vee (b - a)$ |
| (ii) | $ab \equiv a \wedge b$ |
| (iii) | $a^* \equiv a$ |

*Let 0 be the zero element of the lattice. Then with respect to this 0 and these operations, $L$ is a Boolean ring.*

*Proof.* It follows from (ii) and Theorem B2 that R5 and R8 are satisfied. Further, R4† follows by (VC) and (i); also by (∧C) and (ii), multiplication is commutative, and hence, in particular, R7 is a consequence of R6. The monotony of the ring operators follows from that of the lattice operators. The proofs of the other properties are as follows.

*Proof of R1.* By (i),

$$a + (b + c) = (a - (b + c)) \vee ((b + c) - a)$$

---

[1] The distributivity condition was unintentionally left out in [LLA], p. 90, definition 3. See Exercise 3.

† R4 (and also commutativity of multiplication) will be proved redundant in Theorem 5.

Now

$$a - (b + c) = a - ((b - c) \vee (c - b)) \qquad \text{by (i)}$$
$$= (a - (b - c)) \wedge (a - (c - b)) \qquad \text{by (10)}$$
$$= (ac \vee (a - b)) \wedge (ab \vee (a - c)) \qquad \text{by (11)}$$

Multiplying out by the distributive law and noting that by (3)

$$ab(a - b) = ac(a - c) = 0$$

we have

$$a - (b + c) = abc \vee (a - b)(a - c) \qquad (12)$$

On the other hand,

$$(b + c) - a = ((b - c) \vee (c - b)) - a \qquad \text{by (i)}$$
$$= ((b - c) - a) \vee ((c - b) - a) \qquad \text{by (iv) of Theorem C5}$$
$$= (b - (a \vee c)) \vee (c - (a \vee b)) \qquad \text{by (ii) of Theorem C5}$$
$$= (b - a)(b - c) \vee (c - a)(c - b) \qquad \text{by (10)}$$

Combining this with (12), we have

$$a + (b + c) = abc \vee (a - b)(a - c) \vee (b - c)(b - a) \vee (c - a)(c - b) \quad (13)$$

Since this is symmetric in $a$, $b$, $c$, and since addition is commutative, we have

$$a + (b + c) = c + (a + b) = (a + b) + c, \quad \text{Q.E.D.}$$

*Proof of* R2

$$a + 0 = (a - 0) \vee (0 - a)$$

Now    $a - 0 = a$                     by (v) of Theorem C5

and    $0 - a = 0$                     by (vi) of Theorem C5

Hence    $a + 0 = a \vee 0 = a$, Q.E.D.

*Proof of* R3.    By (iii) and (i) and (vi) of Theorem C5,

$$a + a^* = a + a = (a - a) \vee (a - a)$$
$$= 0 \vee 0 = 0, \quad \text{Q.E.D.}$$

*Proof of* R6

$$a(b + c) = a((b - c) \vee (c - b))$$
$$= a(b - c) \vee a(c - b) \qquad \text{by (10) of Sec. B}$$
$$= (ab - ac) \vee (ac - ab) \qquad \text{by (9)}$$
$$= ab + ac \qquad \text{by (i), Q.E.D.}$$

This completes the proof of Theorem 4.

The ob $a + b$ defined by (1) is called by various names, such as the *symmetric difference*, the *exclusive*, or *disjunctive*, *sum*, etc.    In terms of sets, $a + b$ is the set of elements which are in one or the other of $a$, $b$ but not in both.    In the propositional interpretation it represents the exclusive 'or' It, rather than $a \vee b$, is properly called the *disjunction* of $a$ and $b$.

Before proving the converse of Theorem 4 we shall derive some properties of a Boolean ring.    The first two of these properties are consequences of R1 to R3 only (these constitute axiom schemes for a "group"); the third follows from these and R6 only.

**Theorem 5.** *For a Boolean ring R4 is redundant; furthermore, the following are elementary theorem schemes:*

(i)                           $0 + a = a$

(ii)                          $a^* + a = 0$

(iii)                         $a0 = 0$

(iv)                          $ab = ba$

(v)                           $a^* = a$

(vi)                          $a + a = 0$

*Proof.* We prove first (ii), then (i) and (iii), as follows:

$$
\begin{aligned}
a^* + a &= (a^* + a) + (a^* + a^{**}) &&\text{by R2, R3}\\
&= a^* + (a + a^*) + a^{**} &&\text{by R1}\\
&= a^* + a^{**} &&\text{by R3, R2}\\
&= 0 &&\text{by R3}\\
0 + a &= (a + a^*) + a &&\text{by R3}\\
&= a + (a^* + a) &&\text{by R1}\\
&= a &&\text{by (ii), R2}\\
a0 &= a0 + ab + (ab)^* &&\text{by R2, R3, R1}\\
&= a(0 + b) + (ab)^* &&\text{by R6}\\
&= ab + (ab)^* &&\text{by (i)}\\
&= 0 &&\text{by R3}
\end{aligned}
$$

Next, we expand $(a + b)(c + d)$ in two ways, thus.

$$
\begin{aligned}
(a + b)(c + d) &= (a + b)c + (a + b)d &&\text{by R6}\\
&= ac + bc + ad + bd &&\text{by R7}\\
(a + b)(c + d) &= a(c + d) + b(c + d) &&\text{by R7}\\
&= ac + ad + bc + bd &&\text{by R6}
\end{aligned}
$$

Comparing these two and adding $(ac)^*$ on the left and $(bd)^*$ on the right to both sides, we have

$$ad + bc = bc + ad \tag{14}$$

the common value of both sides being

$$(ac)^* + (a + b)(c + d) + (bd)^* \tag{15}$$

All this is true in an arbitrary ring. In a Boolean ring we get R4 from (14) at once by taking $c \equiv b$, $d \equiv a$. On the other hand, if we take $c \equiv a$, $d \equiv b$, then (15) becomes

$$a^* + a + b + b^*$$

which is 0 by (ii) and R3; so that, since this is the same as either side of (14),

$$ab + ba = 0 \tag{16}$$

Here if we take $b \equiv a$, we have (vi) by R7. Then (v) follows by adding $a^*$ to both sides.

Finally, from (16), we have

$$ab = ab + 0 = ab + ab + ba = 0 + ba = ba \qquad \text{by (vi), (i)}$$

This completes the proof of Theorem 5.

**Theorem 6.**    *Let L be a Boolean ring.    Let the lattice operations be defined thus:*

(i)                         $a \vee b = a + b + ab$

(ii)                        $a \wedge b = ab$

(iii)                       $a - b = a + ab$

*Let 0 be the ring 0.    With respect to this 0 and to these operations, L is a classical subtractive lattice.*

*Proof.*    In accordance with Theorem B5, we show that the properties L2 to L4 hold.    So far as the left halves of these postulates are concerned, L2 is true by (iv) of Theorem 5, L3 by R5, and L4 thus:

$$a \wedge (a \vee b) = a(a + b + ab) = aa + ab + ab = aa = a$$

On the right we have L2 thus:

$$\begin{aligned} a \vee b &= a + b + ab \\ &= b + a + ba \qquad \text{by Theorem 5} \\ &= b \vee a \end{aligned}$$

As for L3, we have

$$\begin{aligned} a \vee (b \vee c) &= a \vee (b + c + bc) \\ &= a + b + c + bc + ab + ac + abc \end{aligned}$$

Since this is symmetric and L2 holds,

$$a \vee (b \vee c) = c \vee (a \vee b) = (a \vee b) \vee c$$

Finally, for L4, we have

$$a \vee (a \wedge b) = a \vee ab = a + ab + ab = a$$

Thus $L$ is a lattice.

Let us now examine the properties of $a - b$.    Since

$$\begin{aligned} abc \vee a(b - c) &= abc \vee a(b + bc) & \text{by (iii)} \\ &= abc \vee (ab + abc) & \text{by R6, R5} \\ &= abc + ab + abc + abc + abc & \text{by (i), R6, Theorem 5} \\ &= ab & \text{by (vi) of Theorem 5} \end{aligned}$$

we have (8).    We also have (6) since

$$(a - b)a = a(a - b) = a(a + ab) = a + ab = a - b$$

Finally, we have (3) since

$$b(a - b) = b(a + ab) = ab + ab = 0$$

and 0 is a lattice 0 by (iii) and (iv) of Theorem 5.    In view of Theorem 2, this completes the proof.

Note that we have shown that, in either Theorem 4 or 6, the lattice 0 and the ring 0 coincide.

The following theorem shows, among other things, that the transformations of Theorems 4 and 6 are inverse to one another, in the sense that if (i), (ii), and (iii) of either theorem are taken as definitions, those of the other theorem follow as theorem schemes.

**Theorem 7.** *Let L be a Boolean ring which is also a lattice, with ring multiplication as lattice meet and ring 0 as lattice 0. Then L is a classical subtractive lattice and*

(i) $$a \vee b = a + b + ab$$

(ii) $$a - b = a + ab$$

(iii) $$(a - b) \vee (b - a) = a + b$$

*Proof of* (i).  Let $c \equiv a + b + ab$.  Then

$$ac = a + ab + ab = a$$

Hence $a \leq c$, and similarly, $b \leq c$.  It follows that

$$a \vee b \leq c$$

Conversely,

$$(a \vee b)c = (a \vee b)a + (a \vee b)b + (a \vee b)ab$$
$$= a + b + ab = c$$

Thus $c \leq a \vee b$, and hence (i) holds.

*Proof of* (ii).  If $a - b$ is defined by (ii), then, since (i) of Theorem 6 has just been shown, and (ii) of Theorem 6 holds by hypothesis, $L$ is a classical subtractive lattice by Theorem 6.

*Proof of* (iii).  Using (ii), we have

$$(a - b) \vee (b - a) = (a + ab) \vee (b + ab)$$
$$= a + ab + b + ab + (a + ab)(b + ab)$$
$$= a + b + ab + ab + ab + ab$$
$$= a + b, \text{ Q.E.D.}$$

This completes the proof of Theorem 7.

By virtue of these theorems a classical subtractive lattice is a Boolean ring, and vice versa.  In a Boolean ring we calculate, as with ordinary polynomials, modulo 2, with idempotent multiplication.

**3. Classical implicative lattices.**  We close with a few remarks on the duals of the above theorems.  The duals of the properties (3) to (6) of Sec. 1 are, respectively, (1), (2),

$$(a \supset b) \supset b \leq a \vee b \tag{17}$$

and (i) of Theorem C1.  A theorem dual to Theorem 1 evidently holds, but it will not be stated here.  The property (2) is a form of "Peirce's law."

The dual of the construction in Sec. 2 is of some interest.[1] If we define

$$a \smallfrown b \equiv (a \supset b) \land (b \supset a)$$
$$a \lor' b \equiv a \lor b$$
$$a^* \equiv a$$

then we have a Boolean ring with $\smallfrown$ as addition and $\lor'$ (or $\lor$) as multiplication. The operation $\smallfrown$ corresponds to equivalence in the propositional interpretation; in the class interpretation, $a \smallfrown b$ consists of those elements which are in both $a$ and $b$ or in neither.

Propositional calculus based on $\smallfrown$ as a primitive operation has interested a certain school of logicians, notably Leśniewski, and the Rumanians (e.g., Mihailescu). The interest probably stems from the attitude of Leśniewski in regard to definitions. The dual Boolean ring shows the following properties of obs constructed with $\smallfrown$ alone:

1. In such an ob the terms can be permuted and associated arbitrarily.
2. A necessary and sufficient condition that $a = 1$ hold is that every atom appear an even number of times.

### EXERCISES

1. Show that in a Boolean ring

$$(a \lor b) + ab = a + b$$

(Birkhoff [LTh$_2$, p. 158, exercise 3].)

2. Show that if a distributive lattice with 0 has the property that for all $a$ and all $b \leq a$ there is a $c$ such that $bc = 0$ and $b \lor c = a$, then the lattice is a classical subtractive lattice (with suitable definition of $a - b$). (Birkhoff [LTh$_2$, p. 155, exercise 3].)

3. Exhibit a nondistributive lattice such that $(-)_1$, (3), and (6) hold. (A finite projective plane geometry with seven points will do.)

4. In the proof of Theorem 6, show directly that $(-)_1$ and $(-)_2$ hold, without involving Theorem 2.

### S. SUPPLEMENTARY TOPICS

**1. Historical and bibliographical comment.** This chapter is a revision of [LLA], chap. 3 and part of chap. 4.

For the chapter as a whole, the standard reference is Birkhoff [LTh$_2$]. The encyclopedic character and condensed style of this work make it unsuitable for introductory purposes, but it has an immense amount of information, including copious references.

The algebra of logic, properly speaking, began with Boole in 1847, although there were some sporadic preliminary studies by Leibniz and even earlier. Extensive improvements were made by W. S. Jevons, C. S. Peirce, and E. Schröder. For brief accounts of this early history see Lewis and Langford [SLg, chap. 1] and Whitehead [UAl, pp. 115ff.]; for more details see Lewis [SSL, chap. 1] and Jørgensen [TFL, chap. 3]. These works and those in Sec. 1S3 should be consulted for citations of the early work.

---

[1] These remarks are due to Feys. (See [LLA], p. 95.)

The nineteenth-century form of this algebra reached a certain completion in Schröder's [VAL]. In this treatment the positive operations were introduced first, so that practically all the present Sec. B is contained there (and indeed in Peirce [ALg$_1$]). Schröder's [VAL], however, is extremely long-winded. Couturat [ALg] is an admirable condensation and forms, in my opinion, the best source of information about the "Boole-Schröder algebra." It is still one of the best introductions to the subject matter of Sec. B.

After Schröder, the main logical interest shifted away from the algebraic approach. Improvements in the Boole-Schröder algebra continued to be made, e.g., in Whitehead [UAl] and Huntington [SIP]. But these developments mostly involved considerations of negation and belong properly in Chap. 6. We shall therefore pass over them and turn attention to lattices of more general character.

The first such algebras appear to have been discovered by Schröder himself. The question of the independence of the distributive law interested him, and in appendices (*Anhänge*) 4 and 5 of his [VAL], he produced what are probably the first examples of nondistributive lattices. Other examples of lattices in mathematics were found, notably by Dedekind, Noether, Skolem, Fritz Klein, Bennett, Menger, and Birkhoff (for references, see Birkhoff [LTh, at beginning of chap. 2], to which should be added Skolem [VLt], which gives a summary of the Norwegian paper of 1913 cited by Birkhoff). From these the general conception of a lattice, as given in Secs. A and B, evolved.

As general references on the theory of Secs. A and B, see, besides Birkhoff [LTh], also Glivenko [TGS], Dubreil-Jacotin *et al.* [LTT], Hermes [EVT], and the older encyclopedia report, Hermes and Köthe [TVr]. These consider matters which go far beyond the scope of this book; the shortest and easiest of them, Glivenko [TGS], is more than ample for our purposes here. For a very brief and elementary treatment see, for example, Bennett [SSO].

So much for Secs. A and B in general. There follow a few comments concerning special topics considered in these two sections.

For examples of lattices, in addition to those in Sec. A2, see Birkhoff [LTh, beginning of chap. 2], Bennett [SSO], and Glivenko [TGS, sec. 3].

The following circumstance suggested that it was worthwhile to state and prove Theorem B1. In Whitehead and Russell [PMt] the theory of deduction is presented as an assertional system with primitive operations denoted by the unary prefix '$\neg$' (actually '$\sim$') and the binary infix '$\lor$'; the binary infix '$\supset$' is defined thus:

$$p \supset q \equiv \neg p \lor q$$

As axiom schemes (really axioms with a tacit substitution rule) the authors give

| Taut | $\vdash (p \lor p) \supset p$ |
|------|-------------------------------|
| Simp | $\vdash p \supset (p \lor q)$ |
| Perm | $\vdash (p \lor q) \supset (q \lor p)$ |
| Assoc | $\vdash (p \lor (q \lor r)) \supset (q \lor (p \lor r))$ |
| Sum | $\vdash (q \supset r) \supset ((p \lor q) \supset (p \lor r))$ |

and the rule (modus ponens)

$$p, p \supset q \vdash p$$

Now suppose we introduce the definition schemes

$$p \leq q \rightarrow \vdash q \supset p \qquad p \wedge q \equiv p \vee q$$

Then Taut, Simp, Perm are, respectively, $\Lambda$W, $\Lambda$K, $\Lambda$C, whereas $\Lambda$B follows easily from Sum. Moreover, from Sum, putting $\neg\, p$ for $p$, we have $(\tau)$. Thus, if we omit Assoc, we have a semilattice by Theorem B1. By Theorem B2, Assoc is redundant. Now there is nothing in the proof of Theorems B1 and B2 which is not already implicit in Peirce [ALg$_1$]. Nevertheless, the authors of Whitehead and Russell [PMt], who supposedly based their system on a combination of the Boole-Schröder algebra and the logistic method of Frege and Peano, overlooked this fact, nor did anyone notice it until Bernays [AUA] (cf. [NAL]).

The postulates L1 to L4 in Theorem B5 are those preferred by Birkhoff [LTh].

Theorem B7 is part of Peirce's proof of the distributive law in Huntington [SIP]. Actually, (10) of Sec. B is a lemma there, and (8) of Sec. B is derived from it essentially as here. Since that time this theorem has been rediscovered by a number of other persons. Birkhoff ([LTh$_1$, p. 74, footnote]; cf. [LTh$_2$], p. 134, exercise 3) credits it to an unpublished work of J. Bowden dated 1936; Dubreil-Jacotin *et al.* ([LTT, p. 75]) credit it to M. Molinaro.

At this point it is pertinent to interpolate a historical remark. In 1880, in his paper [ALg$_1$], which deals with "syllogistic logic," Peirce stated that the distributive laws could be "easily proved by (4) and (2), but the proof is too tedious to give" (see Peirce [CPC, sec. 3.200]). Here (4) turns out to be $\Lambda$K, $\Lambda$K', VK, and VK', and (2) to be $\Lambda$S and VS. On this account Birkhoff ([LTh$_2$, p. 133] or [LTh$_1$, p. 74]) states that Peirce "thought that every lattice was distributive." Closer examination shows that Peirce's "syllogistic logic" contained principles of negation which were apparently intended to be used along with "(4) and (2)." When Peirce's original proof was finally published in Huntington [SIP], it turned out to depend on Huntington's postulates 8 and 9; the proof shows, in effect, that any lattice with 0 and 1 is distributive if complements (see Birkhoff, [LTh$_2$, p. 23]) exist such that, if $c$ is any complement of $b$,

$$ab = 0 \rightarrow a \leq c$$

This is quite correct (see Birkhoff [LTh$_2$, p. 171, exercise 6]). Peirce seems to have been in error on two counts: first in that he was misleading about the assumptions other than this "(4) and (2)," and second in that, when he learned of Schröder's independence proof, he put a footnote in his [ALg$_2$] (see his [CPC, sec. 3.384]) saying that Schröder had proved that it was impossible to derive the distributive law in syllogistic logic, whereas Schröder had done nothing of the kind. These facts hardly warrant the conclusion that Peirce thought that every lattice was distributive. Indeed, it is more likely that he did not, as Schröder did, formulate the idea of a lattice at all.

Theorem B9 has been taken from Lorenzen [ALU]. Dubreil-Jacotin *et al.* give Corollary 9.1 as exercise to their chap. 6 and credit it (on p. 75) to M. Molinaro.

For the logic of quantum mechanics see Birkhoff and Von Neumann [LQM]. This has never been followed up. For other references on the same subject, see Fraenkel and Bar-Hillel [FST, p. 230, note 3] and the papers by Rubin, Jordan, Fevrier, and Destouches in Henkin *et al.* [AMS]. The notions of implicative and subtractive lattice are clearly formulated in Skolem [UAK]. I do not know of any earlier formulation. (Inverse operations of subtraction and division were admitted by Boole; Peirce and Schröder later dropped these out as being superfluous. Whether there is any connection of Skolem's operations with these earlier ones I do not know.) When I wrote [LLA], which forms the basis for the present treatment, I was not aware of this work of Skolem, and consequently the only part of the present treatment which is taken directly from Skolem is the distributivity proof in Theorem C3. As stated in [LLA], p. 96, most of the theorems of Sec. C were suggested by Tarski [GZS]. The topological applications have been explored largely by Tarski, Stone, and their coworkers. Theorem C2, in particular, is the dual of McKinsey and Tarski [CEC, Theorem 1.4]. For further information see Birkhoff [LTh$_2$, pp. 147ff., 153ff., 176ff., 195, 204]; McKinsey and Tarski [CEC], [ATp], [TSC]; Stone [RBA]; and works there cited. Such lattices are special cases of "residuated lattices" due to Ward and Dilworth; for citations see Birkhoff [LTh$_2$, pp. 201ff.]. For the terminology in connection with these lattices and some further details, see Sec. 2.

The theory of Boolean rings is due to Stone. The earlier part of Stone [TRB], which contains most of the theorems presented here, is quite readable; see also Stone [RBA]. For complete citations see Birkhoff [LTh$_2$, pp. 153ff.]. Boolean algebras, which are Boolean rings with unit, will not concern us until Chap. 6. Note that Stone's $a + b$ also appeared in Boole, whereas $a \vee b$ was introduced later by Jevons in 1864 and (apparently independently) by Peirce in 1867. There is, however, an important difference. In Boole $a + b$ was regarded as uninterpretable unless $ab = 0$, whereas in Stone $a + b$ is always interpretable; in Boole $a + a$ is meaningless, whereas in Stone $a + a = 0$. There were some anticipations of Stone's idea, notably by Zhegalkin (transliterated into French as "Gegalkine") [ASL]; these are cited in Birkhoff [*loc. cit.*] and Stone [RBA]. See also Herbrand [RTD, chap. 1, sec. 6].

For references on the work on equivalence see Church [IML$_2$, p. 159]. The dual form of Boolean ring is mentioned explicitly in Stone [NFL].

**2. Notes on terminology.** The term 'lattice' was introduced in Birkhoff [CSA] and has taken hold in spite of conflict with a previous use of the same term (which, I understand, is due to Minkowski). The term '*Verband*', used by F. Klein and introduced, according to Hermes [EVT, p. 1], by Dedekind, has become the standard term in German. The accepted term in French now seems to be '*treillis*'; when [LLA] was written this term had not become established.

The term 'semilattice' appeared in Birkhoff [LTh$_2$, pp. 18, 25] after F, Klein's '*Halbverband*'; the term '*groupe logique*', used in [LLA], is from Moisil [RPI].

The names 'meet' and 'join' for the lattice operations are those of Birkhoff. Other terms frequently used instead of 'meet' are 'product', 'intersection' ('*Durchschnitt*'); likewise one finds 'sum' or 'union' ('*Vereinigung*') instead of 'join'. For the operational infixes Birkhoff [LTh] uses '∩' and '∪', and Dubreil-Jacotin *et al.* and Hermes follow him in this; other authors (e.g.,

Glivenko, Schröder) prefer the notation of ordinary algebra, with some form of multiplication infix, or even simple juxtaposition, for meets and '+' for joins—this last, of course, conflicts with the use of '+' for ring addition, for which an infix 'Δ' (symmetric difference) is sometimes used. Still other authors (Ward, Dilworth, Ore, Menger) prefer to use parentheses and brackets, for example, '(a,b)', '[a,b]', and in such cases, on account of duality, it is often not clear which operation is which. For the infix '≤' some authors prefer '⊂', but there is very general agreement on '='. The terms 'cap' and 'cup' were suggested by H. Whitney (see note in Birkhoff [LTh, beginning of chap. 2] as readings for '∩' and '∪'; Dubreil-Jacotin *et al.* suggest reading '∧' as 'inter'.

The infix '⊃' was originally an inverted 'C'. As such it appears in Peano [APN]. Gergonne (see quotation in Bochenski [FLg, sec. 40.12]) had previously used it for a proper inclusion relation between classes. For further details on the history see Church [BSL, subject index, under 'Sign ⊃'].

The term 'Skolem lattice' for implicative and subtractive lattices together is suggested here for the first time. The historical aptness of this term should be clear from what was said in Sec. 1, but it has other advantages. I shall discuss here its relations with 'Brouwer lattice', 'Heyting lattice', 'relatively pseudocomplemented lattice', and 'residuated lattice', all of which have been used in similar connections elsewhere.

The term 'Brouwer algebra' was used in McKinsey and Tarski [CEC] for what is here called an absolute subtractive lattice. The reasons for attaching Brouwer's name to these lattices seem to have been as follows. On the grounds of his intuitionistic philosophy of mathematics, Brouwer denied the validity of certain logical principles, e.g., the law of excluded middle. Up to 1930 only fragmentary treatments of logic which were supposedly more in agreement with this intuitionistic philosophy had appeared (see, for example, Brouwer [OLP], [IZM] 1925; Glivenko [LBr], [PLB]; possibly Kolmogorov [PTN]; cf. Heyting [FMI]). In that year Heyting [FRI] presented a formalization of a propositional algebra and predicate calculus for such a logic. This was an assertional system of a sort which we shall consider in Chap. 5. By methods which we shall learn in that chapter it can be put into a lattice form; this was done, for example, in Ogasawara [RIL]. The lattices considered in Tarski [GZS] are precisely of that kind. As we saw in Sec. C5, the open sets in a topological space form such a lattice. But for topological purposes it is more natural to deal with closed sets and closures than with open sets and interiors. On this account the lattices obtained from the Heyting system were dualized. The use of the term 'Brouwer logic' in Birkhoff [LTh] is similar, the dualization being apparently motivated by the desire to make 0 represent truth. If we accept these writings as fixing the meaning of 'Brouwer lattice', then that term has two characteristics: (1) it is related to an intuitionistic origin which is definitely nonclassical, and (2) it arises in a nonsymmetric situation where the subtractive form is preferred to the implicative.

If 'Brouwer lattice' is to be used in the way just described, it would be appropriate to call an absolute implicative lattice, which is obtained from Heyting's system without dualizing, a Heyting lattice. This term will then have the connotation that the lattice is nonclassical and implicative.

Monteiro [AIA] uses 'Brouwer algebra' in this sense, but later, in his [AHM], prefers 'Heyting algebra'.

As the lattices were introduced in Skolem [UAK], the inverse operations were brought in on formal grounds alone, without any commitment to an intuitionistic or related interpretation. Moreover, the two kinds were introduced simultaneously, in parallel columns. On this account the term 'Skolem lattice' seems well adapted as a term for all the kinds of lattices of Secs. C and D. Skolem lattices thus cross-classify in two ways: on the one hand, into implicative or subtractive (or both); on the other hand, into absolute and classical (with possible intermediate kinds). But to speak of classical Brouwer lattices, or even classical Heyting lattices, would seem rather odd.

Relatively pseudocomplemented lattices are, in Birkhoff's terminology (see his [LTh₂]), the same as absolute implicative lattices. The term seems unnecessarily cumbrous. Moreover, it applies to that particular kind of lattice only.

Residuated lattices, in the sense of Ward and Dilworth (see the citations in Sec. 1), are obtained by adjoining "residuation" operations to "lattice-ordered semigroups," i.e., lattices with a third "multiplication" operation having certain properties. Skolem lattices arise when the third operation is identified with the appropriate lattice operation. Since Skolem lattices are always distributive, and residuated lattices, as lattices, do not even have to be modular, it would be confusing to use the term 'residuated lattice' in the place of 'Skolem lattice'.

**3. Further developments.** It would take us too far afield to mention all the ways in which the theory of lattices has been generalized, specialized, and other-ized in modern mathematics. The extent of the impact of the subject on other branches of mathematics can be glimpsed by looking at the references already cited. Here I shall mention a few further branches which have logical interest. I shall give few references; where they are given I have included the latest publication known to me, as well as the one which seems most comprehensive.

Some study has been made of partially ordered sets in general; also of closure spaces, i.e., spaces with a closure satisfying properties similar to I to III of Sec. 2B3, of semigroups and related algebraic systems, etc. The concerns of quantum mechanics (see Sec. 1) have led Jordan to study lattices which are not commutative; see his [TSV]. Choudbury [BNr] has studied rings in which the associative law may fail. Algebras related to multiple-valued logics in much the same way that Boolean algebra is related to classical logic have been called Post algebras or Moolean algebras; for them see, for example, Rosenbloom [PAI] (also references in his [EML]) and Chang [AAM]. R. L. Foster's 'ring logics' go off at a somewhat different angle in the same general direction.

There are also algebraic systems which superpose on a lattice some additional notions. Among these are the lattice-ordered groups and semigroups. The oldest algebra of this kind is the algebra of relations. This was developed largely by Peirce (the ideas are said to go back to DeMorgan); the third volume of Schröder [VAL] is devoted to it. For a comprehensive recent report see Chin and Tarski [DML]. Boolean algebras with extra operations are considered from a general point of view in Jonsson and Tarski [BAO]. Probability considerations have led A. H. Copeland, Sr., to introduce implication operations of a different sort from those considered here; see, for example, his [IBA], [POP]. For algebras involving modal operations see Chap. 8; for those related to quantification see Chap. 7.

Finally, the application of higher epitheoretic methods to algebras of the sort considered in this chapter leads to interesting results. In particular, the Stone representation theorem (Stone [RBA]), to the effect that any Boolean algebra can be represented as a field of sets, has led to many similar studies for other types of algebra. Such theorems are beyond the scope of this book.

# Chapter 5

# THE THEORY OF IMPLICATION

At this point we make a transition to systems which are assertional in character. This form of system is particularly suitable when the obs are taken as propositions (i.e., to be interpreted as statements). We shall be concerned in this chapter with the simpler forms of such assertional algebras, which involve only the positive operations of Chap. 4. In these systems the principal operation is the ply operation; although the other operations, analogous to join and meet, also play a role, that role is rather minor. On that account this chapter is called the theory of implication. The theory of implicative lattices, considered in Secs. 4C and 4D3, belongs under that heading; however, it is convenient to make the break at the point where we transfer to the assertional form.

We shall start, in Sec. A, with discussion of certain preliminary matters, viz., the nature of the propositional interpretation and of propositions, and the relationship between relational systems, in which the basic predicate is a quasi ordering or an equality whose interpretants are, respectively, relations of implication or equivalence, and the corresponding assertional systems. We shall also consider the possibility of epitheoretic interpretation, in which the propositions we are formalizing are themselves interpretable as elementary statements, or simple forms of epistatements, relating to some more fundamental formal system or theory, so that our systems formalize part of the epitheory of that system. The interpretation of the operations on that basis will lead to the formulation of certain inferential rules, here called T rules, which are similar to the "natural" rules introduced by Gentzen and called by him "N rules."

In Sec. B we shall embark on the formal development of the simplest types of assertional system. There will be two forms of such system, a T form, which is generated by the T rules directly, and a properly assertional form, or H form, in which the only rule (modus ponens) is

$$A, A \supset B \vdash B$$

In contrast to these the lattice form of Chap. 4 will be called the E form. In each form there will be two kinds, called, as in Chap. 4, the absolute and classical kinds. The absolute system of the T form, called henceforth the TA system, will be that generated by the T rules of Sec. A3 alone; it and the corresponding HA system will turn out to be equivalent to an absolute implicative lattice, or EA system.   Systems TC and HC, analogous to the classical implicative lattice, or EC system, will require some further assumptions, just as was the case in Sec. 4D.

The later sections of this chapter will deal with a fourth, or L form, corresponding to the L systems of Gentzen.   This involves passing to a higher stage of formalization.   The elementary statements of the L systems are abstracted from certain epistatements of demonstrability by means of proof trees in a T system.   These L systems give the most profound analysis of implication.   The principal theorem (Gentzen's "*Hauptsatz*") concerning these systems, here called the "elimination theorem," is one of the major theorems of modern mathematical logic.

Throughout most of this chapter we shall be concerned with processes which not only apply to the situation in this chapter, where we have only the finite positive operations, but can be carried out with only minor variations in the succeeding chapters.   Consequently, there has been an attempt to formulate these processes—particularly in the later sections—so that they are capable of application under more general circumstances than those which occur here.

In this chapter capital italic letters will be used for the obs.   This contrasts with the usage in Chap. 4, where small italic letters were used for that purpose in order to conform to standard practice in algebra.   But when we reach Chap. 7, it will be convenient to use capital letters for propositions and small letters for individuals, and it is more advantageous to make the necessary change at this point than it will be then.

## A. GENERAL PRINCIPLES OF ASSERTIONAL LOGICAL ALGEBRA

In this section we discuss principles relating to assertional logical algebras in general.   We begin with more formal considerations, such as the interrelations of relational and assertional systems, and then treat matters of interpretation, such as the nature of propositions and that of the operations connecting them.

**1. Relational and assertional logical algebras.**   In Sec. 2D1 we saw that any formal system could be reduced to an assertional system.   We shall now study this situation more in detail for the case where the original system is a relational algebra of the sort considered in Chap. 4.

Suppose that we have a quasi-ordered system $\Re$ with basic relation $\leq$. All the elementary theorems of $\Re$ are thus of the form

$$A \leq B \tag{1}$$

Let $\mathfrak{S}$ be the assertional system, formed as in Sec. 2D1, whose predicate is indicated by a prefixed '$\vdash$'. Let the infix '$\supset$' indicate the operation which replaces the predicate $\leq$. If $\mathfrak{S}$ is so formed, then its elementary theorems will all be of the form

$$\vdash A \supset B \tag{2}$$

and the equivalence

$$A \leq B \rightleftarrows \vdash A \supset B \tag{3}$$

expresses the fact that (2) holds in $\mathfrak{S}$ when and only when (1) holds in $\mathfrak{R}$. Moreover, the following will be the translation in $\mathfrak{S}$ of the $(\rho)$ and $(\tau)$ in $\mathfrak{R}$:

$$\vdash A \supset A \tag{4}$$

$$A \supset B, B \supset C \vdash A \supset C \tag{5}$$

In the foregoing we have supposed that $\mathfrak{R}$ was given and that $\mathfrak{S}$ was constructed from it. But it is evident that if an assertional system $\mathfrak{S}$ with a ply operation is given such that all the elementary statements of $\mathfrak{S}$ are of the form (2), then we can construe (3) as definition of a relation $\leq$, and this relation will indeed be a quasi ordering if (4) and (5) hold. Thus $\mathfrak{S}$ can be transformed into a relational system $\mathfrak{R}$. Furthermore, the two transformations—from $\mathfrak{R}$ into $\mathfrak{S}$ and from $\mathfrak{S}$ into $\mathfrak{R}$—are reciprocal to one another in the sense that each exactly undoes the work of the other.

None of this is disturbed in any way if the operation $\supset$ already occurs in $\mathfrak{R}$, or in other words, if such an operation already occurs in the construction of the $A$ and $B$ in (1) and (2). For the instance of ' $\supset$ ' occurring in the middle of (2) is the principal instance, in the sense that it indicates the operation forming the bottom node in the construction of $A \supset B$. The transformations in the preceding argument are still uniquely defined.

The situation is different, however, if we start with an assertional system $\mathfrak{S}$ in which not all elementary statements are of the form (2). In that case we proceed as follows. Suppose there is an ob 1 of $\mathfrak{S}$ such that

$$A \vdash 1 \supset A \ \& \ 1 \supset A \vdash A \tag{6}$$

Let $\mathfrak{S}_1$ be the elementary theorems of $\mathfrak{S}$ which are of the form (2). Then there will be a reciprocal relation, defined as before by (3), between $\mathfrak{S}_1$ and a relational system $\mathfrak{R}$, and every elementary theorem of $\mathfrak{S}$ will be equideducible with one of $\mathfrak{S}_1$. Moreover, if (3) holds between $\mathfrak{R}$ and $\mathfrak{S}_1$, we shall have

$$\vdash A \rightleftarrows 1 \leq A \tag{7}$$

in the sense that the left side is an elementary theorem of $\mathfrak{S}$ if and only if the right side is an elementary theorem of $\mathfrak{R}$. This establishes a correspondence between $\mathfrak{S}$ and that part $\mathfrak{R}_1$ of $\mathfrak{R}$ consisting of statements (1) for which $A \equiv 1$.

A sufficient condition for the existence of such a 1 is that

$$A \vdash B \supset A \tag{8}$$

$$A, A \supset B \vdash B \tag{9}$$

be valid rules of $\mathfrak{S}$ and that 1 be an ob [for instance, if (4) holds, the ob $E_1 \supset E_1$] such that

$$\vdash 1 \tag{10}$$

For the left half of (6) follows from (8), and the right half from (9) and (10). Then, in the system $\mathfrak{R}$, we have

$$B \leq 1 \qquad \text{from (10), (8)} \tag{11}$$

$$A \leq B \rightleftarrows 1 \leq A \supset B \qquad \text{by (6), (3)} \tag{12}$$

Suppose now that $\mathfrak{R}$ is a quasi-ordered system with an ob 1 and an operation $\supset$ such that (11) and (12) hold. Adopt (7) as definition of $\vdash$. Then

(10) holds immediately by $(\rho)$. Likewise, we have (3) by (12) and (7). From (3) we have (4) and (5) as before. We can derive (8) as follows:

$$\vdash A \to 1 \leq A \qquad\qquad \text{by (7)}$$
$$\to B \leq 1 \,\&\, 1 \leq A \qquad \text{by (11)}$$
$$\to B \leq A \qquad\qquad\quad \text{by } (\tau)$$
$$\to \vdash B \supset A \qquad\qquad \text{by (3)}$$

Further, we have

$$\vdash A \,\&\, \vdash A \supset B \to 1 \leq A \,\&\, A \leq B \qquad \text{by (7), (3)}$$
$$\to 1 \leq B \qquad\qquad\qquad\qquad \text{by } (\tau)$$
$$\to \vdash B \qquad\qquad\qquad\qquad\quad \text{by (7)}$$

so that (9) holds. From this we have (6) as before.

The foregoing argument establishes that the notions of quasi-ordered system (with 1 and $\supset$) satisfying (11) and (12) and of assertional system (also with 1 and $\supset$) satisfying (4), (5), (8), (9), (10) are equivalent in the sense that, given a system of either type, there is an associated system of the other type, such that either system can be validly interpreted in the other. There are no objective grounds for regarding either type of system as necessarily prior to the other.

It follows that any preference we may have for one type of system over the other must be based on grounds such as utility, naturalness, suggestibility, or the like. But from such points of view both types of system have their advantages. No doubt a unary predicate is a simpler notion than a binary one, and on that account the assertional type has advantages from a foundational standpoint. Also, it is more natural from the standpoint of the propositional interpretation, to be discussed presently at more length. On the other hand, the similarity of the relational approach to that of ordinary mathematics is suggestive. The fullest development of our science is probably best to be achieved by considering both points of view. Furthermore, the interpretation of a relational system in an assertional one seems more natural than that of an assertional one in a relational one. In other words, it is easier to apply the theorems of the relational system in developing the assertional one than the reverse. On this account the relational approach was taken up first in Chap. 4, the assertional approach being postponed to the present chapter.

**2. The propositional interpretation.** From now on until further notice the obs of the various systems being studied will be called *propositions*. This agrees with the tradition according to which we use terms which suggest the intended application—so that our formal discussions may not be an unintelligible jargon—without being so closely tied to it as to contaminate our formal reasoning with contensive considerations which obscure the formal structure. In order to see how this terminology contributes to this objective, a little preliminary discussion will be necessary.

The term 'proposition' is a controversial subject in modern mathematical logic. Some logicians eschew it like poison; they insist on replacing it, in all contexts where it had been used as a matter of course, by the word 'sentence'; others insist on using it, ostensibly on the ground that we need to postulate entities which it can properly denote. The usage here adopted is neutral in regard to this metaphysical controversy.   A proposition is simply an ob; as the term suggests, a particular kind of interpretation is intended for it, but no commitment as to the metaphysical nature of that interpretation is made. Those who object to the term 'proposition' do so, in the main, for two types of reason.   On the one hand, it has metaphysical connotations which they desire to avoid; on the other hand, it is a vague term, whereas 'sentence' is relatively precise.   It will be necessary to consider these objections separately.

The first objection is evidently a matter of one's personal philosophy. Philosophers do use the term in rather mysterious ways.   I admit that I do not understand what their discussion is about.   But it is clear that the discussion has nothing to do with formal structure.   Therefore, so far as that structure is concerned, the term 'proposition' as used there is literally meaningless, and we are free to use it in any way we like.   Mathematicians can and do use, as technical terms, words which are commonly used for other purposes in unrelated contexts.   The use of the term 'proposition' in a mathematical context does not commit one, unless he so chooses, to postulating mysterious entities of an esoteric sort.

In regard to the vagueness, it is pertinent to remark that the term 'sentence' is vague too.   Consider, for example, the following lines of type (not including the numbers written at the extreme right):

$$2 + 3 = 5 \tag{13}$$

$$\text{The sum of 2 and 3 is 5.} \tag{14}$$

$$\text{The sum of 2 and 3 is 5.} \tag{15}$$

$$\text{The sum of 2 and 3 is 5.} \tag{16}$$

$$\textit{The sum of 2 and 3 is 5.} \tag{17}$$

$$\textit{Die Summe von 2 und 3 ist 5.} \tag{18}$$

$$\mathfrak{Die\ Summe\ von\ 2\ und\ 3\ ist\ 5.} \tag{19}$$

$$3 + 2 = 5 \tag{20}$$

$$2 + 3 = 6 \tag{21}$$

How many sentences are there?   Certain logicians would say there were $9n$, where $n$ is the number of extant copies of this book; a typographer would doubtless say eight, identifying (14) and (15); a grammarian would probably go further and identify (14), (15), (16), (17) and also (18) and (19); while a German logician, understanding 'sentence' as a translation of '*Satz*', might identify all the first seven.   Prior to the advent of the logical usage, the term 'sentence' had primarily a grammatical connotation, but that usage has tended to change its meaning.

Because of this ambiguity in the current use of 'sentence', it seems expedient to introduce five related terms, viz., 'inscription', 'sentence', 'statement', 'proposition', and 'clause'.   It will not be possible, at least here, to make the distinction between these perfectly precise, but an approximate explanation can be made as follows.

The term '*inscription*' was introduced in Sec. 2A2 to designate a single concrete instance of a linguistic expression.[1]   Thus in the above table there

---

[1] The term 'expression' is used throughout the present context rather loosely.   It would be more correct, according to Sec. 2A3, to use 'phrase'.

are nine inscriptions in this particular copy of this book; if there are $n$ copies of the book extant, then there are $9n$ inscriptions in the corresponding spaces of all the extant copies collectively. (The term is still a little vague, because a part of an inscription is again an inscription; we might possibly speak of a sentential inscription as one which is an instance of a sentence, or is separated off by periods, etc., like the linguists' 'utterance'.)

The term 'sentence' is used in a grammatical sense. A sentence is an expression (or other unit) of some communicative language which performs a certain communicative function. Such a sentence is a class of inscriptions which may be considered the instances of the sentence; different instances of the same sentence are equiform. I cannot say precisely what the communicative function in question is, and I know of no really satisfactory definition; yet, given a communicative language, there seems to be little disagreement among the users of that language as to which of its expressions are sentences.[1] Thus all the expressions (13) to (17) and (20) and (21) are sentences of mathematical English; the expressions (13) and (18) to (21) are sentences of mathematical German. Likewise, there is some vagueness about the notion of equiformity. Certainly (14) and (15) are equiform, and probably one should consider them both equiform to (16), but whether they are equiform to (17) and whether (18) and (19) are equiform may depend on circumstances.

The sense of the term 'statement' may be vaguely described as the meaning of a sentence. One can associate with this description any metaphysical notion which one pleases, but if one wishes to be objective, one can say that a statement is a sentence considered without regard to certain linguistic features which do not affect meaning. Thus a statement is a class of sentences which are equisignificant. This relation of equisignificance is vague—even more so than equiformity—but this vagueness does not concern us because we do not have to count sentences—let alone statements—in logic. The term 'statement' thus merely indicates in a rough way the level of abstraction. Thus when we speak of "the statement (15)" we imply that what we have to say will apply equally well to any of (13) to (19) and possibly to (20), but when we say "the sentence (15)" we imply that we are including (14) and (16) and possibly (17). We make the same sort of distinctions in ordinary life when we speak of the same object as a fruit, an orange, or a valencia. Admittedly, there are cases where it is not clear which of the two terms is the more appropriate, and one may, if one prefers, identify 'statement' and 'sentence'.

A clause is a linguistic expression which names a statement or sentence as the case may be. All the natural languages—at least the more developed ones—have devices for doing this. None of the expressions (13) to (21) are clauses, but the following are examples in our U language:

$$\text{'The sum of 2 and 3 is 5'} \tag{22}$$

$$\text{That the sum of 2 and 3 is 5} \tag{23}$$

$$\text{The equality of } 2 + 3 \text{ and 5} \tag{24}$$

$$2 + 3 \,\square\, 5 \tag{25}$$

[1] There are certain exceptions. For instance, Chomsky maintains that 'sincerity admires John' is not a grammatical English sentence, whereas I should be inclined to the view that it is a perfectly grammatical sentence, although an absurd one.

Here the quotation marks are an essential part of (22), and the infix '$\square$' is an operator which replaces equality. All of (22) to (25) are clauses; it is natural to think of (22) as naming the sentence (15) and the others as naming the statement.[1]

A *proposition* is the meaning of a clause in the same sense that a statement is the meaning of a sentence; i.e., it is a class of equisignificant clauses. One may reify this notion in a variety of ways. One can indeed suppose that a proposition is a platonistic abstraction, but one does not need to do this. One can also say that a proposition is a sentence in an object language which is being talked about but not used, but one does not need to do that either. There is thus a certain latitude in what one considers a proposition to be. The important point is that a proposition is an object (or fiction) being talked about and that it is in some way related to a certain unique statement.

As stated, the obs of the systems we are about to develop will be called propositions. This means that we have in mind their interpretation as propositions in the sense just discussed. Using the term for the interpretation does no harm and has the advantage mentioned at the beginning of this article. There is some latitude in regard to propositions, and this latitude may be extended to include the possibility of other sorts of interpretations—where the obs are classes, numbers, closed sets, etc., as in Sec. 4A. On the other hand, if the term 'sentence' is used for the obs, certain cautions are necessary. Such a usage is not strictly compatible with the usage just described. For if the formal variables of the theory are interpreted as intuitive variables, they are variables for which nouns, not sentences, can be substituted. Hence if the obs are taken as sentences, they must be sentences which are named, not asserted; i.e., they must be sentences of an O language. The names of these sentences in the U language, not the sentences themselves, are to be substituted for the A nouns. In the present discussion the term 'sentence' is confined in principle to expressions of a language actually used for communication, and it is irrelevant for the formal theory whether the propositions are O sentences in that sense or not.

The use of the term 'proposition' has another advantage which will concern us later. In Sec. C we shall formalize certain types of epistatements, called *compound statements*, concerning a formal system $\mathfrak{S}$. We do this by constructing a formal system $\mathfrak{T}$ whose obs are interpreted as the compound statements of $\mathfrak{S}$. In such a case the term 'proposition' for these obs of $\mathfrak{T}$ is particularly appropriate.

Propositions thus differ from statements in that they are talked about as obs. Now sometimes we have to talk about statements: to say that they are true, that they are asserted, deducible, complex, etc. This is permissible usage according to the habit of the English language; it was already alluded to in Sec. 2A3. We shall adhere to this usage in informal discussions, reserving the term 'proposition' for cases where the procedure is in some degree formalized. Indeed, there is no point to making a distinction between proposition and statement at all except with reference to some formalization. In discussing interpretations, where both terms occur, we shall speak of either propositions or statements as being true, a proposition being

---

[1] The reduction to an assertional system is essentially a systematic transformation of certain A sentences into clauses.

true just when the statement associated with it is also true.  Indeed, we may, in such a context, identify propositions with the statements which are their interpretants.

**3. The interpretation of operations.**  In Sec. 2 we have discussed the nature of the propositional interpretation.  Let us now turn our attention to the interpretation of the operations denoted by the infixes '$\supset$', '$\wedge$', '$\vee$', respectively.  We shall attempt to find principles which will guide us in the choice of postulates to be made later.

For this purpose it is expedient to define a special kind of interpretation called a *normal interpretation*.  We have seen that a proposition is an object being talked about, which is nevertheless associated with a contensive statement in a definite way.  We shall say that an interpretation of a system $\mathfrak{S}$ is a *normal interpretation* just when the proposition $A$ is true when and only when $\vdash A$.  Since the latter statement is true when and only when it is demonstrable, this is a very restricted kind of interpretation.

The operation $\wedge$ is to be the propositional analogue of the conjunction connective, in that $A \wedge B$ is true just when $A$ and $B$ are both true.  Then in a normal interpretation this would mean

$$\vdash A \wedge B \rightleftarrows \vdash A \;\&\; \vdash B \tag{26}$$

and this could be obtained if we had rules

$$\begin{aligned} \vdash A \wedge B &\rightarrow \vdash A \\ \vdash A \wedge B &\rightarrow \vdash B \end{aligned} \tag{27}$$

$$\vdash A \;\&\; \vdash B \rightarrow \vdash A \wedge B \tag{28}$$

Notice that the rule (28) states the circumstances under which we can infer that $\vdash A \wedge B$, whereas (27) states consequences which can be drawn from it. We can call (28) the rule for introducing $\wedge$, while (27) can be called rules for eliminating it.  We shall call the rules (27) and (28) $\wedge$e and $\wedge$i, respectively.

The operation $\vee$ is to be a propositional analogue of the alternation connective.  Now this alternation connective is much less clear to us than conjunction.  However, it is easy to see what rules for introduction and elimination are suitable for it.  For introduction we have the rules

$$\begin{aligned} \vdash A &\rightarrow \vdash A \vee B \\ \vdash B &\rightarrow \vdash A \vee B \end{aligned} \tag{29}$$

For elimination we note that in order to infer $\vdash C$ from $\vdash A \vee B$, we must show that we can infer it both from $\vdash A$ and from $\vdash B$; this gives the rule

$$A \vdash C \;\&\; B \vdash C \;\&\; \vdash A \vee B \rightarrow \vdash C \tag{30}$$

The rules (29) and (30) we call $\vee$i and $\vee$e, respectively.

The operation $\supset$ is to be a propositional analogue of a conditional connective.  Its rule of introduction is to be that we conclude $\vdash A \supset B$ when we can infer $\vdash B$ from $\vdash A$.  If this inference is to be by a formal proof in whatever system we are formalizing, this gives the rule of introduction

$$A \vdash B \rightarrow \vdash A \supset B \tag{31}$$

For the rule of elimination we take the converse of (31), viz.,

$$\vdash A \supset B \rightarrow A \vdash B \tag{32}$$

This is a form of the rule of modus ponens, which is taken as primitive in most systems of propositional algebra.[1] It will be called Pe, whereas (31) will be called Pi. Note that in these rules the conditional connective represented is one of deducibility within a system. On this account the theory in which $\supset$ plays the principal role is appropriately called the *theory of formal deducibility*.

Lorenzen has shown that the rules of elimination are consequences of the rules of introduction in accordance with his "principle of inversion." The idea of his proof is as follows. Suppose that the operations $\supset$, $\wedge$, $\vee$ do not occur in the axioms of a system $\mathfrak{S}$, and the rules are such that none of them other than (28), (29), and (31) has a conclusion of the form $\vdash A \wedge B$, $\vdash A \vee B$, or $\vdash A \supset B$. Then we can derive (27), (30), and (32) as follows. Suppose we have a proof that $\vdash A \wedge B$, and let this proof be exhibited in tree form. Then the bottom node of this tree must be formed by an application of (28). The branches leading to the two nodes constituting the premises will furnish the desired proof that $\vdash A$ and that $\vdash B$, thus establishing (27). Again, assume that we have proofs that $\vdash A \supset B$ and $\vdash A$. Then, since the last step in the first proof must be an application of (31), we must have a proof that $A \vdash B$. From this proof we obtain a proof that $\vdash B$ by putting the proof of $\vdash A$ over $\vdash A$ wherever it occurs as a top node in the proof tree. Finally, suppose we have proofs of the premises of (30). The last step in the third proof must then be an application of (29). The premise must be either $\vdash A$ or $\vdash B$. Suppose it is the former. Let the proof of $A \vdash C$ be exhibited in tree form, and let the proof of $\vdash A$, in tree form, be placed over each occurrence of $\vdash A$ as a premise in the first proof. The result will be the desired proof of $\vdash C$. If the given proof is that of $\vdash B$, we proceed similarly with the second premise of (30).

The argument which we have just been through depends in a subtle way on the normality of the interpretation. To see this, consider the elementary statement

$$\vdash A \supset (B \supset A) \tag{33}$$

We shall see that this is valid in any normal interpretation. From the foregoing it follows that (33) is equivalent to $A \vdash B \supset A$, and this in turn to $A, B \vdash A$; since $A$ is an axiomatic proposition of the indicated extension, this is obviously true. Suppose, however, that the interpretant of $\vdash A$ is a statement to the effect that $A$ is true, while the statement $\vdash A$ itself is true only when it is deducible in a rather restricted system. In such a case (33) may not be valid. For example, let $A \supset B$ be the proposition that $A \vdash B$,[†] where $A$ is the proposition that New York is in North America and $B$ is the proposition that Calcutta is in Africa, and $\vdash$ refers to some system of logic in the ordinary sense; then $\vdash A$ is valid in the interpretation, $\vdash B \supset A$ is not, and hence (33) is invalid. This invalidity arises from the fact that no reasonable system of logic would allow us to derive $\vdash A$ from $\vdash B$.

---

[1] Note that if it is a primitive rule, it can be expressed as $A, A \supset B \vdash B$.

[†] In a normal interpretation this explanation did not have to be made.

**4. Auxiliary interpretations.** In Sec. 2C5 a direct interpretation was
defined as one obtained from a valuation by assigning to each formal predicate a contensive predicate defined over the values. Interpretations defined
in that way[1] in a more or less artificial manner are often an important tool in
the epitheoretical study of propositional algebras. We shall call these
interpretations *auxiliary interpretations*.

A special case of some importance is where the interpretation is such that
obs which are equal (in the sense of Sec. A1) have the same value. Such an
interpretation will be called a *regular interpretation*. The values in a regular
interpretation will be called *elements*. A system with a regular interpretation is an "algebra" in the sense of ordinary mathematics. The usage of
'elements' conforms with general mathematical practice. In case the elements are sets, there may be some confusion, since it may seem more natural
to call the members of these sets elements, rather than the sets themselves;
when such confusion seems dangerous, one can avoid it by calling the members of the elements "members" or "individuals," or by making some specific convention. The methods of modern algebra can evidently be applied
to systems with a regular interpretation.

In connection with regular interpretations it is permissible to use the same
symbol for both the ob and the element. Thus we can speak about "the ob
$A$" and "the element $A$," meaning by the latter the element associated with
the former.

A special case of a regular interpretation is that in which the element
associated with an ob $A$ is the set of all obs $B$ such that

$$A = B$$

The ordinary algebra so obtained is known, especially in the case where the
original system is assertional, as the *Lindenbaum algebra*[2] associated with
the system; from the present standpoint it is appropriate to refer to it as the
*Lindenbaum interpretation*.

In terms of regular interpretations the notion of homomorphism, and
related notions, can be explained. Let $S_1$ and $S_2$ be two given systems.
Then a *homomorphism* of $S_1$ into $S_2$ is a mapping associating to each element
(or ob) $A$ of $S_1$ an element $A^*$ of $S_2$ and to each operation $\omega$ or predicate $\phi$
of $S_1$ an operation $\omega^*$ or predicate $\phi^*$ of $S_2$ of the same degree, such that, for
all obs $A_1, \ldots, A_n$ of $S_1$,

$$\omega(A_1, \ldots, A_n)^* = \omega^*(A_1^*, \ldots, A_n^*)$$
$$\phi(A_1, \ldots, A_n) \to \phi^*(A_1^*, \ldots, A_n^*)$$

An *isomorphism* is a homomorphism which is one-to-one between elements
and works both ways; an *endomorphism* is a homomorphism of a system into
itself; and an *automorphism* is an endomorphism which is also an isomorphism.
These concepts can, at least in part, be defined between systems, but seem
most useful between systems with regular interpretations.

[1] Here it is intended to include not only direct interpretations in the narrow sense of
Sec. 2C5, but also those in which the basic predicate is interpreted as the possession of a
certain property for all admissible valuations. This is admitted explicitly under matrix
interpretations below.

[2] From Adolf Lindenbaum, a Polish logician, who perished in World War II.

Another special kind of interpretation, called a *matrix interpretation*, is defined as follows.[1] Let there be given a certain set α of values and a subset β of values which are called *designated values*. To the operations of an assertional system $S$ let there be assigned functions of the same degree which determine a value in α for each set of arguments in α. Let the function so assigned to implication be $f(-_1, -_2)$. The ordered sequence consisting of α, β, $f$ and the other functions assigned to the operations constitutes a *matrix* for $S$. In a matrix an assignment of values to the atomic obs determines uniquely the value assigned to every ob. An ob $A$ is tautologous for a matrix $M$ if and only if $A$ has a designated value in every possible assignment of values to the atoms. Then ⊢ $A$ is interpreted as saying that $A$ is tautologous. (Validity by 0-1 tables is a special case.)

A matrix is called *normal* just when $y$ is in β whenever $x$ and $f(x,y)$ are both in β. In such a case, if $A_1, \ldots, A_m$ are assigned values in β, the sole rule of inference is (9), and

$$A_1, \ldots, A_m \vdash B$$

then the value assigned to $B$ must be in β. Normal matrices, especially finite ones, are frequently an important tool in investigating questions of consistency and independence.

A matrix interpretation, even a normal one, is not necessarily regular. A sufficient condition that a normal matrix give a regular interpretation is that $f(x,y)$ and $f(y,x)$ be both designated if and only if $x$ and $y$ are the same value.

## B. PROPOSITIONAL ALGEBRAS

In this section we study certain propositional algebras which express the interpretations discussed in Sec. A3. We shall discuss three forms of such algebras and their relations to one another. One of these, called the T form, will be based directly on the rules considered in Sec. A3; the idea is due to Gentzen,[2] who proposed what he called "N rules" (for "natural rules") of this nature. The second form, called the H form, has modus ponens (i.e., the Pe of Sec. A3) as its sole rule, and axiom schemes of the traditional sort. The implicative lattices considered in Secs. 4C and 4D constitute the third form, which will be referred to here as the E form. There will further be two kinds of algebra. The algebra whose asserted propositions consist of those derivable by the rules of Sec. A3 and nothing else will be called the *absolute propositional algebra;* its three forms will be designated TA, HA, and EA, respectively, where it will turn out that EA is precisely the absolute implicative lattice of Sec. 4C3. The algebra formed by adjoining to the absolute algebra just enough so that its E form is a classical implicative lattice will be called the *classical positive propositional algebra;* its three forms will be called TC, HC, and EC, respectively. Later we shall see that the assertions

---

[1] Cf. Sec. 4A2, item 8°; also Łukasiewicz and Tarski [UAK, Definitions 3 and 4].

[2] Gerhard Gentzen (1909–1945), German mathematical logician. He got his doctorate at Göttingen under Hermann Weyl about 1933; died in a prison camp in Prague at the end of World War II. His inferential methods are presented in his [ULS]. Later (in his [WFR], [NFW]) he applied them, in connection with a transfinite induction up to the first ε number, to prove the consistency of a formalization of arithmetic. These have become famous, whereas his thesis has become known only in recent years.

of this classical algebra are precisely those which are tautologies in the ordinary two-valued truth tables.

**1. The system TA.** As stated in the introduction to this section, the system TA is the system based on the rules of Sec. A3. Since these rules are now primitive, they can be expressed as follows:[1]

$$\Lambda e \begin{cases} A \wedge B \vdash A \\ A \wedge B \vdash B \end{cases} \qquad \Lambda i \quad A, B \vdash A \wedge B$$

$$Ve \quad A \vdash C \ \& \ B \vdash C \rightarrow A \vee B \vdash C \qquad Vi \begin{cases} A \vdash A \vee B \\ B \vdash A \vee B \end{cases}$$

$$Pe \quad A, A \supset B \vdash B \qquad\qquad Pi \quad A \vdash B \rightarrow \ \vdash A \supset B$$

The following technique for writing these rules and for exhibiting proofs formed from them was given by Gentzen. It will be noted that two of these rules, namely, Pi and Ve, are not elementary. Suppose we indicate the elementary rules by writing the premises above the horizontal line and the conclusion below it, so that the rules for $\wedge$ become

$$\Lambda e \quad \frac{A \wedge B}{A} \qquad \frac{A \wedge B}{B} \qquad \Lambda i \quad \frac{A \quad B}{A \wedge B}$$

The prefix '$\vdash$' is here superfluous; the rules can be regarded as relations between obs. Proofs formed from these rules alone can then be exhibited in tree form according to the technique of Sec. 2A6, the abbreviations '$\Lambda e$', '$\Lambda i$', etc., being written at the right of the horizontal lines to indicate the rule being applied. A tree so constructed indicates a derivation of the conclusion (i.e., the bottom node) from the premises (i.e., the top nodes); if the top nodes are $A_1, \ldots, A_m$ and the bottom node is $B$, the tree indicates a derivation of

$$A_1, \ldots, A_m \vdash B \tag{1}$$

For example, a derivation of

$$A \wedge (B \wedge C) \vdash (A \wedge B) \wedge C$$

which is a form of the associative law, is given by the tree

$$\frac{\dfrac{A \wedge (B \wedge C)}{A}\Lambda e \qquad \dfrac{\dfrac{A \wedge (B \wedge C)}{B \wedge C}\Lambda e}{B}\Lambda e}{A \wedge B}\Lambda i \qquad \dfrac{\dfrac{A \wedge (B \wedge C)}{B \wedge C}\Lambda e}{C}\Lambda e}{(A \wedge B) \wedge C}\Lambda i$$

To extend this technique so as to include the rules with nonelementary premises (namely, Pi, Ve), Gentzen indicates a premise of the form $A \vdash B$ by writing the subsidiary premise, here indicated by '$A$', in brackets over

---

[1] It will be recalled that $A \vdash B$ is that special case of

$$\vdash A \rightarrow \ \vdash B$$

where the inference is made by application of the rules. This special case will apply if the inference is an instance of a primitive rule.

the subsidiary conclusion, here indicated by '$B$'. Thus the remaining rules, in Gentzen's notation, would be written thus:

$$\text{Ve} \quad \frac{\overset{[A]\ [B]}{C \quad C \quad A \lor B}}{C} \qquad \text{Vi} \quad \frac{A}{A \lor B} \quad \frac{B}{A \lor B}$$

$$\text{Pe} \quad \frac{A \quad A \supset B}{B} \qquad \text{Pi} \quad \frac{\overset{[A]}{B}}{A \supset B}$$

To form the constructions using such rules we need only some way of indicating that the conclusion of an instance of Ve or Pi no longer depends on the bracketed premise. We can do this most simply by numbering the distinct premises, and then marking along with 'Ve' and 'Pi' the number or numbers of the premise(s) to be canceled. At the same time the number of the premise can be canceled in all occurrences, not already canceled, over the inference in question. For example, a proof of (33) of Sec. A can be exhibited thus:

$$\frac{\dfrac{\overset{1}{A} \qquad \overset{2}{B}}{B \supset A} \text{Pi} - 2}{A \supset (B \supset A)} \text{Pi} - 1 \tag{2}$$

Here both of the indicated premises have been canceled; hence the conclusion is an assertion of the absolute propositional algebra.

Other examples of this technique giving, respectively, proofs of

$$\vdash A \supset . B \supset C :\supset: A \supset B . \supset . A \supset C \tag{3}$$
$$A \supset B, B \supset C \vdash A \supset C \tag{4}$$
$$A \land B . \supset C \vdash A \supset . B \supset C \tag{5}$$
$$A \supset B, A \supset C \vdash A \supset B \land C \tag{6}$$
$$A \supset C, B \supset C \vdash A \lor B \supset C \tag{7}$$

are the following:

$$\frac{\dfrac{\dfrac{\overset{3}{A} \quad \overset{1}{A \supset (B \supset C)}}{B \supset C} \text{Pe} \quad \dfrac{\overset{3}{A}, \overset{2}{A \supset B}}{B} \text{Pe}}{\dfrac{C}{A \supset C} \text{Pi} - 3}}{\dfrac{A \supset B . \supset . A \supset C}{A \supset . B \supset C :\supset: A \supset B . \supset . A \supset C} \text{Pi} - 1} \text{Pi} - 2$$

$$\frac{\dfrac{\dfrac{\overset{3}{A} \quad \overset{1}{A \supset B}}{B} \text{Pe} \quad \overset{2}{B \supset C}}{C} \text{Pe}}{A \supset C} \text{Pi} - 3$$

$$
\frac{\dfrac{\overset{2}{A}\quad\overset{3}{B}}{A \wedge B}\ \Lambda\mathrm{i} \qquad \overset{1}{A \wedge B \, . \supset C}}{\dfrac{\dfrac{C}{B \supset C}\ \mathrm{Pi}-3}{A \supset . \, B \supset C}\ \mathrm{Pi}-2}\ \mathrm{Pe}
$$

$$
\frac{\dfrac{\overset{3}{A}\quad\overset{1}{A \supset B}}{B}\ \mathrm{Pe} \qquad \dfrac{\overset{3}{A}\quad\overset{2}{A \supset C}}{C}\ \mathrm{Pe}}{\dfrac{B \wedge C}{A \supset B \wedge C}\ \mathrm{Pi}-3}\ \Lambda\mathrm{i}
$$

$$
\frac{\dfrac{\overset{3}{A}\quad\overset{1}{A \supset C}}{C}\ \mathrm{Pe} \qquad \dfrac{\overset{4}{B}\quad\overset{2}{B \supset C}}{C}\ \mathrm{Pe} \qquad \overset{5}{A \vee B}}{\dfrac{C}{A \vee B \supset C}\ \mathrm{Pi}-5}\ \mathrm{Ve}-3,4
$$

The reader should observe that these proofs may be constructed from the bottom up in a very natural manner.

Let us now establish a connection between the system TA and the system EA.

**Theorem 1.** *If the relation $\leq$ is defined by*

$$A \leq B \rightleftarrows \vdash A \supset B \tag{8}$$

*then the system* TA *is an implicative lattice.*[1]

*Proof.* We note first that, in view of Pe and Pi,

$$A \leq B \rightleftarrows A \vdash B$$

Then $(\rho)$ is obvious; $\Lambda$K, $\Lambda$K′ follow by $\Lambda$e, and VK, VK′ by Vi; $(\tau)$ follows by (4), $\Lambda$S by (6), VS by (7), and $P_2$ by (5); while $P_1$ follows by $\Lambda$e and Pe thus:

$$
\frac{\dfrac{\overset{1}{A \wedge (A \supset B)}}{A}\ \Lambda\mathrm{e} \qquad \dfrac{\overset{1}{A \wedge (A \supset B)}}{A \supset B}\ \Lambda\mathrm{e}}{B}\ \mathrm{Pe}
$$

Thus all the postulates of EA are valid when interpreted by (8) in TA. This completes the proof.

A converse to Theorem 1 will be established in Sec. 3.

**2. The system HA.** Before proceeding to the converse of Theorem 1, we seek a formulation HA of the absolute propositional algebra in which the only rule is Pe (i.e., modus ponens).

It is clear that, in the presence of Pi, any premise of the form $A \vdash B$ in a

---

[1] According to Sec. 4A1 (fifth paragraph) this means that every elementary theorem of an absolute implicative lattice is valid when interpreted in TA by (8). The proof shows this by deductive induction.

nonelementary rule can be replaced by $\vdash A \supset B$; further, in the presence of Pe, any elementary rule of the form

$$A_1, \ldots, A_m \vdash B$$

can be replaced by a statement scheme

$$\vdash A_1 \supset . A_2 \supset . \cdots \supset . A_m \supset B$$

By the first of these principles, Ve can be replaced by the elementary rule

$$A \supset C, B \supset C, A \vee B \vdash C$$

By the second principle, the rule $\Lambda$i will follow from the scheme

$$\vdash A \supset . B \supset . A \wedge B \tag{9}$$

whereas $\Lambda$e, Vi, and Ve will follow, under the same assumptions, from the schemes

| | |
|---|---|
| $\Lambda$K | $\vdash A \wedge B . \supset . A$ |
| $\Lambda$K' | $\vdash A \wedge B . \supset . B$ |
| VK | $\vdash A . \supset . A \vee B$ |
| VK' | $\vdash B . \supset . A \vee B$ |
| VS | $\vdash A \supset C . \supset : B \supset C . \supset : A \vee B . \supset . C$ |

No confusion should arise from the fact that the same abbreviations have been used for these postulates as for corresponding postulates of EA, for the latter are direct translations of the former according to the principles of this article and Sec. 1 (cf. Sec. 3).

This brings us to Pi. We seek axiom schemes from which Pi will follow, in the presence of Pe as sole rule, as an epitheorem. To do this we go through the motions of a proof of Pi by deductive induction; more precisely, of a proof that any ob $C$ such that $A \vdash C$ is such that $\vdash A \supset C$. The basic step of this induction involves two cases, viz., when $C$ is $A$ itself and when $C$ is some other axiom; and the inductive step takes the case where $\vdash C$ is obtained from $\vdash B \supset C$ and $\vdash B$ by Pe. This gives three cases as follows:

CASE 1. $C$ is $A$. The desired conclusion will follow immediately from the scheme

PI                              $\vdash A \supset A$

CASE 2. $C$ is an axiom, so that

$$\vdash C$$

holds. Then, by Pe, we should have the desired result from the scheme

$$\vdash C \supset . A \supset C$$

i.e., from

PK                              $\vdash A \supset . B \supset A$

CASE 3. $C$ is obtained by Pe from $B \supset C$ and $B$. By the hypothesis of the induction,

$$\vdash A \supset . B \supset C$$

and                             $\vdash A \supset B$

Hence the desired conclusion will follow from the scheme

PS $\qquad \vdash A \supset . B \supset C :\supset: A \supset B .\supset. A \supset C$

We thus see that, regardless of what additional axioms or axiom schemes there may be, Pi will be an epitheorem whenever we have PI, PK, and PS as axiom or theorem schemes. Before stating this conclusion formally, it will be expedient to make two remarks.

The first remark is that the scheme PI is a consequence of PK and PS. This was shown in Sec. 3A3.

The second remark is that (9) can be replaced by the weaker principle

ΛS $\qquad \vdash A \supset B .\supset: A \supset C .\supset: A \supset . B \wedge C$

which is analogous to VS. It is perhaps easiest to see this by a direct proof. Let $\vdash 1$, and let '$\rightarrow$' indicate formal deducibility. Then we deduce (9) from ΛS thus:[1] Since, by PK,

$$A \vdash 1 \supset A \ \& \ B \vdash 1 \supset B$$

we have $\qquad \vdash A \ \& \vdash B \rightarrow \vdash 1 \supset A \ \& \vdash 1 \supset B$

$\qquad\qquad\qquad\quad \rightarrow \vdash 1 \supset . A \wedge B \qquad$ by ΛS

$\qquad\qquad\qquad\quad \rightarrow \vdash A \wedge B \qquad\qquad$ since $\vdash 1$

Hence $\qquad\qquad \vdash A \rightarrow \vdash B \supset . A \wedge B \qquad$ by Pi

$\qquad\qquad\quad\ \vdash A \supset . B \supset . A \wedge B \qquad$ by Pi

Conversely, suppose (9) holds. Then

$$\vdash A \supset B \ \& \vdash A \supset C \ \& \vdash A \rightarrow \vdash B \ \& \vdash C$$

$\qquad\qquad\qquad\qquad \rightarrow \vdash B \wedge C \qquad\qquad\qquad$ by (9)

$\vdash A \supset B \ \& \vdash A \supset C \rightarrow \vdash A \supset . B \wedge C \qquad$ by Pi

whence we derive ΛS by two applications of Pi as before.

This discussion can now be formalized in the following definition and theorems. It is convenient to state two different theorems because of the interest in the results.

**Theorem 2. (Deduction theorem.)** *Let $\mathfrak{S}$ be a system of propositional algebra with a ply operation $\supset$. Let Pe be the sole rule of $\mathfrak{S}$, and let the elementary theorems of $\mathfrak{S}$ include all instances of the schemes PK and PS. Then Pi is an epitheorem for $\mathfrak{S}$ or any of its axiomatic extensions.*

DEFINITION. The *system* HA is that system of propositional algebra with operations $\supset$, $\wedge$, $\vee$ which is generated by the axiom schemes PK, PS, ΛK, ΛK′, ΛS, VK, VK′, VS, with Pe as sole deductive rule.[2]

---

[1] In each line, except where Pi is cited, there are one of more applications of Pe; this Pe is not explicitly mentioned.

[2] It is understood that what is important in this definition is that the system is assertional, that the sole rule is Pe, and that the assertions are those generated from the axiom schemes described. Another set of axiom schemes which generated the same set of assertions by the use of the same rule Pe would be considered as another formulation of HA, not as a distinct system. In some contexts, however (e.g., in Sec. E2 below), it is desired to make statements about HA which require that the system be formulated exactly as here described. Since the context will make this clear, it is not necessary to be fussy about this distinction. But when great exactness is desired, the term '*standard formulation of HA*' will be used to describe the formulation given here.

**Theorem 3.** *Every assertion of* TA *is an assertion of* HA; *furthermore, the rules of* TA *are valid as epitheorems of* HA.

*Proof.* The preliminary discussion shows the validity in HA of the rules of TA; hence every assertion of TA is also one of HA.

The converse of Theorem 3 will be established in Sec. 3 (Theorem 5).

**3. The absolute propositional algebra.** We shall now complete the study, begun in Secs. 1 and 2, of the absolute propositional algebra. The first theorem will be concerned with the relation between the systems HA and EA; then the studies of Secs. 1 to 3 will be gathered together in a final statement of equivalence of the three types of systems.

**Theorem 4.** *If* ⊢A *in* HA, *then* $1 \leq A$ *in* EA.

*Proof.* We note first that, by (vi) of Theorem 4C1,

$$1 \leq B \supset C \rightleftarrows B \leq C \tag{10}$$

[so that EA satisfies the condition (12) of Sec. A1]; further, by (3) of Sec. 4C (p. 141),

$$B_1 \leq (B_2 \supset C) \rightleftarrows B_1 \wedge B_2 \leq C \tag{11}$$

Now if $A$ is an ob of the form $B_1 \supset \cdot B_2 \supset \cdots B_m \supset \cdot C$, we say that an elementary theorem of EA is an E transform of $A$ just when it is either

$$1 \leq A$$

or, for some $k \leq m$, it is

$$B_1 \wedge \cdots \wedge B_k \leq B_{k+1} \supset .B_{k+2} \supset \cdots \supset .B_m \supset C$$

Then by virtue of (10) and (11) all these E transforms are equivalent in EA, and any one of them is true if and only if they all are.

Let us say that an ob of HA is EA-valid just when it has an E transform which is true in EA; this amounts to the same thing as saying that all its E transforms are true in EA. We have to show by deductive induction that every assertion of HA is EA-valid.

The inductive step of this induction has already been shown in Sec. A1 [proof of (9)]. To complete the proof it is only necessary to show that each of the axiomatic propositions of HA is EA-valid. This is shown in the following table, in which the numbers of the axiom schemes are listed in the first column, a suitable E transform for an instance of that scheme in the second column, and a reference to a proof of EA validity in the third.

| | | |
|---|---|---|
| PK | $A \leq B \supset A$ | (i) of Theorem 4C1 |
| PS | $A \supset (B \supset C) \leq (A \supset B) \supset (A \supset C)$ | (iii) of Theorem 4C1 |
| ΛK | $A \wedge B \leq A$ | ΛK |
| ΛK′ | $A \wedge B \leq B$ | ΛK′ |
| ΛS | $(A \supset B) \wedge (A \supset C) \leq A \supset (B \wedge C)$ | (iv) of Theorem 4C1 |
| VK | $A \leq A \vee B$ | VK |
| VK′ | $B \leq A \vee B$ | VK′ |
| VS | $(A \supset C) \wedge (B \supset C) \leq (A \vee B) \supset C$ | Theorem 4C3 |

This completes the proof of Theorem 4.

† For $k = m$, the right side is interpreted as $C$.

Combining Theorems 1, 3, and 4 into one statement, we have:

**Theorem 5.** *The systems* TA, HA, *and* EA *are equivalent, in that, given an ob A, the three statements*

$$\vdash A \quad in \ TA \tag{12}$$

$$\vdash A \quad in \ HA \tag{13}$$

$$A \ is \ EA\text{-}valid \tag{14}$$

*are mutually equivalent.*

*Proof.* In view of the fact that Pe and $\vdash 1$† hold in TA, we have (12) from (14) by Theorem 1; (13) follows from (12) by Theorem 3; and (14) follows from (13) by Theorem 4.

**4. The classical positive propositional algebra.** In Sec. 4C5 we saw that the scheme

$$(A \supset B) \supset A \leq A \tag{15}$$

was not an elementary theorem scheme of an absolute implicative lattice, and in Sec. 4D1 a classical implicative lattice was defined, in effect, as an implicative lattice for which (15) holds. This classical implicative lattice is here called the *system EC.*

Acting by analogy with the absolute system, we can define classical positive propositional systems HC and TC by adjoining to HA and TA, respectively, postulates in agreement with (15). The postulate for HC is the scheme

Pc $$\qquad\qquad \vdash A \supset B . \supset A : \supset A ‡$$

which is commonly known as "Peirce's law"; that for TC is the rule

Pk $$\qquad\qquad \frac{[A \supset B]}{\dfrac{A}{A}}$$

The interest of this classical system is that all assertions of HC are obs formed by $\supset$, $\wedge$, $\vee$, which yield tautologies when evaluated by the ordinary two-valued truth tables. This is easily shown by deductive induction. That conversely every such tautology is an assertion of HC, a result which will be called the completeness theorem for HC, cannot be shown conveniently with our present apparatus, but will emerge in due course.

In the meantime, we note simply that the systems TC, HC, EC are equivalent in the same sense that TA, HA, EA were; in fact, we have the following:

**Theorem 6.** *If A is an ob, the three statements*

$$\vdash A \ in \ TC$$

$$\vdash A \ in \ HC$$

$$1 \leq A \ in \ EC$$

*are mutually equivalent.*

---

† This is to take care of the case where the only E transform of $A$ is $1 \leq A$.
‡ The dots are redundant if we use the rule of association to the left.

*Proof.* The proof that (15) is validly interpreted in TC is given by the following diagram.

$$\frac{A \supset B . \supset A \quad \overset{1}{\phantom{x}} \quad A \supset B \overset{\not{\phantom{x}}}{\phantom{x}}}{\dfrac{A}{\dfrac{A}{A} \text{Pk} - 2}} \text{Pe}$$

The rest goes as in Theorems 2 to 5.

## EXERCISES

The following abbreviations are used as names for the indicated statement schemes in all cases.

| | |
|---|---|
| PB | $\vdash B \supset C . \supset : A \supset B . \supset . A \supset C$ |
| PB′ | $\vdash A \supset B . \supset : B \supset C . \supset . A \supset C$ |
| PC | $\vdash A \supset . B \supset C :\supset: B \supset . A \supset C$ |
| PI | $\vdash A \supset A$ |
| PK | $\vdash A \supset . B \supset A$ |
| PS | $\vdash A \supset . B \supset C :\supset: A \supset B_{.}\supset . A \supset C$ |
| PW | $\vdash A \supset . A \supset B :\supset . A \supset B$ |

1. Show directly that PB, PC, PW are true in HA.
2. Show that if PK, PS are replaced by any of the following sets

$$\text{PB, \quad PC, \quad PK, \quad PW}$$
$$\text{PB′, \quad PK, \quad PW}$$

we have a sufficient set of axiom schemes for HA. (The first set was used in Hilbert [GLM]; the second in Hilbert and Bernays [GLM. I]. The latter, pp. 70ff., discusses other axioms.)
3. Show directly that Rp holds in HA. (Use PB, PB′.)
4. Investigate the independence of any of the sets of axiom schemes for HA (Hilbert and Bernays [GLM. I, pp. 72–82]).
5. Put in all details in the above proof of (9), so as to get a proof in the formalism of HA without using Pi.
6. Show that, if (9) is accepted as an axiom scheme, PK becomes redundant if PB′ (or PB, PC) and PW are postulated, but remains independent if only PS is postulated (Church [IML₂, exercise 26.8]).
7. Discuss the following set of axiom schemes for HA (due to Heyting [FRI]).

$$\vdash A \supset . A \wedge A$$
$$\vdash A \wedge B . \supset . B \wedge A$$
$$\vdash A \supset B . \supset . A \wedge C \supset B \wedge C$$
$$\vdash A \supset B . \wedge . B \supset C . \supset . A \supset C$$
$$\vdash B \supset . A \supset B$$
$$\vdash A \wedge . A \supset B . \supset . B$$
$$\vdash A \supset . A \vee B$$
$$\vdash A \vee B . \supset . B \vee A$$
$$\vdash A \supset C . \wedge . B \supset C . \supset . A \vee B \supset C$$

Show that these are equivalent to those here given. (Schröter [UHA]. Show adequacy by reducing to EA, validity by TA.)

**8.** What is the effect on the deduction theorem of the admission of a substitution rule in addition to Pe?

**9.** Formulate a system admitting Pe and, instead of Pi, the rule that if $A \vdash B$, then either $\vdash B$ or $\vdash A \supset B$, but such that PK fails. (Church [WTI].)

**10.** What sort of deduction theorem would hold in a system with PB and PI only; with PB, PC, and PI; and with PB, PC, PW, and PI? ([GDT].)

**11.** Show that a necessary and sufficient condition that $A \vdash B$ be demonstrable in TA with the omission of the rules for P is that $A \leq B$ hold in a general distributive lattice.

**12.** Show that we get a formulation for HC by adjoining either of the schemes

$$\vdash A \vee . \; A \supset B$$
$$\vdash A \supset . \; B \vee C . \supset : A \supset B . \vee C$$

as additional axiom scheme to HA.

**13.** Show that PK, PB′, and Pc are a sufficient set of axiom schemes for pure implication in HC. (Schmidt [VAL, Theorem 56]. The formulation is due to Tarski and Bernays and appeared in Łukasiewicz and Tarski [UAK, Theorem 29].)

**14.** Show that the statements deducible by Pe from the single axiom scheme

$$\vdash A \supset B . \supset C : \supset : E \supset : B \supset . \; C \supset D . \supset . \; B \supset D$$

are precisely the elementary theorems of HA (here 'E' is a U variable). (Meredith [SAP].)

**\*15.** Can the elementary theorems of a lattice, and if so of what sort of lattice, be obtained from the axiom schemes for HA other than those for P by making a suitable definition of $A \leq B$? If not, what addition should be made? [If we use the definition (8), there is trouble with $(\rho)$ and $(\tau)$; if we use that of Exercise 11, with $\Lambda S$ and VS.]

**16.** Show that HA has the following normal matrix interpretation (Sec. A4):

$$\alpha \equiv \{1, 2, \ldots, n\} \qquad \beta \equiv \{1\}$$
$$A \vee B = \min (A, B)$$
$$A \wedge B = \max (A, B)$$
$$A \supset B = \begin{cases} 1 & \text{if } A \geq B \\ B & \text{if } A \leq B \end{cases}$$

Hence show that Pc is not demonstrable in HA. On the other hand, if $A_1, \ldots, A_m$ are distinct atoms, and

$$B_{ij} \equiv A_i \supset A_j . \wedge . \; A_j \supset A_i$$
$$C_k \equiv B_{1k} \vee B_{2k} \vee \cdots \vee B_{k-1,k}$$
$$D_m \equiv C_2 \vee C_3 \vee \cdots \vee C_m$$

then $D_m$ is not demonstrable, even though it always has a designated value in the above matrix if $n < m$. (Gödel [IAK].)

## C. THE SYSTEMS LA AND LC

In this section we shall begin the study of certain systems which Gentzen introduced and called L systems. These differ from the systems considered up to now principally in the following respects. The elementary statements are statements of deducibility; each such statement has a certain conclusion and a set, which may be void, of premises; and it is interpreted as stating the existence of a T proof leading to the stated conclusion and having no uncanceled premises other than some of those stated. But whereas such theorems for a T system would not constitute a deductive theory, at least not one with elementary rules, they do constitute such a theory in the L system.  Furthermore, the rules are such that new combinations can be introduced but not eliminated, and this gives the system a quasi-constructive character which has important consequences.

Along with the L systems another innovation will be introduced at the same time.  It is assumed that the propositions have as interpretants certain statements concerning an underlying formal system $\mathfrak{S}$.  This $\mathfrak{S}$ is subject only to broad restrictions which admit as special case the possibility that $\mathfrak{S}$ may be vacuous.  In that special case the theory of Gentzen—and propositional algebra as ordinarily understood—results.  We thus have here a slight generalization; as a result of it propositional logic becomes part of the methodology of formal systems in general, rather than the theory of a special formal system being studied for its own sake.  This innovation is independent of those mentioned in the preceding paragraph; it could have been introduced in Sec. B or even in Sec. 4C.  It is introduced at this point because it fits in conveniently with the semantical discussion in Sec. 1.

Since the theory of the L systems is rather extensive, it is convenient to divide it between three sections.  The present section will deal with the formulation of the system and of techniques and simple theorems connected with it.  The relations between the different kinds of L systems and other types of systems will concern us in Sec. D, and the further development of the system in Sec. E.  The exercises for all three sections will be found at the end of Sec. E.

This section will lay the foundation, not only for the systems LA and LC, which principally concern this chapter, but for the analogous L systems in the succeeding chapters.

**1. Preliminary study of the absolute system.**  We shall first consider, in a preliminary and partly intuitive manner, the formation of the system $LA(\mathfrak{S})$ based on a given formal system $\mathfrak{S}$.  In this consideration we shall proceed with a certain interpretation of the system and a discussion of the validity of postulates relative to that interpretation, in order to form a basis for the formalization to be introduced later.

The propositions of $LA(\mathfrak{S})$ will have as interpretants certain statements of $\mathfrak{S}$.  We call these statements the *compound statements* of $\mathfrak{S}$.  They form an inductive class generated from the elementary statements of $\mathfrak{S}$ by the positive sentential connectives.  Correspondingly, we speak of the elementary statements of $\mathfrak{S}$ as forming the class $\mathfrak{E}$ of *elementary propositions*; the (compound) propositions are the class $\mathfrak{P}$ generated from the elementary propositions by the operations.  In the present chapter these operations are $\supset$, $\wedge$, and $\vee$.  Both of the classes $\mathfrak{E}$ and $\mathfrak{P}$ are relative to $\mathfrak{S}$; when we wish to make this dependence explicit, we call them $\mathfrak{E}(\mathfrak{S})$ and $\mathfrak{P}(\mathfrak{S})$, respectively.

Among the propositions we postulate two sorts of deducibility relation, symbolized, respectively, by

$$A_1, \ldots, A_m \vdash_0 B \tag{1}$$

$$A_1, \ldots, A_m \Vdash B \tag{2}$$

These forms are admissible for any value of $m \geq 0$; for the case $m = 0$, they are written with nothing to the left of the signs '$\vdash_0$' and '$\Vdash$', respectively.

Of these, (1) will hold only when $A_1, \ldots, A_m$ and $B$ are elementary, and then, for $m > 0$, just when $B$† is a consequence of $A_1, \ldots, A_m$ by virtue of an instance of a deductive rule of $\mathfrak{S}$, and for $m = 0$, just when $B$ is an axiom of $\mathfrak{S}$.‡  The statements (1) are thus specified by a finite list of schemes, and their purpose in the theory is purely auxiliary.  On the other hand, the statements (2) will be the elementary statements, properly speaking, of the system $\mathrm{LA}(\mathfrak{S})$; they will be a deductive class specified in the usual way by axioms and rules.[1]

This study is thus concerned principally with elementary statements of the form (2).  It will be convenient to use German capital letters, generally from the end of the alphabet, to stand for sequences of propositions; thus a typical statement (2) is

$$\mathfrak{X} \Vdash B \tag{3}$$

where $\mathfrak{X}$ is a finite sequence, which may be void, of propositions.

The interpretation we now associate with (3) is that $B$ is a consequence of $\mathfrak{X}$.  We may define this more specifically as saying that $B$ is the conclusion of a T proof having no uncanceled premises except certain members of $\mathfrak{X}$.  Alternatively, we may regard (3) as stating that $\Vdash B$ is demonstrable in a system $L(\mathfrak{S};\mathfrak{X})$ formed by adjoining $\mathfrak{X}$ to $L(\mathfrak{S})$.

For this interpretation (3) will be valid in the following cases:

($a$1) When $B$ is in $\mathfrak{X}$.

($a$2) When $B$ is an axiom of $\mathfrak{S}$.

($a$3) When $B$ is an elementary proposition and $\mathfrak{X}$ contains elementary propositions $A_1, \ldots, A_m$ such that (1) holds.

In the same interpretation the validity of (3) depends only on the class of propositions in $\mathfrak{X}$, and not at all on their order or multiplicity; further, one can add arbitrary propositions to $\mathfrak{X}$ without destroying the validity of (3).  These facts can be expressed by three inferential rules, called *structural rules*.  The first of these, called the *rule of permutation*, says that (3) is derivable from

$$\mathfrak{X}' \Vdash B$$

where $\mathfrak{X}'$ is any permutation of $\mathfrak{X}$; the second, called the *rule of contraction*, says that (3) is derivable from

$$\mathfrak{X}, A \Vdash B \tag{4}$$

provided $A$ is present in $\mathfrak{X}$; and the third, called the *rule of weakening*, says that (4) is a consequence of (3) for any $A$.  It will be convenient to call these three rules C, W, K, after the combinators C, W, and K, respectively.

---

† In discussing interpretations it is permissible to identify propositions with the statements they represent (cf. Sec. A2).

‡ The form (1) can represent only an elementary rule (Sec. 3D3).  Thus our considerations are restricted to systems $\mathfrak{S}$ which have only such rules.  This is a matter of some importance.

[1] The relations $\vdash_0$ and $\Vdash$ are thus predicates of an unspecified number of arguments. This is a possibility not explicitly considered in Sec. 2C.  To bring it under the conception of Sec. 2C, it is necessary to analyze these predicates into an assemblage of predicates each with a fixed number of arguments.  From this point of view, the rules adopted later would be quite complex.

Next we consider the rules related to the operations. We adopt the principle, already suggested in Sec. A3, that the meaning of a concept is determined by the circumstances under which it may be introduced into discourse. For statements of the form (3), introduction can be made either on the left or on the right. This suggests the following notation. Let 'P', 'Λ', 'V' be used, as heretofore, for the operations ⊃, ∧, ∨, respectively, and let 'O' stand for any of these, or indeed, for the name of any operation introduced now or later. Then we indicate a rule of introduction on the right by writing the appropriate one of these symbols with an '∗' on the right; likewise we indicate a rule of introduction on the left by writing such a symbol with an '∗' on the left.[1]

For the rules on the right the situation is essentially the same as in Sec. A3. In fact, the rules P∗, Λ∗, V∗ are analogous to the rules Pi, Λi, Vi, thus:

P∗ $\qquad\qquad \mathfrak{X}, A \Vdash B \to \mathfrak{X} \Vdash A \supset B$

Λ∗ $\qquad\qquad \mathfrak{X} \Vdash A \;\&\; \mathfrak{X} \Vdash B \to \mathfrak{X} \Vdash A \wedge B$

V∗ $\qquad\qquad \mathfrak{X} \Vdash A \to \mathfrak{X} \Vdash A \vee B$

$\qquad\qquad\qquad \mathfrak{X} \Vdash B \to \mathfrak{X} \Vdash A \vee B$

For the rules on the left we have to ask questions analogous to those we asked for the rules Oe in Sec. A3. To justify the introduction of $A \circ B$ to form

$$\mathfrak{X}, A \circ B \Vdash C$$

we have to ask when, in the presence of $\mathfrak{X}$, $C$ is a consequence of $A \circ B$. If we examine the motivation of the rules Oe, we conclude the following. We can infer $C$ from $A \wedge B$ when we can infer it either from $A$ or from $B$, thus giving the rules

∗Λ $\qquad\qquad \mathfrak{X}, A \Vdash C \to \mathfrak{X}, A \wedge B \Vdash C$

$\qquad\qquad\qquad \mathfrak{X}, B \Vdash C \to \mathfrak{X}, A \wedge B \Vdash C$

We infer $C$ from $A \vee B$ just when we can infer it both from $A$ and from $B$, thus giving the rule [cf. (30) of Sec. A]

∗V $\qquad\qquad \mathfrak{X}, A \Vdash C \;\&\; \mathfrak{X}, B \Vdash C \to \mathfrak{X}, A \vee B \Vdash C$

Finally, unless $A$ is present, we can infer nothing from $A \supset B$ that we cannot infer from it as a wholly unanalyzed proposition, but if $A$ is present, we can infer from $A \supset B$ any $C$ which we can infer from $B$ alone, thus giving the rule

∗P $\qquad\qquad \mathfrak{X} \Vdash A \;\&\; \mathfrak{X}, B \Vdash C \to \mathfrak{X}, A \supset B \Vdash C$

---

[1] This notation is a modification of that introduced by Kleene [IMM]. The latter uses '→' for '⊩' and the infixes '⊃', 'Λ', 'V' in place of 'P', 'Λ', 'V'. He indicates the rules by writing '→' so that the operational symbol rather than the '→' indicates the side on which the introduction takes place. The modifications adapt these notations to the general conventions made here and in combinatory logic. In the latter theory 'P', 'Λ', 'V' are prefixes for the operations, and it is essential to have such prefixes which are distinct from the infixes. Other prefixes, 'C', 'K', 'A', respectively, are used by the Polish school, but they are incompatible with the other conventions of combinatory logic.

It will be seen that the rules *O express essentially the same principles as the rules Oe, only now there is no elimination.

To express the fact that the set of propositions $A$ such that $\mathfrak{X} \Vdash A$ holds is closed with respect to the rules of $\mathfrak{S}$, we need a rule

⊢*    If (1) holds and

$$\mathfrak{X} \Vdash A_i \qquad i = 1, 2, \ldots, m$$

then                                    $\mathfrak{X} \Vdash B$

Reciprocal to this, in a certain sense, is a rule

*⊢    If (1) holds and

$$\mathfrak{X}, B \Vdash C$$

then                          $\mathfrak{X}, A_1, \ldots, A_m \Vdash C$

Finally, the rule

$$\mathfrak{X} \Vdash A \ \& \ \mathfrak{X}, A \Vdash B \to \mathfrak{X} \Vdash B \tag{5}$$

appears to be valid on the basis of the interpretation chosen. This rule was called by Gentzen "cut" (*Schnitt*).

The rules so stated are redundant. Thus $(a3)$ expresses intuitively the same principle as ⊢*, and is easily derived from it and $(a1)$ (see Theorem 5 in Sec. 8). Having $(a3)$ and ⊢*, we can then deduce *⊢ by (5). We shall therefore omit the rules $(a3)$ and *⊢. But the most striking redundancy is that of (5). This redundancy constitutes the principal theorem (*Hauptsatz*) of Gentzen's thesis. Here we shall not admit (5) as a primitive rule, but shall prove an epitheorem, called the *elimination theorem*, to the effect that it is an admissible rule on the basis of the others.[1] This profound theorem has important consequences, some of which were mentioned in the introduction to this chapter. Other redundancies will manifest themselves as we proceed.

The postulates $(a1)$, $(a2)$, C, W, K, O*, *O (where O is P, $\Lambda$, or V), and ⊢* constitute the basis we adopt for LA($\mathfrak{S}$); this will be called simply LA when it is not necessary to be explicit about $\mathfrak{S}$. The rules O*, *O will be called its *operational rules*. The system $\mathfrak{S}$ is arbitrary, except that its rules must be of the form (1), so that it is elementary in the sense of Sec. 2D3. The possibility is admitted that $\mathfrak{S}$ may have no axioms and rules, so that $(a2)$ and (1) are vacuous; in that case $\mathfrak{S}$ will be called '$\mathfrak{D}$'. Then $\mathfrak{E}(\mathfrak{D})$ will consist of $E_1, E_2, \ldots$, which function as indeterminates, and LA($\mathfrak{D}$) will be a form of propositional algebra.

The system LA defines a class of propositions $A$ such that

$$\Vdash A \tag{6}$$

These propositions will be called the assertible propositions, or simply the *assertions*, of LA. We shall see in due course that LA($\mathfrak{D}$) has precisely the same assertions as HA.

---

[1] A careful consideration of the principles according to which the rules are set up suggests that this redundancy can be expected a priori. For the structural rules simply explain that the $\mathfrak{X}$ in (3) is to be regarded as a class, whereas the rules O* and *O explain a complex concept in terms of something simpler. The rules $(a2)$, ⊢* show that deducibility in $\mathfrak{S}$ is carried over into LA($\mathfrak{S}$). Intuitively one would feel that these rules suffice to explain the meaning of each statement (3). Thus (5) ought to be an epitheorem, and in fact it is.

Before proceeding to the strict formalization of these ideas, we shall pause to discuss another interpretation of a more classical nature.

**2. The classical system LC.** In the system LA a lack of symmetry between the treatment of the left and right sides is immediately apparent. Suppose now we set up, using formal analogy as a guide, a more extended system in which some, at least, of this lack of symmetry is removed. In this new system the elementary statements replacing (2) are of the form

$$A_1, \ldots, A_m \Vdash B_1, \ldots, B_n \tag{7}$$

i.e., of the form

$$\mathfrak{X} \Vdash \mathfrak{Y} \tag{8}$$

in which $\mathfrak{Y}$ as well as $\mathfrak{X}$ may be a propositional sequence of arbitrary length. Such a statement we accept as axiomatic in the cases:

($a$1)            Some $B_j$ is the same as some $A_i$.

($a$2)            Some $B_j$ is an axiom of $\mathfrak{S}$.

We admit further structural rules affecting $\mathfrak{Y}$ which are the duals, so to speak, of those affecting $\mathfrak{X}$, and following the analogy of the operational rules, we call the former C∗, W∗, K∗, respectively, whereas the latter (those affecting $\mathfrak{X}$) are ∗C, ∗W, and ∗K, respectively. As for the operational rules, we adopt the same rules with the addition of an arbitrary propositional sequence $\mathfrak{Z}$ on the right of all premises and conclusion; thus the rules for P are

∗P            $\mathfrak{X} \Vdash A, \mathfrak{Z} \,\&\, \mathfrak{X}, B \Vdash C, \mathfrak{Z} \to \mathfrak{X}, A \supset B \Vdash C, \mathfrak{Z}$

P∗            $\mathfrak{X}, A \Vdash B, \mathfrak{Z} \to \mathfrak{X} \Vdash A \supset B, \mathfrak{Z}$

In this way we form a system which we call LC [more explicitly LC($\mathfrak{S}$)]. The rules will be written out in full in Sec. 4.

The system LC has been formulated by analogy. It has, however, a valid interpretation in terms of truth-table valuations. Suppose that we call a valuation any assignment of values 0 and 1 to the elementary propositions. We call such a valuation *admissible* just when every axiom has the value 1, and in any instance of (1) in which the premises $A_1, \ldots, A_m$ have the value 1, the conclusion $B$ also has the value 1. Then in any admissible valuation every assertible elementary proposition has the value 1, but an elementary proposition which is not assertible may have either of the values 0 or 1, arbitrarily.[1] Let the values assigned to compound propositions be obtained by the usual two-valued truth tables with 1 taken as truth. Then the interpretant of (7) is true by 0-1 tables just when for every admissible valuation either some $A_i$ has the value 0 or some $B_j$ has the value 1. It is then easy to show, by deductive induction, that the interpretant of (7) is true whenever (7) is derivable in LC.

On the other hand, there are statements (3) which are valid on the interpretation here considered but not on that of Sec. 1. For example, the statement $\Vdash A \lor (A \supset B)$ is true for the interpretation by 0-1 tables, no matter what the system $\mathfrak{S}$ or the propositions $A$, $B$ are. On the other hand, in the system $\mathfrak{S}$ for which $\mathfrak{E}(\mathfrak{S})$ consists of $E_1$, $E_2$, and $E_3$, of which $E_3$ is the sole axiom, and the sole rule of inference is $E_1 \vdash E_2$, consider the

---

[1] Note that there may be a nonconstructive element here.

proposition $E_2 \vee (E_2 \supset E_1)$. Here $E_2$ is not assertible because the only assertible elementary proposition of $\mathfrak{S}$ is $E_3$, and $E_2 \supset E_1$ is not assertible since on adjoining $E_2$ to $\mathfrak{S}$ the only elementary theorems are $E_2$ and $E_3$; hence $\Vdash E_2 \vee (E_2 \supset E_1)$ is not assertible in LA($\mathfrak{S}$) for that $\mathfrak{S}$.†

Thus the rules for LA (as given in Sec. 1) are essentially the same as the rules for LC (as given here), with the additional requirement that $\mathfrak{Y}$ of every statement (8) consist of a single proposition. We can express this by saying that LA is singular, whereas LC is multiple. However, this does not imply that one cannot interpret (7) in terms of the concepts of Sec. 1. In fact, suppose we interpret (7) in LA as meaning that

$$A_1, \ldots, A_m \Vdash B_1 \vee \cdots \vee B_n$$

With this interpretation, as we shall see later, all the rules of LC are valid, with the exception of P∗. Thus there is a multiple formulation of LA; it differs from that of LC in that in P∗ the $\mathfrak{Z}$ must be void.

Again, we may have a singular formulation of LC. We shall see in due course that the assertions [i.e., propositions for which (6) holds] of LC($\mathfrak{D}$) are precisely those of the classical positive propositional algebra. A comparison with Sec. B4 suggests that we can get a singular formulation of LC by adjoining to LA the rule

Px                     $\mathfrak{X}, A \supset B \Vdash A \rightarrow \mathfrak{X} \Vdash A$

We shall use subscripts '1' and '$m$' to distinguish singular and multiple forms. Thus the singular form of LA, described in Sec. 1, is LA$_1$, its multiple form is LA$_m$; the singular and multiple forms of LC are, respectively, LC$_1$ and LC$_m$. When these subscripts are absent, it will be because the context does not require that the form be specified.

The relations between these various formulations, singular and multiple, of LA and LC will concern us later. We shall find that the various relations suggested here informally can be established by rigorous arguments.

**3. Formulation of the morphology.** We now turn to the strict formalization of the systems LA and LC and some related systems. We begin here by formulating the morphology, leaving the theory proper to Sec. 4.

*a. Elementary Propositions.* We start with a class $\mathfrak{E}$ whose members we call elementary propositions. Nothing is postulated concerning $\mathfrak{E}$ except that it is a definite class of formal objects. In the interpretation of Secs. 1 and 2 it corresponds to the elementary statements of the underlying system $\mathfrak{S}$. We reserve the letter '$E$', perhaps with affixes, to designate elementary propositions.[1]

*b. Propositions.* The class $\mathfrak{P}$ of propositions is the monotectonic inductive class generated from the elementary propositions by the operations $\supset$, $\wedge$, $\vee$. We thus have the principles

($b1$)                     $\mathfrak{E} \subseteq \mathfrak{P}$

($b2$)  If $A$ and $B$ are in $\mathfrak{P}$, then $A \supset B$, $A \wedge B$, $A \vee B$ are in $\mathfrak{P}$ but not in $\mathfrak{E}$.

---

† This example is from [TFD], pp. 28ff. One can show rigorously that

$$\Vdash E_3 \vee (E_2 \supset E_1)$$

cannot be obtained by the rules. This is a special case of the general procedure in Secs. 5 and E7.

[1] Whether these symbols are U constants or U variables will be left to be determined from the context.

The propositions are the obs of our system, whereas the elementary propositions are the atoms. We use the letters '*A*', '*B*', '*C*', '*D*', and when necessary other capital italic letters, for propositions.

*c. Prosequences.* A finite sequence of propositions will be called a *prosequence.* We use capital German letters, usually from the end of the alphabet, for prosequences. The propositions which belong to a prosequence will be called its *constituents*; this is to be understood in the sense that when a prosequence contains repetitions of the same proposition, each individual occurrence is a separate constituent. A prosequence may be void, or it may have any finite number of constituents. A prosequence with not more than one constituent will be called *singular*; one with two or more will be called *multiple.* The void prosequence will be indicated either by '0' or by a blank space. A prosequence with a single constituent will not be distinguished notationally from that constituent. When the symbols for two or more prosequences are written one after the other, separated by commas, the complex expression will designate the prosequence formed by concatenation of those indicated; for example,

$$\mathfrak{X}, A_1, \ldots, A_m, \mathfrak{Y}, \mathfrak{X}$$

is the prosequence consisting of the constituents of $\mathfrak{X}$ in order, then $A_1, \ldots,$ $A_m$, then those in $\mathfrak{Y}$, then those in $\mathfrak{X}$ (repeated). Constituents which are occurrences of the same proposition will be said to be *alike.*

*d. Auxiliary Statements.* Among the elementary propositions, we postulate a relation $\vdash_0$ of an unspecified number of arguments, producing elementary statements of the form (1). These statements we call the *auxiliary statements.* The true auxiliary statements are exhaustively specified as the instances of a finite number of statement schemes of the form (1); these schemes we call the *auxiliary postulates.* These auxiliary postulates constitute the basic system $\mathfrak{S}$ so far as our formal theory is concerned. If the set of auxiliary postulates is void, so that there are no auxiliary theorems, the system $\mathfrak{S}$ is designated $\mathfrak{O}$.† An elementary proposition B such that (1) holds for $m = 0$ will be called an *axiom.*

*e. Elementary Statements.* The elementary statements are of the form (8), where $\mathfrak{X}$ and $\mathfrak{Y}$ are prosequences. We call $\mathfrak{X}$ the *left prosequence,* or *antecedent,* of (8); $\mathfrak{Y}$ the *right prosequence,* or *consequent.* The *constituents* of (8) will be precisely the constituents of $\mathfrak{X}$ together with those of $\mathfrak{Y}$; those of $\mathfrak{X}$ will be called the *left,* or *antecedent, constituents,* those of $\mathfrak{Y}$ the *right,* or *consequent, constituents.*

An elementary statement will be called *singular* or *multiple* according as its consequent is singular or multiple. For LA and LC this will mean that (8) is singular just when $\mathfrak{Y}$ has a unique constituent, but in Chap. 6 elementary statements with void consequent will be admitted.

A formulation (or system) will be called *singular* when all its elementary statements are required to be singular; *multiple* when there is no restriction in the rules requiring an elementary statement to be singular; *mixed* when there are no such restrictions for the formulation as a whole, but there are such restrictions on the applicability of certain rules. From now on the letters '$\mathfrak{X}$', '$\mathfrak{Y}$', '$\mathfrak{Z}$' will be used systematically, with or without affixes, as

† The case of general $\mathfrak{S}$ will be reduced to $\mathfrak{O}$ in Theorem E4.

follows: '$\mathfrak{X}$' will always denote an arbitrary, possibly void, prosequence; in
a singular formulation '$\mathfrak{Y}$' will denote a singular prosequence and '$\mathfrak{Z}$' a void
one; in a multiple formulation, on the other hand, '$\mathfrak{Y}$' and '$\mathfrak{Z}$' also denote
unrestricted prosequences; and in a mixed formulation '$\mathfrak{Y}$' will denote a
singular or an unrestricted prosequence and $\mathfrak{Z}$ a void or an unrestricted prosequence according to circumstances stated in the context.   Of the systems
to be defined later, LA$_1$ and LC$_1$ will be singular, LC$_m$ will be multiple, and
LA$_m$ will be mixed.

**4. Theoretical formulation; Formulation I.**   When we come to formulate the theory proper, it turns out that there are several different formulations, each of which has some advantage, and therefore it is necessary
to consider, more or less on a par, a number of variant formulations.   Here
a certain basic formulation, called Formulation I, will be presented; other
formulations will be introduced later as modifications of this one.   The formulation reflects directly the intuitive considerations of Secs. 1 and 2.   The
rules are stated in a table following the preliminary discussion.

The elementary statements which function as axioms in the theory will be
called *prime statements*.   The term 'axiom' will be reserved for the usage of
Sec. 3$d$.   The prime-statement schemes are listed in the table of rules under '$p$'.

The rules proper are to hold for the various systems, with the understanding that LA$_1$ and LC$_1$ are singular systems; LC$_m$ is a multiple system;
LA$_m$ is a mixed system which is singular for P* and unrestricted for all the
others.   The rule Px is postulated (in singular form) for LC$_1$ only; later it
will be shown that the general form is redundant for LC$_m$.

The rules are stated in parallel columns.   The rules for introduction in
the antecedent are on the left; in the consequent on the right.   As in Sec.
B, the almost self-explanatory technique of the Hilbert school is used, the
premises being above the line, the conclusion below.   Supplementary hypotheses are written on the line opposite the word 'if'.

The rules on the left and right are distinguished by using an asterisk on the
side on which the rule is to operate.   Thus P* is the rule for the introduction
of implication on the right.   When it is desired to talk about both rules, the
asterisks are put on both sides.

The rules *C*, *W*, *K* are called *structural rules*; the rules *O*, where
'O' stands for one of 'P', '$\Lambda$', 'V' (and later for names of other operations)
will be called *operational rules;* while $\vdash$* will be called the *rule of $\mathfrak{S}$ derivation*.
It is understood that the structural rules on the right are always inapplicable,
even with $\mathfrak{Z}$ void, in a singular system.

In the following table the letters '$A$', '$B$' are U variables for arbitrary
propositions; '$E_1$', '$E_2$', etc., for elementary propositions; and $\mathfrak{X}$, $\mathfrak{Y}$, $\mathfrak{Z}$ for
prosequences according to the conventions of Sec. 3$e$.

*p. Prime Statements*

($p$1)                                $A \Vdash A$

($p$2)                                $\Vdash E$          where $E$ is an axiom

*C*   *Rules of permutation*
    If $\mathfrak{X}'$ is a permutation of $\mathfrak{X}$,          $\mathfrak{Y}'$ is a permutation of $\mathfrak{Y}$.

$$\frac{\mathfrak{X} \Vdash \mathfrak{Y}}{\mathfrak{X}' \Vdash \mathfrak{Y}} \qquad\qquad \frac{\mathfrak{X} \Vdash \mathfrak{Y}}{\mathfrak{X} \Vdash \mathfrak{Y}'}$$

∗W∗    *Rules of contraction*

$$\frac{\mathfrak{X}, A, A \Vdash \mathfrak{Y}}{\mathfrak{X}, A \Vdash \mathfrak{Y}} \qquad\qquad \frac{\mathfrak{X} \Vdash B, B, 3}{\mathfrak{X} \Vdash B, 3}$$

∗K∗    *Rules of weakening*

$$\frac{\mathfrak{X} \Vdash \mathfrak{Y}}{\mathfrak{X}, A \Vdash \mathfrak{Y}} \qquad\qquad \frac{\mathfrak{X} \Vdash 3}{\mathfrak{X} \Vdash B, 3}$$

∗P∗    *Rules of implication*
If $\mathfrak{Y}$ is singular,[1]

$$\frac{\mathfrak{X} \Vdash A, 3;\ \mathfrak{X}, B \Vdash \mathfrak{Y}, 3}{\mathfrak{X}, A \supset B \Vdash \mathfrak{Y}, 3} \qquad\qquad \frac{\mathfrak{X}, A \Vdash B, 3}{\mathfrak{X} \Vdash A \supset B, 3}$$

∗Λ∗    *Rules of conjunction*

$$\frac{\mathfrak{X}, A \Vdash \mathfrak{Y}}{\mathfrak{X}, A \wedge B \Vdash \mathfrak{Y}};\ \frac{\mathfrak{X}, B \Vdash \mathfrak{Y}}{\mathfrak{X}, A \wedge B \Vdash \mathfrak{Y}} \qquad\qquad \frac{\mathfrak{X} \Vdash A, 3;\ \mathfrak{X} \Vdash B, 3}{\mathfrak{X} \Vdash A \wedge B, 3}$$

∗V∗    *Rules of alternation*

$$\frac{\mathfrak{X}, A \Vdash \mathfrak{Y};\ \mathfrak{X}, B \Vdash \mathfrak{Y}}{\mathfrak{X}, A \vee B \Vdash \mathfrak{Y}} \qquad\qquad \frac{\mathfrak{X} \Vdash A, 3}{\mathfrak{X} \Vdash A \vee B, 3};\ \frac{\mathfrak{X} \Vdash B, 3}{\mathfrak{X} \Vdash A \vee B, 3}$$

⊢∗    *Rule of ⊖-derivation*

$$\text{If } E_1, E_2, \ldots, E_m \vdash E_0$$
$$\frac{\mathfrak{X} \Vdash_0 E_1, 3;\ \ldots;\ \mathfrak{X} \Vdash_0 E_m, 3}{\mathfrak{X} \Vdash E_0, 3}$$

Px    *Peirce rule*

$$\frac{\mathfrak{X}, A \supset B \Vdash A, 3}{\mathfrak{X} \Vdash A, 3}$$

**5. Examples of proof technique.** Before proceeding with further details of the formulation, it will probably be helpful to consider examples of finding proofs in this system. These examples will illustrate an important fact about these systems, viz., that given an elementary statement Γ containing a constituent $M$, there are only a finite number of possibilities for obtaining Γ as conclusion of a rule $R$. It is therefore possible, in principle, to search for a proof starting with the desired conclusion. In due course we shall see that the situation is actually decidable and that one can set up an algorithm for constructing a proof or showing that none exists. But for the illustrative examples presently to be considered, it will merely be shown that proofs can or cannot be found subject to certain reasonable assumptions; if a proof is found, this of course settles the question, but a negative answer depends on verification of the assumptions later.

In these examples we shall consider only the systems $LA_1$ and $LC_m$; these will be called simply LA and LC, respectively. The assumptions are: that we can omit ∗K∗ if we admit as prime statements those of the form $(a1)$ and

---

[1] The rule ∗P, if formulated strictly according to Sec. 2, would have $C$ for $\mathfrak{Y}$. In Chap. 6, we shall need the case where $\mathfrak{Y}$ is void.

(a2) in Secs. 1 and 2, and that $*W*$ can be omitted if we modify the operational rules so as to admit the presence in the premises of a constituent like
the new one being introduced.   In the latter case the modified rule is equivalent to the original rule followed immediately by an application of $*W*$.
Thus for $*P$ the modified rule is

$$\frac{\mathfrak{X}, A \supset B \Vdash A, \mathfrak{Z} \quad \mathfrak{X}, A \supset B, B \Vdash \mathfrak{Y}, \mathfrak{Z}}{\mathfrak{X}, A \supset B \Vdash \mathfrak{Y}, \mathfrak{Z}} \tag{9}$$

This can be obtained from the regular $*P$ thus:

$$\frac{\dfrac{\mathfrak{X}, A \supset B \Vdash A, \mathfrak{Z} \quad \mathfrak{X}, A \supset B, B \Vdash \mathfrak{Y}, \mathfrak{Z}}{\mathfrak{X}, A \supset B, A \supset B \Vdash \mathfrak{Y}, \mathfrak{Z}}*P}{\mathfrak{X}, A \supset B \Vdash \mathfrak{Y}, \mathfrak{Z}}*W$$

Conversely, the regular $*P$ can be obtained from it by using $*K*$ to introduce the extra constituent $A \supset B$.   In these examples $*P$ will be assumed
in the form (9).

*Example 1.*   We seek a derivation in LA($\mathfrak{D}$) of

$\Gamma_1$ $\qquad\qquad\qquad \Vdash A \supset . A \supset B :\supset . A \supset B$

where $A$ and $B$ are elementary.   The only rule of which $\Gamma_1$ can be the conclusion (of an instance of the rule, of course) is $P*$; the premise in that case
would be

$\Gamma_2$ $\qquad\qquad A \supset . A \supset B \Vdash A \supset B$

Here we have two choices, $*P$ and $P*$, for the next preceding rule.   If we
were to use the former in the form (9), we should require as left premise

$\Gamma_3$ $\qquad\qquad A \supset . A \supset B \Vdash A$

but this is false on the 0-1 valuation (cf. Sec. 2), and hence it is not derivable in LC, let alone in LA.   We therefore take the second alternative, $P*$.
In that case the premise would be

$\Gamma_4$ $\qquad\qquad A \supset . A \supset B, A \Vdash B$

Now we have no choice but to use $*P$, for which the premises are

$\Gamma_5$ $\qquad\qquad A \supset . A \supset B, A \Vdash A$
$\Gamma_6$ $\qquad\qquad A \supset . A \supset B, A, A \supset B \Vdash B$

Here $\Gamma_5$ is of type (a1).   As for $\Gamma_6$, if we use $*P$ with $A \supset B$ as principal
constituent,[1] the premises would be

$\Gamma_7$ $\qquad\qquad A \supset . A \supset B, A \supset B, A \Vdash A$
$\Gamma_8$ $\qquad\qquad A \supset . A \supset B, A \supset B, A, B \Vdash B$

Since these are both of type (a1), we have a derivation of $\Gamma_1$.   The derivation in tree form would look like this:

$$\frac{\Gamma_5 \quad \dfrac{\dfrac{\Gamma_7 \quad \Gamma_8}{\Gamma_6}*P}{}}{\dfrac{\dfrac{\Gamma_4}{\Gamma_2}P*}{\Gamma_1}P*}*P$$

---

[1] If we used $A \supset . A \supset B$ again as principal constituent, the right premise would be
the same as $\Gamma_6$.

*Example 2.* Let us investigate the derivability in LA($\mathfrak{O}$) and LC($\mathfrak{O}$) of

$\Gamma_1$  $\qquad\qquad \Vdash A \supset B . \supset A :\supset A$

where $A$ and $B$ are elementary. Suppose, now, that we are in LA. Evidently $\Gamma_1$ can only come from P∗, with the premise

$\Gamma_2$  $\qquad\qquad A \supset B . \supset A \Vdash A$

Then $\Gamma_2$ must come from ∗P with premises

$\Gamma_3$  $\qquad\qquad A \supset B . \supset A \Vdash A \supset B$

$\Gamma_4$  $\qquad\qquad A \supset B . \supset A, A \Vdash A$

Here $\Gamma_4$ is of type $(pl)$. We therefore investigate $\Gamma_3$. The only rules possible are ∗P, P∗. If we used the former, the left premise would have to be identical with $\Gamma_3$. For the latter the premise is

$\Gamma_5$  $\qquad\qquad A \supset B . \supset A, A \Vdash B$

Here the only possibility is ∗P. But in that case the right premise would have to be the same as $\Gamma_5$. Thus the set of statements $\Gamma_1$ to $\Gamma_3$, $\Gamma_5$ does not contain any prime statements, yet is such that none of its members can be the conclusion of a rule instance which does not have some member of the set as premise. It follows that a demonstration of $\Gamma_1$ in LA($\mathfrak{O}$) is impossible.

It is easy, however, to derive $\Gamma_1$ in LC($\mathfrak{O}$). In fact, in LC we could replace $\Gamma_3$ by

$\Gamma_3'$  $\qquad\qquad A \supset B . \supset A \Vdash A \supset B, A$

and this follows by P∗ from the prime statement

$\qquad\qquad A \supset B . \supset A, A \Vdash B, A$

Thus $\Gamma_1$ is derivable in LC($\mathfrak{O}$), but not in LA($\mathfrak{O}$).

In Example 1 we excluded certain possibilities (for example, $\Gamma_2$) by appealing to the 0-1 interpretation of Sec. 2. It was not necessary to do this; we could have handled all the possible alternatives by the method of Example 2 until we reached the proof. On the other hand, if we had made the same appeal in Example 2, we could have observed that $\Gamma_3$ is invalid on that interpretation (viz., where $A$, $B$ have the values 1, 0, respectively) and ended the argument for LA at that point. This shows how the use of known models can shorten the search for a decision. Theorems and techniques to be established later will shorten it still more.

The argument for Example 1 shows that its $\Gamma_1$ could have been derived using the original ∗P—in other words, the result is derivable in Formulation I without ∗W. It will be instructive to consider an example for which that is not true.

*Example 3.* Let us investigate the demonstrability in LA($\mathfrak{O}$) of the distributive law

$\qquad\qquad \Vdash A \wedge (B \vee C) . \supset . (A \wedge B) \vee (A \wedge C)$

Here we adopt the abbreviations

$\qquad\qquad D_1 \equiv A \wedge (B \vee C) \qquad D_2 \equiv (A \wedge B) \vee (A \wedge C)$

so that our thesis is

$$\Gamma_1 \qquad\qquad\qquad \Vdash D_1 \supset D_2$$

This can come only via P* from

$$\Gamma_2 \qquad\qquad\qquad D_1 \Vdash D_2$$

Suppose now that we are using Formulation I without *W*. Then there are two possibilities, *Λ and V*. But for *Λ the premises would have to be one or the other of

$$\Gamma_3 \qquad\qquad\qquad A \Vdash D_2$$
$$\Gamma_4 \qquad\qquad\qquad B \vee C \Vdash D_2$$

whereas for V* we need similarly one or the other of

$$\Gamma_5 \qquad\qquad\qquad D_1 \Vdash A \wedge B$$
$$\Gamma_6 \qquad\qquad\qquad D_1 \Vdash A \wedge C$$

All four of these possible premises are invalid in the 0-1 valuation. It is therefore impossible to derive $\Gamma_1$ in Formulation I without *W. With *W we can proceed thus: $\Gamma_2$ can come via *W from

$$\Gamma_7 \qquad\qquad\qquad D_1, D_1 \Vdash D_2$$

and this in turn by two applications of *Λ successively from

$$\Gamma_8 \qquad\qquad\qquad A, D_1 \Vdash D_2$$
$$\Gamma_9 \qquad\qquad\qquad A, B \vee C \Vdash D_2$$

Here $\Gamma_9$ can come from *V from $\Gamma_{10}$ and $\Gamma_{11}$, namely,

$$\Gamma_{10} \qquad\qquad\qquad A, B \Vdash D_2$$
$$\Gamma_{11} \qquad\qquad\qquad A, C \Vdash D_2$$

For $\Gamma_{10}$ we can use V* and the premise

$$\Gamma_{12} \qquad\qquad\qquad A, B \Vdash A \wedge B$$

which comes via Λ* from the two prime premises

$$A, B \Vdash A \qquad A, B \Vdash B$$

We can prove $\Gamma_{11}$ similarly. Thus $\Gamma_1$ is demonstrable with *W, but not without it.

These examples will suffice to illustrate the technique. We shall return to general considerations.

**6. General properties of the rules; constituents.** Before we proceed to discuss modifications, there will be formulated here certain general conditions on the rules which will turn out to be important in the proofs of theorems. Later, when we meet other systems of similar structure, but differing in the presence of new operations (and perhaps in other respects), the theorems will generalize readily if these new systems satisfy the same conditions.

The *constituents* of an instance of a rule are defined as the constituents of the premises together with those of the conclusion. These constituents are classified into principal, subaltern, and parametric constituents. For the operational rules these are as follows. The *principal constituent* is the new constituent introduced into the conclusion. The *subaltern constituents* (or simply the subalterns) are the constituents in the premises, replicas of which are combined[1] by an operation to form the principal constituent. The *parametric constituents* are those, like the constituents of the various $\mathfrak{X}$'s, $\mathfrak{Y}$'s, and $\mathfrak{Z}$'s, which appear in the premises and are carried unmodified, so to speak, into the conclusion. By analogy we can define such constituents in all the other rules except *C*. Thus, in *W* there are two like constituents in the premise which are replaced by a single constituent, which is like them, in the conclusion; the latter is the principal constituent, and the two former the subalterns. In *K* the new constituent which is introduced in the conclusion is the principal constituent; there are no subalterns. In ⊢* the $E_0$ in the conclusion is the principal constituent; the $E_i$ in the premises are the subaltern constituents. In Px the unique constituent of the consequent of the conclusion is the principal constituent; in the premise the $A \supset B$ in the antecedent and the $A$ in the consequent will be the subalterns. The parametric constituents in all these cases will be those included in $\mathfrak{X}$, $\mathfrak{Y}$, or $\mathfrak{Z}$.

In the sequel there will be occasion for introducing modifications which will make possible situations not considered in the foregoing. For example, the rule (9) is a modification of *P in which there is a copy of the principal constituent in each premise. Such copies of the principal constituent appearing in one or more premises [in (9) it is redundant in the right premise] will be called *quasi-principal constituents* and will be reckoned as special cases of subalterns.

To take care of such eventualities let us formulate the following conditions on the rules:

($r$1) Every constituent in the premises and conclusion is either the principal constituent, a subaltern constituent, or a parametric constituent.

($r$2) The principal constituent, if present, is unique and occurs in the conclusion only.

($r$3) The subaltern constituents, if present, occur in the premises only.

($r$4) There is an equivalence relation among the parametric constituents (p.c.) which we call *congruence*, such that ($a$) congruent p.c. are alike and on the same side; ($b$) every p.c. is congruent to exactly one p.c. in the conclusion; and ($c$) every p.c. is congruent to at least one p.c. in the premises and to at most one in any one premise. A set of mutually congruent p.c. will be called a *parameter*.

($r$5) A correct inference by any rule remains correct if a parameter is deleted.

---

[1] In the case of the left half of V* in Formulation I, the subaltern constituent is $A$ and the principal constituent is $A \lor B$, which seems to involve an extraneous $B$. But $A \lor B$ can be thought of as the result of performing on $A$ the operation $- \lor B$.

($r6$) A correct inference by any rule remains correct if a parameter is changed or a new parameter is inserted, provided the general restrictions of the system[1] and of ($r4$) are satisfied.

($r7$) The principal constituent is composite, and is formed from replicas of the subaltern constituents by an operation.

In regard to ($r4$), the different formulations satisfy more specific requirements. In Formulation I all rules except ∗P satisfy a more stringent form of ($r4$), which will be called ($r4$)′:

($r4$)′ The congruence relation satisfies ($a$), ($b$) of ($r4$), and ($c$′), every p.c. is congruent to a unique p.c. in each premise.

We call a rule which satisfies ($r1$) to ($r7$) a *regular rule*; one which satisfies ($r1$) to ($r5$) and ($r7$), a *semiregular rule*; one which fails to satisfy ($r7$), an *irregular rule*. Thus the operational rules of $LA_1$, $LC_1$, $LC_m$ are regular; the structural rules, ⊢∗, and (in $LC_1$) Px are irregular. If we leave the rules ∗C∗ to one side, all the rules of $LA_1$, $LC_1$, $LC_m$ satisfy ($r1$) to ($r6$), but in $LA_m$ the rule P∗ fails to satisfy ($r6$).†

The term '*L system*' will be applied to any system formed from LA or LC by adjoining additional rules, and possibly imposing additional conditions, such that ($r1$) to ($r5$) are satisfied.[2] Thus $LC_1$ and $LA_m$ are L systems. To such a system the terms 'singular', 'multiple', and 'mixed' may be applied as explained in Sec. 3$e$.

**7. Conventions regarding theorems and proofs.** The use of the rules ∗C∗ in what follows will generally be tacit. Unless there is an express statement to the contrary, these rules are assumed everywhere; rearrangements of prosequences are made in proofs without explicit justification; and in counting steps (for certain inductions) such rearrangements will not be counted as separate steps. In short, prosequences which are rearrangements of one another will be treated as identical.

Capital Greek letters 'Γ', 'Δ', 'Θ' will be used to designate statements or sets of statements.

If Θ is a set of elementary statements, a *deduction* from Θ is a construction (Sec. 2A6) on the basis consisting of prime statements and Θ, using the rules of Sec. 4. In accordance with Sec. 2A6, such a deduction can be represented as a sequence Δ of elementary statements $Γ_1$, $Γ_2$, . . . , $Γ_n$ such that each $Γ_k$ is either (1) in Θ, (2) a prime statement, or (3) derived from one or more of its predecessors by a rule. We can suppose without loss of generality that every $Γ_k$, except the last, is used once and only once as premise for

---

[1] By "general restrictions" I mean those which apply to the system or formulation as a whole, as opposed to those which apply to individual rules. Thus in a singular system no parameter can be added on the right (since we do not, as yet, have the possibility of a void consequent), whereas in a multiple system there is no such restriction. In a mixed system there is no general restriction forbidding the addition of a parameter on the right, but there may be in certain rules; in such a system the condition ($r6$) is not satisfied on the right for those rules.

† See the preceding footnote.

[2] We shall find it expedient to consider systems in which even some of these conditions are modified. We may call these modified L systems. Much of what is said about L systems applies to them with reservations.

inferring some $\Gamma_m$, $m > k$ by a rule $R_m$. In such a case $\Delta$ will be called a⁻ *regular deduction*. A demonstration is a deduction with $\Theta$ void.

An $\mathfrak{S}$ *deduction* is one such that its statements (including those of the basis) are of the form (6), where $A$ is elementary, and the only rule used is ⊢∗. According to the interpretation of Sec. 1, such a deduction can be interpreted as a deduction in the underlying system $\mathfrak{S}$ defined by the postulates (1).

We now consider certain conventions which are suggested by Kleene [PIG].

A constituent in the conclusion of a rule will be called an *immediate descendant* of one in the premises just when either the two are congruent parametric constituents or the former is principal constituent and the latter is a subaltern. The relation of being a *descendant* is defined as the quasi ordering generated by immediate descendance. A constituent will be called an *ancestor* of a second one just when the second is a descendant of the first. Note that descendance and the ancestral relation are both reflexive.[1] In both cases the modifier 'parametric' will mean that only parametric constituents are involved, so that a *parametric ancestor* or *descendant* is one which would still be an ancestor or descendant if the case of a principal constituent vs. a subaltern were omitted from the definition of immediate descendance.

The following theorem, due to Kleene, is valid in any L system.

**Theorem 1.** *If a constituent in the conclusion of a deduction has no ancestor which is constituent of a prime statement, then the deduction remains valid if that constituent and all its ancestors are omitted.*

*Proof.* No prime statement is invalidated by the omission. Likewise, the inferences remain valid, although certain of them may collapse because the premises and conclusion become identical. In that case the conclusion and, if there is more than one premise, the entire derivation of the extra premises can be omitted, Q.E.D.

We may speak of an *ultimate ancestor* (*ultimate descendant*) of a constituent as one which has no further ancestors (descendants). Thus the ultimate ancestors of a constituent which are not constituents of a prime statement are principal constituents of an instance of ∗K∗. The theorem applies when all the ultimate ancestors are of the latter kind. In particular, the theorem applies if there is no parametric ancestor which is constituent of a prime statement or principal constituent of a rule other than ∗K∗.

**8. Modified formulations II, IK, IIK.** As already stated, there exist a number of modifications of the formulation of Sec. 4. Here we shall consider a modification to be called Formulation II, together with certain modifications due to Ketonen [UPK].

Formulation II differs from Formulation I in that the parametric constituents in the different premises are not identified in the rules with more than one premise. The formulation satisfies the following specialization of $(r4)$:

$(r4)''$ The congruence relation satisfies $(a)$, $(b)$ of $(r4)$, and also $(c'')$, every p.c. in the conclusion is congruent to exactly one p.c. in the premises collectively.

---

[1] By definition of 'quasi ordering'.

The rules affected are $*$P, $\Lambda*$, $*$V, and $\vdash*$. For the first, second, and fourth of these the modified rules are:

$*$P
$$\frac{\mathfrak{X}_1 \Vdash A,\, \mathfrak{Z} \qquad \mathfrak{X}_2,\, B \Vdash \mathfrak{Y}}{\mathfrak{X}_1,\, \mathfrak{X}_2,\, A \supset B \Vdash \mathfrak{Y},\, \mathfrak{Z}}$$

$\Lambda*$
$$\frac{\mathfrak{X}_1 \Vdash A,\, \mathfrak{Z}_1 \qquad \mathfrak{X}_2 \Vdash B,\, \mathfrak{Z}_2}{\mathfrak{X}_1,\, \mathfrak{X}_2 \Vdash A \wedge B,\, \mathfrak{Z}_1,\, \mathfrak{Z}_2}$$

$\vdash*$   If $E_1, E_2, \ldots, E_m \vdash_0 E_0$

$$\frac{\mathfrak{X}_i \Vdash E_i,\, \mathfrak{Z}_i \qquad i = 1, 2, \ldots, m}{\mathfrak{X}_1,\, \mathfrak{X}_2,\, \ldots,\, \mathfrak{X}_m \Vdash E_0,\, \mathfrak{Z}_1,\, \mathfrak{Z}_2,\, \ldots,\, \mathfrak{Z}_m}$$

For $*$V the rules for the singular and multiple cases must be stated separately, thus:

$*$V (singular)
$$\frac{\mathfrak{X}_1,\, A \Vdash \mathfrak{Y} \qquad \mathfrak{X}_2,\, B \Vdash \mathfrak{Y}}{\mathfrak{X}_1,\, \mathfrak{X}_2,\, A \vee B \Vdash \mathfrak{Y}}$$

$*$V (multiple)
$$\frac{\mathfrak{X}_1,\, A \Vdash \mathfrak{Y}_1 \qquad \mathfrak{X}_2,\, B \Vdash \mathfrak{Y}_2}{\mathfrak{X}_1,\, \mathfrak{X}_2,\, A \vee B \Vdash \mathfrak{Y}_1,\, \mathfrak{Y}_2}$$

The equivalence of Formulations I and II is shown by the following theorem:

**Theorem 2.** *The rules of Formulation II can be derived from those of Formulation I by the use of $*$K$*$ (and $*$C$*$); the converse derivation can be made with the use of $*$W$*$ (and $*$C$*$).*

*Proof.* Suppose we have an inference by Formulation II; then the parameters can be made the same in the premises and conclusion by using $*$K$*$ to introduce the missing constituents; the conclusion can then be inferred by Formulation I. If we have an inference by Formulation I, then by Formulation II we can draw the same conclusion except that there may be several copies of the same parameter corresponding to the different premises; these copies can then be removed by $*$W$*$.

For example, in the case of $*$P for a multiple system the I-II transformation can be made thus (here $\mathfrak{Y}$ is singular):

$$\frac{\dfrac{\mathfrak{X}_1 \Vdash A,\, \mathfrak{Z}_1}{\mathfrak{X}_1,\, \mathfrak{X}_2 \Vdash A,\, \mathfrak{Z}_1,\, \mathfrak{Z}_2}\ *\text{K}* \qquad \dfrac{\mathfrak{X}_2,\, B \Vdash \mathfrak{Y},\, \mathfrak{Z}_2}{\mathfrak{X}_1,\, \mathfrak{X}_2,\, B \Vdash \mathfrak{Y},\, \mathfrak{Z}_1,\, \mathfrak{Z}_2,}\ *\text{K}*}{\mathfrak{X}_1,\, \mathfrak{X}_2,\, A \supset B \Vdash \mathfrak{Y},\, \mathfrak{Z}_1,\, \mathfrak{Z}_2}\ *\text{P}_\text{I}$$

In the opposite direction, the transformation is made thus:

$$\frac{\dfrac{\mathfrak{X} \Vdash A,\, \mathfrak{Z} \qquad \mathfrak{X},\, B \Vdash \mathfrak{Y},\, \mathfrak{Z}}{\mathfrak{X},\, \mathfrak{X},\, A \supset B \Vdash \mathfrak{Y},\, \mathfrak{Z},\, \mathfrak{Z}}\ *\text{P}_\text{II}}{\mathfrak{X},\, A \supset B \Vdash \mathfrak{Y},\, \mathfrak{Z}}\ *\text{W}*$$

This completes the proof of Theorem 2.

The modifications of Ketonen concern the rules $*$P, $*\Lambda$, V$*$ in the multiple systems. In the case of $*$P, the modification is

$*$P
$$\frac{\mathfrak{X} \Vdash A,\, \mathfrak{Y} \qquad \mathfrak{X},\, B \Vdash \mathfrak{Y}}{\mathfrak{X},\, A \supset B \Vdash \mathfrak{Y}}$$

In the cases $*\Lambda$ and $V*$ the Ketonen rules are

$$*\Lambda \quad \frac{\mathfrak{x}, A, B \Vdash \mathfrak{Y}}{\mathfrak{x}, A \wedge B \Vdash \mathfrak{Y}} \qquad V* \quad \frac{\mathfrak{x} \Vdash A, B, \mathfrak{Z}}{\mathfrak{x} \Vdash A \vee B, \mathfrak{Z}}$$

Of these transformations, the change in $*\Lambda$ can be made even in a singular system. We therefore define *Formulation IK* as that obtained by carrying out the Ketonen modifications in so far as possible—in the singular system leaving $*P$ and $V*$ unchanged. We may similarly define a *Formulation IIK* as that derived from Formulation II by changing $*\Lambda$ and, in so far as possible, $V*$, leaving $*P$ unchanged.

Note that in Example 3 of Sec. 5 the transition from $\Gamma_2$ to $\Gamma_9$ can be made directly by the Ketonen form of $*\Lambda$. We shall see later that the Ketonen rules have generally (with an important exception) the effect of making the rules $*W*$ dispensable.

The equivalence of Formulations I and IK is shown as follows:

**Theorem 3.** *An inference by either of the rules $*\Lambda$ or $V*$ in Formulation I can be obtained by an application of the corresponding Ketonen rule preceded by a single application of $*K*$ on the same side; conversely, an inference by the Ketonen rule can be obtained by two applications of the rule of Formulation I followed by an application of $*W*$ on the same side. In the case of $*P$, an application of $K*$ to the left premise followed by the Ketonen rule will give any inference by $*P$ in Formulation I; the converse requires at most a previous application of $K*$ and a following one by $W*$.*

*Proof.* For the case of $*\Lambda$ the two proofs are as follows:

$$\frac{\dfrac{\mathfrak{x}, A \Vdash \mathfrak{Y}}{\mathfrak{x}, A, B \Vdash \mathfrak{Y}} *K}{\mathfrak{x}, A \wedge B \Vdash \mathfrak{Y}} *\Lambda \qquad \frac{\dfrac{\dfrac{\mathfrak{x}, A, B \Vdash \mathfrak{Y}}{\mathfrak{x}, A \wedge B, B \Vdash \mathfrak{Y}} *\Lambda}{\mathfrak{x}, A \wedge B, A \wedge B \Vdash \mathfrak{Y}} *\Lambda}{\mathfrak{x}, A \wedge B \Vdash \mathfrak{Y}} *W$$

(The other case of the left-hand proof is similar.)

In the case of $V*$, the situation is dual to that of $*\Lambda$.

In the case of $*P$ the derivation of the Formulation I rule from that of Ketonen[1] is

$$\frac{\dfrac{\mathfrak{x} \Vdash A, \mathfrak{Z}}{\mathfrak{x} \Vdash A, \mathfrak{Y}, \mathfrak{Z}} K* \quad \mathfrak{x}, B \Vdash \mathfrak{Y}, \mathfrak{Z}}{\mathfrak{x}, A \supset B \Vdash \mathfrak{Y}, \mathfrak{Z}} *P$$

Conversely, the Ketonen form of $*P$ is identical with that case of $*P$ in Formulation I in which $\mathfrak{Y}$ is void. If void prosequences on the right are not admitted, we take the $\mathfrak{Y}$ of Ketonen's $*P$ to be $C, \mathfrak{Z}$; then the rule can be derived in Formulation I thus:

$$\frac{\mathfrak{x} \Vdash A, C, \mathfrak{Z} \quad \dfrac{\dfrac{\mathfrak{x}, B \Vdash C, \mathfrak{Z}}{\mathfrak{x}, B \Vdash C, C, \mathfrak{Z}} K*}{}}{\dfrac{\dfrac{\mathfrak{x}, A \supset B \Vdash C, C, \mathfrak{Z}}{\mathfrak{x}, A \supset B \Vdash C, \mathfrak{Z}} W*}{}} *P$$

This completes the proof.

---

[1] The proof is for the multiple case only. In the singular case the two formulations of $*P$ are identical.

The Ketonen modification of ∗P has somewhat the opposite effect of that of Formulation II, since it makes the parametric constituents uniform in all the premises, and thus $(r4)'$ is true without exception. Of course, it is only applicable in case ∗P is multiple; in the singular cases we must leave ∗P as in Formulation I, and the exception to $(r4)'$ must stand.

Theorems 2 and 3 use the structural rules ∗K∗ and ∗W∗ to establish the equivalence of Formulations I and II, on the one hand, and of the Ketonen rules and those of Formulation I, on the other. In generalized situations where the ∗K∗ and ∗W∗ do not both hold, these equivalences may fail. In such a case one of the variant formulations may have definite advantages.

**9. Some simple properties.** We shall establish here two rather simple, but unrelated, properties.

**Theorem 4.** *Let the prime statement scheme* $(p1)$ *be replaced by the scheme*

$$(p1)' \qquad\qquad E \Vdash E$$

*where* $E$ *is elementary. Then the general scheme* $(p1)$ *is an elementary theorem scheme of* $LA_1$.

*Proof.* We proceed by structural induction. The basic step, where $A$ is elementary, is precisely $(p1)'$; the inductive step is established using the rules of Formulation II and the Ketonen form of ∗Λ as follows:

$$\frac{\dfrac{A \Vdash A \qquad\qquad B \Vdash B}{A, A \supset B \Vdash B}\;∗\mathrm{P}}{A \supset B \Vdash A \supset B}\;\mathrm{P}∗$$

$$\frac{\dfrac{A \Vdash A \qquad\qquad B \Vdash B}{A, B \Vdash A \wedge B}\;Λ∗}{A \wedge B \Vdash A \wedge B}\;∗Λ \qquad \text{(Ketonen form)}$$

$$\frac{\dfrac{A \Vdash A}{A \Vdash A \vee B}\;\mathrm{V}∗ \qquad \dfrac{B \Vdash B}{B \Vdash A \vee B}\;\mathrm{V}∗}{A \vee B \Vdash A \vee B}\;∗\mathrm{V}$$

This completes the proof.

*Remark.* The second of the above derivations could be replaced by the dual of the third, and the third, if multiple elementary statements are allowed, by the dual of the second. Note that the proof does not require any rules other than those stated.

Statements of the forms $(a1)$ and $(a2)$ in Sec. C2 are obtained from $(p1)$ and $(p2)$ by ∗K∗ only. Such statements will be called *quasi-prime statements*. In such cases the constituent(s) introduced by the original prime statement will be called the *principal constituent(s)*. A prime (quasi-prime) statement will be called *elementary* just when the (corresponding) prime statement is of type $(p1)'$ or $(p2)$. As long as ∗K∗ are accepted, a demonstration starting with quasi-prime statements can be converted trivially into one starting with prime statements.

**Theorem 5.** *Let* $A_1, \ldots, A_m, B$ *be elementary, and let* $B$ *be* $\mathfrak{S}$ *deducible from* $A_1, \ldots, A_m$. *Then* (2) *holds in* $LA_1$.

*Proof.* If $B$ is one of the $A_i$, then (2) holds by $(p1)'$ and $*K$. If $B$ is an axiom, then (2) holds by $(p2)$ and $*K$. Suppose now that $B \equiv B_0$ and

$$B_1, B_2, \ldots, B_n \vdash_0 B_0$$

Suppose that for all $B_j$,

$$A_1, \ldots, A_m \Vdash B_j$$

Then (2) follows by $\vdash_*$. The theorem therefore follows by deductive induction on the given $\mathfrak{S}$ deduction.

## D. EQUIVALENCE OF THE SYSTEMS

The theme of this section is the proof of equivalence of the various formulations of L systems, as given in Sec. C, with each other and with the other types of system formulated in Sec. B. The key theorem for this purpose is the elimination theorem mentioned in (5) of Sec. C. This theorem will be proved in Sec. 2 under circumstances which are generalized so as to be applicable later. After Sec. 2 the term 'L system' will be understood to include the elimination theorem. The inversion theorem of Sec. 1, although not strictly necessary for the elimination theorem, is of a similar nature and allows shortening of the proof of Sec. 2 in a number of special cases and strengthening of it in other cases; since the inversion theorem is needed in Sec. E, it is put at the beginning of this section. In the later articles of this section the elimination theorem is applied to prove the equivalences mentioned, together with some other results which follow more or less immediately. These include, in Sec. 6, the completeness of LC.

**1. Direct inversion of inferences.** It is often desirable to know, when we have a proof whose concluding statement contains a composite constituent which is parametric, whether the proof can be recast so as to end with the introduction of that constituent. This amounts to showing that the rule in question can be inverted. We shall study here a method for carrying out this inversion under certain conditions.

Let $\Gamma$ be an elementary statement containing a compound constituent $M$. Let $\mathfrak{X}$ be the other constituents of $\Gamma$. Let $\Gamma$ be written, without indication of the sides on which the constituents occur, as

$$\mathfrak{X}, M \tag{1}$$

For given $M$ let there be a unique operational rule $R$ which can have $M$ as principal constituent. Let $R$ have $p$ premises, and let the subalterns—supposedly uniquely determined by $M$—in the $i$th premise be $\mathfrak{U}_i$. (The parameters may, of course, be different in the different applications of $R$.) Suppose further that $\mathfrak{X}$ is such that $\Gamma$ could be inferred by $R$ from premises

$$\mathfrak{X}, \mathfrak{U}_i \tag{2}$$

We seek conditions under which we can derive (2) and from it get (1) by $R$.

We consider first the case where there are no rules other than $R$ and $*K*$ which introduce $M$. This requires, in particular, that $M$ be not introduced by $*W*$.

Let $\Delta$ be a proof of $\Gamma$ in tree form. Let $\Delta_1$ be the subtree of $\Delta$ formed by those nodes which contain a parametric ancestor of $M$. At each node of

$\Delta_1$ suppose we delete $M$ and put in $\mathfrak{U}_i$. If all the rules used in the subtree are such that $(r5)$, $(r6)$ hold, then all inferences in which $M$ was parametric will remain valid. We have to consider the top nodes of the subtree. We can exclude the case that such a top node is a prime statement with $M$ as principal constituent by requiring that all such principal constituents be elementary (Theorem C4). The only other possibilities are that $M$ be introduced by $*K*$, or that $M$ be introduced by an application of $R$. In the former case we can introduce $\mathfrak{U}_i$ by $*K*$; this may require the use of $*K*$ on the opposite side to that of its use in introducing $M$. In the latter case, when $M$ is replaced by $\mathfrak{U}_i$, the statement in question becomes the $i$th premise of that application of $R$. Thus, under the assumptions made, the changes convert the given proof of (1) into a proof of the $i$th statement (2). If we do this for all $i = 1, 2, \ldots, p$, then we have (2) and we can apply $R$ to conclude (1).

In the case where $*W*$ occurs, let us define a *quasi-parametric ancestor* as one obtained by modifying the definition of Sec. C7 by admitting the principal constituent of $*W*$ as immediate descendant of either of the subalterns. Then a quasi-parametric ancestor of $M$ will be like $M$ (Sec. C3c). Now carry out the replacement of $M$ by $\mathfrak{U}_i$ in the subtree $\Delta_1$ consisting of all nodes of $\Delta$ containing a quasi-parametric ancestor of $M$. Then an inference in $\Delta_1$ in which $M$ is principal constituent of $*W*$ would be of the form

$$\frac{\mathfrak{Y}, M, M}{\mathfrak{Y}, M}$$

This would be replaced by

$$\frac{\mathfrak{Y}, \mathfrak{U}_i, \mathfrak{U}_i}{\mathfrak{Y}, \mathfrak{U}_i}$$

and can be effected by applications of $*W*$, possibly on the opposite side. The rest of the argument goes through as before, there being simply some additional fuss when there are two or more instances of $M$ being introduced at different places.

This argument establishes the following theorem:

**Theorem 1.** *Let the prime statements of type $(p1)$ be such that the principal constituent is elementary. Let $\Gamma$ be an elementary statement containing a compound constituent $M$. Let $\Delta$ be a derivation in tree form ending in $\Gamma$, and let $\Delta_1$ be the subtree formed by all statements of $\Delta$ which contain a quasi-parametric ancestor of $M$. Then it is sufficient for the existence of a derivation $\Delta'$ of $\Gamma$ in which $M$ is introduced at the end by a single application of $R$ that the following conditions be fulfilled:*

*(a) The $M$ in $\Gamma$ is so placed that $\Gamma$ can be the conclusion of $R$ with $M$ as principal constituent.*

*(b) All the ultimate quasi-parametric ancestors of $M$, except for those introduced by $*K*$, are introduced by $R$ with the same subalterns $\mathfrak{U}_i$ in the $i$th premise.*

*(c) All the rules used in $\Delta_1$ satisfy $(r5)$ and $(r6)$, at least in so far as deleting $M$ and inserting $\mathfrak{U}_i$ is concerned.*[1]

---

[1] That is, whenever $M$ can appear as parameter, the $M$ can be dropped and the $\mathfrak{U}_i$ added.

(d) *If M can be inserted by* $*K*$, *so can any* $\mathfrak{U}_i$.

(e) *If M can be contracted by* $*W*$, *so can all the* $\mathfrak{U}_i$.

Let us consider the significance of the conditions of the theorem in the multiple systems $LA_m$ and $LC_m$.

The condition (a) imposes a restriction only in the case of $P*$ in $LA_m$. In that case where M is, say, $A \supset B$, it must be the sole constituent on the right in (1).

The condition (b) requires that the subalterns be uniquely determined (as to nature and side on which they occur) by M. This condition is not satisfied in the case of $*\Lambda$ and $V*$ of the original Formulation I, for there are then two distinct rules with the same principal constituent. In Formulation IK, however, the condition (b) imposes no restriction on the operational rules proper, but it might interfere with the possibility of applying the theorem if special rules are present. However, there can be no conflict with $\vdash*$ since the principal constituent of $\vdash*$ is elementary.

The condition (c) imposes no restriction in $LC_m$. In $LA_m$, however, there is a restriction in that the rule $P*$ fails to satisfy (r6). If there is an instance of $P*$ in $\Delta_1$, then M must be on the left (since it is parametric); in that case there is no difficulty unless one of the subalterns is on the right. This exceptional situation occurs only when R is $*P$.

The conditions (d) and (e) impose no restriction so long as the rules $*K*$ and $*W*$, if present at all, are assumed without special restrictions.

When the transformation indicated in the theorem can be carried out, we shall say that *a direct inversion of R can be completed*. The demonstration obtained for the ith premise (2) will be called a *direct inversion of R relative to that premise*. Such a relative direct inversion can be carried out whenever the conditions (b) to (e) are satisfied (relative to that premise); in such a case we shall say that R is *directly invertible* (relative to that premise). Thus a rule may be directly invertible with respect to all premises although the direct inversion cannot be completed on account of failure of the condition (a).

COROLLARY 1.1.   *In Formulation IK of* $LC_m$, *all operational rules are directly invertible and a direct inversion can always be completed.*

COROLLARY 1.2.   *In Formulation IK of* $LA_m$, *all operational rules except* $*P$ *are directly invertible. A direct inversion can always be completed with the following exceptions:*

*If R is* $P*$, *then M must be the sole constituent on the right in order for (1) to be obtained from (2) by R, and it may not be possible to invert* $*P$ *if there is an instance of* $P*$ *in* $\Delta_1$.

COROLLARY 1.3.   *If R is directly invertible relative to one of its premises, then the construction of the theorem leads from (1) to a demonstration of the corresponding statement (2) regardless of (a).*

The following are special cases of this corollary.

COROLLARY 1.4.   *In Formulation IK of* $LA_m$, *if*

$$\mathfrak{X} \Vdash A \supset B, \mathfrak{Z}$$

*then*                                $$\mathfrak{X}, A \Vdash B, \mathfrak{Z}$$

COROLLARY 1.5.    *In Formulation* IK *of* $LA_m$, *if*

$$\mathfrak{X}, A \supset B \Vdash \mathfrak{Y}$$

*then*                        $$\mathfrak{X}, B \Vdash \mathfrak{Y}$$

Let us define the *degree* of a proof tree as the total number of its nonstructural inferences, and the *rank* as the maximum number of such inferences in any one branch. Then we have the following:

COROLLARY 1.6.    *When the inversion is possible, the proof of each statement* (2) *has a degree and a rank which are not greater than that of* (1), *and the degree is actually less unless M can be dropped from* (1) *altogether. Further, the new proof does not contain any inference by nonstructural rules which were not used in the proof of* (1); *the same applies to the structural rules except that a rule used in* (1) *on one side may be needed in* (2) *on the other.*

*Proof.* This follows by inspection of the method of proof. The modifications made were changes in parameters, omissions of applications of $R$, and insertions of new structural rules as stated in the corollary. If there are no omissions of $R$, then $M$ can be omitted from (1) by Theorem C1.

The situation in the singular systems, which is evidently more complex, will not be treated here.[1]

The problem of inversion is related to that of permutability of inferences. Suppose that the last inference in $\Delta$ is by a rule $R'$ with $M$ as parameter. Then after the inversion there will be no applications of $R$ to introduce a quasi-parametric ancestor of $M$ before the various applications homologous to $R'$. This is one of the senses in which we can say that $R$ and $R'$ have been interchanged. (For other possible senses see Kleene [PIG].)

In the following examples this idea of interchange is the dominant consideration. It is left to the reader to see how the interchange can be obtained by the argument of Theorem 1. (In the proofs accompanying these examples the top nodes are quasi prime.)

*Example* 1.    $A \supset . A \supset B \Vdash A \supset B$.    (Compare Example 1 in Sec. C5.) If the last step in the demonstration is by P∗, the rest of the demonstration is uniquely determined in $LA_m$, except for a possible quasi-principal constituent in the ∗P, as follows:

$$\frac{A \Vdash A \quad \dfrac{\dfrac{A \Vdash A \quad A, B \Vdash B}{A, A \supset B \Vdash B} \ast P}{\dfrac{A \supset . A \supset B, A \Vdash B}{A \supset . A \supset B \Vdash A \supset B} P\ast} \ast P}{} $$

On the other hand, if the last step in the derivation is by ∗P without quasi-principal constituent, the premise on the left in $LA_m$ would have to be

$$\Vdash A, A \supset B \tag{3}$$

Since the only rule whose conclusion can give (3) is K∗,† and both of the two possible premises are invalid by 0-1 tables, (3) cannot be derived in $LA_m$. If we use the form with quasi-principal constituent, the premise would be

$$A \supset . A \supset B \Vdash A, A \supset B$$

---

[1] See, however, Exercise 13 at the end of Sec. E.

† The case of W∗ can be eliminated by a theorem to be proved later (cf. Corollary E7.1).

in $LA_m$. This is more complex than the original statement. To be sure it is derivable and the interchange may be regarded as satisfied in a trivial sense, yet no significant interchange has been accomplished, and it can be shown that no proof is possible in which an instance of P* is not preceded by one of *P.† In $LC_m$, however, the interchange is possible in a nontrivial sense, for (3) can be derived immediately by P* from the prime statement $A \Vdash A, B$, and the right premise for the final *P is prime.

*Example* 2.   $A \vee B, A \supset B \Vdash B$.   If the last rule applied is *V, the demonstration is as follows:

$$\frac{\dfrac{A \Vdash A, B \quad A, B \Vdash B}{A, A \supset B \Vdash B} \text{ *P} \quad B, A \supset B \Vdash B}{A \vee B, A \supset B \Vdash B} \text{ *V}$$

In reverse order the derivation in $LA_m$ is

$$\frac{\dfrac{A, B \Vdash A, B}{A \vee B \Vdash A, B} \text{ *V} \quad A \vee B, B \Vdash B}{A \vee B, A \supset B \Vdash B} \text{ *P}$$

The interchange is impossible in $LA_1$.

*Example* 3.   The following proof is valid in $LA_1$.

$$\frac{\dfrac{A \Vdash A}{A \Vdash B \vee A} \text{ V*} \quad \dfrac{B \Vdash B}{B \Vdash B \vee A} \text{ V*}}{A \vee B \Vdash B \vee A} \text{ *V}$$

This proof uses the original form of V*, which fails to satisfy the condition (*b*) that the rule is uniquely determined by the principal constituent. The two instances of V* are actually distinct rules. If we use the Ketonen form of the rule, which does satisfy the condition (*a*), the proof becomes the following in $LA_m$.

$$\frac{\dfrac{A \Vdash B, A}{A \Vdash B \vee A} \text{ V*} \quad \dfrac{B \Vdash B, A}{B \Vdash B \vee A} \text{ V*}}{A \vee B \Vdash B \vee A} \text{ *V}$$

In this form the rules can be permuted, thus:

$$\frac{\dfrac{A \Vdash B, A \quad B \Vdash B, A}{A \vee B \Vdash B, A} \text{ *V}}{A \vee B \Vdash B \vee A} \text{ V*}$$

*Example* 4.   Consider the following derivation in $LA_1$.

$$\frac{\dfrac{B \Vdash B}{A \wedge B \Vdash B} \text{ *}\wedge \quad \dfrac{A \Vdash A}{A \wedge B \Vdash A} \text{ *}\wedge}{A \wedge B \Vdash B \wedge A} \text{ }\wedge\text{*}$$

The theorem does not apply to this for the same reason as in Example 3.

† See Exercise 15.

However, in this case the order of the rules can be reversed, using the same rules, thus:

$$\frac{\dfrac{B \Vdash B}{A, B \Vdash B} *K \qquad \dfrac{A \Vdash A}{A, B \Vdash A} *K}{\dfrac{\dfrac{A, B \Vdash B \wedge A}{A \wedge B, B \Vdash B \wedge A} *\Lambda}{\dfrac{A \wedge B, A \wedge B \Vdash B \wedge A}{A \wedge B \Vdash B \wedge A} *W} *\Lambda} \Lambda*$$

In this case an application of $*W$ is necessary at the end. The combination of the two rules $*\Lambda$ and this $*W$ is of course the same as a single application of the Ketonen form of $*\Lambda$.

**2. The elimination theorem.** We shall now turn to the formulation and proof of the theorem of which the singular form was stated in (5) of Sec. C. With the same conventions as to '$\mathfrak{X}$', '$\mathfrak{Y}$', '$\mathfrak{Z}$' as in Sec. C3e, the statement is as follows:

**Elimination theorem.** *If*

$$\mathfrak{X}, A \Vdash \mathfrak{Y} \tag{4}$$

*and*

$$\mathfrak{X}' \Vdash A, \mathfrak{Z} \tag{5}$$

*then*

$$\mathfrak{X}, \mathfrak{X}' \Vdash \mathfrak{Y}, \mathfrak{Z} \tag{6}$$

The statement of the theorem does not specify the type of L system to which it applies. This is therefore not a theorem in the strict sense, but has the character of a theorem scheme. The validity of the theorem for specific L systems and types of systems will be stated in the regularly numbered theorems. The term 'elimination theorem' will be abbreviated 'ET'.

The first proof will be for systems $LA_1$, $LC_1$, and $LC_m$ (with some complication to include $LC_1$). This proof is given in such a way as to make as little use as possible of special and structural rules in the proof itself, so that the proof may apply to generalizations later.

We shall often say that (6) comes from (4) and (5) by the elimination of $A$ and that $A$ is the *eliminated proposition* in the particular instance of the theorem which is under discussion. The indicated occurrences of $A$ in (4) and (5) will be called the *eliminated constituents*. The premises (4) and (5) will sometimes be spoken of as the first and second premises, respectively.

The proof will involve three stages. In Stage 1 we examine the proof of (4). We assume as *hypothesis of the stage* that the theorem is true for the same $A$ and the same second premise whenever the proof of the first premise [corresponding to (4)] ends with the introduction of the eliminated constituent by a regular operational rule, and we show that for this same $A$ and second premise the theorem then holds generally. Since the hypothesis of the stage is vacuously true when $A$ is elementary, the theorem is completely proved in Stage 1 when $A$ is not compound. We therefore assume throughout the rest of the proof that $A$ is compound. In Stage 2 we perform a reduction, analogous to that of Stage 1, but relative to the proof of (5), rather than to that of (4), with the additional restriction that $A$ be compound. Finally, in Stage 3, under the assumption that elimination is possible for any proper component of $A$, we show that the hypotheses of the first two stages hold for a given compound $A$.

The entire argument is thus a complex double induction. The primary induction is a structural induction on the eliminated proposition. The basic step of this primary induction is taken care of incidentally in Stage 1. The rest of the argument may be regarded as the inductive step of the primary induction, but the inductive hypothesis plays no role until we come to Stage 3. Given a fixed compound $A$ and a given second premise, Stage 1 reduces the primary induction step to the special case where the demonstration of the first premise has a special form. Stage 2 does the same for the second premise. Since neither stage requires any restriction on or change in the opposite premise, the two together reduce the primary induction step (for a given fixed $A$) to the case where the demonstration of both of the premises have the special form. Then Stage 3 gives the *coup de grâce*. The argument in Stages 1 and 2 is a secondary deductive induction on the proofs of (4) and (5). These two stages are practically independent of one another; except for the fact that we have to take care of the basic step of the primary induction in Stage 1 and do not allow $A$ to be elementary in Stage 2, the two stages could be carried out in either order.

The proof for Stages 1 and 2 will apply not only to the systems $LA_1$ and $LC_1$ and to $LC_m$, but to L systems of rather general character. The addition of regular rules to these systems hardly affects the proof, but when additional irregular rules are adjoined, modifications may have to be made. Thus the argument of Stage 2 is considerably complicated by the presence of Px; the reader who is not interested in $LC_1$ may ignore these complications.

In case the rule for introducing the eliminated constituent on the left is directly invertible, we may replace Stage 1 by the argument of Sec. 1; similarly, if the rule for introducing it on the right is directly invertible, we can dispense with Stage 2. But the exceptions in Sec. 1 will require the present method, and it is nearly as easy to handle the general case by the present method as well. Thus the present proof, so far as it goes, is independent of Sec. 1. However, the method of Sec. 1 will permit some extensions.

In view of Theorem C2, we may suppose that we are dealing with Formulation II, for which the condition $(r4)''$ is satisfied. Likewise, in view of Theorem C4, we can suppose that the prime statements have only elementary constituents.

*Proof of Stage 1.* Let $\Delta$ be a regular derivation $\Gamma_1, \ldots, \Gamma_n$ of (4). Then, by definition of a regular derivation, each $\Gamma_k$ for $k < n$ is used in $\Delta$ once and only once as premise for deriving a $\Gamma_m$, $m > k$, by a rule $R_m$. Let $\Gamma_k$ be

$$\mathfrak{X}_k, \mathfrak{U}_k \Vdash \mathfrak{B}_k, \mathfrak{Y}_k$$

where $\mathfrak{U}_k$, $\mathfrak{B}_k$ (and hence $\mathfrak{X}_k$, $\mathfrak{Y}_k$) are defined by induction, working backward from $\Gamma_n$, as follows:

(*a*) $\mathfrak{U}_n$ is the indicated occurrence of $A$ in (4), $\mathfrak{B}_n$ is void, $\mathfrak{X}_n$ is $\mathfrak{X}$, and $\mathfrak{Y}_n$ is $\mathfrak{Y}$.

(*b*) If $\Gamma_k$ is used as premise for deriving $\Gamma_m$ by $R_m$, then:

    (*b1*) All parametric constituents of $\mathfrak{U}_m$ ($\mathfrak{B}_m$) which are in $\Gamma_k$ are in $\mathfrak{U}_k$ ($\mathfrak{B}_k$);

    (*b2*) If the principal constituent of an irregular rule is in $\mathfrak{U}_m$ or $\mathfrak{B}_m$, then each of the subaltern constituents is in that one of $\mathfrak{U}_k$, $\mathfrak{B}_k$ which is on the appropriate side.

By the general principle of an inductive definition, viz., that $\mathfrak{U}_k$, $\mathfrak{B}_k$ contain only those constituents which belong by virtue of the stated rules, $\mathfrak{U}_k$ and $\mathfrak{B}_k$ are defined for all $k \leq n$. Since the only irregular rules admitted so far are $*\mathrm{K}*$, $*\mathrm{W}*$, $\vdash*$, Px,† it follows that $\mathfrak{B}_k$ is void for all $k$ and that the constituents of $\mathfrak{U}_k$ are all like $A$. They are actually all the quasi-parametric ancestors (Sec. 1) of $A$.

Let $\Delta_1$ be that part of $\Delta$ in which $\mathfrak{U}_k$ (as well as $\mathfrak{B}_k$) is void, and let $\Delta_2$ be the rest of $\Delta$. Then since $\mathfrak{U}_k$ is void [under case $(b)$] whenever $\mathfrak{U}_m$ is void, all premises used to derive members of $\Delta_1$ are also in $\Delta_1$.

With each $\Gamma_k$ we now associate a statement $\Gamma'_k$ as follows:

$$\mathfrak{X}_k, (\mathfrak{X}') \Vdash \mathfrak{Y}_k, (\mathfrak{Z}) \qquad (3)$$

Here '$(\mathfrak{X}')$' indicates a set (possibly void) of replicas of $\mathfrak{X}'$, there being one such replica for every occurrence of $A$ in $\mathfrak{U}_k$; likewise, '$(\mathfrak{Z})$' indicates a set of replicas of $\mathfrak{Z}$, one for each occurrence of $A$ in $\mathfrak{U}_k$. Note that if $\Gamma_k$ is in $\Delta_1$, then $\Gamma'_k$ is the same as $\Gamma_k$, and $\Gamma'_n$ is precisely the statement (6).

We show by an induction on $k$—which is a deductive induction with reference to the fixed deduction $\Delta$—that if the hypothesis of the stage holds, every $\Gamma'_k$ is derivable. There are five cases to be considered, which will be indicated by the Greek letters '$\alpha$' to '$\epsilon$', as follows:

($\alpha$) $\Gamma_k$ is in $\Delta_1$. Then $\Gamma'_k$ is the same as $\Gamma_k$.

($\beta$) $\Gamma_k$ is prime and in $\Delta_2$. Since $\mathfrak{U}_k$ is then nonvoid, $\Gamma_k$ cannot be of type $(p2)$; hence $\Gamma_k$ must be

$$A \Vdash A$$

where the $A$ on the left is the unique constituent of $\mathfrak{U}_k$. Then $\Gamma'_k$ is precisely the statement (5) and hence is demonstrable. (By the restriction on the prime statements, $A$ is then elementary.)

($\gamma$) $\Gamma_k$ is in $\Delta_2$ and is derived by a rule $R_k$ for which all the constituents of $\mathfrak{U}_k$ are parametric. Let the premises be $\Gamma_i$, $\Gamma_j$, .... By the hypothesis of the deductive induction, $\Gamma'_i$, $\Gamma'_j$, ... are demonstrable. By $(r5)$, $(r6)$, $\Gamma'_k$ is derivable from $\Gamma'_i$, $\Gamma'_j$, ... ; moreover, $(r6)$ is applied on the right only if $\mathfrak{Z}$ is nonvoid. This shows that $\Gamma'_k$ is demonstrable in the appropriate system.

($\delta$) $\Gamma_k$ is obtained by an irregular rule whose principal constituent is in $\mathfrak{U}_k$. The only possible irregular rules are $*\mathrm{K}$ and $*\mathrm{W}$. Let the premise be $\Gamma_i$, where $\Gamma'_i$ is derivable by the hypothesis of the deductive induction. If the rule is $*\mathrm{W}$, then $\Gamma_k$ differs from $\Gamma_i$ only in that it contains one less instance of $A$ in $\mathfrak{U}_k$ and $\Gamma'_k$ differs from $\Gamma'_i$ only in that it contains one less replica of $\mathfrak{X}'$ on the left and, if $\mathfrak{Z}$ is nonvoid, one less replica of $\mathfrak{Z}$ on the right; thus $\Gamma'_i$ can be transformed into $\Gamma'_k$ by a succession of applications of $*\mathrm{W}$ and $\mathrm{W}*$, the latter being used only if $\mathfrak{Z}$ is nonvoid. If the rule is $*\mathrm{K}$, there is one additional instance of $A$ in $\mathfrak{U}_k$, and $\Gamma'_k$ can be obtained from $\Gamma'_i$ by using a succession of applications of $*\mathrm{K}$ to insert an additional replica of $\mathfrak{X}'$ on the left, and applications of $\mathrm{K}*$ to insert an additional replica of $\mathfrak{Z}$ on the right, the latter being necessary only if $\mathfrak{Z}$ is nonvoid. Thus $\Gamma'_k$ is derivable from $\Gamma'_i$.‡

---

† The proof of Stage 1 applies to $\mathrm{LC}_1$ without change.

‡ If we were to entertain systems in which one of the rules $*\mathrm{W}$ or $*\mathrm{K}$ was not accepted, then one or the other of the subcases could not occur. However, if we admit either of these rules and also allow nonvoid $\mathfrak{Z}$, then we must admit the corresponding rule on the right also.

($\epsilon$) $\Gamma_k$ is obtained by a regular operational rule whose principal constituent is in $\mathfrak{U}_k$. Let $\Gamma_k''$ be derived from $\Gamma_k$ by replacing all parametric constituents of $\mathfrak{U}_k$ by replicas of $\mathfrak{X}'$ and adding replicas of $\mathfrak{Z}$ on the right. Then we can derive $\Gamma_k''$ by the same argument as in ($\gamma$). We derive $\Gamma_k'$ from $\Gamma_k''$ by the hypothesis of the stage.

Since case $\epsilon$ cannot arise when $A$ is elementary, the proof of the elimination theorem is complete for that case. If $A$ is compound, the theorem is reduced to proving the hypothesis of Stage 1 for the particular $A$ and second premise concerned.

*Proof of Stage 2.* Stage 2 of the proof is rather similar to Stage 1; it is therefore permissible to be more brief and to restrict attention principally to those points in which the two treatments differ. Those differences arise from the singularity restrictions and from the existence of certain irregular rules with principal constituent on the right. Among these latter the rules ⊢∗ and Px have no analogues in Stage 1. Complications in regard to ⊢∗ are avoided by the exclusion of an elementary eliminated constituent, but those for Px have to be taken into account if the system $LC_1$ is to be treated at all.

Let $\Delta$ be a derivation $\Gamma_1, \ldots, \Gamma_n$ of (5). For $\Gamma_k$ we take the form

$$\mathfrak{X}_k', \mathfrak{U}_k \Vdash \mathfrak{B}_k, \mathfrak{W}_k$$

For the determination of $\mathfrak{U}_k$, $\mathfrak{B}_k$, we take the following rules:

(a) $\mathfrak{U}_n$ is void, $\mathfrak{B}_n$ is the indicated eliminated constituent in (5), $\mathfrak{X}_n'$ is $\mathfrak{X}'$, $\mathfrak{W}_n$ is $\mathfrak{Z}$.

(b) If $\Gamma_k$ is used as premise for deriving $\Gamma_m$,

(b1) The parametric constituents of $\mathfrak{B}_m(\mathfrak{U}_m)$ which are in $\Gamma_k$ are in $\mathfrak{B}_k(\mathfrak{U}_k)$.

(b2) If $R_m$ is an irregular rule whose principal constituent is in $\mathfrak{U}_m$ or $\mathfrak{B}_m$, then the subaltern constituents are in that one of the $\mathfrak{U}_k$, $\mathfrak{B}_k$ which is on the appropriate side.

(b3) If $R_m$ is a rule ∗P whose principal constituent is in $\mathfrak{U}_m$, let the left premise be $\Gamma_k$ and the right premise be $\Gamma_i$; then the subaltern constituent of $\Gamma_k$ is in $\mathfrak{B}_k$, but no specification is made as to the subaltern constituent in $\Gamma_i$.

This constitutes an inductive definition of the $\mathfrak{U}_k$, $\mathfrak{B}_k$, and hence of the $\mathfrak{X}_k'$, $\mathfrak{W}_k$. By induction, working from $\Gamma_n$ backward, we shall see that all constituents of $\mathfrak{B}_k$ are like $A$, and all those of $\mathfrak{U}_k$ are of the form $A \supset C$ (the $C$'s not necessarily the same). This is evidently true for $\Gamma_n$. Suppose it true for $\Gamma_m$. Then it is certainly true for the parametric constituents which are in $\Gamma_k$ by virtue of (b1). For the irregular rules entering under (b2), the possibilities are ∗K, ∗W on the left, and K∗, W∗, ⊢∗, and Px on the right; of these ⊢∗ is excluded by the requirement that the eliminated constituent be compound; in all the other cases the constituents introduced into $\Gamma_k$ are as stated. Finally, the constituent introduced into $\mathfrak{B}_k$ by virtue of (b3) is like $A$, which completes the induction. Note that $\mathfrak{U}_k$ is void, except in the case of $LC_1$, which is a singular system.

For the case where Px is not admitted, $\Gamma_k'$ will be the statement

$$\mathfrak{X}_k', (\mathfrak{X}) \Vdash (\mathfrak{Y}), \mathfrak{W}_k$$

where $(\mathfrak{X})$ is a set of replicas of $\mathfrak{X}$, and $(\mathfrak{Y})$ a set of replicas of $\mathfrak{Y}$, there being one replica of each for each occurrence of $A$ in $\mathfrak{B}_k$. If Px is admitted, we have necessarily a singular case,[1] in which $\mathfrak{Y}$ has a single constituent $B$; then $\Gamma'_k$ is obtained from $\Gamma_k$ by replacing a constituent $A \supset C$ of $\mathfrak{U}_k$ by $B \supset C$ and a constituent $A$ of $\mathfrak{B}_k$ by $B$ and by adjoining on the left a replica of $\mathfrak{X}$ for each constituent of $\mathfrak{U}_k$ or $\mathfrak{B}_k$. Then $\Gamma'_k$ is the same as $\Gamma_k$ for $\Gamma_k$ in $\Delta_1$ (i.e., when $\mathfrak{U}_k$ and $\mathfrak{B}_k$ are void); also $\Gamma'_n$ is precisely (6).

The proof that every $\Gamma'_k$ is demonstrable proceeds as in Stage 1. The added parametric constituents on the right come from replicas of $\mathfrak{Y}$, rather than of $3$, and those on the left from replicas of $\mathfrak{X}$, rather than of $\mathfrak{X}'$. In the singular cases $\mathfrak{Y}$ is singular, and either $\mathfrak{B}_k$ is singular and $\mathfrak{W}_k$ void or vice versa. If Px is admitted, there is one additional case, here called case $\zeta$. The details are as follows:

($\alpha$) $\Gamma_k$ is in $\Delta_1$. Then $\Gamma'_k$ is the same as $\Gamma_k$.

($\beta$) $\Gamma_k$ is in $\Delta_2$ and is prime. This case is impossible by the restrictions on the prime statements, since $A$ is compound.

($\gamma$) $\Gamma_k$ is in $\Delta_2$ and is derived by a rule $R_k$ for which all constituents in $\mathfrak{U}_k$ and $\mathfrak{B}_k$ are parametric. Then $\Gamma'_k$ is derivable by the same argument as in Stage 1.

($\delta$) $\Gamma_k$ is obtained by a rule $R_k$ satisfying the conditions ($b2$). If $R_k$ is one of $*$K$*$ or $*$W$*$, the situation is again analogous to that of Stage 1. If the rule is Px, leading from $\Gamma_i$ to $\Gamma_k$ by dropping a constituent $A \supset C$ from $\mathfrak{U}_i$, then the inference from $\Gamma'_i$ to $\Gamma'_k$ can be made by dropping an instance of $B \supset C$ by the same rule Px and dropping a superfluous replica of $\mathfrak{X}$ by successive applications of $*$W. It is necessary to postulate $*$W in this case.

($\epsilon$) $\Gamma_k$ is obtained by a rule $R_k$ for which the principal constituent is in $\mathfrak{B}_k$. This situation is analogous to that in Stage 1 using the hypothesis of the stage. Note that there can be parametric constituents in $\mathfrak{B}_k$, only in the multiple case. In the singular case we apply the hypothesis of the stage at once.

($\zeta$) $\Gamma_k$ is obtained by a rule $R_k$ satisfying the conditions ($b3$). Let the premises be $\Gamma_i$, $\Gamma_j$ where $\Gamma'_i$ and $\Gamma'_j$ are derivable. Then we derive $\Gamma'_k$ from $\Gamma'_i$ by the same rule.

*Proof of Stage* 3. We now suppose that the derivations of both (4) and (5) terminate in the introduction of the eliminated constituent $A$ on the left and right, respectively, by the appropriate operational rules, and that the elimination has been established if the eliminated proposition is a proper component of $A$. There are three cases corresponding to the three operations so far introduced.

In these cases it will be convenient to use '$\mathfrak{X}_1$', '$\mathfrak{X}_2$' for '$\mathfrak{X}$', '$\mathfrak{X}''$, and where these are subdivided, to understand that $\mathfrak{X}_{11}$, $\mathfrak{X}_{12}$ constitute $\mathfrak{X}_1$, etc.

CASE 1, $A \equiv B \supset C$. Then (4) comes from the premises

$$\mathfrak{X}_{11} \Vdash B, 3_1 \tag{7}$$

$$\mathfrak{X}_{12}, C \Vdash \mathfrak{Y}_1 \tag{8}$$

---

[1] If a multiple case were needed, it would not be difficult to treat it. One would simply replace $A \supset C$ by $\mathfrak{Y} \supset C$, where $\mathfrak{Y} \supset C$ is a prosequence whose constituents are $B \supset C$, with $B$ an element of $\mathfrak{Y}$. Such a case occurs in Chap. 6, but not in connection with a system of primary interest.

where $\mathfrak{Y}$ is $\mathfrak{Y}_1$, $\mathfrak{Z}_1$.  The premise for (5) is

$$\mathfrak{X}_2, B \Vdash C, \mathfrak{Z}_2 \qquad (9)$$

Then from (8) and (9), by elimination of $C$, we have

$$\mathfrak{X}_{12}, \mathfrak{X}_2, B \Vdash \mathfrak{Y}_1, \mathfrak{Z}_2$$

and from this and (7), by elimination of $B$, we have

$$\mathfrak{X}_{11}, \mathfrak{X}_{12}, \mathfrak{X}_2 \Vdash \mathfrak{Y}_1, \mathfrak{Z}_1, \mathfrak{Z}_2$$

which is (6).

CASE 2, $A \equiv B \wedge C$.  If we use the Ketonen form of $*\Lambda$, the premise for (4) is

$$\mathfrak{X}_1, B, C \Vdash \mathfrak{Y} \qquad (10)$$

The premises for (5) must be

$$\mathfrak{X}_{21} \Vdash B, \mathfrak{Z}_1 \qquad (11)$$

$$\mathfrak{X}_{22} \Vdash C, \mathfrak{Z}_2 \qquad (12)$$

From (10) and (12) and elimination of $C$ we have

$$\mathfrak{X}_1, \mathfrak{X}_{22}, B \Vdash \mathfrak{Y}, \mathfrak{Z}_2$$

From this and (11), eliminating $B$, we have

$$\mathfrak{X}_1, \mathfrak{X}_{21}, \mathfrak{X}_{22} \Vdash \mathfrak{Y}, \mathfrak{Z}_1, \mathfrak{Z}_2$$

which is (6).

Without the use of the Ketonen rule the situation would be dual to that considered in the next case.

CASE 3, $A \equiv B \vee C$.  In the multiple case, with the use of the Ketonen form for $\vee *$, the situation is dual to that in Case 2.  In the singular case we can argue as follows.  The premises for (4) must be

$$\mathfrak{X}_{11}, B \Vdash \mathfrak{Y} \qquad (13)$$

$$\mathfrak{X}_{12}, C \Vdash \mathfrak{Y} \qquad (14)$$

The premise for (5) must be one or the other of

$$\mathfrak{X}_2 \Vdash B \qquad \mathfrak{X}_2 \Vdash C \qquad (15)$$

Supposing it is the first of (15), then, eliminating $B$ with (13), we should have

$$\mathfrak{X}_{11}, \mathfrak{X}_2 \Vdash \mathfrak{Y}$$

This is not quite the same as (6); to get (6) we need to postulate $*K$ in order to get $\mathfrak{X}_{12}$ on the left.

This completes the proof of the elimination theorem so far as LA, LC, $LC_1$ are concerned.  Thus we have the following:

**Theorem 2.** *The elimination theorem holds for* $LA_1$, $LC_1$, $LC_m$. *It further holds for any L system formed by adding regular operational rules to one of these, provided that the argument of Stage 3 holds for the new operations.*

*Proof for* $LA_m$.†  The proof fails to go through for $LA_m$ because $P*$ fails to satisfy $(r6)$.  However, it can be proved for $LA_m$ by modifying the argument.  The crucial point is that when $\mathfrak{Z}$ is void, Stage 1 requires $(r6)$ only

† An alternative method of extending the proof to $LA_m$ is suggested in Sec. 5 (see Exercise 17 at the end of Sec. E).

on the left and therefore goes through for $LA_m$ for void $3$. In particular, ET is true for $A$ elementary and $3$ void.

If $A$ is compound and not of the form $B \supset C$, then we can eliminate Stages 1 and 2 by Theorem 1. Hence we have only to consider the cases where $A$ is elementary or $A$ is of the form $B \supset C$. In these we begin by modifying the proof of Stage 2. Since $\mathfrak{B}_k$ now consists of the quasi-parametric ancestors of the eliminated constituent, and since there are such in all[1] statements of $\Delta_2$, there can be no application of P* within $\Delta_2$. Therefore the part of the proof of Stage 2 under cases $\alpha$, $\gamma$, and $\delta$ stands without change and case $\zeta$ is vacuous. It is only necessary to consider case $\epsilon$ for the case where $A$ is $B \supset C$; in the case where $A$ is elementary, we have to reconsider not only ($\beta$) but the possibility that ⊢* may occur.

If $A$ is $B \supset C$, then case $\epsilon$ can occur only when the right side is singular. The Stage 2 induction is completed by direct application of the hypothesis of the stage. It remains to verify that hypothesis. But, under the conditions of that hypothesis we have a case in which $3$ is void, and therefore Stages 1 and 3 go through without trouble. ET is therefore proved provided it holds when $A$ is elementary.

If $A$ is elementary and case $\beta$ occurs, then again we have a void $3$ at that point, and the case has already been covered. Thus ET is proved in full for any elementary $A$ such that there is no instance of ⊢* introducing $A$ into $\Delta_2$.

If some parametric ancestor of the eliminated constituent is introduced into $\Delta_2$ by ⊢*, we use a tertiary induction[2] on the number of such applications of ⊢*. The preceding argument takes care of the basic step. Suppose then that $\Gamma_k$ is

$$\mathfrak{X}_k \Vdash A, 3_k$$

(where $3_k$ contains $\mathfrak{W}_k$ and parametric instances of $\mathfrak{B}_k$) and that $\Gamma_k$ is obtained by ⊢* from

$$\mathfrak{X}_{ki} \Vdash B_i, 3_{ki} \qquad i = 1, 2, \ldots, m \qquad (16)$$

where

$$\mathfrak{X}_k \equiv \mathfrak{X}_{k1}, \ldots, \mathfrak{X}_{km} \qquad 3_k \equiv 3_{k1}, \ldots, 3_{km}$$

and

$$B_1, \ldots, B_m \vdash_0 A$$

Then by Theorem C5,

$$B_1, \ldots, B_m \Vdash A$$

From this and (4), by cases already proved, we have

$$\mathfrak{X}, B_1, \ldots, B_m \Vdash \mathfrak{Y} \qquad (17)$$

After the transformations described in Stage 2, let (16) become

$$\mathfrak{X}''_{ki} \Vdash B_i, 3'_{ki}$$

----

[1] Except the last. It is assumed in this context that $k < n$.

[2] Alternatively, one may eliminate ⊢* by proving Theorem E4 below by a direct induction which does not involve ET. This is the only place where the presence of ⊢* causes any trouble.

From these and (17), eliminating the $B_i$ by the tertiary inductive hypothesis, we have $\Gamma'_k$. Thus the proof of ET for this $A$ and (5), and so for $LA_m$ generally, is complete.

This establishes the following:

**Theorem 3.**   *The elimination theorem holds for* $LA_m$.

There is a certain amount of interest in ET for modified L systems in which the structural rules are not postulated in their full strength. These structural rules have entered into the foregoing proofs in the following ways: (1) in case $\delta$ of Stage 1 (and also dually for Stage 2), it is required that if $*W*$ or $*K*$ holds on one side, it does on the other; (2) in case $\delta$ of Stage 2, we need $*W$ in order to take account of Px; (3) in Case 3 of Stage 3, we need $*K$ in the singular systems.[1]   Thus the proof of the theorem is not complete for generalizations in which some of these structural rules are denied.   However, in the multiple systems with Ketonen rules, ET, as here stated, does not require that either $*K*$ or $*W*$ be postulated.[2]

In case these principles are postulated, we have the following:

**Theorem 4.**   *If* $*W*$ *both hold, then* ET *entails the following* ET′:

ET′ *If*

$$\mathfrak{X}, A \Vdash \mathfrak{Y}$$

*and for some* $\mathfrak{Z}$ *which is a part* (*or the whole*) *of* $\mathfrak{Y}$

$$\mathfrak{X} \Vdash A, \mathfrak{Z}$$

*then*

$$\mathfrak{X} \Vdash \mathfrak{Y}$$

*Further, if* $*K*$ *both hold,* ET′ *entails* ET.   *Thus* ET *and* ET′ *are equivalent for* L *systems.*

The proof is analogous to that of Theorem C2.

Henceforth the definition of 'L system' (Sec. C6) will be understood to include the validity of ET.

**3. The replacement theorem.**   As a tool for later use it is expedient to formulate a theorem which has a relation to the theory of L systems similar to that which the replacement theorem of Sec. 3B had to the algebraic systems of Chap. 4.   Indeed, by passing to a higher stage of epitheory, it may be regarded as a modified form of that theorem; it will also be called the replacement theorem and abbreviated Rp when there is no likelihood of confusion.

Let $B$ be a proposition which contains a proposition $A$ as component. Then the composition from $A$ to $B$ is accomplished by a series of applications of unary operations of the forms

$$C \supset -, \; - \wedge C, C \wedge -, \; - \vee C, C \vee -, \; - \supset C$$

Here the dash indicates the position of the argument, while '$C$' indicates a parameter which may be different in the different steps of the composition, but is independent of the argument.   Now, on the hypothesis that

$$A \Vdash A' \tag{18}$$

---

[1] This is not surprising if we request that the rules $V*$ (and also the original form of $*\Lambda$) contain principles with approximately the same intuitive meaning as $K*$ ($*K$).

[2] This has not been checked for the case where there are inferences by $\vdash*$.

one can easily establish that

$$C \supset A \Vdash C \supset A'$$
$$A \wedge C \Vdash A' \wedge C$$
$$C \wedge A \Vdash C \wedge A'$$
$$A \vee C \Vdash A' \vee C$$
$$C \vee A \Vdash C \vee A'$$
$$A' \supset C \Vdash A \supset C$$

These show that if we interpret the relation $R$ of Sec. 3B as $\Vdash$, then the first five operations in the above list are directly monotone, while the sixth one is inversely monotone. Let us say that $A$ is *positive* or *negative* in $B$ according as the number of inversely monotone operations in the composition from $A$ to $B$ is even or odd. Now the relation $\Vdash$ is a quasi ordering by virtue of the elimination theorem. Hence by Theorem 3B1, if $B'$ is the result of replacing $A$ by $A'$ in that particular occurrence, the left-hand one of the two statements

$$B \Vdash B' \qquad B' \Vdash B \qquad\qquad (19)$$

will hold if $A$ is positive in $B$, while the right-hand one will hold if $A$ is negative in $B$.

Now let $\Gamma$ be an elementary statement in which there is a constituent $B$ containing an occurrence of $A$. Let us say that $A$ is positive in $\Gamma$ if $B$ is in the consequent of $\Gamma$ and $A$ is positive in $B$, or $B$ is in the antecedent of $\Gamma$ and $A$ is negative in $B$; and that $A$ is negative in $\Gamma$ if $B$ is in the consequent of $\Gamma$ and $A$ is negative in $B$, or $B$ is in the antecedent of $\Gamma$ and $A$ is positive in $B$. Let $\Gamma'$ be obtained from $\Gamma$ by putting $A'$ in the place of $A$ at that occurrence. Then we have the following:

**Theorem 5.** *In any* L *system with monotone operations, we can infer from* (18) *that*

$$\Gamma \to \Gamma' \qquad or \qquad \Gamma' \to \Gamma \qquad\qquad (20)$$

*respectively, according as $A$ is positive or negative in $\Gamma$.*

*Proof.* Let $B$ be on the right in $\Gamma$, so that $\Gamma$ is

$$\mathfrak{X} \Vdash B, \mathfrak{Y}$$

Then $\Gamma'$ is

$$\mathfrak{X} \Vdash B', \mathfrak{Y}$$

By the elimination theorem we have that one of the relations (20) which is on the same side as that occupied by the true statement in (19). This will be the left one if $A$ is positive in $\Gamma$ and the right one if $A$ is negative in $\Gamma$. This completes the proof for this case.

Let $B$ be on the left in $\Gamma$, so that $\Gamma$, $\Gamma'$ are, respectively,

$$\mathfrak{X}, B \Vdash \mathfrak{Y} \qquad \mathfrak{X}, B' \Vdash \mathfrak{Y}$$

Then the elimination theorem allows us to infer that one of (20) which is on the opposite side to that occupied by the true statement in (19). This will be the left side if $A$ is positive in $\Gamma$ (hence negative in $B$) and the right side if $A$ is negative in $\Gamma$. This completes the proof.

**4. Equivalence of T and singular L.** From the preliminary discussion in Secs. C1 and C2 it would be expected that the systems LA and LC would be equivalent to the corresponding T systems. That this is indeed so, and in exactly what sense, is now to be shown.

We first extend the T systems so as to include a rule analogous to $\vdash\ast$. This rule will be called $\vdash$i; its formulation is

$\vdash$i  If $E_1, \ldots, E_m \vdash_0 E_0$ (in the sense of Sec. C3$d$), then

$$\frac{E_1, \ldots, E_m}{E_0}$$

DEFINITION. The statement

$$\mathfrak{X} \Vdash^T B \tag{21}$$

relative to one of the systems TA or TC will mean that in the T system under discussion there is a derivation of $B$ such that all uncanceled premises are propositions in $\mathfrak{X}$.

Our task now is to show that (21) relative to TA (TC) is equivalent to

$$\mathfrak{X} \Vdash B$$

relative to LA$_1$ (LC$_1$). For the sake of explicitness, the latter of these relations will now be written

$$\mathfrak{X} \Vdash^L B \tag{22}$$

This investigation has one difficulty at the very beginning, in that the relation (21) is less rigorously formalized than (22). The statement (21) is to be verified by inspection of a tree diagram; we have to see that all the uncanceled premises of the proof are in $\mathfrak{X}$. We have to take it as intuitively evident that any true instance of (21) can be derived from the prime statements

($p$1)           $A \Vdash^T A$

($p$2)           $\Vdash^T E$      if $E$ is an axiom of $\mathfrak{S}$

by means of the rules analogous to $\ast$C, $\ast$W, and $\ast$K and the following rules:

Pe  $\dfrac{\mathfrak{X} \Vdash^T A \quad \mathfrak{X} \Vdash^T A \supset B}{\mathfrak{X} \Vdash^T B}$         Pi  $\dfrac{\mathfrak{X}, A \Vdash^T B}{\mathfrak{X} \Vdash^T A \supset B}$

$\Lambda$e  $\dfrac{\mathfrak{X} \Vdash^T A \wedge B \quad \mathfrak{X} \Vdash^T A \wedge B}{\mathfrak{X} \Vdash^T A \qquad \mathfrak{X} \Vdash^T B}$         $\Lambda$i  $\dfrac{\mathfrak{X} \Vdash^T A \quad \mathfrak{X} \Vdash^T B}{\mathfrak{X} \Vdash^T A \wedge B}$

Ve  $\dfrac{\mathfrak{X} \Vdash^T A \vee B \quad \mathfrak{X}, A \Vdash^T C \quad \mathfrak{X}, B \Vdash^T C}{\mathfrak{X} \Vdash^T C}$         Vi  $\dfrac{\mathfrak{X} \Vdash^T A}{\mathfrak{X} \Vdash^T A \vee B} \qquad \dfrac{\mathfrak{X} \Vdash^T B}{\mathfrak{X} \Vdash^T A \vee B}$

$\vdash$i  If $E_1, \ldots, E_m \vdash E_0$,

$$\frac{\mathfrak{X} \Vdash^T E_i \qquad i = 1, 2, \ldots m}{\mathfrak{X} \Vdash^T E_0}$$

Pk  $\dfrac{\mathfrak{X}, A \supset C \Vdash^T A}{\mathfrak{X} \Vdash^T A}$

With this matter clarified, we proceed to the following theorem.

**Theorem 6.** *In order that* (22) *hold, it is necessary and sufficient that* (21) *hold.*

*Proof of Necessity.* (Proof does not require ET.) The result will follow by deductive induction on the proof of (22) as soon as we show that the prime statements and rules of the L system are valid when $\Vdash^L$ is interpreted as $\Vdash^T$.

For the prime statements this is clear.   It is also clear for the structural rules.   For the other rules the proof is as follows:

P∗   By hypothesis there is a derivation of $B$ whose uncanceled premises are in the prosequence $\mathfrak{X}$, $A$.   By Pi there is a derivation of $A \supset B$ whose uncanceled premises are in $\mathfrak{X}$.

$\Lambda$∗   By hypothesis there is a derivation from $\mathfrak{X}$ of $A$ and also one for $B$. By $\Lambda$i there is a derivation of $A \wedge B$ from the same premises.

V∗, ⊢∗   Similarly using Vi, ⊢i.

∗P   By the first hypothesis there is a derivation of $A$ from uncanceled premises in $\mathfrak{X}$.   By Pe there is therefore a derivation of $B$ from $\mathfrak{X}$ and $A \supset B$. By the second hypothesis there is a derivation of $C$ from $\mathfrak{X}$ and $B$.   Over each occurrence of $B$ as premise for this derivation place the established derivation of $B$ from $\mathfrak{X}$ and $A \supset B$.   The result is a derivation of $C$ from $\mathfrak{X}$ and $A \supset B$.

∗$\Lambda$   By hypothesis there is a derivation of $C$ from $\mathfrak{X}$ and either $A$ or $B$ or both (for the Ketonen form).   Over each occurrence of one of these premises put the derivation of that premise from $A \wedge B$ by a single application of $\Lambda$e.   The result is a derivation of $C$ from $\mathfrak{X}$ and $A \wedge B$.

∗V   By hypothesis there is a derivation $\Delta_1$ of $C$ from $\mathfrak{X}$ and $A$ and a derivation $\Delta_2$ of $C$ from $\mathfrak{X}$ and $B$.   If we adjoin the premise $A \vee B$, then we can conclude $C$ by Ve and cancel the premise $A$ over $\Delta_1$ and the premise $B$ over $\Delta_2$.   The result is a derivation of $C$ whose uncanceled premises are either $A \vee B$ or in $\mathfrak{X}$.

Px   By hypothesis there is a derivation of $A$ from $\mathfrak{X}$ and $A \supset B$.   By Pk we can cancel the premise $A \supset B$.   The result is a derivation of $A$ from $\mathfrak{X}$, Q.E.D.

*Proof of Sufficiency.*   (Using ET.)   It is enough to show that the rules of the T system are valid when $\Vdash^T$ is interpreted as $\Vdash^L$.   This is clear for ($p$1), ($p$2) and the rules analogous to ∗C, ∗W, and ∗K.   Likewise, it is true for the rules Pi, $\Lambda$i, Vi, ⊢i, and Pk, since the interpretation carries these directly into P∗, $\Lambda$∗, V∗, ⊢∗, and Px, respectively.   For the remaining rules the validity is shown by the following schemes, in which 'H1', 'H2', etc., represent the hypotheses of the rule in question, and 'ET' signifies an application of the elimination theorem, possibly along with structural rules:

$$
\text{Pe} \qquad \dfrac{\dfrac{\overset{\text{H1}}{\mathfrak{X} \Vdash A} \quad \overset{(p1)}{\mathfrak{X}, B \Vdash B}}{\mathfrak{X}, A \supset B \Vdash B} \ast\text{P} \qquad \overset{\text{H2}}{\mathfrak{X} \Vdash A \supset B}}{\mathfrak{X} \Vdash B} \ \text{ET}
$$

$$
\Lambda e_1 \qquad \dfrac{\overset{\text{H1}}{\mathfrak{X} \Vdash A \wedge B} \quad \dfrac{\overset{(p1)}{A, B \Vdash A}}{A \wedge B \Vdash A} \ast\Lambda}{\mathfrak{X} \Vdash A} \ \text{ET}
$$

$$\Lambda e_2 \qquad \cfrac{\mathrm{H1}\quad \cfrac{\mathrm{H1}\quad A, B \Vdash B}{A \wedge B \Vdash B}\ *\Lambda}{\mathfrak{X} \Vdash A \wedge B \qquad A \wedge B \Vdash B}\ \mathrm{ET}$$

$$\mathrm{Ve} \qquad \cfrac{\mathrm{H1}\quad \cfrac{\mathrm{H2}\qquad\qquad \mathrm{H3}}{\mathfrak{X}, A \Vdash C \qquad \mathfrak{X}, B \Vdash C}\ *\mathrm{V}}{\mathfrak{X} \Vdash A \vee B \qquad \mathfrak{X}, A \vee B \Vdash C}\ \mathrm{ET}$$

This completes the proof of Theorem 6.

In connection with the theorems of Sec. B we have the following corollaries:

COROLLARY 6.1.    *The following statements are all equivalent:*

$$\begin{array}{ll}\Vdash A & in\ \mathrm{LA}(\mathfrak{D})\\ \vdash A & in\ \mathrm{TA}(\mathfrak{D})\\ \vdash A & in\ \mathrm{HA}\\ 1 \le A & in\ \mathrm{EA}\end{array}$$

*Likewise, the statements obtained by substituting* $\mathrm{LC_1}$, $\mathrm{TC}$, $\mathrm{HC}$, *and* $\mathrm{EC}$, *respectively, for* $\mathrm{LA}$, $\mathrm{TA}$, $\mathrm{HA}$, *and* $\mathrm{EA}$.

COROLLARY 6.2.    *The following statements are also equivalent*

$$\begin{array}{ll}A_1, \ldots, A_m \Vdash B & in\ \mathrm{LA}(\mathfrak{D})\\ A_1, \ldots, A_m \Vdash B & in\ \mathrm{TA}(\mathfrak{D})\\ \vdash A_1 \supset. A_2 \supset. \cdots \supset. A_m \supset B & in\ \mathrm{HA}\ (or\ \mathrm{TA})\\ A_1 \wedge A_2 \wedge \cdots \wedge A_m \le B & in\ \mathrm{EA}\end{array}$$

*Likewise the statements arising from the same substitutions as in Corollary 6.1.*

## 5. The equivalence of singular and multiple systems.

We now attack the problem of showing that the singular systems $\mathrm{LA_1}$ and $\mathrm{LC_1}$ are equivalent, respectively, to the multiple systems $\mathrm{LA}_m$ and $\mathrm{LC}_m$. We shall do this first for the absolute systems $\mathrm{LA_1}$ and $\mathrm{LA}_m$; then the argument will be extended to the classical systems $\mathrm{LC_1}$ and $\mathrm{LC}_m$.

The following conventions will be understood. All systems are formulated in Formulation IK, that is, with the Ketonen forms of $*\mathrm{P}$ and $*\Lambda$. When explicitness is desired, '$\Vdash_1$' will be written for '$\Vdash$' in connection with a singular system, whereas '$\Vdash_m$' will be so used in connection with a multiple system. It will be supposed further that $\mathfrak{Y}$ is $C_1, \ldots, C_p$ and that

$$C \equiv C_1 \vee C_2 \vee \cdots \vee C_p \tag{23}$$

with association to the right; similarly, that $\mathfrak{Y}'$ is $C_1', \ldots, C_q'$ and $\mathfrak{Z}$ is $D_1, \ldots, D_r$ with $C'$ and $D$ defined analogously to $C$. Given a statement

$$\mathfrak{X} \Vdash_m \mathfrak{Y} \tag{24}$$

the statement

$$\mathfrak{X} \Vdash_1 C \tag{25}$$

will be called its *singular transform*; likewise, we say that (24) is a *multiple transform* of (25).   The singular transform is unique, but since there may be several decompositions (23), the multiple transform is not necessarily so; however, if $C$ is not an alternation (i.e., if it does not have $\lor$ as its outside operation), $\mathfrak{Y}$ is singular and (24) and (25) are identical.   The inference

$$\frac{\mathfrak{X} \Vdash A}{\mathfrak{X} \Vdash B}$$

can always be made by Rp and Corollary 6.2 in case

$$A \leq B$$

is true in EA; in such a case the inference will be said to be made by EA. Thus the various ways of associating in (23) are equivalent by EA, so that the agreement as to association was a pure technicality.   Finally, in proofs indicated by the tree diagrams, I shall use 'ET' to indicate an application of the elimination theorem and 'EA' to denote an inference by EA.   The diagrams are thus not demonstrations, but indications of how demonstrations may be constructed; the reader is advised to read these from the bottom upward and to construct those parts of them which are demonstrations by the techniques of Sec. C5.

**Theorem 7.**   *A necessary and sufficient condition that a statement of form* (24) *hold in* $\mathrm{LA}_m$ *is that its singular transform* (25) *hold in* $\mathrm{LA}_1$.

*Proof of Necessity.*   We use a deductive induction on the proof of (24).

If (24) is a prime statement, then $\mathfrak{Y}$ is singular and so (24) and (25) are identical.

If (24) is derived by any of the rules $*$C, $*$K, $*$W, $*\Lambda$, $*$V, all of which leave the right side completely unchanged, then the same rule will allow us to derive (25) from the singular transform(s) of the premise(s).[1]

Suppose that (24) is derived by $*$P.   Then $\mathfrak{X}$ is $\mathfrak{X}'$, $A \supset B$, and the premises are

$$\mathfrak{X}' \Vdash_m A, \mathfrak{Y} \qquad \mathfrak{X}', B \Vdash_m \mathfrak{Y}$$

The singular transforms may be written, by EA, as

$$\mathfrak{X}' \Vdash A \lor C \qquad \mathfrak{X}', B \Vdash C \tag{26}$$

From these premises we argue as follows.   We construct first a proof of $A \supset B, A \lor C \Vdash B \lor C$, thus:

$$\frac{\dfrac{\dfrac{A \Vdash A \qquad B \Vdash B}{A \supset B, A \Vdash B}\,*\text{P}}{A \supset B, A \Vdash B \lor C}\,\text{V}* \qquad \dfrac{\dfrac{(p1)}{A \supset B, C \Vdash C}}{A \supset B, C \Vdash B \lor C}\,\text{V}*}{A \supset B, A \lor C \Vdash B \lor C}\,*\text{V}$$

From this conclusion and the left-hand premise (26) we have, by ET,

$$\mathfrak{X}', A \supset B \Vdash B \lor C \tag{27}$$

---

[1] The only case of multiple premises is $*$V.   In that case it is important that we are using Formulation I, so that the right sides in the two premises are the same.

On the other hand, using the right-hand premise (26), we have

$$\frac{(p1)}{\mathfrak{X}', B \Vdash C \quad \mathfrak{X}', C \Vdash C}{\mathfrak{X}', B \vee C \Vdash C} \ *V$$

Hence, eliminating $B \vee C$ with (27), we have

$$\mathfrak{X}', A \supset B \Vdash C$$

which is (25).

Next, suppose (24) is derived from

$$\mathfrak{X} \Vdash \mathfrak{Y}' \tag{28}$$

by one of the rules C*, K*, W*. Then we derive (25) from the singular transform of (28) by EA.

If (24) is derived by P*, then $\mathfrak{Y}$ is singular and so is the premise from which (24) is derived. By the hypothesis of the induction, since $p = 1$, that premise is also derivable in $LA_1$; hence so is (24). In such a case, (24) and (25) are identical.

If (24) is obtained by $\Lambda$*, then $\mathfrak{Y}$ is $A \wedge B$, $\mathfrak{Z}$ and the premises are

$$\mathfrak{X} \Vdash A, \mathfrak{Z} \quad \mathfrak{X} \Vdash B, \mathfrak{Z}$$

The singular transforms may be written, in view of the association to the right of $\vee$, in the form

$$\mathfrak{X} \Vdash A \vee D \quad \mathfrak{X} \Vdash B \vee D$$

From these by $\Lambda$* we have in $LA_1$

$$\mathfrak{X} \Vdash (A \vee D) \wedge (B \vee D)$$

On the other hand, since

$$(A \vee D) \wedge (B \vee D) \leq (A \wedge B) \vee D$$

holds in EA, we have by EA

$$\mathfrak{X} \Vdash (A \wedge B) \vee D$$

which is (25) for this case.

Finally, suppose (24) is obtained from V*. Then $\mathfrak{Y}$ is $A \vee B$, $\mathfrak{Z}$. Let the singular transform of the premise be

$$\mathfrak{X} \Vdash C'$$

Then since

$$C' \leq C$$

holds in a lattice, we have (25) by EA.

If (24) is obtained by $\Vdash$*, then the inference is

$$\frac{\mathfrak{X} \Vdash_m E_i, \mathfrak{Z} \quad i = 1, 2, \ldots, m}{\mathfrak{X} \Vdash_m E_0, \mathfrak{Z}}$$

By the inductive hypothesis,

$$\mathfrak{X} \Vdash_1 E_i \vee D \quad i = 1, 2, \ldots, m \tag{29}$$

By Theorem C5,

$$E_1, E_2, \ldots, E_m \Vdash_1 E_0$$

and hence by V*,

$$E_1, E_2, \ldots, E_m \Vdash E_0 \vee D$$

This is the case $k = 0$ of

$$E_1 \vee D, \ldots, E_k \vee D, E_{k+1}, \ldots, E_m \Vdash E_0 \vee D \qquad (30)$$

To show that this holds for all $k \leq m$ it is sufficient to give the induction step for induction on $k$. This is obtained from (30) and the conclusion of

$$\frac{D \Vdash_1 D}{D \Vdash_1 E_0 \vee D} \; \mathrm{V}*$$

by $*\mathrm{V}$ in Formulation II (which in turn follows from Formulation I by Theorem C2). Thus (30) holds for $k = m$. From this and (29) we have (25) by successive applications of ET.

This completes the proof of necessity.

*Proof of Sufficiency.* Since all the rules of $\mathrm{LA}_1$ are valid in $\mathrm{LA}_m$,† (25) is valid in $\mathrm{LA}_m$. From this we obtain (24) by Theorem 1, inverting with respect to the join operations on the right.

This completes the proof of Theorem 7.

COROLLARY 7.1. *The statements of the form*

$$\mathfrak{X} \Vdash B$$

*which are valid in* $\mathrm{LA}_m$ *are the same as those which are valid in* LA.

*Proof.* This is the case where $p = 1$.

Since ET for $\mathrm{LA}_m$ was not used in the proof of Theorem 7, one can use Theorem 7 to give an alternative proof of ET for $\mathrm{LA}_m$, namely, by deducing it from ET for $\mathrm{LA}_1$.‡

The following corollary generalizes Corollary 6.2.

COROLLARY 7.2. *The following statements are equivalent*:

$$A_1, \ldots, A_m \Vdash B_1, \ldots, B_n \qquad\qquad in\ \mathrm{LA}_m$$

$$A_1, \ldots, A_m \Vdash B_1 \vee B_2 \vee \cdots \vee B_n \qquad\qquad in\ \mathrm{TA}$$

$$\vdash A_1 \supset. A_2 \supset \cdots \supset. A_m \supset. B_1 \vee B_2 \vee \cdots \vee B_n \qquad\qquad in\ \mathrm{HA}$$

$$A_1 \wedge A_2 \cdots \wedge A_m \leq B_1 \vee B_2 \cdots \vee B_n \qquad\qquad in\ \mathrm{EA}$$

So much for the absolute systems. We turn now to the classical ones.

**Theorem 8.** *A necessary and sufficient condition that* (24) *hold in* $\mathrm{LC}_m$ *is that* (25) *hold in* $\mathrm{LC}_1$.

*Proof of Necessity.* It is only necessary to add to the cases considered in Theorem 7 the case of the only inference possible in $\mathrm{LC}_m$ but not in $\mathrm{LA}_m$, namely, that by the unrestricted rule P*. If (24) is so obtained, then $\mathfrak{Y}$ is $A \supset B$, $\mathfrak{Z}$; the premise is

$$\mathfrak{X}, A \Vdash B, \mathfrak{Z}$$

The singular transform, which is valid in $\mathrm{LC}_1$ by the inductive hypothesis, is

$$\mathfrak{X}, A \Vdash B \vee D$$

---

† One might think that $*\mathrm{P}$ was an exception, but the $\mathrm{LA}_1$ form of $*\mathrm{P}$ is valid in $\mathrm{LA}_m$.

‡ See Exercise 16. Instead of using Theorem 1 in the sufficiency proof we could have used a deductive induction directly.

From this we derive (25) as follows:

$$\frac{A, B \Vdash B}{\frac{B \Vdash A \supset B}{\frac{B \Vdash A \supset B . \vee D}{\frac{B \vee D \Vdash A \supset B . \vee D}{\frac{\mathfrak{X}, A \Vdash A \supset B . \vee D}{\frac{\mathfrak{X}, A, A \supset B . \vee D . \supset B \Vdash B}{\frac{\mathfrak{X}, A \supset B . \vee D . \supset B \Vdash A \supset B}{\frac{\mathfrak{X}, A \supset B . \vee D . \supset B \Vdash A \supset B . \vee D}{\mathfrak{X} \Vdash A \supset B . \vee D} \text{Px}} \text{V*}} \text{P*}}} } } } }$$

*Proof of Sufficiency.* Here again the only case to consider is that where the inference is made by Px. The inference in $\text{LC}_1$ would then be of the form

$$\frac{\mathfrak{X}, C \supset B \Vdash C}{\mathfrak{X} \Vdash C}$$

The corresponding inference in $\text{LC}_m$ can be justified as follows:

$$\frac{\frac{(p1)}{\frac{C_i \Vdash B, \mathfrak{Y} \quad i = 1, 2, \ldots, p}{C \Vdash B, \mathfrak{Y}} \text{*V}}{\frac{\mathfrak{X}, C \supset B \Vdash \mathfrak{Y} \quad \Vdash C \supset B, \mathfrak{Y}}{\mathfrak{X} \Vdash \mathfrak{Y}} \text{ET}} \text{P*}}$$

This completes the proof of Theorem 8.

COROLLARY 8.1. *The statements of the form*

$$\mathfrak{X} \Vdash B$$

*which are true in* $\text{LC}_m$ *are the same as those which are true in* $\text{LC}_1$.
*Proof.* This is the case where $p = 1$.
The following corollary connects Theorem 8 with Corollary 6.2.

COROLLARY 8.2. *The statements of Corollary 7.2 remain equivalent if the systems* $\text{LA}_m$, TA, HA, *and* EA *are replaced, respectively, by* $\text{LC}_m$, TC, HC, *and* EC.

COROLLARY 8.3. *The rule analogous to* Px, *namely,*

$$\mathfrak{X}, A \supset B \Vdash A, \mathfrak{Z} \to \mathfrak{X} \Vdash A, \mathfrak{Z}$$

*is redundant in* $\text{LC}_m$.
*Proof.* The sufficiency proof of the theorem holds whenever $C$ is in $\mathfrak{Y}$.

COROLLARY 8.4. *Let* $\text{LC}_1'$ *be the same as* $\text{LC}_1$ *except that* Px *is subject to the additional restriction that* B *be a component of* A. *Then any elementary theorem of* $\text{LC}_1$ *is demonstrable in* $\text{LC}_1'$ *and vice versa.*
*Proof.* The necessity proof for Theorem 8 used only such instances of Px

as satisfied the additional restriction. (The only use of Px was in validating inferences by P∗.) Hence if $\Gamma$ is demonstrable in $LC_1$, it is derivable in $LC_m$ by Corollary 8.3, and hence, by the necessity part of Theorem 8, in $LC_1'$. The converse is clear since any demonstration in $LC_1'$ is a fortiori one in $LC_1$.

**6. Completeness of LC.**  We saw in Sec. C2 that any elementary statement of LC was valid in an interpretation by 0-1 tables. The completeness theorem for LC is the converse statement for the case where $\mathfrak{S}$ is $\mathfrak{D}$. Since validity by 0-1 tables may be regarded as a fifth form of formulation, parallel to the L, T, H, and E formulations it is appropriate to prove this theorem in the present section. The proof is, in principle, due to Ketonen [UPK].

**Theorem 9.**  *A necessary and sufficient condition that*

$$\mathfrak{X} \Vdash \mathfrak{Y} \tag{31}$$

*hold in* $LC(\mathfrak{D})$ *is that it be valid relative to every evaluation by* 0-1 *tables in the sense of Sec. C2.*

*Proof.*  The necessity of this condition [even for $LC(\mathfrak{S})$] is shown by deductive induction. It is hardly necessary to give the details explicitly. It is enough to state the following facts. A prime statement of type $(p1)$ has a common constituent on both sides, and the statement is valid for either of the two possible values of this common constituent. In the inductive step it is sufficient to consider the case that all parametric constituents on the left have the value 1 and those on the right have the value 0, since otherwise the conclusion is valid; under these circumstances the validity of the premises restricts the values for the subalterns, and for these values the ordinary truth tables assign a value to the principal constituent such that the conclusion is valid.

It remains to prove the sufficiency. Suppose then that (31) is valid as stated. We construct a tree from the bottom up by applying the operational rules of $LC_m$, formulation IK, in the inverse direction. Since each premise of such an operational rule contains fewer operations than the conclusion, the process must eventually terminate in the construction of a tree $\mathfrak{D}$ in which the top nodes contain only elementary constituents. If the statements at these top nodes are demonstrable, then from $\mathfrak{D}$ we can construct a proof tree for (31).

Now the inverted rules of $LC_m$ also preserve the property of being valid by truth tables, as we can show by working the inductive step of the necessity proof in the opposite direction. Hence all the nodes of $\mathfrak{D}$, and therefore the top nodes, will be tautologous by truth tables. But if (31) is tautologous and has only elementary constituents, then there must be a common constituent in $\mathfrak{X}$ and $\mathfrak{Y}$; otherwise there would be a valuation which would give all constituents of $\mathfrak{X}$ the value 1 and all those of $\mathfrak{Y}$ the value 0. Therefore all the top nodes of $\mathfrak{D}$ are quasi prime (Sec. C9) and $\mathfrak{D}$ can be trivially converted into a demonstration of (31), Q.E.D.

Since the demonstration so obtained from $\mathfrak{D}$ contains no instances of ∗W∗, the proof shows that the contraction rules are superfluous. Likewise the uses of ∗K∗ are specialized. Generalizations of these results will concern us in Sec. E.

*Remark.* The sufficiency argument will go through in any case in which all the operational rules are directly invertible and preserve truth-table validity when inverted.

## E. L DEDUCIBILITY

This section is devoted to the development of theorems and techniques concerned with what can or cannot be done in the operation of the L systems. These systems have two characteristics: the first is that inferences by the rules generally increase the complexity, so that a demonstration is a process of synthesis; the second is that one can (with certain exceptions) tell by inspection whether an elementary statement can be the conclusion of a given rule, so that there is the possibility of a corresponding analysis.

The section begins with the study of a "composition property," which expresses in a refined way the synthetic characteristic. It then proceeds, in Secs. 2 to 4, to study properties which are more or less immediate consequences of the composition property and to express broad necessary conditions for deducibility. These include the separation property, to the effect that rules for an operation are relevant only when that operation actually occurs; the conservation properties, which concern relations to the underlying system $\mathfrak{S}$; and the alternation property, which in the simplest case says that $A \vee B$ is assertible in the absolute propositional algebra only when one or the other of $A$, $B$ is. Then in Secs. 5 to 6 the structural rules are studied with a view to restricting or eliminating them, and thus reducing the number of alternatives in the analytic process; in particular, it is shown in Sec. 6 that in slightly modified formulations, called Formulations III and IV, the rules *W* are redundant. This paves the way for the decidability theorem of Sec. 7. Finally, Sec. 8 is devoted to a "tableau" method, essentially due to Beth, which simplifies, and to some extent mechanizes, the decision process.

**1. The composition property.** The most striking property of the operational rules for the L systems is that they form new combinations, but they do not allow combinations to drop out unless they are repetitions of combinations already present. We proceed to formulate this property precisely. We shall use the term 'component' in the sense of Sec. 3B1. That definition determines when a proposition $A$ is a component of another proposition $B$. The possibility that $A$ is the same as $B$ is included. Then a proposition $A$ will be said to be a *component of an elementary statement* $\Gamma$ just when it is a component of a proposition occurring as a constituent in $\Gamma$.

A rule $R$ will be said to have the '*composition property*'[1] just when every subaltern constituent is like a component of the principal constituent. A system will be said to have the composition property just when every one of its rules does. A rule or system will be said to have the *composition property for compound constituents* just when the conditions of the above definition are satisfied for all compound subaltern constituents, but not necessarily for elementary ones.

Thus all the rules, regular or irregular, of the systems $LA_1(\mathfrak{D})$, $LA_m(\mathfrak{D})$,

---

[1] The term 'subformula property' is beginning to be standard for this property, but it does not agree so well with the terminology of this book as the term used in the text.

and $LC_m(\mathfrak{O})$ satisfy the composition property, and consequently these systems do. If $\mathfrak{S}$ is such that the auxiliary statements (Sec. C3$d$) are non-vacuous, then the rule $\vdash_*$ does not satisfy the composition property, but since the subaltern constituents of $\vdash_*$ are necessarily elementary, it does satisfy the composition property for compound constituents; therefore all the systems $LA(\mathfrak{S})$, $LC(\mathfrak{S})$, $LA_m(\mathfrak{S})$ do also. Finally, the rule Px does not satisfy the composition property, not even for compound constituents, and therefore the same is true for all systems $LC_1$.

Since all the constituents in the premises of a rule are either subaltern or parametric, and the latter are like some constituent in the conclusion, every constituent in the premises of the rule satisfying the composition property will be like some component of a constituent in the conclusion. The same will be true under the restriction to compound constituents if the rule satisfies the composition property for compound constituents. Thus we have, either with the omission of all the words in parentheses or with the inclusion of all of them, the following theorem:

**Theorem 1.** *If an* L *system satisfies the composition property (for compound constituents), then every (compound) constituent of an elementary statement in a regular demonstration is like a component in the final result.*

*Proof.* Every descendant of the constituent in question will contain it as a component, and there will be such a descendant in the final result, Q.E.D.

**2. Separation property.** A system will be said to have the *separation property* (with respect to a set of operations $\Omega$) just when, for every operation $\omega$ (of $\Omega$), every elementary theorem $\Gamma$ in which $\omega$ does not occur is demonstrable without using postulates related to $\omega$. For L systems the postulates related to $\omega$ (unless there is some auxiliary explanation)[1] consist of those rules in which $\omega$ is the main operation in the principal constituent. In other cases the relationship is left a little vague, but is understood to be that which is made apparent in the formulation of the system.

**Theorem 2.** *If an* L *system has the composition property for compound constituents, then it has the separation property with respect to all operations.*

*Proof.* If $\Gamma$ is an elementary theorem, it has a demonstration; hence it has a regular demonstration. By Theorem 1 none of the deleted rules is used in the regular demonstration, Q.E.D.

Since all the rules of $LA_1$, $LA_m$, $LC_m$ have the composition property for compound constituents, we have at once:

COROLLARY 2.1. *The systems* $LA_1$, $LA_m$, *and* $LC_m$ *have the separation property for all operations.*

To derive the separation property for $LC_1$ requires a little thought. We note first:

COROLLARY 2.2. *The elementary theorems of* $LC_1$ ($LC_m$) *which do not contain implication are the same as those of* $LA_1$ ($LA_m$).

*Proof.* If $\Gamma$ is an elementary theorem of $LC_1$, then by Theorem D8 it is demonstrable in $LC_m$. If $\Gamma$ does not contain implication, it has a demonstration in $LC_m$ not using the P rules (by Corollary 2.1). Since the operational rules of $LC_m$ other than the P rules are the same as in $LA_m$, $\Gamma$ is demonstrable

---

[1] As in Corollary 2.3.

in $LA_m$. If it is singular, it is demonstrable in $LA_1$ by Theorem D7. Conversely, any elementary theorem of $LA_1$ ($LA_m$) is clearly demonstrable in LC, Q.E.D.

Now suppose that $\Gamma$ is an elementary theorem of $LC_1$ which involves implication. By Theorem D8 and Corollary 2.1, $\Gamma$ has a demonstration in $LC_m$ using only rules for operations which actually occur in $\Gamma$. This may be transformed into a demonstration in $LC_1'$ by Theorem D8 and Corollary D8.4. All the rules of $LC_1'$, except Px, have the composition property, and Px can drop out only the constituent $A \supset B$, which, since $B$ is a component of $A$, contains no operations except implication and operations in $A$. Hence each node of the demonstration contains no operations, except possibly implication, which are not in the node just below it. Since implication occurs in $\Gamma$, no node contains operations not in $\Gamma$, and hence no rules corresponding to operations not in $\Gamma$ can occur.

Since this is also true if $\Gamma$ does not contain implication by Corollaries 2.1 and 2.2, we have:

COROLLARY 2.3.   *The system $LC_1$ has also the separation property for all operations, even if Px is added to the list of rules related to implication.*

The equivalence theorems of Sec. D allow the separation property to be extended to other types of formulation. This requires a detailed examination of the equivalence proofs. In the equivalence between the T and singular L systems in Theorem D6, the justification of a rule in either system required only the use of a corresponding rule for the same operation in the other system; consequently, an analogue of Theorem 2 holds without modification for TA and TC. But the proofs of equivalence between the E, H, and T systems in Sec. B did not separate the operations in that way. Thus the reduction of TA to HA in Theorem B3 required properties of implication for all the T rules, and the reduction of EA to HA in Theorem B4 required properties of conjunction as well. Hence the conclusion to be drawn from Theorem 2 is the following:[1]

COROLLARY 2.4.   *The systems TA, TC have the separation property for all operations; the systems HA, HC in standard formulation for all except implication. Further, an elementary theorem of EA which does not contain the operation $\vee$ is a theorem of an implicative semilattice.*

There does not seem to be much point to investigating the situation with respect to EC.

**3. Conservation property.** An L system over $\mathfrak{S}$ will be said to have the *conservation property* relative to $\mathfrak{S}$ just when every elementary theorem with only elementary constituents has some proposition in the consequent which is $\mathfrak{S}$-deducible (Sec. C7) from those in the antecedent.

**Theorem 3.** *If an L system over $\mathfrak{S}$ is such that all rules other than $\vdash\!*$ and the structural rules are related to some operation and the separation property holds for all operations, then the L system has the conservation property relative to $\mathfrak{S}$.*

*Proof.* Let $\Gamma$ be an elementary statement all of whose constituents are elementary. Let $\Delta$ be a regular demonstration of $\Gamma$. By the separation property no operational rules are used in $\Delta$, and hence all constituents in $\Delta$

---

[1] Cf. Exercises 11 and 15 of Sec. B.

are elementary. We use a deductive induction with respect to $\Delta$. If $\Gamma$ is prime, then either some proposition of the consequent of $\Gamma$ is the same as one in the antecedent or some proposition in the consequent is an axiom; in either case the thesis of our theorem is verified. If $\Gamma$ is derived by a rule other than ⊢*, then there is one premise, and the propositions appearing on either side in the premise appear also on the same side in the conclusion; since, by the inductive hypothesis, our thesis holds for the premise, it holds for the conclusion also. Finally, suppose $\Gamma$ is obtained by rule ⊢*. Let the inference be

$$\frac{\Gamma i \qquad \mathfrak{X}_i \Vdash E_i, \mathfrak{Z}_i \qquad i = 1, 2, \ldots m}{\Gamma \qquad \mathfrak{X} \Vdash E_0, \mathfrak{Z}}$$

where all the propositions in any of the $\mathfrak{X}_i$ are contained in $\mathfrak{X}$ and all in any of the $\mathfrak{Z}_i$ are contained in $\mathfrak{Z}$.† If a proposition in $\mathfrak{Z}_i$ is $\mathfrak{S}$-derivable from those in $\mathfrak{X}_i$, then a fortiori some proposition of $\mathfrak{Z}$ is $\mathfrak{S}$-derivable from those in $\mathfrak{X}$. By the inductive hypothesis the only remaining possibility is that, for every $i$, $E_i$ be derivable from the propositions in $\mathfrak{X}_i$; then $E_0$ is $\mathfrak{S}$-derivable from $E_1, \ldots, E_m$ and hence in turn from those in $\mathfrak{X}$, Q.E.D.

The following theorem, which reduces LA($\mathfrak{S}$) and LC($\mathfrak{S}$) to LA($\mathfrak{D}$) and LC($\mathfrak{D}$), may also be considered as a kind of conservation theorem.

**Theorem 4.** *Let $\Delta$ be a demonstration in* LA($\mathfrak{S}$) *or* LC($\mathfrak{S}$) *of*

$$\mathfrak{X} \Vdash \mathfrak{Y} \tag{1}$$

*Let $\mathfrak{M}$ be the (finite) set of all propositions $M$ which appear as axioms [i.e., constituents of prime statements of type $(p2)$] in $\Delta$, and let $\mathfrak{N}$ be the set of all propositions $N$ of the form*

$$N_1 \supset. \; N_2 \supset \cdots N_n \supset N_0 \tag{2}$$

*where the auxiliary statement*

$$N_1, \ldots, N_n \vdash_0 N_0 \tag{3}$$

*is used as justification for an inference by* ⊢* *in $\Delta$. Then*

$$\mathfrak{M}, \mathfrak{N}, \mathfrak{X} \Vdash \mathfrak{Y} \tag{4}$$

*is demonstrable in the corresponding L system over $\mathfrak{D}$.*

*Proof.* By Theorems D7 and D8 there is a proof tree in the corresponding singular system of

$$\mathfrak{X} \Vdash C$$

where $C$ is as in Sec. D5. By Theorem D6 there is a T proof ending in $C$, all of whose uncanceled premises are in $\mathfrak{X}$. In this T proof each $M$ in $\mathfrak{M}$ will appear at a top node, and inferences by ⊢* will become inferences by ⊢i. Now let the members of $\mathfrak{M}$ at the top nodes be taken as additional premises, and where there is an inference by ⊢i justified by (3), let the inference be made by successive applications of Pe with (2) as additional premise. Then we shall have a T proof relative to $\mathfrak{D}$ of

$$\mathfrak{M}, \mathfrak{N}, \mathfrak{X} \Vdash C$$

This can be converted into a proof of (4) by Theorems D6 to D8, Q.E.D.

† This treats Formulations I and II simultaneously.

**4. Alternation property.** This is the property of absolute propositional algebra, not possessed by classical algebra, to the effect that $A \lor B$ is assertible only if either $A$ or $B$ is assertible.

To formulate this precisely, we assume that we are dealing with the system $LA_1$. We say that an operation is *nondilemmatic* if its rule for introduction on the left satisfies $(r6)$ on the right and either has one premise or, if it has more than one, the right constituent in the conclusion is congruent to the right constituent in exactly one of the premises. Thus $\Lambda$ is nondilemmatic because $*\Lambda$ has only one premise; P is nondilemmatic in $LA_1$ because the right constituent of the conclusion of $*P$ is congruent to that of the right premise only; but $\lor$ is dilemmatic because the right constituent of $*\lor$ is parametric in both premises. In Chap. 7 we shall meet a case where an operation is dilemmatic because it fails to satisfy $(r6)$.

**Theorem 5.** *Let $\mathfrak{L}$ be a singular L system formed by adjoining to $LA_1$ at most semiregular rules for additional operations. Let $\mathfrak{X}$ be constructed from elementary propositions at most by nondilemmatic operations. Let $A$ and $B$ be propositions not necessarily elementary. Let $\Gamma$, viz.,*

$$\mathfrak{X} \Vdash A \lor B \tag{5}$$

*be demonstrable in $\mathfrak{L}$. Then one or the other of*

$$\mathfrak{X} \Vdash A \qquad \mathfrak{X} \Vdash B \tag{6}$$

*is demonstrable in $\mathfrak{L}$.*

*Proof.* Let $\Delta$ be a regular derivation of (5), and let it be exhibited in tree form. Since the operation $\lor$ does not occur on the left in $\Gamma$ and the right side is not elementary, $\Gamma$ is not quasi prime. Then the bottom node of $\Delta$ must be obtained by an inference. This inference cannot be by $\Vdash*$, since the right side is not elementary, nor can it be obtained by an operational rule on the right other than $\lor*$. If this bottom inference is not by $\lor*$, it must therefore be by a structural rule or by an operational rule on the left. In either case, by the restrictions of the theorem, there will be a unique premise whose consequent is congruent to that of (5). We pass up the tree to that premise. We continue climbing the tree in this way until we reach a node $\Gamma'$ which has been introduced by $\lor*$. Then $\Gamma'$ must be of the form

$$\mathfrak{X}' \Vdash A \lor B$$

and the premise immediately above it must be one or the other of

$$\mathfrak{X}' \Vdash A \qquad \mathfrak{X}' \Vdash B \tag{7}$$

Now the rules along the branch from $\Gamma'$ to $\Gamma$ all satisfy $(r6)$ on the right; hence we can replace $A \lor B$ by whichever of $A$ or $B$ is in the consequent of the premise for $\Gamma'$. This will convert $\Delta$ into a proof of one or the other of (6), Q.E.D.

**Corollary 5.1.** *The rule $*P$ in Formulation IK of $LA_m$ cannot be inverted as to its left premise.*

*Proof.* (Cf. Sec. D1, Example 1.) In LA we have

$$A \supset B \Vdash A \supset B$$

If *P could be inverted, the left premise would be

$$\Vdash A, A \supset B$$

and from this we should have by V*

$$\Vdash A \lor . A \supset B$$

By Theorem D7 this could be derived in $LA_1$. But this contradicts Theorem 5 since both of the possibilities (6) are invalid by truth-table valuation.

**5. Restriction of \*K\*.** We now, investigate to what extent the rules *K* can be restricted without loss of deductive power.

Suppose we have a rule $R$ which is regular (Sec. C6). If such a rule is followed by an application of *K*, then by virtue of $(r6)$ the inference would still be valid if the new constituent is introduced as an additional parameter before applying the rule $R$. Thus if all the rules satisfy $(r1)$ to $(r6)$, as in the case with $LA_1$, $LC_1$, $LC_m$, the applications of *K* can be pushed upward (on the proof tree) until they are made immediately following the introduction of the prime statements. In such a case let us say that *K* is applied only *initially*. If one were to allow starting with quasi-prime statements (Sec. C9), one would not need *K* any further. But in cases where there are some rules, like the P* in $LA_m$, which do not satisfy $(r6)$, it is not possible to push an application of *K* on the relevant side up the proof tree beyond such a rule; therefore we have to admit the possibility of an application of K* or *K, as the case may be, immediately following such a rule.

Next let us consider the nature of the principal constituent introduced by *K*. If all the rules are regular, then we can use an argument similar to the proof of Theorem C4[1] to show that we can restrict *K* to have the principal constituent elementary. But in the general case there will be certain exceptions. In those formulations, considered in Sec. 6, in which certain rules are required to have a quasi-principal constituent, then a first instance of that constituent may have to be introduced by *K*. Again, in $LA_m$ an extra constituent of the form $A \supset B$ cannot be introduced on the right by P*.

Let us call an operational rule *unrestricted* if it is regular and if, whenever a constituent of the form of the principal constituent can be introduced by *K*, that constituent can also be introduced by first using *K* to bring in the subalterns and then using the operational rule. Then we do not need to postulate *K* for principal constituents for which an unrestricted operational rule is present.

The upshot of this discussion is the following:

**Theorem 6.** *The rules* *K* *can be restricted, without loss of generality, to be made initially or immediately[2] after a rule which fails to satisfy $(r6)$ on the same side and to have a principal constituent which is elementary or for which no unrestricted operational rule for introducing it is present.*

COROLLARY 6.1. *In the Formulation I of* $LA_1$, $LC_1$, $LC_m$, *K* *can be restricted to be made initially, with principal constituents elementary.*

---

[1] That is, we can introduce the subalterns first and then use the operational rule in question.

[2] There may of course be several successive applications of *K* in such positions.

COROLLARY 6.2. *In Formulation* I *of* LA$_m$, *$*$K can be restricted to be made initially and with principal constituents elementary;* K$*$ *can be restricted to be made initially or after rule* P$*$, *with principal constituents which are either elementary or of the form* $A \supset B$.

Further exceptions must be made in case quasi-principal constituents are required (see Corollary 7.3).

For some purposes one may ask under what circumstances an application of $*$K$*$ is moved down the tree, rather than up. We shall not go into this question.

**6. Reduction of $*$W$*$; Formulation III.** The next problem is the reduction or elimination of the rules of contraction. This is especially important in the case of a decision procedure because these rules increase the number of alternatives to be considered.

In Sec. D6 we noticed that in LC$_m$, where all the operational rules are directly invertible, the rules $*$W$*$ are superfluous. This suggests that the invertibility of the rules is the essential factor. We can accomplish this for such a rule as $*$P, which is not invertible in LA$_m$ with respect to its left premise, by requiring that there be a quasi-principal constituent in the left premise, so that the rule becomes

$$\frac{\mathfrak{X}, A \supset B \Vdash A, \mathfrak{Y} \quad \mathfrak{X}, B \Vdash \mathfrak{Y}}{\mathfrak{X}, A \supset B \Vdash \mathfrak{Y}} \tag{8}$$

Here the left premise can be derived from the conclusion by K$*$. Since the quasi-principal constituent can be introduced into the premise by $*$K, the rule is not weaker than $*$P, and $*$P can be derived from it by $*$W (cf. Sec. C8).

These two forms of inversion have an important feature in common, viz., that the proofs of the premises are not essentially more complex than the original proof of the conclusion. However, there are differences between them, so that it seems best not to subsume them under a unified concept of invertibility.

In the proof of the following theorem the *degree* of a demonstration is to be understood in the sense of the remark preceding Corollary D1.6. Also an application of one of $*$W$*$ will be called a *contraction*, and its subalterns, the *contracted constituents*.

**Theorem 7.** *Let an* L *system satisfy the following conditions:* (a) *all rules satisfy* (r4)$'$ *except that those which are singular on the right do not have to satisfy it on that side;* (b) *for every nonstructural rule and every premise, either the premise contains a quasi-principal constituent or the rule is directly invertible with respect to that premise. Then the rules* $*$W$*$ *are redundant.*

*Proof.* Suppose that there is a contraction leading from $\Gamma$ to $\Gamma'$. Suppose further that there is a demonstration $\Delta$ of degree $n$ and not containing any contraction, leading to $\Gamma$. We shall see that there is then a demonstration $\Delta'$, also of degree $\leq n$ and not containing any contraction, leading to $\Gamma'$. This will be proved as a lemma by induction on $n$.

Let $\Delta$ be exhibited in tree form. Since all prime statements are singular on both sides, $\Gamma$ is not a top node of $\Delta$. Hence there is a rule $R$ such that $\Gamma$ is the conclusion of an instance $R_1$ of $R$ in $\Delta$. If $R$ is singular on the right, then the contracted constituents cannot be on the right. Hence such of these constituents as are parametric appear in all premises of $R_1$ by (r4)$'$.

If both the contracted constituents are parametric, then they both appear in all the premises. If the premises can be contracted, then by $(r5)$ the rule $R$ will lead from the contracted premises to $\Gamma'$. If $R$ is a nonstructural rule, then the contraction can be accomplished without using $*W*$ by the inductive hypothesis, and we have the $\Delta'$ sought. Otherwise we simply cut the final $*K*$ off of $\Delta$ and establish the existence of a $\Delta'$ for the $\Delta$ so shortened.

This reduces our lemma to the case where one of the contracted constituents is the principal constituent of $R_1$. If $R$ is structural, then it is one of $*K*$. In this case the premise $\Gamma_1$ of $R_1$ is identical with $\Gamma'$, and the part of $\Delta$ leading to $\Gamma_1$ is the $\Delta'$ sought. Since this exhausts the possibilities if $n = 0$, the basic step of our induction is complete.

Suppose now that $R$ is nonstructural, and that the contracted constituents are the principal constituent $M_1$ of $R_1$ and a like parametric constituent $M_2$. Let the premises of $R_1$ be $\Gamma_1, \Gamma_2, \ldots, \Gamma_p$. Let $\Gamma'_1, \ldots, \Gamma'_p$ be obtained by dropping (the constituent congruent to) $M_2$ from $\Gamma_i$. The inference from $\Gamma'_1, \ldots, \Gamma'_p$ to $\Gamma'$ is, by $(r5)$, an instance of $R$. Hence if we show that there are contraction-free demonstrations $\Delta'_1, \ldots, \Delta'_p$, each of degree $\leq n - 1$, for $\Gamma'_1, \ldots, \Gamma'_p$, respectively, we shall have the $\Delta'$ sought.

Let $\Gamma_i$ have a quasi-principal constituent. Then this constituent and $M_2$ can be contracted to give $\Gamma'_i$. Since $\Gamma_i$ has a demonstration in $\Delta$ which contains no contraction and is of degree $\leq n - 1$, the existence of $\Delta'_i$ follows by the inductive hypothesis.

Let $R$ be directly invertible with respect to its $j$th premise. In a notation similar to that of Sec. 1, let $\Gamma_j$ be

$$\mathfrak{X}, \mathfrak{U}_j, M_2 \tag{9}$$

This can be the conclusion of an instance $R_2$ of $R$ with $M_2$ as principal constituent. By Corollary D1.3 we can invert with respect to the $j$th premise of $R$, obtaining

$$\mathfrak{X}, \mathfrak{U}_j, \mathfrak{U}_j \tag{10}$$

from which $\Gamma'_j$ follows by a series of contractions. Now (9) has a proof in $\Delta$ of degree $\leq n - 1$ and not containing any contractions; hence, by Corollary D1.6, (10) has a proof as required by the inductive hypothesis. By successive applications of that hypothesis we have the required $\Delta'_j$.

This completes the proof of the lemma. But that lemma shows that the rules $*W*$ are eliminable rules—one can eliminate them from any demonstration by starting at the top and working downward, Q.E.D.

This suggests a new formulation for $LA_m$ and $LC_m$ which will be called *Formulation* III. It differs from Formulation IK only in the following respects: $(a)$ the rules $*W*$ are omitted; $(b)$ in $LA_m$, $*P$ is taken as (8), but in $LC_m$, as in Formulation IK; $(c)$ $\vdash*$, if admitted, has a quasi-principal constituent in every premise.

COROLLARY 7.1. *In Formulation III of* $LA_m$ *and* $LC_m$, *the rules* $*W*$ *are epitheorems. For these systems Formulation* III *is equivalent to Formulation* I.

*Proof.* The rules of Formulation I are epitheorems of Formulation III. This was shown for $*W*$ in the theorem; for the other rules this follows by an argument similar to that of Sec. C8 [for $*P$ see the text after (8) of this article]. The converse argument follows along the lines of Sec. C8.

COROLLARY 7.2.    *Peirce's law, viz.*,

$$A \supset B \;.\supset A \;.\supset A$$

*is not assertible in* LA.

*Proof.* The argument of Example 2 of Sec. C5 is quite rigorous in Formulation III.

COROLLARY 7.3.   *In those cases of Formulation* III *in which there are rules with a quasi-principal constituent on the left, we must allow applications of* ∗K *in which the principal constituent has the form of such a quasi-principal constituent; we should need similar cases on the right if we should adjoin rules with a quasi-principal constituent on that side.*

The foregoing theorem gives circumstances under which the rules ∗W∗ are redundant. This helps with finding a decision process in Sec. 7. But they do not exclude the possibility of repeated constituents. The following example, using the Ketonen form of ∗Λ, shows this:

$$\frac{\dfrac{A,\, A \;\Vdash\, A}{A \wedge A \;\Vdash\, A}\;\ast\Lambda}{\Vdash A \wedge A \;.\supset.\, A}\;\text{P}\ast$$

We shall now seek a formulation such that repeated occurrences of the same constituent do not need to occur. This is rather a technical question.

The new formulation will be called Formulation IV. It will consist of Formulation III plus some additional rules. Demonstrations in it will be called *IV-demonstrations*, whereas those in Formulation III will be called *III-demonstrations*. Then our thesis is the following. Let $\Gamma'$ be obtained from $\Gamma$ by contracting until there are no repetitions. Let there be a III-demonstration of degree $n$ of $\Gamma$. Then we seek a IV-demonstration of degree $\le n$ of $\Gamma'$. This will be proved possible by induction on $n$.

The process of eliminating contractions used in the proof of Theorem 7 did not increase the degree. By that process there will be a III-demonstration, call it $\Delta$, of $\Gamma'$. We may therefore suppose that $\Gamma'$ is the same as $\Gamma$.

If there are structural rules at the end of $\Delta$, let $\Gamma_1$ be such that $\Gamma$ is obtained from $\Gamma_1$ in $\Delta$ by applications of ∗K∗, whereas $\Gamma_1$ is not obtained in $\Delta$ by structural rules (the only structural rules are ∗K∗). Then $\Gamma_1$ contains no repetitions and has a III-demonstration of degree $n$. Further, if we have a IV-demonstration of $\Gamma_1$, these same applications of ∗K∗ will give a IV-demonstration of $\Gamma$. We can therefore suppose that $\Gamma$ is the same as $\Gamma_1$, that is, that it does not end with applications of structural rules.

If $n = 0$, $\Gamma$ must now be prime. Our thesis is trivially satisfied, and the basic step of the induction is complete.

We now suppose that $\Delta$ ends with a nonstructural rule $R$ whose premises, let us say, are $\Gamma_1, \ldots, \Gamma_p$. These have III-demonstrations of degree $\le n - 1$. By the inductive hypothesis there are IV-demonstrations of $\Gamma'_1, \ldots, \Gamma'_p$. We need rules to validate the transition from $\Gamma'_1, \ldots, \Gamma'_p$ to $\Gamma$. To get them we take a look at the contractions which may take place in $\Gamma_1, \ldots, \Gamma_p$. There are three kinds conceivable: (*a*) between two parameters, (*b*) between two subalterns, and (*c*) between a parameter and a subaltern.

Contractions between two parameters are impossible, since there would then be repetitions in $\Gamma$.

Contractions between two subalterns can indeed arise, as the above example shows. We need additional rules to take care of this eventuality. It can occur only when there are two subalterns on the same side in the same premise, viz., in $*\Lambda$ and $V*$.

Contractions between a subaltern and a parameter can also occur. We therefore need rules which allow the same constituent to act as both parameter and subaltern. However, these rules are trivial. If the rule has a quasi-principal constituent, it may happen that one of the premises is identical with the conclusion. In all other cases, if we have unrestricted $*K*$, we can drop the parameter and restore it after the inference by $*K*$.† Thus rules of this character are needed only when $*K*$ is restricted.

In view of this discussion *Formulation IV* is defined as follows. The elementary statements are restricted to have no repeated constituents in any prosequence (so that a prosequence is simply a class of propositions). As additional rules we have

$$\frac{\mathfrak{X}, A \Vdash \mathfrak{Y}}{\mathfrak{X}, A \wedge A \Vdash \mathfrak{Y}} \qquad \frac{\mathfrak{X} \Vdash A, \mathfrak{Z}}{\mathfrak{X} \Vdash A \vee A, \mathfrak{Z}}$$

**Theorem 8.** *Let $\Gamma$ be an elementary statement of Formulation III, and let $\Gamma'$ be obtained by contracting it until there are no repetitions. Then a necessary and sufficient condition that $\Gamma$ be a theorem of Formulation III is that $\Gamma'$ be a theorem of Formulation IV.*

*Proof.* The preliminary discussion shows the necessity. The sufficiency follows, since the missing constituents can be reinstated by $*K*$.

**7. Decidability.** The composition property, in combination with the fact that the operational rules (in certain formulations) are such that the rule and the premises are uniquely determined once the conclusion and the principal constituent are given, suggests that after trial of a finite number of alternatives we ought to be able to ascertain whether a proposed elementary statement of such a formulation is demonstrable or not. Since there is only a finite number of choices for the principal constituent, it should be possible to discover a derivation of a given elementary statement, or to prove that none exists, by starting with the desired conclusion and using the rules backward, taking into account all possible alternatives. The examples of Sec. C5 are illustrations of decisions as to demonstrability which are reached by just such a process. The results of Secs. 5 and 6 considerably decrease the number of alternatives.

That such a process must eventually lead to a decision may be seen as follows. Let $\Gamma$ be the elementary statement to be tested, and let $n$ be the number of distinct components in $\Gamma$. Suppose we use Formulation IV, where prosequences consist of unlike constituents. Then the number of possible prosequences is $2^n$, and the number of distinct elementary statements which satisfy the condition of Theorem 8 is $2^{2n}$ in the multiple cases and $n \cdot 2^n$ in the singular cases. Let this number be $m$. If $\Gamma$ is demonstrable,

---

† The following is an example:

$$\frac{\mathfrak{X}, A \Vdash B}{\mathfrak{X}, A \Vdash A \supset B}$$

Here one can perform an ordinary $P*$ followed by a $*K$.

the demonstration can be exhibited as a finite sequence of such statements, and hence as an initial segment of one of the $m!$ permutations of them. Given any such permutation, it is a definite question whether an initial segment of it constitutes a demonstration of $\Gamma$. This argument applies when the system has the composition property in the full sense. Further, we need the fact that $*W*$ are redundant. Hence we have the following:

**Theorem 9.** *If an L system has the composition property and satisfies the conditions a and b of Theorem 7, then it is decidable.*

Since the hypotheses are satisfied for $LA_m(\mathfrak{O})$ and $LC_m(\mathfrak{O})$, we have the following:

COROLLARY 9.1.    *The systems* $LA_m(\mathfrak{O})$ *and* $LC_m(\mathfrak{O})$ *are decidable.*

The method used in the proof of Theorem 9 does not give a practical decision process. Thus for Example 1 of Sec. C5, where $n = 5$, there are 160 possible statements. But at least it shows that a decision is possible, and any systematic search must eventually reach it.

**8. Proof tableaux.** Although a theoretical solution of the decision problem was presented in Sec. 7, yet the procedures for carrying out the decision are often tedious. This is true for two principal reasons: on the one hand, there are a large number of alternatives to be investigated; and on the other hand, since the changes from one step to another involve only a small number of constituents (the subalterns and the principal constituent), the same expressions have to be written down repeatedly. There is need of a method for making the decision which will eliminate some of the excess labor. We shall study here such a method, which is a modification of one due to E. W. Beth.

There are two aspects to the problem as stated, viz., the description of the procedure and the notational devices for recording it. These will be treated separately.

It will be convenient to describe the notation first. The basic idea is as follows. Each step in a proof cancels certain constituents (the subalterns) and inserts the principal constituent. If the rules are reversed, as is natural when we are seeking a decision for a given statement, the principal constituent is dropped and the subalterns are added; where there is a quasi-principal constituent, the changes consist solely of additions. Accordingly, we arrange the analysis in the form of a tableau, consisting of two columns, in which we make entries as follows. We write the antecedent of the statement to be tested on the first line of the left column; the consequent of that same statement in the first line of the right column. Supposing, for the moment, that we are dealing with cases where there is only one premise and that one is uniquely determined, we write the subaltern(s) in the appropriate column(s) of a new line; we may or may not cancel the principal constituent, depending on which formulation we are concerned with. It is then understood that uncanceled constituents on the line above are included among the constituents of the statement represented on the line being written. The process continues until we reach a line with a common constituent on both sides, or the tableau reaches a point where no new constituents can be added. In the former case we say the tableau *closes;* in the latter, that it *fails to close.* If a tableau closes, the statement which has been reached is quasi prime (Sec. C9), and the tableau read backward furnishes an abbreviated proof of the statement on its first line; if none of the possible ways of developing the tableau leads to a closed tableau, then (assuming invertible rules) such a proof is impossible. For example, the proof of

$$\Vdash A \supset. B \supset. C \vee A$$

is given by the tableau[1]

$$
\begin{array}{c|l}
 & A \supset., B \supset. C \vee A \\
A & B \supset. C \vee A \\
B & C \vee A \\
 & C, A
\end{array}
$$

Here the fourth line represents a prime statement, and the tableau closes.

The basic principle must be modified in case we take into account rules with multiple premises or a choice of rules with the same principal constituent. In such a case the tableau will split into two or more subtableaux. The splitting may be conjunctive, in the sense that closure of the whole tableau is equivalent to closure of all the subtableaux, or it may be alternative, in the sense that closure of the whole tableau is equivalent to closure of at least one subtableau. The conjunctive case occurs when there is a single rule with multiple premises; the alternative case, when two or more rules are applicable. The splitting of a tableau will be indicated here by choosing one of the subtableaux, usually that corresponding to the left premise, to be written as a continuation of the main tableau, whereas the others are carried out as independent side tableaux. The presence of such a side tableau is indicated by the appearance, at the end of the relevant line in the main tableau, of the appropriate one of the signs '&', 'or', followed by the first line of the side tableau. In case we are dealing with a system which is not decidable (as will be the case later), it would be necessary to operate the main and all side tableaux simultaneously, so as to discover a proof or counterproof as quickly as possible; but when the situation is decidable, as we know it is here from Sec. 7, we may continue with the main tableau until we are through with it, and then pick up the side tableaux later. This enables us to write the whole tableau in a single pair of columns (aside from the incidental indications). The repetition of the constituents at the top of a side tableau, perhaps in abbreviated form, avoids a difficulty which arises when cancellations are allowed, viz., that when a constituent in the main tableau above the split is canceled in one of the subtableaux, these cancellations may not occur at the same time in the other subtableaux.

In the system LC one can interpret a tableau as an attempt to construct a counterexample for the statement on the first line. If that statement is false (by 0-1 tables), then all the propositions in the right column must have the value 0, those in the left the value 1. The rules are such that this holds throughout the tableau. If the tableau closes, then the same constituent must have both values; since this is impossible, the statement on

---

[1] In each line the consequent of the preceding line is canceled. A technique for indicating this will be introduced shortly.

the first line must be valid by 0-1 tables.[1] Because of this interpretation, Beth speaks of 'semantic tableaux'. For the absolute system Beth found an interpretation by means of treelike models. But although this interpretation is both ingenious and interesting, it seems artificial from the semantic point of view of Secs. A3 and C1. For this reason the tableaux are called here simply 'proof tableaux'.

Before giving examples of tableaux it will be necessary to explain the conventions in regard to them. In accordance with Theorem 8, we avoid repeated occurrences of the same proposition in the same prosequence. Likewise, when a rule would merely transform a line above into an identical line below, the rule is regarded as not applicable. A horizontal line drawn across a subtableau indicates that we are through with that subtableau, and nothing above the line is to be regarded as part of the subtableau which follows. The horizontal line is marked with 'T', 'F', '∼', or 'Δ' to indicate the disposition of the subtableau above. The letter 'T' indicates closure of the subtableau; 'F' or '∼' indicates failure to close; 'Δ' indicates that the subtableau has no further interest because the question of derivability has already been settled. The letter 'F' indicates failure to close because the final statement cannot be reduced further, or is false by 0-1 tables or previously established results, etc. The sign '∼' is used when a statement has already occurred farther up; such a recurrence does not necessarily mean failure to close, because the rules of the algorithm may be such that a new rule is applicable at that point, in which case the analysis proceeds with that rule; but where a statement recurs and no rule not previously applied to that statement is applicable, further prolongation of the reduction can lead to no new statements, and so the subtableau fails to close; this is the situation which is indicated by '∼'. A horizontal dashed line drawn across one side of a tableau indicates cancellation of everything on that side above it.

To get all this information down in handy form it is convenient to use an arrangement with six columns. The steps are numbered in the first column. The second column gives the justification for the step. The third and fourth columns are the two columns of the main tableau as above described. The fifth column indicates the beginning of a side tableau. The sixth column indicates the disposition of the side tableau; usually this is simply the line of the main tableau where that side tableau is taken up.

In the second column the following abbreviations are used. A notation such as 'L6.1' means that the principal constituent is the first constituent on the left side of line 6; likewise 'R2' indicates it is the constituent on the right of line 2, etc.; in cases like these, the principal constituent determines the rule. An '&' or 'or' followed by a numeral indicates the beginning of the subtableau whose existence was signalized at the line cited. The notation '(PK)∗' will be explained in the next paragraph. In some complex cases (not illustrated here) I have found it expedient to write in this second column an indication, using '=' and a line number, that the statement in question has previously occurred.

By virtue of Theorem 6 we need to consider only applications of ∗K which are made initially and those of K∗ which are made initially or after an

---

[1] Cf. Theorem D9. The tableau method is indeed a very efficient method of carrying out the test by truth tables.

instance of P∗. The initial applications of ∗K∗ can be eliminated if we allow starting with quasi-prime statements. Those after P∗ can be avoided if we replace P∗ by the complex rule

$$\text{(PK)}∗ \qquad \frac{\mathfrak{X},\, A \Vdash B}{\mathfrak{X} \Vdash A \supset B,\, \mathfrak{Z}}$$

in which $A \supset B$ will be called the principal constituent. Now if, in the tableau, we come across a statement of the form of the conclusion of (PK)∗, we cannot infer that it was obtained by (PK)∗ by the particular premise indicated. In this case it is necessary to split the tableau alternatively, thus:

$$\mathfrak{X},\, A \Vdash B \qquad \text{or} \qquad \mathfrak{X} \Vdash \mathfrak{Z},\, A \supset B$$

where the right alternative is to make the rule invertible (in the side tableau we can explore any other possibilities). The shifting of the '$A \supset B$' will remind us that this particular possibility has been considered. This eventuality is indicated by writing '(PK)∗' in the second column.

We now proceed to two examples. Example 1, which closes, shows that PS is assertible in LA. Example 2, which fails to close, shows that Peirce's law is not assertible in LA.†

*Example* 1

| 1 | | | $A \supset. B \supset C :\supset: A \supset B .\supset. A \supset C$ | | | |
|---|---|---|---|---|---|---|
| 2 | R1 | $A \supset. B \supset C$ | $A \supset B .\supset. A \supset C$ | | | |
| 3 | R2 | $A \supset B$ | $A \supset C$ | | | |
| 4 | R3 | $A$ | $C$ | | | |
| 5 | L2 | | $A$ | & | L3, $A, B \supset C \Vdash C$ | 6 |
| | | | —————————T | | | |
| 6 | &5 | $A \supset B, A, B \supset C$ | $C$ | | | |
| 7 | L6.1 | | $A$ | & | $A, B \supset C, B \Vdash C$ | 8 |
| | | | —————————T | | | |
| 8 | &7 | $B \supset C, B, A$ | $C$ | | | |
| | | | $B$ | & | $B, A, C \Vdash C$ | |
| | | | —————————T | | —————————T | |

*Example* 2

| 1 | | $A \supset B .\supset A$ | $A$ | | | |
|---|---|---|---|---|---|---|
| 2 | L1 | | $A \supset B$ | & | $A \Vdash A$ | |
| | | | —————————————— | | —————T | |
| 3 | (PK)∗ | | $A \supset B$ | or | L1 $\Vdash A, A \supset B$ | |
| | | | —————————————— | | ——————————∼ | |
| 4 | R3 | $A$ | $B$ | | | |
| 5 | L1 | | $A \supset B$ | & | $A \Vdash B$ | |
| | | | —————————————△ | | —————F | |

So much for the tableaux themselves. We now turn to the question of constructing a systematic procedure or algorithm—in the general sense of an effective process, not necessarily the specific sense of Sec. 2E—for applying the rules efficiently. Except for certain details, such an algorithm is

† Cf. Example 2 of Sec. C5 and Corollary 7.2.

given for the classical system in the proof of Theorem D9; the present discussion will therefore have primary reference to the absolute system.

Let us first discuss some general principles. It will be an advantage to use rules which are reversible, because we then know that we never have to go back to the beginning, and if the tableau closes we have a demonstration. Again it is advisable to begin with the rules which do not require splitting of the tableau, because that avoids a certain amount of repetition. Another principle is that we should apply first rules which may be expected to cut down the total number of distinct components and postpone till last those which leave this unchanged. There is some empirical evidence that it is, on the whole, better to apply rules on the right first. The algorithm given here is constructed according to these principles, one stated earlier having precedence over one stated later.

Again it is necessary to have some indication as to which of several possible candidates is the principal constituent. Beth does this by the device of cycling used above in the discussion of (PK∗). In his algorithm the principal constituent is always the leading (i.e., leftmost) constituent of its prosequence; after execution of the rule the subalterns are placed last (i.e., on the extreme right). Although this does keep track of what constituent is principal, I have found it a nuisance in practice; besides, it requires supplementary treatment for certain cases where the leading constituent is elementary or is to be passed over. It is here used only in the cases (like the above PK∗) where the principal constituent does not disappear. In other cases the following device is used. The letters '𝔛', '𝔜' will denote arbitrary prosequences; '𝔍', prosequences which are void in LA, arbitrary in LC. One of these letters with a subscript 1 will denote a prosequence containing no constituent of the same type as the principal constituent (which immediately follows it); one with subscript 2, or one without a subscript, is not so restricted. Then the principal constituent is simply the first (leftmost) constituent which is eligible.

In accordance with these principles, an algorithm will be proposed below. In this proposal the following conventions are supposed to be understood. The rules are stated as rules for entry in a tableau and are thus the inverses of rules intended for proof construction. For this reason the terms 'premise' and 'conclusion' may be misleading; the upper line of a rule will therefore be called its *datum*, the lower line its *result*.[1] The terms 'subaltern' and 'principal constituent' shall then be understood, in the same sense as previously, as if the rule were being used in the order of deduction, i.e., from the result to the datum. A rule is applicable just when (1) the datum has the form indicated in the rule and (2) the same rule has not previously been applied to the same datum.[2] The rules are given in order; as usual in the case of an algorithm, the rule actually to be applied is the first one, in the order given, which is applicable.

---

[1] When the result in this sense is compound, we shall sometimes apply the word 'result' to the elementary statements of which it is composed. In such a case there will be more than one result.

[2] We should perhaps add here "with the same principal constituent." However, the formulation in the text is such that the datum and the rule together determine the principal constituent uniquely.

Subject to these conventions, the rules for $LA_m$ are as follows.[1]

I
$$\frac{\mathfrak{X} \Vdash \mathfrak{Z}_1, A \supset B, \mathfrak{Z}_2}{\mathfrak{X}, A \Vdash \mathfrak{Z}_1, B, \mathfrak{Z}_2}$$

II
$$\frac{\mathfrak{X} \Vdash \mathfrak{Y}_1, A \vee B, \mathfrak{Y}_2}{\mathfrak{X} \Vdash \mathfrak{Y}_1, A, B, \mathfrak{Y}_2}$$

III
$$\frac{\mathfrak{X}_1, A \wedge B, \mathfrak{X}_2 \Vdash \mathfrak{Y}}{\mathfrak{X}_1, A, B, \mathfrak{X}_2 \Vdash \mathfrak{Y}}$$

IV
$$\frac{\mathfrak{X} \Vdash \mathfrak{Y}_1, A \wedge B, \mathfrak{Y}_2}{\mathfrak{X} \Vdash \mathfrak{Y}_1, A, \mathfrak{Y}_2 \ \& \ \mathfrak{X} \Vdash \mathfrak{Y}_1, B, \mathfrak{Y}_2}$$

V
$$\frac{\mathfrak{X}_1, A \vee B, \mathfrak{X}_2 \Vdash \mathfrak{Y}}{\mathfrak{X}_1, A, \mathfrak{X}_2 \Vdash \mathfrak{Y} \ \& \ \mathfrak{X}_1, B, \mathfrak{X}_2 \Vdash \mathfrak{Y}}$$

VI
$$\frac{\mathfrak{X} \Vdash A, \mathfrak{Y}}{\mathfrak{X} \Vdash A \ \text{or} \ \mathfrak{X} \Vdash \mathfrak{Y}, A}$$

VII
$$\frac{\mathfrak{X}_1, A \supset B, \mathfrak{X}_2 \Vdash \mathfrak{Y}}{\mathfrak{X}_1, \mathfrak{X}_2, A \supset B \Vdash \mathfrak{Y}, A \ \& \ \mathfrak{X}_1, \mathfrak{X}_2, B \Vdash \mathfrak{Y}}$$

In the algorithm for $LC_m$, rule VI is deleted and the quasi-principal constituent is left out of VII.

**Theorem 10.** *A necessary and sufficient condition that*

$$\mathfrak{X} \Vdash \mathfrak{Y} \tag{11}$$

*be derivable in* $LA_m(\mathfrak{O})$ *or* $LC_m(\mathfrak{O})$ *is that the corresponding algorithm lead to a closed tableau.*

*Proof of Sufficiency.* We need to show only that inverses of the algorithmic rules, i.e., the inferences from result to datum, are valid as inferences in the pertinent L system. The inverses of the rules I, II, III, IV, V, VII are rules of $LA_m$; so are those for I and VII in $LC_m$. It remains only to consider the rule VI in $LA_m$. The inference from the left result to the datum can be made by K*, while that from the right result is a specialization of C*.

*Proof of Necessity.* For $LC_m$ this follows by Theorem D9. We therefore confine attention to $LA_m$. We suppose we have a formulation of type IV, with elementary prime statements, and *K* applied only initially or as part of a (PK)*. (For some details see Theorem 7B11.)

By hypothesis there exists a demonstration $\Delta$ of (11). Let $n$ be the number of its operational steps. If $n = 0$, then (11) is quasi prime and the tableau closes at the very beginning. It therefore suffices to prove the necessity for a given value of $n$ on the assumption that it holds for any smaller value of $n$.

Suppose that (11) contains a constituent of one of the first five types. By Corollary D1.6 (adapted to Formulations III and IV) there will be demonstrations of the results[2] of the first step in the tableau, and these demonstrations

---

[1] From correspondence with S. Kripke in the summer of 1958, I understand that he had found independently a similar, but apparently not quite identical, algorithm.

[2] There may, of course, be only one result.

will have fewer than $n$ operational steps.[1]   By the inductive hypothesis their subtableaux all close.   Hence the tableau for (11) closes in that case.

In any other case the only rules which can terminate $\Delta$ are K$*$ and $*$P, and the algorithm passes through to VI.   Let $\mathfrak{Y}$ be $C_1, C_2, \ldots, C_m$.   Then VI is equivalent to a rule giving as result the alternation

$$\mathfrak{X} \Vdash C_1 \text{ or } \mathfrak{X} \Vdash C_2 \text{ or } \cdots \text{ or } \mathfrak{X} \Vdash C_m \text{ or } \mathfrak{X} \Vdash \mathfrak{Y} \tag{12}$$

The tableau will close if that for

$$\mathfrak{X} \Vdash C_k \tag{13}$$

closes for some value of $k$; otherwise we go through a complete cycle and pass on to VII.

Suppose the last step in $\Delta$ is by K$*$.   This cannot be an initial K$*$ since $n > 0$; it must therefore be part of an inference by (PK)$*$.   Let $C_k \equiv A \supset B$ be the principal constituent of that (PK)$*$.   Then application of I to (13) will give the premise of that (PK)$*$.   The part of $\Delta$ over that premise will have $n - 1$ operational inferences.   Hence the subtableau for that premise closes by our inductive hypothesis.   The tableau for (11) therefore closes in that case.

If the last step in $\Delta$ is by $*$P, then we choose the alternative in VI which carries us on into VII and examine its subtableaux.   (We may, of course, be able to close the tableau in VI, but that is irrelevant.)   Let

$$A_1 \supset B_1, \ldots, A_m \supset B_m \tag{14}$$

be all the constituents of the type of the principal constituent in VII in the order in which they occur in $\mathfrak{X}$.   Suppose that the premises of the last inference in $\Delta$ are

$$\mathfrak{X}', A_k \supset B_k \Vdash \mathfrak{Y}, A_k, \qquad \mathfrak{X}', B_k \Vdash \mathfrak{Y} \tag{15}$$

where $\mathfrak{X}', A_k \supset B_k$ is a permutation of $\mathfrak{X}$.   These have demonstrations (in a secondary $\Delta$) with fewer than $n$ operational steps in the two together.   We show by induction on $k$ that the tableau will close.   In fact, if $k = 1$, the premises (15) are precisely the result of the next step; hence the tableau closes.   For $k > 1$, suppose that $\mathfrak{X}$ is $\mathfrak{X}_1, A_1 \supset B_1, \mathfrak{U}, A_k \supset B_k, \mathfrak{X}_2$.   Then the result of the next algorithmic step is the conjunction of

$$\mathfrak{X}_1, \mathfrak{U}, A_k \supset B_k, \mathfrak{X}_2, A_1 \supset B_1 \Vdash \mathfrak{Y}, A_1 \tag{16}$$

$$\mathfrak{X}_1, \mathfrak{U}, A_k \supset B_k, \mathfrak{X}_2, B_1 \Vdash \mathfrak{Y} \tag{17}$$

Now (16) can be inferred by $*$P from the conjunction of

$$\mathfrak{X}_1, \mathfrak{U}, A_k \supset B_k, \mathfrak{X}_2, A_1 \supset B_1 \Vdash \mathfrak{Y}, A_1, A_k \tag{18}$$

$$\mathfrak{X}_1, \mathfrak{U}, B_k, \mathfrak{X}_2, A_1 \supset B_1 \Vdash \mathfrak{Y}, A_1 \tag{19}$$

---

[1] In that phase of Corollary D1.6 where $M$ drops out because it is introduced solely by $*$K$*$, we employ a secondary induction.   Let $M$ be of order $m \geq 0$ if it is formed by $m$ uses of $\wedge$, $\vee$ from obs which are elementary or of the form $A \supset B$.   The case of order 0 cannot occur here because the omission of $M$ would leave a theorem with void consequent.   Thus the results of the first step of the algorithm will have the same $n$, but the total order of all the constituents introduced by $*$K$*$ will be less.

and similarly (17) can be inferred from

$$\mathfrak{X}_1, \mathfrak{U}, A_k \supset B_k, \mathfrak{X}_2, B_1 \Vdash \mathfrak{Y}, A_k \tag{20}$$

$$\mathfrak{X}_1, \mathfrak{U}, B_k, \mathfrak{X}_2, B_1 \Vdash \mathfrak{Y} \tag{21}$$

Now (18) can be derived from the left premise of (15), and (19) from the right premise of (15), by the unrestricted K∗; hence these have demonstrations with the same number of operational steps as in the demonstration of that premise.    Similarly, (20) and (21) can be derived from the premises of (15) by the unrestricted K∗ and Corollary D1.5, and the number of operational inferences will not be greater by Corollary D1.6.    Thus (16) and (17) will have demonstrations with not more than $n$ operational steps; they also have a smaller value of $k$.    By the inductive hypothesis on $k$ their subtableaux will close, and hence that for (11) will.[1]

This completes the proof of Theorem 10.

COROLLARY 10.1.    *In* $LA_m$ *and* $LC_m$ *the uses of* ∗C∗ *can be confined to cyclic permutations.*

*Proof.*    As already mentioned in the introductory discussion, the rules II to V can be given an alternative form using a cycling rule of the form,

$$\frac{\mathfrak{X} \Vdash A, \mathfrak{Z}}{\mathfrak{X} \Vdash \mathfrak{Z}, A}$$

to be used whenever $A$ is not of the appropriate form.    One can then require that the principal constituent can always be the first constituent in the prosequences.

*Remark.*    It is not claimed that the algorithm gives the shortest possible derivations.    I have no reason, except purely heuristic ones, for preferring it to any of various conceivable modifications.    It has the advantage that similar steps can be telescoped.    Thus by combining several applications of rule I we can make inferences of the form

$$\frac{\mathfrak{X} \Vdash A_1 \supset. A_2 \supset. \cdots. \supset. A_m \supset B}{\mathfrak{X}, A_1, \ldots, A_m \Vdash B}$$

Similarly, one can combine several steps of rule II to clear the right prosequence of all constituents of the form $A \lor B$, etc.    In rule VI, one could theoretically restrict attention to the case "$A$ is of the form $B \supset C$," but this seems to complicate matters.    When one comes to rule VII, the mechanical operation of the algorithm can be very tedious.    If one is alert, one can shorten the process considerably by arranging the constituents (14) in a suitable order; thus, if we have on the left, along with (14), an instance of $A_k$, it is an advantage to begin with $A_k \supset B_k$, for then the left result is quasi prime and the right result is simpler.    But if one wants a purely algorithmic procedure—e.g., if one wishes to have a machine to do the job—the algorithm is, so far as I know, as effective as any other.

[1] In the proof of Theorem 7B11 there is an alternative method of treating this phase of the algorithm.

**EXERCISES**

**1.** Show directly that the axiomatic propositions of HA are assertible in LA.

**2.** Prove that the following are assertible in LA:

$$A \supset B .\supset. A \supset C :\supset: A \supset. B \supset C$$
$$A \supset B .\supset A :\supset: A \supset B .\supset B$$
$$A \supset B .\wedge. C \supset D :\supset: A \wedge C .\supset. B \wedge D$$

What about the converses (i.e., those obtained by interchanging the two sides of the main ply operation)?

**3.** Show that the following are assertible in LC($\supset$) but not in LA($\supset$):

$$A \supset B .\supset B :\supset: B \supset A .\supset A$$
$$A \supset B .\supset: A \supset C .\supset B .\supset B$$

**4.** Show that

$$A, A \supset B \Vdash B$$
$$A \wedge B .\supset C \Vdash A \supset. B \supset C$$
$$A \supset. B \supset C \Vdash A \wedge B .\supset. C$$

are demonstrable in LA($\supset$) using Formulation I without \*W, but not

$$A \wedge. A \supset B \Vdash B$$
$$A \supset B .\wedge. B \supset C \Vdash A \supset C$$
$$A \vee. A \supset B :\supset B \Vdash B \qquad \text{(Kripke)}$$
$$\Vdash P \supset B .\supset B \qquad \text{(Wajsberg)}$$

where

$$P \equiv A \supset B .\supset A .\supset A$$

**5.** Show that, without change in the elementary theorems of an L system, the rule \*P could be stated in the form: if $\mathfrak{Z}$ is a part of $\mathfrak{Y}$, then

$$\frac{\mathfrak{X} \Vdash A, \mathfrak{Z} \quad \mathfrak{X}, B \Vdash \mathfrak{Y}}{\mathfrak{X}, A \supset B \Vdash \mathfrak{Y}}$$

**6.** Formulate systems dual to L systems in which implication is replaced by subtraction, and develop some of their properties. (Feys [SLP].)

**7.** Show that if prosequences are restricted to be singular and nonvoid on both sides, we have a formulation of a lattice. (Lorenzen [ALU].)

**\*8.** What sort of system should we get from Formulation I of LA (presumably LA$_m$) if \*W were denied, or if the quasi-principal constituent were left out of \*P in Formulations III and IV? What would be the associated H and E forms? (Cf. Exercises 3, 4, and 7.)

**\*9.** What sort of system should we get if the rules were restricted to be singular on the left but not on the right?

**\*10.** What sort of multiple system should we get if we replaced P\* by a rule

$$\frac{\mathfrak{X}, A \Vdash B_1, B_2, \ldots, B_n}{\mathfrak{X} \Vdash A \supset B_1, A \supset B_2, \ldots, A \supset B_n}$$

How would it be related to LA and LC?

**\*11.** What sort of L system would one get if one interpreted $\mathfrak{X} \Vdash B$ as meaning that there is a demonstration of $B$ using as uncanceled and unremovable premises (a) exactly those propositions which appear in $\mathfrak{X}$; (b) some of those in $\mathfrak{X}$ with not greater than the multiplicity with which they occur there; (c) both of these restrictions at once?

**12.** Show that a necessary and sufficient condition that

$$\mathfrak{X}, A_1 \supset B_1, A_2 \supset B_2, \ldots, A_n \supset B_n \Vdash \mathfrak{Y}$$

be true in HA is that all the $2^n$ statements be true which are formed from

$$\mathfrak{X}, A_1 \supset B_1, \ldots, A_n \supset B_n \Vdash, \mathfrak{Y} A_1, \ldots, A_n$$

by dropping out any subset (including the null set) of the $A_i$ on the right and changing the corresponding $A_i \supset B_i$ to $B_i$ on the left.

**13.** Under what conditions are inferences directly invertible in $LA_1$? (Cf. Kleene [PIG].)

**14.** Show that the invertibility (but not the direct invertibility) of all operational rules in $LC_m$ and in Formulation III (Sec. E6) of $LA_m$ can be derived from the elimination theorem.

**15.** Prove in detail that PW cannot be derived in LA without having an instance of $*$P above one of P$*$.

**16.** Construct demonstrations in HC of the theorems that the propositions of Exercise 2 are assertible.

**17.** Derive the elimination theorem for $LA_m$, using the arguments of Sec. D5, from that theorem for $LA_1$. What is the situation with respect to $LC_1$ and $LC_m$?

**18.** If $\mathfrak{Z}$ is $D_1, D_2, \ldots, D_n$ as in Sec. D5, show that in LC

$$\mathfrak{X} \Vdash {}_m A, \mathfrak{Z} \rightleftarrows \mathfrak{X}, D_1 \supset A, D_2 \supset A, \ldots, D_n \supset A \Vdash A$$

(Cf. [TFD, Theorems II16, II17].)

**19.** Show that, in $LA(\mathfrak{D})$, $LC(\mathfrak{D})$, if a component has a positive or negative occurrence (Sec. D3) in any elementary statement of a demonstration, then it has an occurrence of the same sign in the conclusion.

**20.** Let $E_1, \ldots, E_n$ be elementary, and let none of them occur as component in $\mathfrak{X}$ or $\mathfrak{Y}$. In $LA(\mathfrak{D})$ or $LC(\mathfrak{D})$ show that if

$$\mathfrak{X}, C_1, \ldots, C_m \Vdash \mathfrak{Y}$$

where

$$E_1, \ldots, E_n \Vdash C_1 \wedge C_2 \cdots \wedge C_m$$

then

$$\mathfrak{X} \Vdash \mathfrak{Y}$$

(Use Exercise 19. The theorem generalizes [SLD], Theorem 1.)

**\*21.** Is it true that in $LC_1$ all applications of Px can be made last? In particular, if $Q$ is assertible in $LC_1$, is there a C such that in LA

$$Q \supset C \Vdash Q$$

**22.** Complete the proof of Theorem E6 by adding details of reduction to special forms of principal constituents.

**23.** The algorithm of Beth [SCI] has the following peculiarities: (a) the quasi-principal constituent is left out of $*$P; (b) the rules II to V and VII are stated with the principal constituent on the extreme left in the datum and the subalterns at the extreme right in the results; (c) one must begin with VI and then apply a rule on the left, then VI, then a rule on the right, then VI again, etc., in cyclic order. Show that the algorithm fails to close in the following cases:

$$A, A \supset B \Vdash B$$
$$B \supset C, A \supset B \Vdash A \supset C$$

although these are demonstrable in LA.

## S. SUPPLEMENTARY TOPICS

**1. Historical comment.** A reader who has some familiarity with current logical literature will notice at once two peculiarities of the present approach. The first of these is the complete separation of the finite positive operations, treated in this chapter, from negation and quantification, treated in Chaps. 6 and 7. The second is the emphasis on the absolute system, which plays here the leading role, whereas the classical system is carried along by an analogy. These peculiarities arose from the interpretation from which we started out. This interpretation was described in Sec. A (especially Secs. A2 and A3), which, in turn, is based on Sec. 3A2 and amplified in Sec. C1.

The following are the essential facts in the history of the present treatment. In the study of combinatory logic (cf. Sec. 3D5) it was necessary to use epitheoretical methods of some complexity and at the same time to adhere strictly to the constructive point of view. Accordingly, the question naturally came up as to how to define the logical connectives and to formulate propositional logic so as to be applicable to statements which are generated by inductive definitions. When Gentzen's thesis [ULS] became available, it suggested a method of attacking this problem. From this the semantical point of view of Secs. A3 and C1 evolved. A preliminary manuscript (never published) was written in 1937 (for an abstract see [PFD]); the manuscript of [TFD] was written in 1948 after a long gap. The second edition of [TFD] contains a preface, written in September, 1955, giving an account of the history up to that time. A discussion of the principles of the interpretation, written primarily for philosophers, appears in [IFI]. The treatment in Secs. A2, A3, and C1 is a revision of [IFI].

The absolute propositional algebra, arrived at in the way just described, coincides exactly with the propositional algebra arrived at independently by other authors starting from similar but yet significantly different interpretations. Two such systems which are particularly interesting are Heyting's formulation of the intuitionistic propositional algebra (see Sec. 4S2, under Brouwer algebra, and Exercise B7, also below in the discussion of H systems) and Lorenzen's *"Konsequenzlogik"* (see below). Both of these systems emphasize a constructive point of view, and they both start (this is explicit with Lorenzen and plausible with Heyting) with an interpretation similar to that with which we began here. However much they may differ from one another and from the present treatment in the details of their interpretations, they agree both in the kind of logical algebra with which they emerge and in the separation of negation. All three treatments therefore exemplify an approach to logical calculus characterized by interpretations leading to these two features; it is appropriate to call this the *constructive approach*.

In contrast to the constructive approach, the more usual procedure, which I shall call the *traditional approach*, has the following features. The principal interpretation is that of validity as tautologies in evaluations by 0-1 tables. This leads to primary emphasis on the classical systems; if nonclassical systems are mentioned at all, they are usually considered as artificial constructs whose interpretation is more or less mysterious. Furthermore, there is no semantical reason for deferring the treatment of negation, and since it, being unary, is regarded as simpler than the binary operations, it is often used as a basis for eliminating some of the latter as primitive operations (see Chap. 6).

Now there is a tendency, even in theories motivated by the traditional approach, to evolve in the direction of a rapprochement with the constructive approach. In particular, the separation of the operations may be motivated, as in mathematics generally it often is, by considerations of formal structure. Thus the operations were separated in Chap. 4. This separation goes back in principle to Schröder, who developed the properties of a lattice first and then superposed on them the properties of negation. This separation did not occur in Boole and was only imperfect in Peirce (see the remark in Sec. 4S1). Now it seems likely that Schröder had no nonclassical interpretation (the classical one then being item 1° in Sec. 4A2) in mind; on the contrary, he sought for and found such an interpretation afterward. Many other such interpretations were found later; e.g., Dedekind found the interpretation 5° of Sec. 4A2 and treated it at some length in his [ZZG]. The separation of positive operations from negation is commonplace in modern lattice theory. In assertional propositional algebra the situation is more complex. In the system of Frege there was a partial separation; it is now known that Frege's schemes for pure implication suffice for the absolute properties of implication, but not for the classical nonabsolute ones. The formulation of Whitehead and Russell [PMt] completely ignored this separation. The separation of the classical properties of various operations is a prominent motive in various Polish investigations reported in Łukasiewicz and Tarski [UAK] (these were mostly due to Tarski), Wajsberg [MLB], etc., but these authors paid no attention to separation of the absolute properties (although Wajsberg did in later papers). The treatment of Hilbert and Bernays [GLM], which forms the basis of most more recent treatments, is thorough in the separation of the absolute properties, but considers only incidentally that of the classical properties. This treatment is to be reckoned as traditional because the interpretations of the absolute system, which Hilbert and Bernays call "positive Logik" (see [GLM.II], supplement III), seem a bit forced and to be dictated primarily by a search for an interpretation to fit the formal structure. That there are certainly similarities in their supplement III to the present interpretation is interesting as showing that one is approaching the same answer from both directions. Their system is thus intermediate in character between the classical and constructive approaches. To what extent it was influenced by the prior existence of Heyting's system I do not know.

In the work of Gentzen the constructive and traditional approaches are considered strictly in parallel. The treatment given here is based on his. Indeed the systems TA, $LA_1$, and $LC_m$ of this chapter are obtained from Gentzen's NJ, LJ, and LK, respectively, by dropping out everything connected with negation; the other inferential systems, TC, $LA_m$, $LC_1$, are modifications introduced later.

From the foregoing it should be clear that the indebtedness of the present treatment to Gentzen's [ULS] is very great. For references on the Gentzen technique in general, see Sec. 1S5 and [IAL]. For the present context it is worthwhile to point out that Gentzen was apparently influenced by Hertz (see his [ASB] and other papers cited in Church [BSL]), and in his very first paper, [EUA], he showed that a complex rule (called "*Syll*") of Hertz' could be replaced by Gentzen's "*Schnitt*." This throws some light on the role of "*Schnitt*" in the Gentzen system. There is no trace in Gentzen [ULS] of the idea that the propositions represent the statements related to an underlying system $\mathfrak{S}$, but the idea is germane to the work of Hertz. Thus the idea is not completely foreign to Gentzen, and perhaps he did not consider it important because he foresaw, in some way, the result of Theorem E4.

There is also some indebtedness to Lorenzen, at least for a parallel and independent development of related ideas. He starts with an idea of a "calculus," which is essentially a syntactical system with elementary rules. (On the relation of the two ideas, see [CFS].) He defines an implication, symbolized by '$\rightarrow$', taken, like the present '$\Vdash$', as a functor with any number of arguments on the left and one argument on the right; propositions formed with this function are to be interpreted as rules concerning a calculus which are valid just when they are admissible (see Sec. 3A2). He presents a number of "principles" which are essentially epitheorems giving methods of establishing such admissibility; one of these, the inversion principle (*Inversionsprinzip*), was mentioned in Sec. A3. His "*Konsequenzlogik*," which appeared in his [KBM] and was incorporated in his book [EOL], secs. 6 and 7, aims to include as assertions just those propositions which are admissible for every calculus. It can be shown that under a suitable translation the assertions of *Konsequenzlogik* become the same as those of LA. This seems to mean (the situation is not quite clear) that it does not make any difference whether one takes implication in the sense of deducibility or of admissibility so long as the underlying calculus is completely unspecified. All this development is certainly completely independent of [TFD], and vice versa. Curiously, Lorenzen shares in the German tendency to shy away from Gentzen; the latter's [ULS] is not listed in the bibliography, and there is no reference to it in the index; the rules of *Konsequenzlogik* are strongly reminiscent of the very rules of Hertz which Gentzen simplified in his [EUA], and the proof of a certain decidability theorem, which would follow immediately from Gentzen's results, is deduced instead from Wajsberg [UAK]. The treatment of Lorenzen [EOL] is, moreover, not free from error (e.g., see [EOL], top of p. 56). Hermes [IPO] shows that the statement of the inversion principle needs correction. Naturally, the development of mathematics in the later portions of Lorenzen [EOL] goes beyond the limits of this book.

So much for this chapter in general and the semantical discussion in Sec. A. The following remarks relate to matters discussed in the separate sections from Sec. B on.

Of the systems considered in Sec. B, the oldest are the H systems. They correspond to the LH systems which Gentzen exhibited in his [ULS], sec. V2, as "*einen dem Hilbertschen Formalismus angeglichenen Kalkül*." Here the 'L' in the names of these and similar systems is stricken out, and the term 'H system' is extended to include not only the particular systems which Gentzen described, but any assertional system based on axiom schemes and modus ponens. Thus the 'H' stands for 'Hilbert'. The designation is apt in so far as Hilbert in his [NBM] and [GLM] considered formulations very similar to those which Gentzen exhibited.

On the history of these H systems see Church [IML$_2$, sec. 29]. To this I shall add only the following remarks concerning details. The earliest such system is that of Frege (P$_F$ in Church). For information about this system, see also Hermes and Scholz [NVB]. From there one learns that the primitive operations were implication and negation; the axiom schemes (speaking of it as if it were a modern system—actually the notion of axiom scheme had not then been formulated) were PK, PS, and PC, of which PC was redundant, as Łukasiewicz first showed. Thus PK and PS, which form an adequate pair of axiom schemes for absolute implication, appeared already in this very first system. The system employed by Hilbert (*loc. cit.*) used PB, PC, PK, and PW as axiom (schemes) for implication, whereas Hilbert and Bernays [GLM] replaced PB and PC by PB′; with respect to other positive operations, both of these systems used the same axioms as the present formulation of HA, except that Hilbert [GLM] used (9) of Sec. B instead of ΛS (cf. Church [IML$_2$, p. 160, note 267]). Gentzen's system LHJ was actually not taken directly from Hilbert, but (so far as HA is concerned) from Glivenko [PLB] (which Gentzen cites); in it PI, PB′, PC, ΛS are all postulated, but not PB or (9) of Sec. B. This system is formed from that of Glivenko [LBr], which contains "*principes connus de la logique*," all of which, except the law of excluded middle, are intuitionistically acceptable, by adding certain new axioms, for some of which the intuitionistic acceptability was credited to Heyting. The positive part of the formulation of Heyting [FRI$_1$] also constitutes a formulation of HA (see Exercise B7); for a detailed proof of its equivalence to that of Hilbert and Bernays [GLM] and some comments about its history and significance, see Schröter [UHA] (this represents a study made in 1937 at the suggestion of Scholz; it is curious, in view of the existence of the Glivenko papers, that it had not been made earlier). The present formulation of HA is the same as that in Wajsberg [UAK], which he credited to the Münster school.

For an extended study of the system HA see Hilbert and Bernays [GLM. II, supplement III] and Wajsberg [UAK]. Many properties, here obtained by the Gentzen formulations, are there obtained by direct consideration of the H formulation.

All the above systems concern HA only; they were obtained by dropping out negation (and in some cases quantification) from systems in which the positive postulates are all absolute. So far as HC is concerned, the first studies concern the system with implication as the only operation. These are due to Tarski, but the realization that by adjoining Pc to HA with implication only we get the corresponding part of HC seems to be due to Bernays (see Łukasiewicz and Tarski [UAK, sec. 4, especially Theorem 29]). The theorem to the effect that PK, PB′, and Pc form a sufficient set of axiom schemes for HC is known as the "theorem of Tarski-Bernays." For further information about HC see Church [IML$_2$, sec. 26], Wajsberg [MLB], Schröter [VIE]. From a historical point of view, Peirce [ALg$_2$] is not without interest; that is the point of appearance of Peirce's law; for a discussion of it see Prior [PAP].

The T systems of this book correspond to the "N systems" (for "natural systems") of Gentzen. The change from 'N' to 'T' was made in [TFD] to forestall possible confusion with 'N' standing for negation. The system TA is the positive part of Gentzen's NJ; the system TC was proposed in [TFD] on the same motivation as that given here. Rules similar to the T rules of Gentzen appear in Jaśkowski [RSF]. There is also some similarity to Tarski [FBM], where Pe and Pi were postulated; it is easy to show that Tarski's postulates are satisfied if $Fl(X)$ [in the English translation $Cn(X)$] is the set of all $B$ such that $X \Vdash^T B$.

Kneale [PLg] gives a multiple form of TC which bears much the same relation to $LC_m$ that the present TC does to $LC_1$.

The equivalence of the H and T systems established in Sec. B, and of the L and T systems in Sec. D4, was proved in Gentzen [ULS, sec. V]. He showed directly that the postulates of his H system could be established in the T system; those of the T system in the L system; and finally, those of the L system in the H system. The latter reduction in particular was rather involved. Some minor improvements in this proof of equivalence were proposed in [NRG], from which the present treatment evolved through [TFD].

The deduction theorem of Sec. B2 has a curious history. Apparently it occurred independently to a number of different workers. It is included, for example, in [PEI], Theorem 13. It is, of course, also implicitly contained in the equivalence proof cited in the preceding paragraph. The name 'deduction theorem' appears to be due to Hilbert and Bernays [GLM.I], where the theorem is stated and proved on pp. 155ff.; it is my understanding that this was also an independent discovery. In all probability this is the publication which made the theorem well known. The theorem appeared earlier, however, in Herbrand [RTD, p. 61]; for an even earlier appearance in a publication of Herbrand see Church [IML$_2$, p. 164]. Herbrand appears to be the first to publish a proof of the theorem. As already noted, Tarski stated Pi as a postulate in his [FBM$_1$]. For his claim to priority in the discovery of the theorem see the note to the paper just cited, in Tarski and Woodger [LSM, p. 32].

Formulation I of the systems LA$_1$ and LC$_m$ was taken directly from Gentzen [ULS], as already explained. The system LC$_1$ was proposed in [SLD], p. 42, and again in [ETM], p. 263. The system LA$_m$ is related to the LJ' of Maehara [DIL] (see below), much as LA$_1$ is to Gentzen's LJ; as presented here, however, it is the development of an idea in [NRG], sec. 6.

The modifications of Ketonen are taken from his [UPK].

The inversion theorem of Sec. D1 is in principle due to Schütte [SWK]. At least there is in that paper the fundamental suggestion that inversion could be established in that way; the details are more easily worked out directly than transferred from his point of view to the present one. Schmidt [VAL] makes extensive use of this idea. Ketonen [UPK] showed the invertibility of his classical propositional rules, of which Formulation IK of LC$_m$ constitutes the positive part, by using the elimination theorem, but the Schütte method gives more information. Kleene's [PIG] has also been suggestive. The notions of ancestor and descendant in Sec. C7 are Kleene's terms, but the same idea occurs in Schütte (and also in Schmidt), where it is ascribed to Gentzen [NFW]. The examples at the end of Sec. D1 are taken from the counterexamples given by Kleene as illustrations of cases where permutation is not possible in LA$_1$.

The elimination theorem is the principal theorem ("*Hauptsatz*") of Gentzen [ULS]. He formulated his systems at first with the cut rule ("*Schnitt*") as a primitive rule—which is natural in view of the way in which his [EUA] indicates it originated (see above); he then showed that uses of this rule could be eliminated from a demonstration. The theorem is therefore often called the "cut-elimination theorem;" but from the present point of view, it is the constituent $A$, rather than the cut, which is eliminated. Gentzen's original proof involved a double induction. The primary induction was, as here, on the number of operations in $A$, the secondary induction on the number of steps in demonstrating the premises. There were a large number of special cases treated independently. The present proof was worked out for [TFD], revised in [ETM], adapted for a different purpose in [CLg], sec. 9F4, and finally revised for this book. Anyone who will compare this proof with Gentzen's will probably find that it is essentially only another variant of the same thing. The extension to include $LA_m$ is new here. It was not ready when [IAL] was written, and at first the theorem was proved in connection with the theorem of Sec. D5. In Umezawa [IPL, p. 22, footnote 3] there is a statement to the effect that Ohnishi proved ET for the system LJ' of Maehara, which includes $LA_m$.

The proof of the equivalence of singular and multiple systems is to be compared with that of Maehara [DIL] (Umezawa [IPL, p. 22, footnote 3] claims independent discovery). Of course, for LC, the equivalence is, in principle, already shown in [TFD].

The various theorems of Sec. E are proved here in ways which involve some generalization over [TFD]. The composition property is of course the main and most obvious characteristic of the Gentzen L rules. Gentzen used this property to prove decidability; this proof is somewhat simpler here on account of the restriction on the structural rules in Secs. E5 and E6.

Some interest attaches to the extension of the separation property to the H systems. This theorem is also, in principle at least, contained in Gentzen's proof of equivalence between the systems, but he was not interested in the question and did not make it explicit. Apparently the first explicit statement and proof were in [NRG]. Wajsberg, in his [UAK], which appeared when [NRG] was in press, stated the theorem and claimed a proof; I have never examined this proof, but I understand from Bernays and from an errata sheet (received January, 1957) to Church [IML$_2$] that it contains an error.

The alternation property for the H system is due to Gödel [IAK].

For references on proof tableaux see Sec. 1S5. The topic was the subject of some correspondence with Kripke, who verified the existence of errors in Beth [SCI]. A correction, Dyson and Kreisel [ABS], has appeared, but it is concerned mostly with the completeness question for the intuitionistic predicate calculus, which goes beyond the scope not only of this chapter, but of this book.

**2. Weakened implications.** The absolute implication of the foregoing theory is related to formal deducibility in that

$$\vdash A \supset B \tag{1}$$

expresses essentially that $B$ is assertible in any extension (of the basic system $\mathfrak{S}$) in which $A$ is assertible. It is characterized by the deduction theorem and the rule of modus ponens, and is the minimal implication which has those two properties. It is an objective notion and has significance for formal deducibility regardless of one's philosophy. But it does not, and was never intended to, satisfy the demands of those who wish to formalize an implication such as we express by conditional sentences in ordinary language. We shall consider here briefly some formalized implications which do better in that respect. It will be convenient to use the term '*entailment*' for relations so expressed in ordinary language, and to call implications which formalize them *entailment implications*. These entailment implications will be *weak implications* in the sense that they are weaker than the absolute.

If (1) expresses an entailment, then most persons would agree that it holds only when there is some sort of logical connection[1] between $A$ and $B$. From this standpoint the statement

$$\vdash A \supset . \ B \supset B \tag{2}$$

which is derivable in HA for arbitrary $A$ and $B$, is not acceptable because there need not be any connection, logical or otherwise, between $A$ and $B \supset B$. Now of course the notion of logical connection is rather vague, and its explication would seem—for the reasons discussed in Chap. 1—to require some sort of previous formalization. Yet, apart from this, it makes sense to seek for a formalized implication which, in one way or another, is reasonable from the standpoint of this requirement of logical connection. The extent of the current discussion concerning contrary-to-fact conditionals, disposition statements, and the like, suggests that such a formalized entailment implication may be useful.

The problem of formulating entailment implication has interested philosophers for a long time.[2] Curiously, most of these attempts have been by way of modal logics, i.e., logics containing notions of necessity or possibility, or the like. The consideration of these is postponed until Chap. 8. In recent years, however, there have been a number of attempts to formulate systems of entailment implication directly. Among these are Church [WTI], Ackermann [BSI], Schmidt [VAL], and a series of papers by Anderson, Belnap, Wallace, and their coworkers at Yale University.[3] The paper by Anderson and Belnap [PCE] explains the motivation of these papers and discusses also the work of Ackermann and Church. The bibliography of Anderson and Belnap [PCE] cites most of the preceding work on this topic, and the more important items are actually discussed in the text. The reader is referred to these sources for the details, but it will be appropriate to add here a few general remarks.

The principal bone of contention is the scheme PK. It is sometimes said that this scheme represents the principle that if $B$ is true, then any $A$ implies it. This is not quite correct. For in order to derive (1) from PK by Pe, one must first have

$$\vdash B \tag{3}$$

---

[1] Ackermann's "*logischer Zusammenhang.*"

[2] Thus Aristotle had a modal logic, and various sorts of implication appear to have been formulated by the Stoics. Cf. Bochenski [AFL] and [FLg].

[3] To some extent based on Fitch [SLg].

and this is true, not when the interpretant of $B$ is true, but when (3) is demonstrable in the system. Thus if the interpretation is such that (3) holds only when $B$ is logically valid (in some sense), then (1) holds only for such $B$. Statement (2) is an example. This point is elaborated in [IFI].

It follows from this that absolute implication is not incompatible with an interpretation of (1) as stating that $B$ is a logical consequence of $A$, at least in one sense in which those words are commonly used. For according to that sense, $B$ is a consequence of certain premises whenever $B$ is a consequence of some subset of those premises. Any mathematician would include the null set of premises as a possible subset. But assuming $\mathfrak{S}$ is a suitable logical system, (3) holds just when $B$ is a logical consequence of the null set of premises; then if $A$ is the conjunction of all the premises, $B$ is a logical consequence of $A$, so that (1) is valid.

The protagonists of entailment implications argue that such words as 'consequence', 'deducibility', and 'implication' do not properly apply when a null set of premises is involved. Yet they accept such principles as $\Lambda K$, $\Lambda K'$, the associativity of $\Lambda$, the transitivity of 'entails', etc., and from these it follows that they accept for all nonvoid subsets the property of logical consequence mentioned in the preceding paragraph. They are thus arguing that ordinary language makes a special exception of the null set. This is probably true. Mankind went for thousands of years without a notation for zero. The logic of Aristotle did not know the null class. The fact that these innovations came at such a late stage is evidence that there is indeed something sophisticated about them. It happens in mathematics again and again that we find it expedient to extrapolate to zero a situation which is given naively only for nonzero values of some parameter (we shall meet this in Chap. 7 for "vacuous quantification"), and in such cases not only is the null situation often unnatural, but the most advantageous extrapolation is sometimes difficult to discern. Thus it is to be expected that ordinary language should have, at least in nontechnical use, a connotation that the situation present is nonnull.

The statements of null character are often unnatural in ordinary language for another reason, viz., that they are trivial. Ordinary language is used for purposes of human communication; one does not ordinarily attempt to communicate trivialities, and if one did so without some indication, one would probably be misunderstood.[1] Nevertheless, the extrapolation to zero mentioned in the preceding paragraph is often a great advantage. The invention of zero is generally regarded as one of the major steps in human progress. It thus frequently happens that we get a more acceptable theory by including trivial special cases than otherwise, the exclusion of trivial cases then being left to common sense. However, this is not necessarily the case. There are theories of some significance in which certain null cases are excluded. Examples are the formulation of Aristotelian logic in Łukasiewicz [ASS] and the ontology of Leśniewski (see Sec. 1S4). Philosophers of nominalistic tendency are said to be interested in such theories.

It would seem that the most obvious way to get an entailment implication would be to modify the system TA by limiting Pi to cases where the canceled premise is actually necessary for the derivation of the conclusion.

[1] Amusing examples of this are given in Anderson and Belnap [PCE].

A corresponding modification of LA would be to star constituents all of whose ancestors are introduced by $*K*$ and to limit $P*$ to unstarred subalterns. But for the system arrived at in that naive fashion, ET will fail, for

$$\Vdash A \wedge B . \supset B$$

$$A \wedge B . \supset B \Vdash . A \supset . B \supset B$$

both hold, but not

$$\Vdash A \supset . B \supset B$$

It seems likely that there should be some modification of $*P$.

Another possibility is to deny $*K*$. If the rules $*\Lambda$ and $V*$ are taken in the original form of Formulation II, we should have a system similar to the foregoing; the interrelations of the two systems are unknown. The proof here given of the elimination theorem fails for it. If the Ketonen rules for $*\Lambda$ and $V*$ were used, we should have a system of much more restricted character, in which even $\Lambda K$ would not be derivable. In this case the ply operation would be more in the nature of an equivalence than an entailment.

The problem of entailment implication is still under active investigation. The various solutions which have been proposed do not quite agree with one another. For some recent ideas see Kripke [PEn].

Besides the systems already mentioned, other forms of weakened systems are conceivable. There would be some interest in knowing what would happen if the quasi-principal constituent were left out of $*P$ in Formulation IV. This possibility has not been explored. Church, in his abstracts [MiL], calls attention to weakened systems and ventures the opinion that there will be one which is minimal. Each of these systems will have a weakened form of the deduction theorem; for systems of pure implication this is studied in [GDT]. The considerations adduced there cast doubt on the idea of a minimal system in Church's sense. Formulations without the distributive law are considered in Lorenzen [ALU] and Moisil [LPs].

**3. Extension of the alternation property.** Since the main text was written, it has been brought to my attention that Harrop and T. Robinson have, in principle, extended Theorem E5 to certain cases where $\mathfrak{X}$ contains dilemmatic operations; viz., to those where each such operation is interior to a component $C$ which is in negative position in $\mathfrak{X}$ (hence positive in $\Gamma$), and all operations whose arguments include $C$ are regular. The proof of Theorem E5 is easily extended to this case. For such a $C$ can only become a principal constituent when it is on the right (cf. Exercise E19), and this can happen only in one of the branches which is by-passed in the process of tree climbing.

# Chapter 6
# NEGATION

The usual practice in mathematical logic is to introduce negation along with the operations considered in the previous chapter. From the semantic point of view of Sec. 5A3, however, negation is an operation of essentially different character, and it can be constructively defined only when the underlying formal system is rather special. For this reason the study of negation was deferred in Chap. 5. The time has now come to take it up, and this chapter is devoted to it.

The chapter begins, in Sec. A, with a discussion of the semantics of negation, in the same sense that Sec. 5A3 discussed the semantics of the ply, meet, and join operations. We shall then proceed with the development of various formal systems for negation. In this development it will be convenient to reverse the order used in the study of the positive operations. That study really began with the relational systems of Chap. 4, proceeded through the T and H systems of Sec. 5B, and reached a climax with the L systems of Secs. 5C to 5E. This arrangement had the advantage that the study began with systems which, on the one hand, are simple and natural from the viewpoint of ordinary mathematics and, on the other hand, have applications in other fields; it ended with systems which are rather abstruse from that point of view, but are very natural from the semantic point of view of Secs. 5A3 and 5C1. Now that this semantic point of view has become familiar, it agrees with the theme of this book to take it as fundamental. Accordingly, Sec. B will be devoted to the L systems for negation. The T, H, and E (i.e., lattice) systems for nonclassical negation, together with the analogous properties of classical negation, will be taken up in Sec. C; the special properties of classical negation in Sec. D. The latter will include a study of the techniques of Boolean algebra.

### A. THE NATURE OF NEGATION

In this section we discuss the meaning of negation more or less intuitively, so as to have a motivation for the formal developments which follow. We begin, in Sec. 1, with an explanation of different kinds of negation as applied to elementary propositions. This is illustrated by the examples considered in Sec. 2. In Sec. 3 we proceed to a formalization in terms of singular L systems. We end with five kinds of formal negation, of which that called simple refutability is fundamental.

1. **Preliminary analysis.** Our first step will be to ask what is meant by the negation of an elementary statement of a formal theory (or system) $\mathfrak{S}$. This amounts to inquiring under what circumstances we say that such a statement is false, for the negation of a statement is then a second statement to the effect that the first one is false.

Since an elementary statement of a deductive theory $\mathfrak{S}$ is true just when there is a demonstration of it according to the deductive postulates of $\mathfrak{S}$, one would most naturally say that such a statement is false just when no such demonstration exists. Negation defined in this way will be called *non-demonstrability*.[1] From a platonistic standpoint it answers all questions. But we saw in Sec. 3A2 that from a constructive standpoint it does nothing of the kind. Furthermore, it contrasts sharply with the positive operations in its behavior with respect to extensions of the basic system. For let $\mathfrak{S}'$ be an extension of $\mathfrak{S}$. Then if a positive compound statement is true for $\mathfrak{S}$, it will remain true for $\mathfrak{S}'$, but since a statement which is nondemonstrable for $\mathfrak{S}$ may become demonstrable for $\mathfrak{S}'$, this invariance will not hold for negation in the sense of nondemonstrability.

There are, however, notions which correspond partially to our intuitions of negation and nevertheless can be treated by methods analogous to those of Chap. 5. We shall consider here two sorts of such notions; these will be called absurdity and refutability, respectively.

For the first notion, an elementary statement is defined to be *absurd* just when the system $\mathfrak{S}'$ formed by adjoining it to $\mathfrak{S}$ is inconsistent—in other words, when every elementary statement is $\mathfrak{S}$-derivable from it. Then we can define negation by taking falsity in the sense of absurdity. This negation evidently has the same invariance properties relative to an extension (provided the extension does not change the elementary statements) that the positive connections do.

The second notion, that of *refutability*, can be defined as follows. Suppose we have a new kind of theory in which, besides the conventions defining the subclass $\mathfrak{T}$ of the elementary theorems (the true elementary statements), there are additional conventions defining a second subclass $\mathfrak{F}$ (viz., the refutable ones). The definition of $\mathfrak{F}$ is also inductive. The basis is a definite class of elementary statements called the *counteraxioms*. The generating specifications state that if $A$ and $B$ are elementary statements such that $B$ is refutable and is at the same time deducible from $A$ by the deductive rules for $\mathfrak{S}$, then $A$ is refutable. If we interpret falsity as refutability, then we again have a negation with the same invariance with respect to extension as the positive connections.

These two kinds of negation are intrinsic. Other kinds may be defined with reference to an interpretation.

With each of these kinds of negation there is associated a kind of consistency and completeness. A theory will be said to be *consistent* with respect to a certain definition of negation just when no elementary statement is both true and false; it will be called *complete* with respect to that definition if every nondemonstrable elementary statement is false. Then consistency in the sense of Post, as defined in Sec. 2B1, is the same as consistency with

---

[1] Called "invalidity" in [TFD].

respect to absurdity, and an analogous identity holds for completeness in the sense of Post, which was defined in Sec. 2B3.

**2. Examples from number theory.** In order to get examples we set up a rudimentary system of number theory; this is a modification of the system of sams in Sec. 2C2, Example 2. The primitive frame for this system is as follows (the single rule of the system is here designated 'Rule 1' because further rules will be introduced afterward).

OBS. These are generated from a single atom, $0$, by one unary operation, indicated by a postfixed accent. Thus the obs are those in the sequence

$$0, 0', 0'', \dots$$

We shall treat these as the natural integers $0, 1, 2, \dots$, and the ordinary numerals and arithmetical notations will be used for them, including the sign for addition. As U variables for obs, the letters '$a$', '$b$', '$c$' will be used.

ELEMENTARY STATEMENTS. These are of the form

$$a = b$$

where $a$ and $b$ are obs.

AXIOM.    $0 = 0$.

RULE 1.    If $a = b$, then $a' = b'$.

In this system the elementary theorems are precisely those elementary statements of the form

$$a = a$$

i.e., those of the form $a = b$ in which $a$ and $b$ are the same ob. The following rules are admissible in the sense (introduced in Sec. 3A2) that the adjunction of any or all of them does not add any elementary theorems to the system.

RULE 2.    If $a = b$, then $b = a$.

RULE 3.    If $a = b$ and $b = c$, then $a = c$.

RULE 4.    If $a' = b'$, then $a = b$.

RULE 5.    If $0 = a'$, then $0 = a$.

These rules, however, have a very decided effect on the properties of negation. Let us consider the effect of adding them one by one, on the assumption that $1 = 7$ and $4 = 6$ are counteraxioms. Then, after each rule is adjoined, the following become refutable or absurd:

After Rule 1: $0 = 2, 1 = 3, 2 = 4, 3 = 5, 0 = 6$ become refutable.

After Rule 2: $2 = 0$, $3 = 1$, $4 = 2$, $5 = 3$, $6 = 4$, $6 = 0$, $7 = 1$ become refutable.

After Rule 3: $0 = 1$ and $1 = 0$ become absurd (and hence refutable); in addition, $1 = 2, 2 = 3, 3 = 4, 4 = 5, 0 = 3, 1 = 4$, and their converses become refutable.

After Rule 4: $a = a'$ and $a' = a$ become absurd for all $a$; $a = a''$, $a = a + 3$, $a = a + 6$, and their converses become refutable for all $a$.

After Rule 5: All nonderivable elementary statements become absurd. The system is then complete in the Post sense.

The example may be modified to give an illustration of a system which is complete with respect to refutability, but not with respect to absurdity, by taking as counteraxioms all those of the form $a = a + a$, where $a \neq 0$, and using Rules 1 to 4.[1]

The example just given shows that the generating principle for refutability, which is a rule with a relation of deducibility as a premise, cannot be replaced by more elementary rules derived from the given rules for $\mathfrak{S}$ by a process of simple contraposition. By this is meant taking each rule of deduction and replacing it by a set of rules to the effect that, if all the premises except one are true and the conclusion is refutable, then the remaining premise is refutable. Thus, if we use the infix ' $\neq$ ' to indicate the refutability of the corresponding statement with ' $=$ ' in the place of ' $\neq$ ', then the following rules are obtained by simple contraposition from Rules 1 to 3:

$$a' \neq b' \to a \neq b$$
$$a \neq b \to b \neq a$$
$$a \neq c \,\&\, a = b \to b \neq c$$
$$a \neq c \,\&\, b = c \to a \neq b$$

These rules do not suffice to deduce $4 \neq 5$ from $4 \neq 6$, although such a conclusion is obtained easily as follows:

$$\cfrac{4 = 5 \qquad \cfrac{\cfrac{4 = 5}{5 = 6}\,1}{4 = 6}\,3 \qquad 4 \neq 6}{4 \neq 5}$$

### 3. Formalization of negation.

Up to the present we have proceeded in an informal and more or less intuitive fashion. We now move toward a formalization of negation. It will be appropriate to do this in several steps.

In the·first step we reduce the theory $\mathfrak{S}$ to a system with two unary predicates. For this purpose we take the elementary statements of $\mathfrak{S}$ as propositions, i.e., as obs, and the interpretants of the two unary predicates will be the classes $\mathfrak{T}$ and $\mathfrak{F}$. To designate these predicates the prefixes '⊢' and '⊣', respectively, will be used. The elementary statements of the new system will thus be of one of the forms

$$\vdash A \qquad \dashv A \qquad\qquad (1)$$

Of these the left-hand one will mean that the statement which is associated with $A$ is in $\mathfrak{T}$, while that on the right will mean that this statement is in $\mathfrak{F}$. If $\mathfrak{S}$ is already a system, transformations analogous to those in Sec. 2D1 will be involved. Thus, if $\mathfrak{S}$ is the system in Sec. 2, we replace the elementary statement $a = b$ (as in Sec. 2D1) by the proposition $a \,\square\, b$; then the statements

$$\vdash a \,\square\, b \qquad \dashv a \,\square\, b$$

will represent, respectively, the $a = b$ and $a \neq b$ of Sec. 2.

The second stage will be the application of the reduction of Sec. 2D1 to reduce the system to an assertional one. This means that we introduce a new operation, designated by the prefix '$\neg$', to replace the predicate prefix

---

[1] For this system Rule 3 is unnecessary. If it is omitted, no elementary statement is absurd.

'⊣', and then in place of the second form of statement in (1) we have the form

$$\vdash \neg A \tag{2}$$

At this point let us pause to make two observations. The first is that the new type of theory mentioned in Sec. 1, in which there are counteraxioms and a notion of refutability, is not a radical innovation as opposed to the type of theory considered in Chap. 2, but is rather a specialization of it. The second is that the reduction of these two stages, although apparently tied to the notion of refutability, applies to the notion of absurdity equally well. Indeed it applies to any situation where we have a theory in which more than one subclass of the class of elementary statements is considered. This will hold for absurdity if we simply take $\mathfrak{F}$ to be the class of absurd statements.

After the first two steps negation has become an operation parallel to those of Chap. 5. The third step is to state postulates expressing the properties which characterize the different species of negation. The explanations will be in terms of a singular L formulation, because that formulation is most explicit in regard to the deducibility relationships. Inasmuch as the other propositional operations may be introduced at the same time, the explanation will include the negations of compound propositions. The statements

$$\mathfrak{X} \Vdash A$$

will be taken for the time being intuitively, i.e., as indicating that $A$ is deducible from $\mathfrak{X}$ in some sense, not completely specified as yet, for which the elimination theorem (ET) holds.

Let us first consider refutability. Let the counteraxioms be $F_1, F_2, \ldots$, and let notations like '$F_i$' indicate an unspecified one of these. Then the basis of the definition of refutability is the set of statements

$$\Vdash \neg F_i \tag{3}$$

The generating specification is a rule which may be stated thus:

$$\frac{\mathfrak{X}, A \Vdash B \quad \mathfrak{X}, A \Vdash \neg B}{\mathfrak{X} \Vdash \neg A} \tag{4}$$

where the antecedents $\mathfrak{X}, A$ in the right premise are inserted, by a weakening principle, to preserve symmetry. If $B$ is some $F_i$, then the right premise of (4) is derived by weakening from (3), so that (4) becomes

$$\frac{\mathfrak{X}, A \Vdash F_i}{\mathfrak{X} \Vdash \neg A} \tag{5}$$

This principle includes (3) (for $\mathfrak{X}$ void and $A \equiv F_i$). On the other hand, if negation is regarded as introduced solely by (3), (4), an induction on the number of applications of (4) shows that (5) suffices in the place of (4), in the sense that whenever the conclusion of (5) holds, the premise will hold for some $F_i$; in fact, by the hypothesis of the induction, the right premise of (4) can be replaced by $\mathfrak{X}, A, B \Vdash F_i$; then by ET one can get the premise of (5). Thus (5) is sufficient for the introduction of negation on the right.

From the formal point of view, at least, the rule (5) can be split into two rules, F* and N*, as follows:

$$\text{F*} \qquad \frac{\mathfrak{X} \Vdash F_i}{\mathfrak{X} \Vdash F}$$

$$\text{N*} \qquad \frac{\mathfrak{X},\, A \Vdash F}{\mathfrak{X} \Vdash \neg A}$$

The proposition $F$ thus introduced can be interpreted as a statement to the effect that some counteraxiom is demonstrable, and hence that the system is inconsistent. If there is only a finite number of counteraxioms, such an $F$ can be defined in terms of alternation. Otherwise we take $F$ as a new primitive. We can so take it even for the void system $\mathfrak{O}$. Thus N* is, so to speak, a universal rule which is valid for any system; on the other hand F*, like $\vdash$*, represents the special conventions of the ground theory $\mathfrak{S}$.

The discussion of the next to the last paragraph suggests that the rule N* should be reversible. That circumstance suggests further that negation be defined thus:

$$\neg A \equiv A \supset F \qquad (6)$$

With such a definition N* becomes a special case of P*. The following rule

$$\text{*N} \qquad \frac{\mathfrak{X} \Vdash A}{\mathfrak{X},\, \neg A \Vdash F}$$

is then a special case of *P. We adopt it as the rule for introduction of negation on the left.

The last three paragraphs may be summed up as follows. Refutability may be characterized by the rule F* and the definition (6) or by the rules F*, N*, and *N. The negation so defined is known as *minimal negation*; it may also be called the system of *simple refutability*. It was studied by Johansson.[1] Its L formulation will be called, after Johansson, LM.

Let us now turn to absurdity. If we here introduce a proposition $F$, whose interpretation is a statement that the system is inconsistent in the Post sense, then (6) will define absurdity. By definition of $F$ the rule[2]

$$\text{Fj} \qquad \frac{\mathfrak{X} \Vdash F}{\mathfrak{X} \Vdash A}$$

will be valid. The rules N* and *N will be admissible since they follow from (6). The rule F* may be vacuous, i.e., $F$ may be the only element of $\mathfrak{F}$, but we may have cases, like that of Sec. 2, in which one establishes the absurdity of certain statements (for example, $0 = 1$ after Rule 3) by an epitheoretic argument not formalizable in the L system, and in such a case these can be taken as counteraxioms. We thus have a system formed by adjoining Fj to LM. It is the system adopted by the intuitionists and therefore is appropriately called *intuitionistic negation*; another name is the

[1] Norwegian mathematician, professor at the University of Oslo. The paper on the minimal algebra is his [MKR]. Cf. Kolmogorov [PTN]. See also Sec. S1.

[2] This rule goes slightly beyond the preliminary discussion (Sec. 1) in that $A$ is not necessarily an elementary proposition. The possibilities in regard to restriction of $A$ in Fj are left open; they are evidently similar to those for K*.

system of *simple absurdity*. Its L formulation will be called, after Gentzen, LJ.

Besides the types of negation which have just been considered, there is some interest in types which are applicable when the underlying system is complete. In Sec. 1 a complete system was defined as one in which every elementary statement is true or false; here we are concerned with a stronger form of completeness in which this is true for every compound proposition. We are thus dealing with systems with a law of excluded middle. This principle may be expressed by saying that any antecedent of the form $A \vee \neg A$ may be dropped; in view of ET and the rule $*V$, this may in turn be expressed by the rule

$$\text{Nx} \qquad \frac{\mathfrak{X}, A \Vdash B \quad \mathfrak{X}, \neg A \Vdash B}{\mathfrak{X} \Vdash B}$$

which has some analogies to (4). When $B$ is $A$, the left premise is prime; hence we have in that case

$$\frac{\mathfrak{X}, \neg A \Vdash A}{\mathfrak{X} \Vdash A} \qquad (7)$$

The rule (7) is a special case of Px if (6) is adopted.[1] But we shall take Nx in the above more general form. To get it from (7) requires ET, whereas the converse argument is immediate.

If we adjoin Nx to LM, the system so obtained will be called the theory of *complete refutability*, or *strict negation*.[2] Its L formulation is called LD.

If we adjoin Nx to LJ, we obtain a system LK which will be called *classical negation*, or the theory of *complete absurdity*.

Up to the present nothing has been said about the nature of the deducibility expressed by '$\Vdash$'. Let it now be specified that for all the systems LM, LJ, LD, LK, the rules for F and N are to be adjoined to LA. It is then natural to ask what would be the effect of adjoining these same rules to LC. Since we are supposing that (6) holds in interpretation, and thus Nx, in principle, is a special case of Px, the system so formed will necessarily be complete. Furthermore, one can show that LK includes LC as follows:

$$\frac{\dfrac{\dfrac{\dfrac{A \Vdash A}{A, \neg A \Vdash F} *N}{A \neg A \Vdash B} \text{Fj}}{\dfrac{\neg A \Vdash A \supset B}{} \text{P}*} \quad \mathfrak{X}, A \supset B \Vdash A}{\dfrac{\mathfrak{X}, \neg A \Vdash A}{\mathfrak{X} \Vdash A} (7)} \text{ET}$$

Thus the only new system is that formed from LC as LM is from LA. Kripke [SLE][3] has studied this system and called it LE. From the present point of view it may be called by the name he proposed, viz., *classical refutability*.

---

[1] Actually, Nx was proposed first (in [TFD]) in the form (7), the '$x$' suggesting 'excluded middle'; later Px was named from its resemblance to this Nx.

[2] This term was proposed in [TFD] on account of a remark in Johansson [MKR] to the effect that the system might be suitable as a theory of strict implication. The term is hardly apt, but no better short name has been proposed.

[3] For anticipations see Sec. S1.

The result of all this is that we have five kinds of constructive negation whose characteristics are as follows:

LM   Minimal negation, or simple refutability, formed by adding to LA the rule F∗ and either the definition (6) or the rules ∗N∗

LJ   Intuitionistic negation, or simple absurdity, formed by adjoining Fj to LM

LD   Strict negation, or complete refutability, formed by adjoining Nx to LM

LE   Classical refutability, formed by adjoining Px to LM; includes LD

LK   Classical negation, or complete absurdity, formed from LJ by adjoining Nx and from LD or LE by adjoining Fj

For all these systems we distinguish an *F formulation*, in which only F rules and (6) are adopted [Nx being formulated in terms of (6)], an *N formulation*, in which (6) is abandoned in favor of ∗N∗, and an *FN formulation*, in which F and ⌐ are both taken as primitives and (6) becomes a theorem. These formulations for LM will be called LMF, LMN, and LM(FN), respectively, with a similar convention for the others. This convention will be used only when there is some reason for distinguishing the formulations.

In regard to the rule ∗N, the rule proposed above has the technical disadvantage that it fails to satisfy condition (r2) of Sec. 5C6. This is on account of the constituent F in the conclusion, which would have to be considered a second principal constituent. But if we take the B in ∗P to be F, the rule we come up with is

$$\frac{\mathfrak{X} \Vdash A \quad \mathfrak{X}, F \Vdash C}{\mathfrak{X} \ \lnot A \Vdash C}$$

This rule is regular. From it we get the above ∗N by taking C to be F (the right premise being then quasi prime). This explains the formulation in the next section.

One should remark that refutability is the fundamental kind of negation, the others being defined by additions to it. This is true for nondemonstrability with respect to a fixed system; one can then take the nondemonstrable elementary statements as counteraxioms.

## B. L SYSTEMS FOR NEGATION

The preceding section was concerned with the motivation and justification of the L rules. In this section these rules will be formulated and various consequences derived from them. The object will be to extend the theorems of Secs. 5C to 5E to the systems of negation and to derive analogous theorems about the relationships between different types of L formulation. The argument is abstract and formal; reference is made to Sec. A only for motivation.

**1. Formulations of L systems for negation.** The conventions for the positive L systems as given in Sec. 5C3 are to apply to the L systems for negation except for certain changes which will be stated presently. In particular, the conventions for prosequences, auxiliary statements, and prime statements are to be taken over without change. The changes in the other conventions will be different for the different formulations. However, the changes will be listed systematically, with indication at the appropriate place of the types to which they apply.

PROPOSITIONS. The following changes will be made in the definition of $\mathfrak{E}$ and $\mathfrak{P}$.

1. A subclass $\mathfrak{F}$, called the *counteraxioms*, is defined in $\mathfrak{P}$. The letters '$F_1$', '$F_2$', '$F_i$', '$F_j$', etc., will stand for counteraxioms. The class $\mathfrak{F}$ may be void, and will always be void when $\mathfrak{S}$ is $\mathfrak{O}$.

2. In some formulations a unary negation operation, indicated by the prefix '$\neg$', will be adjoined to those used in generating $\mathfrak{P}$. Then if $A$ is in $\mathfrak{P}$, $\neg A$ will be in $\mathfrak{P}$ but not in $\mathfrak{E}$. The letter 'N' will sometimes be used, parallel to 'P', '$\Lambda$', 'V', to designate this negation.

3. In some formulations a fixed proposition $F$ will be adjoined to $\mathfrak{E}$. This $F$ is an indeterminate so far as $\mathfrak{S}$ is concerned. In terms of $F$ the negation operation may be defined thus:

$$\neg A \equiv A \supset F \tag{1}$$

In this case also, for any $A$ in $\mathfrak{P}$, $\neg A$ will be in $\mathfrak{P}$ but not in $\mathfrak{E}$.

ELEMENTARY STATEMENTS. The only change from Sec. 5C3e is that in the N formulation there may be a void prosequence on the right.

RULES. The following new rules will be added to those in Sec. 5C4, subject to special restrictions in the different systems and formulation types as explained later.

F   *Rules of direct refutability*
F∗   If $F_i$ is a counteraxiom,

$$\frac{\mathfrak{X} \Vdash F_i, \mathfrak{Z}}{\mathfrak{X} \Vdash F, \mathfrak{Z}}$$

Fj

$$\frac{\mathfrak{X} \Vdash F, \mathfrak{Z}}{\mathfrak{X} \Vdash A, \mathfrak{Z}}$$

N   *Rules of negation*
∗N   (multiple)                              N∗   (all formulations)

$$\frac{\mathfrak{X} \Vdash A, \mathfrak{Y} \quad \mathfrak{X}, F \Vdash \mathfrak{Y}}{\mathfrak{X}, \neg A \Vdash \mathfrak{Y}} \qquad \frac{\mathfrak{X}, A \Vdash F, \mathfrak{Z}}{\mathfrak{X} \Vdash \neg A, \mathfrak{Z}}$$

∗N   (singular)

$$\frac{\mathfrak{X} \Vdash A \quad \mathfrak{X}, F \Vdash B}{\mathfrak{X}, \neg A \Vdash B}$$

$$\text{Nx} \quad \frac{\mathfrak{X}, A \Vdash \mathfrak{Y} \quad \mathfrak{X}, \neg A \Vdash \mathfrak{Y}}{\mathfrak{X} \Vdash \mathfrak{Y}} \quad \text{(for LD only)}$$

FORMULATION TYPES. As already mentioned in Sec. A3, there will be three formulation types: the F formulation in which negation is defined in terms of $F$ by (1); the N formulation in which $\neg$ is taken as primitive and $F$ is not postulated; and the FN formulation in which both $F$ and $\neg$ are postulated [and (1) will turn out to hold in the sense of a provable equivalence].

The rules just formulated are those for the FN formulation. In the F formulation the rules ∗N∗ are omitted as primitive rules, but are valid as specializations of $*P*$, respectively. In the N formulation $F$ is simply omitted wherever it occurs in the above rules except in $*N$; the latter becomes

$$\frac{\mathfrak{X} \Vdash A, \mathfrak{Z}}{\mathfrak{X},\ \neg A \Vdash \mathfrak{Z}} \tag{2}$$

The singular and multiple systems are to be subject to the same conventions as in Sec. 5C. It is understood that in the multiple formulations of certain systems particular rules may be restricted to be singular; i.e., they may be mixed systems in the sense of Sec. 5C3e. It is further to be understood that structural rules on the right, in particular K$*$, are inapplicable in singular systems even where a void consequent is admitted.

F and N formulations will be distinguished, when that is expedient, by writing the letters 'F', 'N', respectively, after the abbreviated name for the system, and for its FN formulation we shall write '(FN)'; thus LMF, LMN, LM(FN). Singular and multiple formulations will be distinguished by subscripts as in the case of LA in Sec. 5C.

DEFINITIONS OF THE SYSTEMS. The singular FN formulations are made by adjoining rules to LA or LC as shown by the following table:

$$\begin{aligned}
\text{LM} &= \text{LA} + *\text{N}* + \text{F}* \\
\text{LJ} &= \text{LA} + *\text{N}* + \text{F}* + \text{Fj} = \text{LM} + \text{Fj} \\
\text{LD} &= \text{LA} + *\text{N}* + \text{F}* + \text{Nx} = \text{LM} + \text{Nx} \\
\text{LE} &= \text{LC} + *\text{N}* + \text{F}* \\
\text{LK} &= \text{LC} + *\text{N}* + \text{F}* + \text{Fj} = \text{LE} + \text{Fj}
\end{aligned} \tag{3}$$

In the multiple formulations Px is not postulated, but in those cases where Px does not hold in the singular systems, the rules N$*$ and P$*$ are to be singular. Thus N$*$ and P$*$ are both restricted to be singular in $\text{LM}_m$, $\text{LJ}_m$, and $\text{LD}_m$, and there are no singularity restrictions in $\text{LE}_m$ or $\text{LK}_m$.

In the system $\text{LD}_m$ analogy would suggest that we should simply omit Nx and allow N$*$ to be multiple. However, in the system so formed

$$\Vdash A, A \supset \neg A$$

would not be demonstrable, whereas its singular transform

$$\Vdash A \vee. A \supset \neg A$$

is demonstrable in $\text{LD}_1$, using the weak form [(7) of Sec. A] of Nx, thus:

$$\cfrac{\cfrac{\cfrac{\cfrac{\cfrac{\cfrac{A \Vdash A}{A \Vdash A \vee. A \supset \neg A}\text{V}* \qquad F \Vdash F}{A,\ \neg(A \vee. A \supset \neg A) \Vdash F}\text{*N}}{A,\ \neg(A \vee. A \supset \neg A) \Vdash \neg A}\text{N}*,\ *\text{K}}{\neg(A \vee. A \supset \neg A) \Vdash A \supset \neg A}\text{P}*}{\neg(A \vee. A \supset \neg A) \Vdash A \vee. A \supset \neg A}\text{V}*}{\Vdash A \vee. A \supset \neg A}\text{Nx}$$

It is therefore necessary to postulate the multiple form of Nx. The form taken here for Nx we shall find easier to work with than the form (7) of Sec.

A. We can then restrict N∗ to be singular, as it would be if it were a special case of P∗.

The F and N formulations are obtained from the FN formulation as above described, but the following peculiarities about their multiple forms should be noted. Since N∗ is the special case of P∗ where the $B$ (in the formulation of the rule in Sec. 3C4) is $F$, $LD_m$ does not require P∗ to be singular in that special case. Again the void consequent admitted in the N system brings with it automatically the "*ex falso quodlibet*," hereafter abbreviated "efq," thus:

$$\frac{A \Vdash A}{A \Vdash A, B} \text{K}*$$
$$\frac{A, \neg A \Vdash B}{} *\text{N}$$
$$\frac{}{\neg A \Vdash A \supset B} \text{P}*$$

Thus, in principle, we cannot have multiple N formulations without Fj, and N forms of $LM_m$, $LD_m$, and $LE_m$ are not defined.

MODIFIED FORMULATIONS. One can state alternative formulations as in Sec. 5C8, but these will not be considered here further. The rules just stated are suitable for adjunction to Formulation IK. From now on the rules ∗K∗ and ∗W∗ will be assumed to be valid, not necessarily as primitive rules, without restriction, and we are therefore free to change back and forth between the different modified formulations as we please.

**2. The inversion theorem.** The proof of the inversion theorem in Sec. 5D1 involved two major steps. The first step was the examination of the top nodes of $\Delta_1$; the second, the validation of the inferences in $\Delta_1$. The first step was taken care of by the assumption, (*b*), that the only rules which could have $M$ as principal constituent were $R$, with the $\mathfrak{U}_i$ then uniquely determined, and the structural rules ∗K∗ and ∗W∗; assumptions (*d*) and (*e*) took care of the structural rules. The second step required the assumption (*c*).

In the present situation an instance of $M$ may also be introduced by Fj. In the N formulation Fj is essentially a special case of K∗, and there cannot be any difficulty as long as ∗K∗ holds without restriction. In the other formulations an $M$ introduced by Fj will be on the right; if in the $i$th premise of $R$, one of the $\mathfrak{U}_i$ is on the right,[1] that subaltern can be introduced by Fj, the rest by ∗K∗. This takes care of the first major step. In the second major step we may have the same difficulty as before with respect to assumption (*c*) in case $R$ has a principal constituent on the left, one or more of its premises has a subaltern on the right, and there are rules in $\Delta_i$ which are singular on the right. Otherwise the proof of Sec. D1 goes through. We thus have the following:

**Theorem 1.** *The inversion theorem (Theorem 5D1) remains true if the condition (b) is modified to the condition (b′) and a new condition (f) is imposed, as follows:*

(*b′*). *The only rules which can have $M$ as principal constituent are $R$ (with the $\mathfrak{U}_k$ then uniquely determined) and* ∗K∗, ∗W∗, Fj.

(*f*). *If $M$ is introduced into $\Delta_1$ by Fj, then the $\mathfrak{U}_i$ can be introduced at that point by Fj and* ∗K∗.

---

[1] This condition is fulfilled for all rules so far. In fact, in all cases where $R$ has a principal constituent on the right, every premise has at least one subaltern on the right.

Since the conditions are fulfilled in the cases which interest us here, we have the following corollaries:

COROLLARY 1.1.    *In the systems* $LE_m$ *and* $LK_m$ *all rules are directly invertible with respect to all premises. Moreover, the direct inversion can always be completed.*

In the system $LD_m$ there appears to be trouble on account of the presence of the rule Nx. That rule has no principal constituent, and hence cannot introduce an instance of $M$ as such. Thus this rule does not cause any exception in regard to condition $(b')$. The exceptional cases in regard to condition $(c)$ are *P and *N on the left. However, *N can be inverted on the left, but not directly, as follows:

$$(p1) \qquad \frac{\dfrac{\mathfrak{X}, \;\neg A \Vdash \mathfrak{Y}}{\mathfrak{X}, A \Vdash A, \mathfrak{Y} \quad \mathfrak{X}, \;\neg A \Vdash A, \mathfrak{Y}} K* }{\mathfrak{X} \Vdash A, \mathfrak{Y}} Nx$$

We thus have the following:

COROLLARY 1.2.    *In the system* $LD_m$ *all rules are directly invertible, except that* *N *and* *P *are not directly invertible with respect to their left premises (or the only premise in the* N *formulation of* *N*). The inversion of* *N *with respect to its left premise can be carried out, but not directly.*

COROLLARY 1.3.    *In the systems* $LM_m$ *and* $LJ_m$ *all rules are directly invertible except that neither* *P *nor* *N *can be inverted with respect to the left premise (or the only premise in certain forms of* *N*).*

COROLLARY 1.4.    *The conclusions of Corollary 5D1.6 also hold if the rule* Fj *and (in* $LD_m$*)* Nx *are reckoned as structural rules.*[1]

**3. The elimination theorem.** Using a method similar to that of Sec. 2, let us examine the proof of the elimination theorem to see what modifications are necessary to accommodate the presence of negation. In view of the remark on modified formulations at the end of Sec. 1, no further attention will be paid to the formulation types of Chap. 5 or to uses of the structural rules.

In Stage 1 the only new possibilities are the cases under $(\gamma)$, in which $R_k$ is one of the rules *N*, Nx, Px, F*, Fj, and the case under $(\epsilon)$, where $R_k$ is *N. Note that the rule Nx causes no difficulty since all the constituents in the conclusion are parametric, and this is true for F* and Fj on the left. The changes made in passing from $\Gamma_k$ to $\Gamma'_k$ are dropping out parametric constituents of $A$ on the left and adding replicas of $\mathfrak{X}'$ on the left and of $\mathfrak{Z}$ on the right. These changes do not affect the validity under $(\gamma)$ of any of the rules in the above list, nor the new case under $(\epsilon)$. Hence Stage 1 goes through for all unmixed cases, and also for the mixed cases where $\mathfrak{Z}$ is void.

In Stage 2 we have the same new possibilities under $(\gamma)$; also F*, Fj, Nx under $(\delta)$ and Nx under $(\zeta)$. The changes made in passing from $\Gamma_k$ to $\Gamma'_k$ consist in dropping parametric instances of $A$ on the right and adding constituents of $\mathfrak{X}$ and $\mathfrak{Y}$ on the left and right, respectively; if Px is present, there are additions of others on the left. In the unmixed systems, where $A$ is composite, none of these changes invalidates an inference by any of the

---

[1] If all quasi-parametric ancestors of $M$ are admitted by K* or Fj, Fj can be taken as $R$, and hence as nonstructural, in that particular case.

new rules.    Further, the case of F∗ under (δ) is impossible since $A$ cannot be
$F$.    The case of Fj under (δ) is similar to that of K∗.    Let the instance of
Fj in question lead from $\Gamma_i'$ to $\Gamma_k'$.    Then if $\mathfrak{Y}$ is void, we have necessarily
an N formulation and the passage from $\Gamma_i'$ to $\Gamma_k'$ can be made by ∗K only;
otherwise it can be made by Fj and ∗K∗.    The case of Nx under (δ) and (ζ)
again causes no difficulty since it has no principal constituent.[1]    Thus the
proof of Stage 2 also goes through for the unmixed systems.

Let us now look at Stage 3.    In the F formulation there is nothing to
prove.    In the other formulations let $A$ be $\neg B$, and suppose that ET holds
for $B$.    The premises of the case of ET under discussion are

$$\mathfrak{X}, \neg B \Vdash \mathfrak{Y} \tag{4}$$

$$\mathfrak{X} \Vdash \neg B, \mathfrak{Z} \tag{5}$$

We suppose that (4) has been obtained by ∗N from the premises [in the N
formulation without $F$ in (7)]

$$\mathfrak{X} \Vdash B, (\mathfrak{Y}) \tag{6}$$

$$\mathfrak{X}, F \Vdash \mathfrak{Y} \tag{7}$$

whereas (5) comes by N∗ from [without $(F)$ in the N formulation]

$$\mathfrak{X}, B \Vdash (F), \mathfrak{Z} \tag{8}$$

From (6) and (8), eliminating $B$, we have

$$\mathfrak{X} \Vdash (F)_{\mathfrak{f}}, (\mathfrak{Y}), \mathfrak{Z}$$

This is the desired conclusion in the N formulation.    In the FN formulation
we eliminate $F$ with (7).

This completes the proof of ET for all the unmixed systems, so that we
have the following:

**Theorem 2.**    *The elimination theorem holds without qualification for all
formulations of the systems* $LM_1$, $LJ_1$, $LD_1$, $LE_1$, $LK_1$, $LE_m$, $LK_m$.

Now let us consider the extension of ET to the mixed systems using the
method of Sec. 5D2.    As there, we have only to consider the cases where
$A$ is $B \supset C$, $\neg B$ (in $LM_m$ and $LJ_m$)[2], or is elementary.    The case where $A$
is $B \supset C$ can be handled as in Sec. 5D2, and the case of $\neg B$ can be handled
analogously, so that we are reduced to the case where $A$ is elementary.    In
the N formulation of $LJ_m$, the rest of the argument can be carried through as
in Sec. 5D2.    But in the F and FN formulations, we have to consider the
possibility that $A$ is $F$.

---

[1] If we were to take Nx in the form

$$\frac{\mathfrak{X}, \neg A \Vdash A, \mathfrak{Z}}{\mathfrak{X} \Vdash A, \mathfrak{Z}}$$

we should be able to carry through this part of ET, by using a method similar to that used
for Px in Sec. 5D2.    But there would be difficulty with the inversion theorem, and this
would cause trouble below for $LD_m$.

[2] In $LD_m$, since ∗N and N∗ can be inverted, we can use that inversion to carry out the
elimination, and it makes little difference whether N∗ is singular or not.    But if we used
the form of Nx in the preceding footnote, then we must have N∗ singular and include
the case $\neg B$.

If $A$ is $F$, it may happen that such an $F$ is introduced by F∗. In that case it seems to be necessary to impose further restrictions. The difficulty cannot arise if the class of counteraxioms is void. There need not be any loss in generality in this if the original counteraxioms were elementary. For then we could take $F$ as single counteraxiom and adjoin to the auxiliary statements the following:

$$F_i \vdash_0 F \tag{9}$$

Then the rule F∗ becomes superfluous since the only inference possible by it has premise and conclusion identical.

In the remaining cases the argument can be carried through as in Sec. 5D2. Thus we have the following:

**Theorem 3.** *The elimination theorem holds without qualification in the N formulation of* LJ$_m$. *For the other formulations of* LJ$_m$ *and the formulations of* LM$_m$ *and* LD$_m$, *it holds if the counteraxioms are elementary.*

*Remark 1.* Instead of the rule F∗, one might introduce a third type of prime statement formed by weakening from

$$F_i \Vdash F \tag{10}$$

This would cause difficulty with the elimination theorem. Statements of the form (10) can be introduced by F∗ thus:

$$\frac{F_i \Vdash F_i}{F_i \Vdash F} \; \text{F∗}$$

This F∗ can be further reduced, as explained in the proof, if the $F_i$ are elementary.

**4. Equivalence between formulation types.** Before we go further it is important to see that the three formulation types—the F, N, and (FN) formulations—are mutually equivalent in so far as they are all defined. This will be shown here under the hypothesis that the elimination theorem holds for all the formulations concerned. The proof of the main theorem is preceded by the definitions of certain transformations and by the proofs of four lemmas.

We first define a *G transformation* from statements of an N formulation to those of the FN formulation as follows:

$$(\mathfrak{X} \Vdash)^G \rightleftharpoons \mathfrak{X} \Vdash F$$
$$(\mathfrak{X} \Vdash \mathfrak{Y})^G \rightleftharpoons \mathfrak{X} \Vdash \mathfrak{Y}$$

Further, if $R$ is a rule of an N formulation, $R^G$ will be the same rule in the corresponding FN formulation in which all premises and conclusion are replaced by their G transforms.

**LEMMA 1.** *If $\Gamma$ is demonstrable in an N formulation, then $\Gamma^G$ is demonstrable in the corresponding FN formulation.*

*Proof.* Let $\Delta$ be a demonstration $\Gamma_1, \ldots, \Gamma_n$ of $\Gamma$ in the N formulation. We show that $\Gamma_k^G$ is demonstrable in the corresponding (FN) formulation on the hypothesis that every $\Gamma_i^G$ for $i \leq k$ is so demonstrable. It suffices to consider the following six cases.

CASE 1. $\Gamma_k$ is prime. Then the consequent is not void. Hence $\Gamma_k^G$ is the same as $\Gamma_k$ and is prime.

CASE 2.   $\Gamma_k$ is derived from $\Gamma_i$, $\Gamma_j$, ..., by a rule $R$ involving neither $F$ nor negation, and the consequent of $\Gamma_k$ is not void.[1]   Then none of the premises has a void consequent; $\Gamma_i^G$, $\Gamma_j^G$, ..., $\Gamma_k^G$ are the same, respectively, as $\Gamma_i$, $\Gamma_j$, ..., $\Gamma_k$, and $R^G$ the same as $R$.   Hence $\Gamma_k^G$ is obtainable in the (FN) formulation from the same premises by the same rule.

CASE 3.   $\Gamma_k$ is obtained from $\Gamma_i$, $\Gamma_j$, ..., by a rule as in Case 2, but $\Gamma_k$ has a void consequent.   Then the rule cannot be a rule with principal constituent on the right.   If the rule is $*$C, $*$K, $*$W, $*\Lambda$, $*$V, or Nx, then the right side is parametric and the inference remains correct if the void consequent is replaced by $F$, giving $\Gamma_k^G$ as consequence of $\Gamma_i^G$, $\Gamma_j^G$, ..., by $R^G$. If the rule is $*$P, let the original inference be

$$\frac{\mathfrak{X} \Vdash A \qquad \mathfrak{X},\, B \Vdash}{\mathfrak{X},\, A \supset B \Vdash}$$

The transformed inference is

$$\frac{\mathfrak{X} \Vdash A \qquad \mathfrak{X},\, B \Vdash F}{\mathfrak{X},\, A \supset B \Vdash F}$$

This is correct by $*$P.

CASE 4.   $\Gamma_k$ is obtained by a rule $R$, namely, Fj or N$*$, which requires, in the (FN) formulation, a constituent $F$ on the right in some premise.   All such rules have one premise, which we call $\Gamma_i$, and a conclusion with a nonvoid consequent.   Then $\Gamma_k^G$ is $\Gamma_k$.   Let $\Gamma_i$ be

$$\mathfrak{X} \Vdash \mathfrak{Y}$$

and let $\Gamma_i'$ be

$$\mathfrak{X} \Vdash F,\, \mathfrak{Y}$$

Then $\Gamma_k$ follows from $\Gamma_i'$ in the (FN) formulation by the same rule.   If $\mathfrak{Y}$ is void, then $\Gamma_i'$ is $\Gamma_i^G$; if not, it is obtainable from $\Gamma_i^G$ by K$*$.   Thus in any case $\Gamma_k^G$ is derivable from $\Gamma_i^G$ in the FN formulation.

CASE 5.   $\Gamma_k$ is obtained by a rule $R$, namely, F$*$, which introduces an $F$ into the conclusion.   In this case the premise $\Gamma_i$ has necessarily a nonvoid consequent; hence $\Gamma_i^G$ is the same as $\Gamma_i$.   Let $\Gamma_k$ be

$\Gamma_k$ $\qquad\qquad\qquad\qquad$ $\mathfrak{X} \Vdash \mathfrak{Y}$

The result of applying the same rule to $\Gamma_i^G$ in the (FN) formulation will be $\Gamma_{k'}$, namely,

$\Gamma_k'$ $\qquad\qquad\qquad\qquad$ $\mathfrak{X} \Vdash F,\, \mathfrak{Y}$

If $\mathfrak{Y}$ is void, this will be precisely $\Gamma_k^G$.   Otherwise, let $B$ occur as constituent in $\mathfrak{Y}$.   Since we have a multiple N formulation, Fj will be postulated.   Then we can argue from $\Gamma_k'$ thus:

$$\frac{\dfrac{\mathfrak{X} \Vdash F,\, \mathfrak{Y}}{\mathfrak{X} \Vdash B,\, \mathfrak{Y}}\text{Fj}}{\mathfrak{X} \Vdash \mathfrak{Y}}\text{W}*$$

The conclusion is $\Gamma_k$, which is the same as $\Gamma_k^G$.   Thus $\Gamma_k^G$ is derivable in the corresponding (FN) formulation.

---

[1] The argument also applies to Nx, which can occur only in $LD_1$.

CASE 6. $\Gamma_k$ is obtained by $*$N. Then the inference is

$$\frac{\mathfrak{X} \Vdash A, \mathfrak{Z}}{\mathfrak{X}, \, \neg A \Vdash \mathfrak{Z}}$$

In the case of $\mathfrak{Z}$ void, the transformed inference can be made thus:

$$\frac{\mathfrak{X} \Vdash A \quad \mathfrak{X}, F \Vdash F}{\mathfrak{X}, \, \neg A \Vdash F} *N$$

If $\mathfrak{Z}$ is nonvoid, it can be made thus:

$$\frac{\dfrac{\mathfrak{X} \Vdash A, \mathfrak{Z}}{\mathfrak{X} \Vdash A, F, \mathfrak{Z}} K* \quad \mathfrak{X}, F \Vdash F, \mathfrak{Z}}{\mathfrak{X}, \, \neg A \Vdash F, \mathfrak{Z}} *N$$

From this point we can argue as in Case 5. The right-hand premise, in either case, is quasi-prime.

This completes the proof of Lemma 1.

Let us now define an $F$ *transformation* associating to each proposition $A$ of an (FN) formulation that proposition $A^F$ of the corresponding F formulation which is obtained by replacing, from within outward, every component of the form $\neg B$ by $B \supset F$. This is a special case of a definitional reduction in the sense of Sec. 3C1; it can be expressed in the form of a definition by structural induction as follows:

a. If $A$ is elementary, $A^F$ is $A$.

b. The transformation is a homomorphism relative to operations other than negation.

c. If $A$ is $\neg B$, $A^F$ is $B^F \supset F$.

The transformation may be extended to prosequences, elementary statements, and rules by the requirement that it be a homomorphism, i.e., that all constituents be replaced by their F transforms.

LEMMA 2. *If $\Gamma$ is an elementary statement of an F formulation, then $\Gamma$ is demonstrable in that formulation if and only if it is demonstrable in the corresponding FN formulation.*

*Proof.* If $\Gamma$ is demonstrable in an F formulation, then it is a fortiori demonstrable in the corresponding FN formulation. The converse follows immediately if the formulation is one for which the composition property holds for compound constituents. But even without that property, we can conclude the converse by deductive induction. For let $\Gamma_1, \ldots, \Gamma_n$ be a derivation $\Delta$ of $\Gamma$. If $\Gamma_k$ is a prime statement, $\Gamma_k^F$ is also a prime statement. Also, if $R$ is a rule of an FN formulation, $R^F$ is a valid inference in the corresponding F formulation. It follows that every $\Gamma_k^F$, and hence $\Gamma^F$, is derivable in the F formulation. Since $\Gamma^F$ is the same as $\Gamma$, this completes the proof.

LEMMA 3. *In an FN formulation for which the elimination theorem holds, an elementary statement $\Gamma$ is demonstrable if and only if $\Gamma^F$ is.*

*Proof.* The proof schemes

$$\frac{\dfrac{A \Vdash A}{A, \, \neg A \Vdash F} *N}{\neg A \Vdash A \supset F} P* \qquad \frac{\dfrac{A \Vdash A \quad F \Vdash F}{A, A \supset F \Vdash F} *P}{A \supset F \Vdash \neg A} N*$$

show that $\neg A$ and $A \supset F$ are interchangeable. The lemma therefore follows by Rp.

Next we define an $S$ *transformation* from an FN formulation to an N formulation. This consists in eliminating $F$ by the definition

$$F \equiv \neg T$$

where $T$ is some assertible proposition, say, $E_1 \supset E_1$. The definition by structural induction is as follows:

a. If $A$ is elementary and distinct from $F$, then $A^S$ is $A$; $F^S$ is $\neg T$.

b. With respect to all operations, the transformation is a homomorphism.

This transformation may also be extended to prosequences and elementary theorems by the requirement that it be a homomorphism.

LEMMA 4. *Let the elimination theorem hold for a certain* (FN) *formulation and also for the corrresponding* N *formulation. Then a necessary and sufficient condition that* $\Gamma$ *be demonstrable in the* (FN) *formulation is that* $\Gamma^S$ *be demonstrable in the* N *formulation.*

*Proof of Necessity.* For each rule $R$ in the FN formulation, let $R^N$ be the corresponding rule in the N formulation and $R^S$ the S transform of $R$. We show that $R^S$ is a correct inference in the N formulation. This is clear if $R$ does not postulate an occurrence of $F$ in either premise or conclusion. If it postulates $F$ on the right in the premise, then, since we have in LMN

$$\frac{\Vdash T}{\neg T \Vdash} *\mathrm{N} \tag{11}$$

we can infer a premise of $R^N$ from that of $R^S$ by the elimination theorem, and then draw the desired conclusion by $R^N$. If $R$ postulates an $F$ on the right in the conclusion, then the premise is the same as in the N formulation, and we can insert the required $\neg T$ in the conclusion thus:

$$\frac{\dfrac{\mathfrak{X} \Vdash 3}{\mathfrak{X}, T \Vdash 3} *\mathrm{K}}{\mathfrak{X} \Vdash \neg T, 3} \mathrm{N}*$$

If $R$ is $*\mathrm{N}$ in the singular case, let the inference be

$$\frac{\mathfrak{X} \Vdash A \quad \mathfrak{X}, F \Vdash B}{\mathfrak{X}, \neg A \Vdash B} *\mathrm{N}$$

The transformed inference can be obtained thus:

$$\frac{\dfrac{\dfrac{\dfrac{\mathfrak{X}^S \Vdash A^S}{\mathfrak{X}^S, \neg A^S \Vdash} *\mathrm{N}}{\mathfrak{X}^S, T, \neg A^S \Vdash} *\mathrm{K}}{\mathfrak{X}^S, \neg A^S \Vdash \neg T} \mathrm{N}* \quad \mathfrak{X}^S, \neg T \Vdash B^S}{\mathfrak{X}^S, \neg A^S \Vdash B^S} \mathrm{ET}$$

In the multiple case the original inference is

$$\frac{\mathfrak{X} \Vdash A, \mathfrak{Y} \quad \mathfrak{X}, F \Vdash \mathfrak{Y}}{\mathfrak{X}, \neg A \Vdash \mathfrak{Y}} *\mathrm{N}$$

The transformed inference is

$$\frac{\mathfrak{X}^S \Vdash A^S, \mathfrak{Y}^S}{\mathfrak{X}^S, \; \neg A^S \Vdash \mathfrak{Y}^S} \, *\mathrm{N}$$

The transform of the right premise is irrelevant.    Thus in all cases $R^S$ gives a correct inference in the N formulation.

Suppose we have a derivation $\Delta$ of $\Gamma$ in the FN formulation.    Since the S transformation carries a prime statement into a prime statement and, as we have just shown, an inference by a rule $R$ into a valid N inference, we see that applying the S transformation to all the statements will produce a derivation of $\Gamma^S$.    This proves the necessity.

*Proof of Sufficiency.*    If $\Gamma^S$ is derivable in the N formulation, then $\Gamma^{SG}$ is derivable in the (FN) formulation by Theorem 1.    Since the consequent of $\Gamma^S$ is not void, $\Gamma^{SG}$ is the same as $\Gamma^S$.    Thus $\Gamma^S$ is derivable in the FN formulation.    The rest follows by Rp, since we have

$$\frac{\Vdash T}{\neg T \Vdash F} \, *\mathrm{N} \qquad \frac{F, T \Vdash F}{F \Vdash \neg T} \, \mathrm{N}* \qquad (12)$$

This completes the proof.

We may summarize this as follows:

**Theorem 4.**    *If the relevant formulations are defined and the elimination theorem holds for them, then we have the following:*

(i) *If $\Gamma$ is an elementary statement of an N formulation, then $\Gamma$ is a theorem of that formulation if and only if $\Gamma^G$ is a theorem of the corresponding FN formulation.*

(ii) *If $\Gamma$ is an elementary statement of an F formulation, then $\Gamma$ is a theorem for that formulation if and only if it is a theorem for the corresponding FN formulation.*

(iii) *If $\Gamma$ is an elementary statement of an FN formulation, then $\Gamma$ is equivalent in that formulation to each of $\Gamma^S$ and $\Gamma^F$; further, its truth in the FN formulation is equivalent to that of $\Gamma^F$ in the corresponding F formulation and also to that of $\Gamma^S$ in the corresponding N formulation.*

*Proof.*    The only-if part of (i) follows by Lemma 1; the if part, by Lemma 4, since $\Gamma^{GS}$ differs from $\Gamma$ only in the possible presence of an extra constituent $\neg T$ on the right, and this can be dropped by (11) and the elimination theorem.

The statement (ii) is the same as Lemma 2.

The part of (iii) relating to $\Gamma^F$ follows from Lemmas 3 and 2.    By Lemma 4, $\Gamma$ holds in the FN formulation if and only if $\Gamma^S$ holds in the N formulation; since $\Gamma^{SG}$ is the same as $\Gamma^S$, we conclude by (i) that $\Gamma^S$ holds in the N formulation if and only if it holds in the FN formulation.

This completes the proof.

**5. Singular and multiple formulations.**    We now attack the question of equivalence between the singular and multiple systems.    This was handled for the positive systems in Sec. 5D5.    It will be shown here that it holds, in so far as the multiple formulations are defined in Sec. 1 and ET holds, for all formulations of negation.    In view of the results of Sec. 4, the number of types to be considered may be cut down, but it seems advisable to make this subsection independent of Sec. 4.

The proof in Sec. 5D5 made use of the fact, established in Sec. 5D4, that a statement $A \le B$ in EA is equivalent to $A \Vdash B$ in LA.   Since LA is included in all the systems, we may use inferences by EA here just as we did in Sec. 5D5.

We adopt the same conventions in regard to '$\mathfrak{Y}$', '$\mathfrak{Y}''$, '$\mathfrak{Z}$', '$C$', '$C''$', '$D$', '$\Vdash_m$', '$\Vdash_1$' as in Sec. 5D5.

**Theorem 5.**   *For any formulation of LM, LJ, LD, LE, or LK for which both singular and multiple forms are defined and ET holds, a necessary and sufficient condition that*

$$\mathfrak{X} \Vdash_m \mathfrak{Y} \tag{13}$$

*hold in the multiple formulation is that*[1]

$$\mathfrak{X} \Vdash_1 C \tag{14}$$

*hold in the corresponding singular formulation.*

*Proof of Necessity.*   We have only to add to the cases arising in the proof of Theorems 5D7 and 5D8 those in which the inference leading to (13) is by one of the rules F∗, Fj, ∗N, N∗.   The proofs for thse cases, which will follow presently, depend on the following observations.   The first is that the inference

$$\frac{A,\, B \Vdash C}{A,\, B \vee D \Vdash C \vee D} \tag{15}$$

can be justified in any of the systems thus:

$$\frac{\dfrac{A,\, B \Vdash C}{A,\, B \Vdash C \vee D}\,\text{V}_* \quad \dfrac{A,\, D \Vdash D}{A,\, D \Vdash C \vee D}\,\text{V}_*}{A,\, B \vee D \Vdash C \vee D}\,{}_*\text{V}$$

The second is that where $C$ is $F$, the analogous inference in an N formulation, viz.,

$$\frac{A,\, B \Vdash}{A,\, B \vee D \Vdash D} \tag{16}$$

can be made, since Fj holds in any multiple N formulation, as follows:

$$\frac{\dfrac{A,\, B \Vdash}{A,\, B \Vdash D}\,\text{Fj} \quad A,\, D \Vdash D}{A,\, B \vee D \Vdash D}\,{}_*\text{V}$$

The third observation is that we can suppose that the $\mathfrak{Y}$, $\mathfrak{Z}$ entering in the following proofs are nonvoid, since otherwise the situation is trivial.

After these observations we proceed to the proofs for the new cases as follows.   Expressions in parentheses are to be omitted in dealing with N formulations.

Suppose first that (13) is obtained by F∗.   Then $\mathfrak{Y}$ is $F$, $\mathfrak{Z}$ and the premise is

$$\mathfrak{X} \Vdash_m F_i,\, \mathfrak{Z}$$

---

[1] In case the consequent of (13) is void, that of (14) is to be void also.

By the hypothesis of the induction,

$$\mathfrak{X} \Vdash_1 F_i \vee D$$

Now $F_i \Vdash (F)$ follows by F∗ from the prime statement $F_i \Vdash F_i$; hence by (15) or (16) we have

$$F_i \vee D \Vdash (F \vee)D$$

By the elimination theorem we can therefore obtain from the transformed premise the conclusion

$$\mathfrak{X} \Vdash (F \vee)D$$

which is (14).

Next suppose (13) is obtained by Fj. Then the original inference is

$$\frac{\mathfrak{X} \Vdash (F), \mathfrak{Z}}{\mathfrak{X} \Vdash A, \mathfrak{Z}}$$

If $F$ is present, the transformed inference is

$$\frac{\mathfrak{X} \Vdash F \vee D}{\mathfrak{X} \Vdash A \vee D}$$

Now the statement $F \Vdash A$ follows from the prime statement $F \Vdash F$ by Fj. Hence, by (15), we have

$$F \vee D \Vdash A \vee D$$

and therefore the transformed inference can be obtained by the elimination theorem. If $F$ is not present, the transformed inference is a special case of V∗.

If (13) is obtained by ∗N, the original inference in the FN formulation is

$$\frac{\mathfrak{X} \Vdash A, \mathfrak{Y} \quad \mathfrak{X}, F \Vdash \mathfrak{Y}}{\mathfrak{X}, \neg A \Vdash \mathfrak{Y}}$$

The transformed inference is

$$\frac{\mathfrak{X} \Vdash A \vee C \quad \mathfrak{X}, F \Vdash C}{\mathfrak{X}, \neg A \Vdash C}$$

This can be obtained thus:

$$\frac{\mathfrak{X} \Vdash A \vee C \quad \dfrac{\dfrac{\dfrac{A \Vdash A}{A, \neg A \Vdash F} \ast\text{N} \quad \mathfrak{X}, F \Vdash C}{\mathfrak{X}, A, \neg A \Vdash C} \text{ET} \quad \mathfrak{X}, C, \neg A \Vdash C}{\mathfrak{X}, A \vee C, \neg A \Vdash C} \ast\text{V}}{\mathfrak{X}, \neg A \Vdash C} \text{ET}$$

In the N formulation the inference is

$$\frac{\mathfrak{X} \Vdash A, \mathfrak{Y}}{\mathfrak{X}, \neg A \Vdash \mathfrak{Y}}$$

and the transformed inference is

$$\frac{\mathfrak{X} \Vdash A \vee C}{\mathfrak{X}, \neg A \Vdash C}$$

The demonstration is—since Fj holds whenever there is a multiple N system—

$$\frac{\dfrac{A \Vdash A}{A, \,\urcorner A \Vdash} *N}{\mathfrak{X}, A, \,\urcorner A \Vdash C} \text{ Fj}, *K$$

From this point on the argument is the same as before.

If (13) is obtained by N*, the original inference is

$$\frac{\mathfrak{X}, A \Vdash (F), \mathfrak{Z}}{\mathfrak{X} \Vdash \urcorner A, \mathfrak{Z}}$$

If $F$ is present, the transformed inference is

$$\frac{\mathfrak{X}, A \Vdash F \vee D}{\mathfrak{X} \Vdash \urcorner A \vee D}$$

Now $F \Vdash \urcorner A$ can be obtained by N* from the quasi-prime statement $A$, $F \Vdash F$; hence, by (15),

$$F \vee D \Vdash \urcorner A \vee D$$

From this and the transformed premise we obtain, by the elimination theorem,

$$\mathfrak{X}, A \Vdash \urcorner A \vee D \tag{17}$$

If $F$ is absent, we obtain (17) from the transformed premise by V*. Again, from a prime statement and V*, we have

$$\mathfrak{X}, \urcorner A \Vdash \urcorner A \vee D \tag{18}$$

By the restrictions on N*, $\mathfrak{Z}$ is nonvoid only in systems where Nx is present.[1] Therefore, from (17) and (18), we have

$$\mathfrak{X} \Vdash \urcorner A \vee D$$

which is (14) for this case.

If (13) is obtained by Nx, the original inference is

$$\frac{\mathfrak{X}, A \Vdash \mathfrak{Y} \quad \mathfrak{X}, \urcorner A \Vdash \mathfrak{Y}}{\mathfrak{X} \Vdash \mathfrak{Y}}$$

The transformed inference is

$$\frac{\mathfrak{X}, A \Vdash C \quad \mathfrak{X}, \urcorner A \Vdash C}{\mathfrak{X} \Vdash C}$$

which is again valid by Nx.

This completes the proof of necessity.

*Proof of Sufficiency.* As in Sec. 5D5, it suffices to show that if (14) holds in the singular system, then it holds in the multiple system, for the rest will follow by the inversion theorem. In the case of LM, LJ, and LD, where there are no rules peculiar to the singular systems, this finishes the argument. In case the rule Px is postulated, the corresponding multiple inference can be made in $LC_m$, and hence in $LE_m$ and $LK_m$, by the proof of Theorem 5D8.

This completes the proof of the theorem. Note that the necessity proof

---

[1] It is left as an exercise to show that Nx is a derived rule in $LE_1$, even though it is not postulated (see Exercise 4).

requires ET for the singular system; the sufficiency proof uses the inversion theorem and, in case Px is present, also ET for the multiple system.

**6. Deducibility questions.** The question now to be considered is the extent to which the deducibility theorems and other theorems of Sec. 5E, as well as some in Secs. 5C and 5D which have not already been dealt with, extend to the systems involving negation. Many of these theorems were stated as general theorems applicable to L systems satisfying broad general conditions. These theorems extend at once to systems with negation. Others extend with slight changes.

In the following, not a great deal of attention is paid to LD. Details about LD are, for the most part, left to the exercises.

Of the theorems in Sec. 5C, Theorems 5C1, 5C2, 5C3, and 5C5 extend without significant change. Theorem 5C4 requires only the additional case

$$\frac{A \Vdash A}{\frac{A, \neg A \Vdash F}{\neg A \Vdash \neg A}\; \text{N}*} *\text{N}$$

Thus it is trivial that the theorems of Sec. 5C hold for all the formulations.[1]

The method of proof of the completeness theorem of Sec. 5D6 can be applied to either of the systems $\text{LE}_m(\mathfrak{D})$ or $\text{LK}_m(\mathfrak{D})$, leading to the following:

**Theorem 6.** *A necessary and sufficient condition that*

$$\mathfrak{X} \Vdash \mathfrak{Y} \tag{19}$$

*be an elementary theorem of* $\text{LK}_m$ *is that it be tautologous with respect to all valuations by 0-1 tables in which F is assigned the fixed value 0, and hence* $\neg A$ *has the value opposite to that of A; for (19) to be an elementary theorem of* $\text{LE}_m$, *it is necessary and sufficient that it be tautologous with respect to all 0-1 valuations in which F is assigned either of the values 0 or 1, and hence* $\neg A$ *has either the value 1 or the value opposite to A.*

The theorems as to the restriction of *K* and the elimination of *W* were formulated in general terms in Chap. 5 (Theorems 5E6 to 5E8). They need no further attention here.[2] For Formulation III we need a quasi-principal constituent for *P and *N in $\text{LM}_m$, $\text{LJ}_m$; for *P, but not for *N, in $\text{LD}_m$; and for none of the rules in $\text{LE}_m$ or $\text{LK}_m$. The rule Fj can be made reversible by having a quasi-principal constituent. The rule F* does not concern us when $\mathfrak{S}$ is void.

More interesting is the question concerning the composition property and properties connected with it. The possibilities in regard to dropping out of components are as follows. Elementary constituents can be dropped by $\vdash*$, counteraxioms by F*, instances of $F$ by Fj, and compound constituents by Px, Nx. This gives us the situation shown by the following theorem and corollaries:

**Theorem 7.** *Let the counteraxioms be elementary. Then all L formulations of Sec. 1 which do not postulate either* Nx *or* Px *have the composition property with respect to compound constituents.*

---

[1] Corresponding to the different forms of *P, there are, of course, variants of *N. These we shall not stop to investigate.

[2] For LD some slight changes in the proof of Theorem 5E7 are necessary. It is necessary to treat Nx as a structural rule.

*Proof.* The only constituents which can be dropped out by any rule are *F*, subalterns for an instance of ⊢∗, and counteraxioms. Since these are all elementary, the theorem is clear.

COROLLARY 7.1.    *If the counteraxioms are elementary, all the* L *systems of Sec.* 1 *which do not postulate* Px *or* Nx *have the separation property. If there are no counteraxioms, they have also the conservation property.*

*Proof.* This follows by Theorem 7 and Theorems 5E2 and 5E3.[1]

COROLLARY 7.2.    *The* N *formulations of the systems* $LM_1(\mathfrak{D})$, $LJ_1(\mathfrak{D})$ *and all the multiple systems over* $\mathfrak{D}$ *have the composition property without restriction.*

*Proof.* The exceptional constituents which can be dropped out, as listed in the proof of Theorem 7, do not exist in those formulations.

COROLLARY 7.3.    *The* N *formulations of* $LM_1(\mathfrak{D})$, $LJ_1(\mathfrak{D})$, $LJ_m(\mathfrak{D})$, *and* $LK_m(\mathfrak{D})$ *are decidable.*

*Proof.* This follows from Corollary 7.2 and Theorem 5E9.

For formulations involving *F*, we can go a little further, as shown by the following theorem:

**Theorem 8.**    *Let the formulation be one containing* F *in which neither* Px *nor* Nx *is postulated, and let the counteraxioms be elementary. If* Fj *is postulated, let there be an effective process associating with each positive proposition* A *and each counteraxiom* $F_i$, *a deduction justifying the inference*

$$\frac{\mathfrak{X} \Vdash F_i, \mathfrak{Z}}{\mathfrak{X} \Vdash A, \mathfrak{Z}} \tag{20}$$

*for arbitrary parametric* $\mathfrak{X}$ *and* $\mathfrak{Z}$; *further, let no true auxiliary statement contain* F *as constituent. Let* $\Delta$ *be a regular demonstration* $\Gamma_1, \Gamma_2, \ldots,$ $\Gamma_n$ *such that neither* F *nor negation occurs in any component of* $\Gamma_n$. *Then there is a demonstration* $\Delta'$ *terminating in* $\Gamma_n$ *such that neither* F *nor negation occurs in any statement of* $\Delta'$.

*Proof.* If negation occurred in any statement of $\Delta$, then it would occur in $\Gamma_n$ by Corollary 7.1.    Hence no negation occurs in $\Delta$, and we have only to consider the case that an *F* so occurs.

If an *F* occurred on the left in some $\Gamma_k$, then it could never drop out later and hence would occur in $\Gamma_n$.    Hence any such *F* must appear on the right. In that case it could only be removed later by N∗ or Fj.    Of these, the former is impossible, since it would introduce a negation.    We may therefore suppose that $\Gamma_k$ immediately precedes an application of Fj.

Let us consider the first such $\Gamma_k$.    Let the inference from $\Gamma_k$ to $\Gamma_{k+1}$ be

$$\frac{\mathfrak{X}_k \Vdash F, \mathfrak{Z}_k}{\mathfrak{X}_k \Vdash A, \mathfrak{Z}_k}$$

This *A* is positive, since if it contained any negation or a proper component *F*, that would appear in $\Gamma_n$.    Let us look at the uppermost parametric

---

[1] If F∗ is replaced by (9), the restriction in regard to the counteraxioms for the conservation property can be dropped.    Thus the argument applies in principle to cases where the counteraxioms are elementary.    Otherwise

$$F_i \Vdash F$$

is a counterexample for the conservation property.

ancestors of this $F$. Such an ancestor might conceivably be introduced as principal constituent of some prime statement, by ⊢∗, by K∗, or by F∗. A prime statement of type $(p1)$ is impossible since there would have to be an $F$ on the left. A prime statement of type $(p2)$, or an introduction by ⊢∗, is impossible, since $F$ cannot be the consequent of an auxiliary statement. Therefore all ancestors of the indicated $F$ must be introduced by K∗ or F∗. In the former case the $F$ can be replaced by $A$; this is true in the latter case also since the inference (20) can be effectively justified. Thus all the ancestors of the indicated $F$ can be replaced by $A$.

This process can be continued until we arrive at a $\Delta'$ not containing any $F$'s, Q.E.D.

*Remark* 1. The hypotheses made concerning the counteraxioms are consistent with the interpretation discussed in Sec. A3. For if Fj is postulated, we have to do with absurdity. Ordinarily one would not expect counteraxioms in such a case; but if they are present, it is because we have some way of knowing their absurdity, perhaps by an epitheoretic argument not formulated in the system. (Compare the examples in Sec. A2.) Thus (20) is a way of expressing the requirement that the counteraxioms be themselves absurd. Again the requirement that no true auxiliary statement contain $F$ is essentially a requirement that $F$ be an indeterminate, as was intended in Sec. 1.

*Remark* 2. The requirement in regard to $F$ may seem to contradict the suggestion made in the proof of ET (Sec. 2) that rule F∗ could be replaced by an auxiliary statement of the form

$$F_i \vdash_0 F \qquad (21)$$

Actually, the two situations do not conflict because there we were talking about $LD_m$ and here about a situation in which Fj holds. However, the present theorem would hold if we replaced F∗ by (21), provided that (21) were the only auxiliary statements in which $F$ is a conclusion, and (20) holds for all $F_i$.

COROLLARY 8.1. *Let the counteraxioms satisfy the conditions of Theorem 8, and let $\Gamma$ be an elementary statement not containing negation nor any instance of $F$. Then, if $\Gamma$ is demonstrable in any formulation containing $F$ of LM, LJ, or LD, it is demonstrable in LA; if it is demonstrable in such a formulation of LE or LK, it is demonstrable in LC.*

*Proof.*[1] If $\Gamma$ is demonstrable in any formulation, it is demonstrable in a multiple formulation by Theorem 5. In that formulation there is a demonstration not containing any $F$ or any negation by Theorem 8. Hence the demonstration is valid in $LA_m$ or $LC_m$ as the case may be, Q.E.D.

COROLLARY 8.2. *Let the counteraxioms satisfy the conditions of Theorem 8, and let $\Gamma$ be an elementary statement all of whose constituents are elementary and none of them is $F$. Then $\Gamma$ is demonstrable in one of the systems of Sec. 1 containing $F$ if and only if some constituent in the consequent of $\Gamma$ is ⊝-deducible from some of those in the antecedent.*

*Proof.* By Corollary 8.1 and Theorem 5E4.

---

[1] The proof does not consider LD, but the corollary is true. See [SLD] and Exercise 7 at the end of this section.

*Remark* 3. For N formulations the conclusions of these corollaries do not hold. In fact,

$$F_i \Vdash$$

is a simple counterexample. But those conclusions hold, in particular, if $\mathfrak{S}$ is $\mathfrak{O}$.

These theorems extend the separation and conservation theorems to systems with negation. The following theorem is the analogue of Theorem 5E4.

**Theorem 9.** *Let the counteraxioms be elementary, and let the formulation contain* $F$. *Let* $\Gamma$ *be an elementary theorem* (19) *of some formulation of Sec.* 1. *Let* $\Delta$ *be a demonstration of* $\Gamma$. *Let* $\mathfrak{M}$ *be the set consisting of all the constituents as in Theorem 5E4 and also of all propositions* $\neg F_i$ *where* $F_i$ *is a counteraxiom used in an application of* $\mathbf{F}*$ *in* $\Delta$. *Let* $\mathfrak{N}$ *be as in Theorem 5E4. Then*

$$\mathfrak{X}, \mathfrak{M}, \mathfrak{N} \Vdash \mathfrak{Y}$$

*is an elementary theorem of the corresponding formulation over* $\mathfrak{O}$.

The proof will be left as an exercise. It may be expedient to employ the techniques of Sec. C.

In regard to proof tableaux, it seems expedient to use the multiple F formulations as a basis.[1] Since we are considering the systems over $\mathfrak{O}$, the only new rule will then be Fj, the distinctions between the systems having mostly the nature of the singularity restrictions on $\mathbf{P}*$ and its specializations. These can be taken care of by making modifications of I as in Sec. 5E8. If Fj is present, we need only add a rule for adding an $F$ to the right side of the datum. (For certain details see Sec. 7B6.)

**EXERCISES**

**1.** For what systems of negation is each of the following assertible?

(a) $\qquad A \supset . \neg\neg A$

(b) $\qquad \neg A \supset A . \supset . \neg\neg A$

(c) $\qquad \neg A \supset A . \supset A$

(d) $\qquad \neg\neg A \supset . A$

(e) $\qquad \neg\neg A \supset . \neg A \supset A$

(f) $\qquad A \supset . \neg A \supset A$

(g) $\qquad \neg\neg\neg A \supset \neg A$

(h) $\qquad \neg\neg A \supset : \neg A \supset A . \supset A$

(i) $\qquad \neg\neg(\neg\neg A \supset A)$

**2.** Answer the same question for the following:

(a) $\qquad \neg A \wedge \neg B . \supset . \neg(A \vee B)$

(b) $\qquad \neg(A \wedge B) . \supset . \neg A \vee \neg B$

(c) $\qquad \neg(\neg A \wedge \neg B) . \supset . A \vee B$

(d) $\qquad \neg(A \vee B) . \supset . \neg A \wedge . \neg B$

---

[1] The argument of Sec. 5E8 depended, at one point, on the impossibility of a void consequent. This could probably be avoided by complicating the induction somewhat, but this is not necessary in an F formulation.

(e) $\qquad\qquad \neg\neg(A \supset B). \supset: \neg\neg A. \supset \neg\neg B$

(f) $\qquad\qquad \neg\neg A \supset \neg\neg B. \supset: \neg\neg(A \supset B)$

(g) $\qquad\qquad A \supset \neg A. \supset B. \supset: A \supset B. \supset B$

**3.** Consider the following properties in $LD_m$:

(a) $\qquad \mathfrak{X}, A \Vdash \mathfrak{Y} \ \& \ \mathfrak{X}, \neg A \Vdash \mathfrak{Y} \to \mathfrak{X} \Vdash \mathfrak{Y}$

(b) $\qquad \mathfrak{X}, \neg A \Vdash \mathfrak{Y} \to \mathfrak{X} \Vdash A, \mathfrak{Y}$

(c) $\qquad \mathfrak{X}, \neg A \Vdash A, 3 \to \mathfrak{X} \Vdash A, 3$

(d) $\qquad \mathfrak{X}, A \Vdash \neg B, 3 \to \mathfrak{X} \Vdash A \supset \neg B, 3$

[Here (a) is Nx.] On the basis of LM show that all the properties follow from (a), that (b) and (c) are equivalent, and that (d) follows from (b), all without use of ET; further, that (a) follows from (b) or (c) by ET. *What would be the situation in regard to ET for LD if Nx were postulated in the form (b), (c), or (d) instead of (a)? (Cf. footnotes in Sec. 3; also Kripke [DCn].)

**4.** Show that Nx is a derived rule in $LE_1$.

**5.** Discuss in detail the analogues of Theorem 8 and its corollaries for the N formulation.

**6.** Complete the proof of Theorem 9.

**7.** (Generalized Glivenko theorem.) If

$$\mathfrak{X} \Vdash \mathfrak{Y}$$

holds in LD (LK), then[1]

$$\mathfrak{X}, \neg \mathfrak{Y} \Vdash \mathfrak{Y}$$

holds in LM (LJ). What can you say about

$$\mathfrak{X} \Vdash \neg\neg \mathfrak{Y}$$

([SLD] gives the proof in the singular case. For the original Glivenko theorem see Glivenko [PLB]. See also Kripke [DCn].)

**8.** Show that if $\mathfrak{X}, \mathfrak{Y}$ do not contain implication and

$$\mathfrak{X} \Vdash \mathfrak{Y}$$

holds in LE, then it holds in LD.

**9.** Let $A$ be a proposition formed from the elementary propositions by $\wedge$ and $\neg$ only. Show that $A$ is assertible in $LK(\mathfrak{D})$ only if it is assertible in $LM(\mathfrak{D})$. Hence show that if $A$ is any proposition of $LK(\mathfrak{D})$ and $A^*$ is a proposition obtained from $A$ by replacing—from within outward as in a definitional reduction—components of the forms $B \supset C$, $B \vee C$, respectively, by $\neg(B \wedge \neg C)$, $\neg(\neg B \wedge \neg C)$, then

$$\Vdash A \supset A^* \quad \text{and} \quad \Vdash A^* \supset A \quad \text{in } LK(\mathfrak{D})$$
$$\Vdash A \quad \text{in } LK(\mathfrak{D}) \rightleftarrows \Vdash A^* \quad \text{in } LM(\mathfrak{D})$$

(Gödel [IAZ]; cf. Schmidt [VAL, sec. 131], who refers to Kolmogorov [PTN].)

**\*10.** What property analogous to the Glivenko theorem will hold for the system LE? (Cf. Exercises 5E21 and 7B8.)

**\*11.** Does the situation in Exercise 9 generalize to the case of statements

$$\mathfrak{X} \Vdash \mathfrak{Y}$$

**\*12.** Suppose that in $LE_m$ the $B$ in multiple cases of P* were restricted to be $\neg A$. What would be the relation of this system to LD? Would ET hold for it?

**\*13.** Suppose that an elementary statement $\Gamma$ holds in both LD (or perhaps LE) and LJ. Does it necessarily hold in LM?

---

[1] Here $\neg \mathfrak{Y}$ is the prosequence formed from $\mathfrak{Y}$ by negating all its constituents.

**14.** Show that in LJ there are no "quasi definitions" (cf. Sec. 6D) of the form

$$A \circ B \Vdash C \ \& \ C \Vdash A \circ B$$

where $A$ and $B$ are elementary; $\circ$ is one of the operations $\supset$, $\wedge$, and $\vee$; and $C$ is a proposition constructed from the remaining operations; likewise, if $C$ is positive, there is no elementary statement of the form

$$C \Vdash \neg A$$

(Wajsberg [UAK, sec. 10]; McKinsey [PIP].)

## C. OTHER FORMULATIONS OF NEGATION

This section will be devoted to the study of the T and H systems of negation and to the algebraic properties of nonclassical negation.  Properties peculiar to the classical system are deferred to Sec. D.  Since the preceding formulations solve the problem of discovering demonstrations, the emphasis here is on the axiomatization.

**1. The T formulations of negation.**  The T formulations of the various systems of negation are defined as those formed by adjoining to TA or TC the appropriate ones of the following rules.

Ne $\quad \dfrac{A \quad \neg A}{F}$

Ni $\quad \dfrac{[A]}{\dfrac{F}{\neg A}}$

Fi $\quad \dfrac{F_{\,\ast}}{F}$

Nj $\quad \dfrac{F}{A}$

Nd $\quad \dfrac{[\neg A]}{\dfrac{A}{A}}$

Nk $\quad \dfrac{\neg \neg A}{A}$

The different systems are defined according to the following scheme:

$$
\begin{aligned}
\text{TM} &= \text{TA} + \text{Ne} + \text{Ni} \\
\text{TJ} &= \text{TM} + \text{Nj} \\
\text{TD} &= \text{TM} + \text{Nd} \\
\text{TE} &= \text{TC} + \text{Ne} + \text{Ni} \\
\text{TK} &= \text{TM} + \text{Nj} + \text{Nd} = \text{TM} + \text{Nk} = \text{TE} + \text{Nj}
\end{aligned}
\tag{1}
$$

Here the first formulation of TK is taken as definition, and the others will be shown to be equivalent to it.

It will now be shown that these T systems are equivalent to the corresponding L systems of the singular FN formulations.  In this we use the same devices for translating from the L systems to the T systems, and vice versa, as in Sec. 5D4.  The statement

$$\mathfrak{X} \Vdash^T B \tag{2}$$

is to mean that there is a T proof of $B$ all of whose uncanceled premises appear as constituents in $\mathfrak{X}$. The T rules can then be interpreted as permitting inferences between statements of the form (2) by supposing that the uncanceled premises over the inference consist of the propositions of $\mathfrak{X}$ plus those indicated; thus any premise $A$ represents $\mathfrak{X} \Vdash^T A$ and any premise

$$[A] \atop B$$

represents $\mathfrak{X}, A \Vdash B$. To distinguish from (2) the elementary statements of the L system, the latter are written, when such explicitness is necessary, thus:

$$\mathfrak{X} \Vdash {}^L B \tag{3}$$

We regard (3) as the L transform of (2) and (2) as the T transform of (3). Likewise, the L transform of a T rule is the rule obtained by replacing all its statements (premises and conclusion) by their L transforms, and the T transform of an L rule is defined similarly.

**Theorem 1.** *If* (3) *holds in an L system, then* (2) *holds in the corresponding* T *system.*

*Proof.* It is only necessary to add to the proof of Sec. 5D4 a demonstration that the T transforms of the L rules F*, Fj, *N*, and Nx are valid inferences of the T system. Now the T transforms of the rules F*, Fj, N*, and Nx[1], respectively, are the T rules Fi, Nj, Ni, and Nd, respectively, and hence they are valid in the appropriate T system. The same is shown for *N as follows:

$$\frac{\dfrac{\mathfrak{X}}{A}\ \mathrm{Hp} \qquad \neg A}{F}\ \mathrm{Ne}$$

This completes the proof.

**Theorem 2.** *If* (2) *holds in a T system, then* (3) *holds in the corresponding* L *system.*

*Proof.* As in Theorem 1, it is here only necessary to add to the proof of Theorem 5D6 the induction step for the L transforms of Fi, Fj, Ne, Ni, and Nd. But the L transforms of Fi, Nj, Ni, and Nd are precisely the rules F*, Fj, N*, and Nx, respectively. The L transform of Ne is established thus:

$$\frac{\dfrac{\mathfrak{X} \Vdash A}{\mathfrak{X}, \neg A \Vdash F}\ {*}\mathrm{N} \qquad \mathfrak{X} \Vdash \neg A}{\mathfrak{X} \Vdash F}\ \mathrm{ET}$$

This completes the proof.

**2. Alternative formulations.** Besides the standard T formulations given in Sec. 1, there are certain variants of these formulations which raise questions of some interest. Some of these will be considered here; others are given in the exercises.

The first such question, already mentioned in Sec. 1, is the existence of various alternative forms of TK.

[1] Here and in proof of Theorem 2, this is the old form of Nx as in Exercise 6B 3(b).

**Theorem 3.**  *As addition to* TM, *the rule* Nk *is equivalent to the conjunction of* Nj *and* Nd *and entails* Pk.

*Proof.*   Derivation of Nk:

$$
\frac{\dfrac{\overset{1}{\neg\,\neg A} \quad \overset{2}{\neg A}}{F}\ \text{Ne}}{\dfrac{A}{A}\ \text{Nj}}
$$
$$
\frac{A}{A}\ \text{Nd-2}
$$

Derivation of Nj:

$$
\frac{\dfrac{F}{\neg\,\neg A}\ \text{Ni}\dagger}{A}\ \text{Nk}
$$

Derivation of Nd:

$$
\frac{\dfrac{\dfrac{[\neg A]}{A}}{\neg A \supset A}\ \text{Pi} \quad \overset{1}{\neg A}}{\dfrac{A}{F}\ \text{Ne}}\ \text{Pe}
$$
$$
\frac{\neg\,\neg A}{A}\ \text{Ni-1}
$$
$$
\frac{}{A}\ \text{Nk}
$$

Derivation of Pk:

$$
\frac{\dfrac{\dfrac{\overset{\cancel{1}}{\neg A} \quad \overset{\cancel{2}}{A}}{F}\ \text{Ne}}{\dfrac{B}{A \supset B}\ \text{Nj}}\ \text{Pi-2}}{\dfrac{A}{A}\ \text{Nd-1}}\ \text{Hp of Pk}
$$

The result about Pk also follows from the fact that Pk is the T translation of Px, and Px is valid in LC, and hence in LK, by Theorem 5D8.

Another result of this nature concerns the formulation of negation without $F$. The principal difficulty here is that there is no T transform of an L statement with void consequent. However, we saw in Sec. B3 that the statements

$$
\mathfrak{X} \Vdash \qquad \mathfrak{X} \Vdash \neg T
$$

where $T$ is some assertible proposition, are equivalent. Thus we can replace $F$ as consequent by $\neg T$, and $F$ as premise by $T$, $\neg T$. This, after some modification, leads to the following:

**Theorem 4.**  *The rules* Ne *and* Ni *together are equivalent to the single rule*

Nm

$$
\frac{B \quad \overset{[A]}{\neg B}}{\neg A}
$$

---

† We here use a weakened form of Ni in which the premise $[\neg A]$, which is discharged, is not actually used in the proof.

*Proof.*  Derivation of Nm:

$$
\begin{array}{c}
1 \\
A \\
\hline
\begin{array}{cc} B & \neg B \end{array} \quad \text{Hp} \\
\hline
F \quad \text{Ne} \\
\hline
\neg A \quad \text{Ni-1}
\end{array}
$$

Derivation of Ne:

$$
\begin{array}{c}
\quad\quad 2 \quad\quad 3 \\
\quad\quad \neg B \quad T \\
1 \\
\begin{array}{cc} B & \quad \neg B \end{array} \\
\hline
\neg T \quad \text{Nm-3}
\end{array}
$$

Derivation of Ni:

$$
\begin{array}{c}
4 \\
A \\
\hline
1 \\
\begin{array}{cc} T & \quad \neg T \end{array} \\
\hline
\neg A \quad \text{Nm-2}
\end{array}
$$

Here the demonstration of $T$ can be placed over $T$, and the result gives $\neg A$ as consequence of $A \Vdash^{T} \neg T$.

**3. H systems.** Following the pattern of Sec. 5B let us define an H system as an assertional system admitting the propositions 𝔅 (Sec. 5C3$b$) as obs and having Pe (i.e., modus ponens) as its sole inferential rule. For the assertion predicate (the '$\vdash$' of Sec. 2D1) of an H system, we may use, when great explicitness is desired, the prefix '$\vdash^{H}$', and we may extend to that prefix the secondary use of '$\vdash$' to indicate deducibility as in Sec. 2D3. Associated with each such H system is a corresponding lattice, or E system; this association is defined in Sec. 5A1. The H system whose assertions are the same as those of LM will be called the system HM, and its corresponding lattice will be EM; likewise, the H and E systems associated with LJ will be HJ, EJ, respectively; etc.

A set of prime assertions for an H system will be a set of assertions of the system such that every assertion can be obtained from those in the set by a deduction using Pe. They correspond to the axiomatic propositions of the H system in the sense of Sec. 2D1, but the new term is used to avoid confusion with the axioms of Sec. 5C3$d$. A set of prime assertions will be said to be *separated* if every assertion $B$ of the system can be obtained by Pe from the prime assertions which contain, besides implication, only operations actually present in $B$.

If we have a system with negation, a separated set of prime assertions will consist of a separated set for the positive part, HA or HC, together with certain schemes which contain only implication and negation (including $F$ as a form of negation). Conversely, any set of propositions of that character will form, when so adjoined to a separated set for the positive system, a separated set for the system in question, provided only that they are assertions of the latter system and are sufficient, in combination with the former system, to validate the T rules of Secs. 2 and 3. In fact, by Theorems B7 and B8, any assertion can be established in the corresponding L systems using only rules concerning the operations which actually occur. The equivalence theorems show that the translation of this L demonstration into an H demonstration can be made by the use of prime assertions satisfying the conditions of separation.

In the discussion of H systems it will suffice, in principle, to consider only the case where the underlying system $\mathfrak{S}$ is $\mathfrak{O}$. This is the case of primary interest; besides, the general case can be reduced to this one by Theorem B9.

**4. The system HM.** If we consider the F formulation of LM, it is clear that its rules, since F∗ is now vacuous, are precisely the same as those of LA. Negation is defined by (1) of Sec. B, so that HM will now be simply a definitional extension of HA. Since we have Rp in LA, this definitional extension is the same as what we should get by adjoining to HA, as prime assertions schemes,

$$\vdash \neg A . \supset . A \supset F$$

$$\vdash A \supset F . \supset . \neg A$$

(4)

Thus we should get a separated set of prime assertions by adjoining (4) to a separated set for HA.

For the N formulation we have seen that special interest attaches to propositions containing only implication and negation. Every such assertion is obtained from an assertion of pure implication of HA by specializing one of its atoms to be $F$ and applying (4). It is convenient to use 'NB', 'NC', etc., for propositions formed in this way from 'PB', 'PC', etc. The following are some examples (using the definition of $\leq$ in Sec. 5A1)[1]:

| | |
|---|---|
| NB | $\neg B \leq A \supset B . \supset . \neg A$ |
| NB′ | $A \supset B \leq \neg B \supset \neg A$ |
| NC | $A \supset \neg B \leq B \supset \neg A$ |
| NS | $A \supset \neg B \leq A \supset B . \supset \neg A$ |
| NW | $A \supset \neg A \leq \neg A$ |
| NI′ | $A \leq \neg \neg A$ |
| NK$_{(1)}$ | $\neg A \leq A \supset \neg B$ |

All of these are true statements concerning LM.

Clearly, NC in combination with Pe is sufficient to justify the rule Nm. Further, since by PK

$$B \leq A \supset B$$

one can derive NC from NS by Rp; hence the same is true for NS.

This discussion proves the following:

**Theorem 5.** *A separated set of prime statements for* HM *can be formed by adjoining to such a set for* HA *either the schemes* (4) *or* NC *or* NS.

---

[1] All but the last two of these are in the list preceding the exercises to Sec. 5B. The last two are related to the combinators I′ ≡ CI, K$_{(1)}$ ≡ BK; the corresponding P propositions are

| | |
|---|---|
| PI′ | $A \supset : A \supset B . \supset B$ |
| PK$_{(1)}$ | $A \supset C . \supset . A \supset . B \supset C$ |

The formulation obtained by adjoining NC to the standard formulation of HA will be taken as the *standard formulation* of HM.

Various other possibilities for prime propositions for HM are given in the exercises. Note that since HM is a subset of HE, the necessary condition for validity in HE can be applied to HM (see Sec. 6).

The scheme NB' is not sufficient alone to generate HM.† Hence the algebra formed by adjoining this scheme to HA would be of some interest. The scheme expresses the fact that negation is a dual endomorphism with respect to inclusion. There should be interesting applications for this weaker algebra. But we do not reach it by the semantical approach of Sec. A3.

**5. The intuitionistic propositional algebra HJ.** The system LM becomes the intuitionistic system LJ if we adjoin Fj. This, in turn, is equivalent to the T rule Nj, which could be justified by

$$F \leq A \tag{5}$$

This in turn is equivalent to efq (the *"ex falso quodlibet"*)

$$\neg A \leq A \supset B \tag{6}$$

**Theorem 6.** *A separated set of prime assertions for HJ is formed by adjoining to a similar set for HM either of the schemes* (5), (6).

The proof is similar to that of Theorem 5.

The set formed by adjoining (6) to the standard formulation of HM will be called the *standard formulation of HJ.*

The property (5) shows that $F$ is a zero element in the lattice EJ. It is therefore appropriate to use '0' for '$F$'. Then (6) gives, by $P_1$,

$$A \wedge (\neg A) \leq 0 \tag{7}$$

Thus $\neg A$ is sometimes called a "pseudocomplement" of $A$, and EJ is called a "pseudocomplemented lattice." A pseudocomplement is not a true complement because of failure of the law

$$B \leq A \vee \neg A$$

which is a form of the law of excluded middle. The lattice EJ has topological applications where the propositions are open sets and the pseudocomplement is the interior of the complement. The dual lattice is a lattice of closed subsets of a fixed "universe."

The following half of the "definition" of material implication of "Principia Mathematica" is true in EJ:

$$\neg A \vee B \leq A \supset B \tag{8}$$

The converse does not hold.

**6. The systems HD and HE.** The strict system HD is formed by adjoining to HM a law of excluded middle. This, of course, can be formulated as

$$\vdash A \vee \neg A \tag{9}$$

† See Exercise 4

Since $A \vee \neg A$ contains alternation, this cannot fulfill the requirement of separateness. However, the scheme

$$\neg A \supset A \leq A \tag{10}$$

justifies the rule Nd, and hence is sufficient in combination with HM to give all assertions of HD. If it is adjoined to a standard formulation of HM, the formulation will be called the *standard formulation of* HD.

**Theorem 7.** *A separated set of prime assertions for the system* HD *is obtained by adjoining* (10) *to such a set for* HM.

In the system HD the converse of (8), viz.,

$$A \supset B \leq \neg A \vee B \tag{11}$$

is valid.

No applications are known for HD, and the system has been little studied. Johansson, who was the first to consider it, suggested that it formed a natural system of strict implication, but this has not been worked out.

The system HE is related to HC in the same way as HM is to HA. Hence, if we adjoin Pc to a separated set for HM, we shall have a set which is still a separated set, since Pc involves no operations except implication and is sufficient for HC. Such a set may also be formed by adjoining any of the sets of Theorem 5 to HC. The *standard formulation of* HE will be taken as that formed by adjoining Pc to the standard HM or NC to the standard HC.

Note that both HM and HD are subsystems of HE. The criterion of Theorem B6, which can sometimes show nondemonstrability in HE very quickly, will then show nondemonstrability in these other systems also.

**7. The system HK.** It follows by Theorem 3 that in the system HK we have the features of HJ and HD simultaneously. We can therefore form a separated set of prime statements for HK by adjoining both (6) and (10) to such a set for HM. Such a formulation would have the property that if we omitted (6) we should have HD, and if we omitted (10) we should have HJ, whereas if we omitted both we should have HM. The formulation so obtained from a standard formulation of HM will be called the *standard formulation of* HK.

There are, however, many other ways of formulating HK. Reserving until Sec. D a study of the algebraic nature of HK, we note here some of the simpler properties which are parallel to the developments treated elsewhere in this section.

In the system HK we have both (8) and (11), and hence, using '=' as in Sec. 5A1, we have

$$A \supset B = \neg A \vee B \tag{12}$$

Furthermore, since Nk is derivable in TK, the statement

$$\neg \neg A \leq A \tag{13}$$

is a theorem of HK; since the converse holds even in HM (by NI'), we have

$$\neg \neg A = A \tag{14}$$

Again consider the four principles of contraposition:

$$A \supset B \le \neg B \supset \neg A \qquad (15)$$
$$A \supset \neg B \le B \supset \neg A \qquad (16)$$
$$\neg A \supset B \le \neg B \supset A \qquad (17)$$
$$\neg A \supset \neg B \le B \supset A \qquad (18)$$

These are all equivalent in HK by virtue of (14). But whereas (15) is NB' and (16) is NC, both of which hold in HM, (17) and (18) fail in HE. The statement (13) justifies the rule Nk of Sec. 1, and hence by Theorem 3 is sufficient to generate HK if adjoined to HM. The statement (18) is sufficient to generate the whole of HK when adjoined to HA. This can be shown as follows:

$$\neg A \le \neg B \supset \neg A \qquad \text{by PK}$$
$$\le A \supset B \qquad \text{by (18)}$$

Thus we have (6). Hence, putting $\neg A$ for $A$ and $\neg B$ for $B$,

$$\neg \neg A \le \neg A \supset \neg B$$
$$\le B \supset A \qquad \text{by (18)}$$

Here if we take $B$ to be $\neg \neg A$, we have (13) by PW. Applying PB' to (13), we have

$$A \supset \neg B . \le . \neg \neg A \supset \neg B$$
$$\le B \supset \neg A \qquad \text{by (18)}$$

which is NC. Thus we have NC and (13), and hence all of HK.

Summing up, we have the following:

**Theorem 8.** *The system* HK *contains all the statements* (12) *to* (18) *and includes* HC. *A set of prime assertions for it can be formed by adjoining* (6) *and* (10) *to* HM, (13) *to* HM, *or* (18) *to* HA. *A separated set can be formed by adjoining* (6) *to one for* HE.

*Proof.* It is only necessary to add to the preliminary discussion a proof of the inclusion of HC. This follows from Theorem 3, but a direct proof that Pc holds is easy, as follows:

$$A \supset B. \supset A \le \neg A \supset A \qquad \text{by (6), Rp}$$
$$\le A \qquad \text{by (10)}$$

## EXERCISES

1. Show that any of the following schemes or combinations of schemes are equivalent, on the basis of HA, to NC, and thus suffice to generate HM from HA:

(a)    NB', NW
(b)    NB', NI'
(c)    NW, $NK_{(1)}$

([LLA], Theorem IV 1.)

2. Show that NW and efq [that is, (6)] suffice to generate HJ from HA. ([LLA], Theorem IV 2.)

3. Show that

$$\vdash \neg \neg A \supset . \neg A \supset A$$

is equivalent, in the presence of HM, to efq. Hence it suffices, when adjoined to HM, in order to get HJ. (Schmidt [VAL, p. 345]. For the history see Hermes and Scholz [MLg, p. 37]; the result is attributed to Bernays.)

**4.** Find independence examples showing that neither NB′ nor NW alone is sufficient to generate HM. The same is true for $NK_{(1)}$ since it is a consequence of NB′. ([TFD], Theorem 9, for NW; for NB′ see Hilbert and Bernays [GLM.I], p. 76.)

**5.** Verify the statements in the text concerning (8) and (11).

**6.** The generalized Glivenko theorem (Exercise B5) can be stated in terms of H systems thus: $A$ is an assertion of HD (or HK) only if $\neg A \supset A$ is an assertion of HM (or HJ). Prove this by deductive induction in the H systems. ([LLA], sec. IV5.)

**7.** Discuss the results of Exercises B7 to B9 from the standpoint of H systems.

**8.** Show that PB′, (10), and

$$\vdash A \supset. \ \neg A \supset B$$

form a sufficient set of axiom schemes for the part of HK which contains implication and negation only. [Church [IML$_2$, exercise 29.2]; originally due to Łukasiewicz. I do not know what the state of affairs is if the new axiom scheme is replaced by (6).]

**9.** Show that the scheme NC in the standard formulation of HK is redundant, but that if (10) is replaced by either of the schemes

(a)       $\vdash A \supset \neg A .\supset A .\supset A$
(b)       $\vdash A \supset \neg A .\supset B .\supset. \ A \supset B .\supset B$

the resulting set of schemes gives an independent (but not separated) set of prime assertions for HK. Show further that (a) is equivalent to (10), relative to HM, but (b) is not. (On the history of these schemes see Hermes and Scholz [MLg, footnote 53]. On the independence, cf. Exercise 11.)

**10.** Show that the standard formulation of HE is such that by adjoining efq one has an independent and separated set for HK. (Kanger [NPP].)

**\*11.** What sort of systems would one get if one admitted the possibility that $\neg A$ was always false, in the same sense that LM, LD, LE admit the possibility (see Theorem B6) that $\neg A$ is always true? In such a case NB′ and (17) would hold, but not (16) or (18). What if both of these possibilities were admitted? Would the system then coincide with that formed by adjoining NB′ alone to some positive system? How would such systems be formulated in the various formulations? What could you say if $\neg A$ has always the same value as $A$?

**12.** Suppose we have an H system, call it HS, which is an (axiomatic) extension of HM such that whenever $A$ is an assertion of HK, $\neg A \supset A$ is an assertion of HS. Show that HS is an extension of HJ. (Porte [PCP], but his proof is erroneous—see the review in *Journal of Symbolic Logic*. The theorem as here stated is constructive.)

**13.** Show directly that

$$A_1, \ldots, A_m \Vdash B_1, \ldots, B_n \qquad \text{in LK}$$

is equivalent to

$$\vdash \neg A_1 \vee \cdots \vee \neg A_m \vee B_1 \vee \cdots \vee B_n \qquad \text{in HK}$$

## D. TECHNIQUE OF CLASSICAL NEGATION

The system of classical negation (HK) is very much further developed than the other systems. Some of the features of this development will be considered in this section. The treatment will center around the concept of a Boolean algebra, which will be defined as a complemented distributive lattice. This notion will be introduced in Sec. 1 and shown to be equivalent to related notions developed previously, viz., classical subtractive and implicative lattices, Boolean rings, the system EK, and tautological interpretations. In Sec. 2, entitled "Quasi definitions," will be considered elementary theorems which have the effect of decreasing the number of primitive operations, together with formulations based on reduced sets of such operations. The remaining subsections will develop the standard technique of Boolean algebra, including representations, normal forms, elimination, solution of equations, etc. However, in keeping with the aim of this book, only finitary properties will be treated; the rather extensive nonfinitary theory of Boolean algebras, including the general case of the "representation theorem," is beyond the scope of this book.

The notation of this section will be modified to fit standard procedures. Thus negation will generally be designated by priming, and the meet operations by simple juxtaposition; and lower-case letters will occasionally be used for obs.

**1. Boolean algebras.**   Given a lattice $L$ with a zero element 0 and a unit element 1, an ob $A'$ will be said to be a *complement* of an ob $A$, just when

$$A \wedge A' = 0 \tag{1}$$

$$A \vee A' = 1 \tag{2}$$

are both true.   A lattice in which every ob has at least one complement will be called a *complemented lattice*.   A complemented distributive lattice will be called a *Boolean algebra*.

**Theorem 1.**   *A classical subtractive lattice with unit in which negation is so defined that*

$$A' = 1 - A \tag{3}$$

*is a Boolean algebra.   So also is a classical implicative lattice with zero and negation such that*

$$A' = A \supset 0 \tag{4}$$

*Proof.*   Let $L$ be a classical subtractive lattice with unit.   Then $L$ is distributive by Theorem 4C5.   From (3) we have (1) and (2) thus:

$$AA' = A(1 - A) = 0 \qquad \text{by (3) of Sec. 4D}$$
$$1 \leq A \vee (1 - A) = A \vee A' \qquad \text{by } (-)_1$$

This proves the first part of the theorem.   The second part follows by duality.

COROLLARY 1.1.   *The algebra EK is a Boolean algebra.*

*Proof.*   In the F formulation of LK, negation is defined by

$$\neg A \equiv A \supset F$$

Since $F$ is the zero of the system by Fj, this is the same as (4).   Thus EK can be generated by adding 0 and (4) to EC, which is a classical implicative lattice.   Thus the theorem applies, Q.E.D.

**Theorem 2.**  *In every Boolean algebra the complement is unique. Moreover,
the following hold for all obs A, B:*

$$A'' = A \tag{5}$$
$$A = AB \vee AB' \tag{6}$$
$$A \leq B \rightleftarrows AB' = 0 \tag{7}$$
$$A \leq B \rightleftarrows B' \leq A' \tag{8}$$

*Proof.*  In a distributive lattice, complements are unique, by Corollary
4B9.1.  For (1) and (2), one concludes by ΛC and VC that $A$ is a complement
of $A'$.  Since $A''$ also is, we have (5).  Again, by (2) and the distributive law,

$$A = A(B \vee B') = AB \vee AB'$$

which proves (6).  To prove (7), note first that, by Rp and (1),

$$A \leq B \rightarrow AB' \leq BB' \leq 0$$

Conversely, by (6),

$$AB' \leq 0 \rightarrow A = AB$$
$$\rightarrow A \leq B \qquad \text{(by } \Lambda K')$$

From this we derive (8) thus:

$$A \leq B \rightarrow AB' = 0 \qquad \text{by (7)}$$
$$\rightarrow A''B' = 0 \qquad \text{by (5)}$$
$$\rightarrow B' \leq A' \qquad \text{by (7)}$$

**Theorem 3.**  *Let L be a lattice with 0, 1 and having a negation operation such
that (1) and*

$$A \leq AB \vee AB' \tag{9}$$

*are satisfied.  Let subtraction be defined in L in such a way that*

$$A - B = AB' \tag{10}$$

*Then L is simultaneously a Boolean algebra and a classical subtractive
lattice.  Moreover, (3) holds.*
*Proof.*  The property $(-)_1$ follows at once from (9).  We also have

$$A - B \leq A \qquad \text{by } \Lambda K$$
$$B(A - B) = 0 \qquad \text{by (1)}$$

By (9) we have 4D(7), and hence $(-)_2$ as in proof of Theorem 4D2.  Then (3)
holds since it is a special case of (10).  The rest follows by Theorem 1.

COROLLARY 3.1.  *A Boolean algebra with a subtraction satisfying (10) is a
classical subtractive lattice with unit, such that (3) holds.*
*Proof.*  The condition (9) follows from (6) by ΛK' and Rp.

COROLLARY 3.2.  *A Boolean algebra with an implication satisfying*

$$A \supset B = A' \vee B \tag{11}$$

*is a classical implicative lattice in which (4) holds.*
*Proof.*  This is the dual of Corollary 3.1.

**Theorem 4.**  *If a Boolean algebra is also a subtractive lattice, then it is a
classical subtractive lattice in which (3) and (10) hold.  Likewise, if a Boolean*

*algebra is an implicative lattice, then it is a classical implicative lattice in which* (4) *and* (11) *hold.*

*Proof.* Since the two halves of this theorem are dual to each other, it will suffice to prove the first half.

Suppose that $L$ is a Boolean algebra which is also a subtractive lattice. Let $C \equiv AB'$. Then

$$BC = 0 \qquad\qquad \text{by (1)}$$
$$A = AB \vee C \leq B \vee C \qquad \text{by (6), } \Lambda\text{K}', \text{Rp}$$
$$A - B \leq C \qquad\qquad \text{by } (-)_2$$

Conversely, since $C \leq A$, $(-)_1$ holds, the lattice is distributive, and $BC = 0$,

$$C = CA \leq C(B \vee (A - B)) = C(A - B) \leq A - B$$

Thus (10) holds. The rest follows by Corollary 3.1.

**Theorem 5.** *A Boolean algebra becomes a Boolean ring with unit if ring addition is defined by*

$$A + B = AB' \vee A'B \tag{12}$$

*and ring multiplication and the ring* 0, 1 *are identified with the lattice meet* 0, 1, *respectively. Conversely, a Boolean ring with unit becomes a Boolean algebra if*

$$A \vee B = A + B + AB \tag{13}$$
$$A' = 1 + A \tag{14}$$

*Proof.* This theorem is a composite of Theorem 1, Corollary 3.1, and Theorem 4D4. If we adjoin (10), then by the corollary we have a classical subtractive lattice, and (12) becomes condition (i) of Theorem 4D4.[1] Conversely, if we have (13) and a suitable definition of $A - B$, we have a classical subtractive lattice in which (3) is the same as (14); we then have a Boolean algebra by Theorem 1, Q.E.D.

**Theorem 6.** *A necessary and sufficient condition that*

$$A = B \tag{15}$$

*hold in a general Boolean algebra is that A and B have the same value in every evaluation by* 0-1 *tables; the same condition for*

$$A \leq B \tag{16}$$

*is that B have the value* 1 *in every evaluation in which A has the value* 1.

*Proof.* If (15) and (16) hold in a general Boolean algebra, then they hold in $\text{EK}(\mathfrak{O})$ by Corollary 1.1; the converse is clear since no special assumptions are made in $\text{EK}(\mathfrak{O})$. Thus the theorem is a consequence of Theorem B6 and the equivalence between LA and EA (Secs. C, 5D4, 5B5), Q.E.D.

Although this theorem gives a theoretical solution to the decision problem for a general Boolean algebra, it is not always the fastest method. If the number of indeterminates is $n$, then there are $2^n$ possibilities to be considered.

---

[1] The condition (iii) of Theorem 4D4 can be ignored here. It was introduced in Sec. 4D2 solely to subsume the notion of Boolean ring under the general concept of a ring.

The method of reduction by translating into a Boolean ring with unit, multiplying out, and adding modulo 2 is faster; so also is Beth's method of proof tableaux. The 0-1 tables for the Boolean operations considered here are given in Table 1.

TABLE 1. TRUTH TABLES FOR VARIOUS FUNCTIONS

| $A$ | $B$ | $A \wedge B$ | $A \vee B$ | $A - B$ | $A \supset B$ | $A + B$ | $A \backsim B$ | $A \mid B$ | $A'$ |
|---|---|---|---|---|---|---|---|---|---|
| 1 | 1 | 1 | 1 | 0 | 1 | 0 | 1 | 0 | 0 |
| 1 | 0 | 0 | 1 | 1 | 0 | 1 | 0 | 0 | 0 |
| 0 | 1 | 0 | 1 | 0 | 1 | 1 | 0 | 0 | 1 |
| 0 | 0 | 0 | 0 | 0 | 1 | 0 | 1 | 1 | 1 |

**2. Quasi definitions.** In (10), (11), (12), and (13) of Sec. 1 we have examples of equations of the following type:

$$A \circ B = f(A,B) \qquad (17)$$

in which the infix '$\circ$' stands for a binary operation and '$f(-_1, -_2)$' for a construction which is independent of that operation. In (3), (4), and (14) we have equations of similar character except that the operation in question is a unary operation. Such equations will be called *quasi definitions;* a particular quasi definition will be said to be a quasi definition of the principal operation on the left, and that operation will be said to be *quasi-definable* by the equation. In partially ordered presentations a quasi definition consists, of course, in the conjunction of the elementary statements; e.g., for (17) these would be

$$A \circ B \leq f(A,B)$$
$$f(A,B) \leq A \circ B$$

When none of the operations of a certain set of operations is quasi-definable in terms of the others, the operations will be said to be *independent*.

When an operation is quasi-definable, one would expect that the operation could be eliminated in the sense that to every elementary theorem containing it there would be an equivalent elementary theorem which was free from it. However, to draw this conclusion from the existence of a quasi definition requires Rp. In the case of a definition (in which case the infix '$\equiv$' would ordinarily be used), Rp is satisfied automatically, but in other cases it has to be established. In the cases considered up to now, and in most of those considered later, Rp holds, and consequently the distinction between a definition and a quasi definition can be ignored; but there are exceptional cases in which the derivation of Rp is an essential difficulty.

In some, at least, of the previous systems the operations were independent.[1] But this is not the case in HK, or even in HC. The rest of this subsection will be centered on such quasi definitions for the operations of Boolean algebra in terms of one another or in terms of extraneous operations. There will be included a sketch of ways in which Boolean algebra can be formulated in terms of a restricted list of primitive operations.

[1] See Exercise B14.

We may as well begin with a recapitulation,[1] from the present point of view, of the results of Sec. 1.

**Theorem 7.** *With respect to the quasi definitions in Table 2, the following systems are equivalent: Boolean algebra, a classical subtractive lattice with unit, a classical implicative lattice with zero, a Boolean ring with unit, and a system formed by adding a zero to the dual of a Boolean ring.*

TABLE 2. QUASI DEFINITIONS†

| Operation | Boolean algebra | Subtractive lattice | Implicative lattice | Boolean ring | Dual Boolean ring |
|---|---|---|---|---|---|
| $A \wedge B$ | | | | | $A \smile B \smile (A \vee B)$ |
| $A \vee B$ | | | | $A + B + AB$ | |
| $A - B$ | $AB'$ | | $[(A \supset B)']$ | | $A \vee B \smile B \smile 0$ |
| $A \supset B$ | $A' \vee B$ | $[(A - B)']$ | | $1 + A + AB$ | $A \vee B \smile B$ |
| $A + B$ | $AB' \vee A'B$ | $(A - B) \vee (B - A)$ | $[(A \smile B)']$ | | $A \smile B \smile 0$ |
| $A \smile B$ | $(A \vee B')(A' \vee B)$ | $[(A + B)']$ | $(A \supset B)(B \supset A)$ | $1 + A + B$ | |
| $A'$ | | $1 - A$ | $A \supset 0$ | $1 + A$ | $A \smile 0$ |

† Those in square brackets are not given in full, but require reference to other lines in the same column.

Let us now turn to the proper business of this subsection.

The property (8), in combination with (5), expresses the fact that negation is a one-to-one mapping of a Boolean algebra on its dual. Such a correspondence is called a *dual automorphism*. The equations of the following theorem are known as *DeMorgan formulas*.[2]

**Theorem 8.** *In a Boolean algebra negation is a dual automorphism. The following are theorem schemes:*

$$(A \wedge B)' = A' \vee B' \tag{18}$$

$$(A \vee B)' = A' \wedge B' \tag{19}$$

*and the quasi definitions*

$$A \wedge B = (A' \vee B')' \tag{20}$$

$$A \vee B = (A' \wedge B')' \tag{21}$$

*also hold.*

*Proof.* These follow immediately from the automorphism property by methods of modern algebra. They can also be verified by Theorem 6. Perhaps quicker than this is the method of transforming into a Boolean ring, thus:

$$A' \vee B' = (1 + A) + (1 + B) + (1 + A)(1 + B)$$
$$= A + B + 1 + A + B + AB = 1 + AB = (AB)'$$
$$A' \wedge B' = (1 + A)(1 + B) = 1 + A + B + AB = (A \vee B)'$$

Of course, (20) and (21) follow from (18) and (19) by negating both sides and using (5).

COROLLARY 8.1. *In a Boolean algebra one can form from an ob $A$ an ob equal to $A'$ as follows. One forms first the ob $A^D$, which corresponds to $A$ by*

---

[1] The recapitulation misses certain details which appear in Sec. 1 and adds some quasi definitions not considered there. Thus Theorem 7 is not quite the same as the conjunction of Theorems 1 to 5.

[2] The term is customary despite its historical inaccuracy. According to Bochenski [FLg], the formulas were known in the Middle Ages.

*duality (interchanging* ∧ *and* ∨, 0 *and* 1), *and then in* $A^D$ *replaces each atom E* (*other than* 0 *and* 1) *by E'*.

This is established by structural induction, moving the negation toward the interior by (18) and (19).

This theorem and corollary show that the principle of duality holds for Boolean algebra in a stronger sense than for a lattice. In the latter case, if $\Gamma$ is an elementary theorem and $\Gamma^D$ its dual, then $\Gamma \rightarrow \Gamma^D$ is an admissible rule, but here one can pass from $\Gamma$ to $\Gamma^D$ provided one can substitute for the indeterminates. For let $\Gamma$ be

$$A \leq B$$

Then by (8) we have

$$B' \leq A'$$

By Corollary 8.1 one can convert $A'$, $B'$, respectively, into $A^D$, $B^D$ by substitutions of the indeterminates and, possibly, applications of (5).

The following theorem shows that one can define all operations in terms of ⊃ and negation.

**Theorem 9.**[1]    *The quasi definitions*

$$A \wedge B = (A \supset B')' \tag{22}$$
$$A \vee B = A \supset B \,.\supset B \tag{23}$$

*hold in* EK; *the second one, even in* EC.

*Proof.* This follows from the following theorems:

$$A \wedge B \Vdash \neg (A \supset \neg B)$$
$$\neg (A \supset \neg B) \Vdash A \wedge B$$
$$A \vee B \Vdash A \supset B \,.\supset B$$
$$A \supset B \,.\supset B \Vdash A \vee B$$

These hold in LM, LK, LA, and LC, respectively.

These theorems show that in HK all operations are quasi-definable in terms of negation and any one of the three binary operations ∧, ∨, ⊃. It is possible to define all these operations in terms of a single binary operation. This is shown as follows:

**Theorem 10.**    *If*

$$A \mid B \equiv (A \wedge B)' \tag{24}$$

*then the following quasi definitions in terms of this operation as the sole primitive operation hold:*

$$\begin{aligned} A' &= A \mid A \\ A \wedge B &= (A \mid B)' \\ A \vee B &= A' \mid B' \\ A \supset B &= A \mid B' \end{aligned} \tag{25}$$

*Proof.* See Theorem 6.

This theorem is attributed to Sheffer [SFI], and the new operation is called the *Sheffer stroke function*. Someone has recently discovered that the idea occurs in the work of Peirce, but this apparently was not known when Sheffer's paper appeared. (See Church [IML₂], note 207.) Some logicians

---

[1] [LLA], Theorem IV 10.

consider the idea a great discovery (e.g., see the introduction to Whitehead and Russell [PMt.I$_2$]), others a mere curiosity (e.g., Hilbert and Ackermann [GZT], sec. I2).

The existence of quasi definitions indicates the possibility of formulating the algebra in terms of a reduced set of operations, using the quasi definitions to introduce the others. There is a great variety of such formulations, and it is not possible to consider them all here.

That a formulation in terms of P and N alone is possible may be seen as follows. The axiom schemes for $\Lambda$ and V in HK remain tautologous when the operations are replaced by their quasi definienda by (22) and (23). By the separation theorem, these are deducible in that part of HC which contains only implication. Hence one needs to adjoin to HC axiom schemes of the type considered in Sec. C7. As shown there, one can make such additions to absolute implication. The very first assertional formulation, that of Frege, was of that character. He had six axiom schemes; three of these, of which one was redundant, were for absolute implication; to these were added NB', NI', and (13) of Sec. C. This has since been abbreviated. A system of three axiom schemes is formed by adjoining to PK, PS the scheme

$$\vdash \neg A \supset \neg B . \supset . B \supset A$$

Another similar system is that of Łukasiewicz, consisting of PB' and

$$\vdash \neg A \supset A . \supset A$$
$$\vdash A \supset . \neg A \supset B$$

Systems with a single axiom scheme are known.

The formulation of Whitehead and Russell [PMt] was in terms of V and N. After eliminating a redundancy (cf. the remark in Sec. 4S1), the axiom schemes are (see also Sec. 4S1):

| | |
|---|---|
| Taut | $\vdash A \supset . A \vee A$ |
| Perm | $\vdash A \vee B . \supset . B \vee A$ |
| Simp | $\vdash A \supset A \vee B$ |
| Sum | $\vdash A \supset B \supset: C \vee A . \supset . C \vee B$ |

These together with the quasi definitions (11) and (20) suffice for the whole of HK. Note that this gives eight axiom schemes; further, it is necessary to establish Rp. A lattice formulation in terms of these primitives is given in Huntington [SIP], [NSP]; these take equality and Rp for granted.

The schemes[1]

$$\vdash A \supset AA$$
$$\vdash AB \supset A$$
$$\vdash (AB)' \supset (BA)'$$
$$\vdash A \supset B . \supset: (CB)' \supset (CA)'$$
(26)

with the quasi definition (21) and

$$A \supset B = (AB')'$$
(27)

---

[1] [LLA], p. 114.

form a set in terms of $\Lambda$ and $N$. The first formulation of that sort was given by Sobocinski. Rosser [LMt, chap. 4] introduces a twist into the fourth axiom by which he is able to dispense with the third. For other formulations of this sort see Porte [SCP].

**3. Finite interpretations.** The preceding subsection has been concerned with formulations of Boolean algebra in general. We shall now deal with regular interpretations (Sec. 5A4) of Boolean algebras; this will cause us to be interested in Boolean algebras of a more special nature.

In this connection there is some conflict of terminology between that which has been used here up to now and that which is current in regard to Boolean algebra. It is therefore necessary to devote some space to clarification of the terminological questions.

The notion of "regular interpretation" was introduced in Sec. 5A4, and also the idea of "element" in connection with it. It is important not to confuse this notion of element with that of ob. Thus, if $E_1$ and $E_2$ are primitive obs, the obs $E_1 \supset E_2$, $E_1' \vee E_2$, $E_2 \vee E_1'$, $E_2' \supset E_1'$, $(E_1 E_2')'$ constitute five distinct obs, but they all correspond to the same element. It is nevertheless permissible (cf. Sec. 5A4) to use the same symbol for an element as for the ob to which it is associated, and thus to say that the above five obs *are* the same element; when we do this, and say, for example, "the element $E_1 \supset E_2$," what we have to say will apply equally well to $E_1' \vee E_2$, but when we say "the ob $E_1 \supset E_2$," it will not.

In Sec. 2C3 the term 'atom' was used for the primitive ob of an ob system. This term was appropriate because these obs precede in the order of construction. This usage, however, conflicts with one which has become common in connection with Boolean algebra. In that usage an atom is a nonnull element which is preceded (in the ordering $\leq$ of the algebra) by no elements distinct from itself except the 0 element. In order to avoid using 'atom' in two senses, I shall, following Birkhoff [LTh], use 'point' in the sense of 'atomic element'. Thus, if the infix '$\neq$' means that the associated elements are (contensively) distinct, a *point* is an element $U$ such that

$$U \neq 0 \tag{28a}$$

and for all $A$,

$$A \neq 0 \ \& \ A \leq U \rightarrow A = U \tag{28b}$$

A *counterpoint* of a Boolean algebra is now defined as an element which is the dual of a point; i.e., it is an element $U$ such that

$$U \neq 1 \tag{29}$$

and for all $A$,

$$U \leq A \ \& \ A \neq U \rightarrow A = 1 \tag{30}$$

A *finite Boolean algebra* is one with a finite number of elements.

In a finite Boolean algebra every nonzero element contains at least one point and is the join of all the points which it contains. For let $A$ be an element. If among the finitely many elements distinct from $A$ and 0 there is none $\leq A$, then $A$ is a point. If not, let $A_1 < A$, $A_1 \neq A$, $A_1 \neq 0$. Then start in again with $A_1$, and so on. Eventually we must reach a point. Let $U_1, \ldots, U_n$ be all the points $U_i$ such that $U_i \leq A$. Let

$$B \equiv U_1 \vee U_2 \vee \cdots \vee U_n$$

Then $$B \leq A$$

Hence $$A B = B$$

If $A - B \neq 0$, then it, and hence $A$, will contain a point distinct from $U_1, \ldots, U_n$, which is impossible.  Hence

$$A - B = 0$$

i.e., $$A = B$$

Thus an element can be identified with the set of points which it contains. This proves the following:

**Theorem 11.** *In a finite Boolean algebra every nonzero element contains at least one point and every element is the union of all the points which are included[1] in it.  The algebra is isomorphic to the system whose elements are all possible sets of its points; if the number of points is $m$, the number of elements is $2^m$.*

A special case of such a finite Boolean algebra is the case where $m = 1$. In that case the unit 1 is itself a point, and the algebra consists of two elements 0 and 1.  Validity with respect to such an interpretation corresponds to validity by 0-1 tables.

Theorem 11 is a special case of the "Stone representation theorem," which says that every Boolean algebra is isomorphic to a field of sets.  This means, in the present terminology, that every Boolean algebra has an interpretation in which the elements are subsets of a fixed set, meet and join are set intersection and set union, respectively, and negation is complementation with respect to the fixed set.  Theorem 11 is the special case where the fixed set is finite.  The proof of the general theorem requires transfinite methods and is therefore beyond the scope of this book.  Of course, the term 'representation' has a different meaning in this connection from the one it had in Sec. 2C.

**4. Developments and bases.**  If $U_1, U_2, \ldots, U_n$ are all the points of a Boolean algebra $L$, then, by Theorem 11, for each element $A$ there exist $a_1, \ldots, a_n$, where $a_i$ is 1 if $U_i \leq A$ and $a_i$ is 0 otherwise, such that

$$A = a_1 U_1 \vee \cdots \vee a_n U_n \qquad (31)$$

It is convenient to write (31) as

$$A = \bigvee_{i=1}^{n} a_i U_i$$

This is a special case of the following situation.  Let $L$ be a Boolean algebra and $K$ be a subalgebra of $L$; that is, the elements of $K$ are some subset of the elements of $L$ and are combined in the same way by the operations. We use lower-case letters for obs and elements of $K$.  Then a set of $L$-elements $U_1, \ldots, U_n$ will be called a *basis* of $L$ relative to $K$ just when the following conditions are satisfied:

*a.* For every $A$ in $L$ there exist $a_1, \ldots, a_n$ in $K$ such that (31) is satisfied.

*b.* For $i \neq j$, $U_i U_j = 0$.

*c.* If $a$ is in $K$ and $aU_i = 0$, then $a = 0$.

---

[1] Here the relation $\leq$ is taken as inclusion, so that, for example, '$A \leq B$' is to be read "$A$ is included in $B$."

These conditions are satisfied if the $U_1, \ldots, U_n$ are points and $a_1, \ldots, a_n$ are either 0 or 1. Another possibility, to be considered later, is that $L$ is an extension of $K$ by adjoining certain indeterminates and the $U_i$ are certain combinations of these indeterminates.

The dual of this situation is occasionally of some interest. In this case the set $U_1, \ldots, U_n$ will be called a *counterbasis* of $L$ relative to $K$; the conditions analogous to $a$ to $c$ are:

$a'$. For every $A$ in $L$ there exists $a_1, \ldots, a_n$ in $K$ such that

$$A = (a_1 \vee U_1)(a_2 \vee U_2) \cdots (a_n \vee U_n) \qquad (\equiv \bigwedge_{i=1}^{n} (a_i \vee U_i)) \qquad (32)$$

$b'$. For $i \neq j$, $U_i \vee U_j = 1$.

$c'$. If $a$ is in $K$ and $a \vee U_i = 1$, then $a = 1$.

The right side of (31) will be called the *alternative development* of $A$ with respect to the basis $U_1, \ldots, U_n$. The separate $a_i U_i$ will be called the *terms* of the development, and the $a_i$ the *coefficients*. The dual development (32) will be called the *conjunctive development* of $A$ relative to $U_1, \ldots, U_n$, with "terms" and "coefficients" defined analogously.

**Theorem 12.** *Let $L$ be a Boolean algebra, $K$ a subalgebra of $L$, and $U_1, \ldots, U_n$ a basis of $L$ relative to $K$. Then the coefficients satisfying (31) are uniquely determined when $A$ is given, and for every $i$ the correspondence*

$$A \sim a_i \qquad (33)$$

*is a homomorphism.*

*Proof.* Let (31) hold, and let

$$B = \bigvee_{i=1}^{n} b_i U_i \qquad C = \bigvee_{i=1}^{n} c_i U_i$$

Suppose now that $A = B$; then

$$A + B = \bigvee_{i=1}^{n} (a_i + b_i) U_i = 0$$

Hence                    $(a_i + b_i) U_i \leq A + B = 0$

Therefore, by condition $c$,

$$a_i + b_i = 0$$
$$a_i = b_i$$

This shows that the $a_i$ are unique as elements.

Now suppose

$$C = A \vee B$$

Then by the distributive law and the uniqueness just shown,

$$c_i = a_i \vee b_i$$

Hence, in the correspondence (33),

$$A \vee B \sim a_i \vee b_i \qquad (34)$$

Again, let

$$C = AB$$

Then
$$C = \left( \bigvee_{i=1}^{n} a_i U_i \right) \left( \bigvee_{j=1}^{n} b_j U_j \right)$$

$$= \bigvee_{i=1}^{n} a_i b_i U_i$$

by the distributive law and condition $b$. Hence, by the uniqueness

$$c_i = a_i b_i$$

and therefore in the correspondence (33),

$$AB \sim a_i b_i \tag{35}$$

Now it is evident that in the correspondence (33)

$$0 \sim 0 \qquad 1 \sim 1$$

since for every $A$ we have

$$0U_i \vee \cdots \vee 0U_n \leq A \leq U_1 \vee \cdots \vee U_n$$

Hence, if

$$B = A'$$

then
$$AB = 0 \qquad A \vee B = 1$$

It follows, by what we have already obtained, that

$$a_i b_i = 0 \qquad a_i \vee b_i = 1$$

Therefore

$$b_i = a_i'$$

and thus in (33)

$$A' \sim a_i' \tag{36}$$

From (34), (35), (36), and the uniqueness result, the correspondence (33) is a homomorphism, Q.E.D.

COROLLARY 12.1. *If $K$ is a finite Boolean algebra with $k$ elements, then $L$ is a finite Boolean algebra with $k^n$ elements.*

A special case of the notions of basis and counterbasis occurs if $L$ is formed by adjoining certain indeterminates $E_1, \ldots, E_m$ to $K$. Then consider the $2^m$ elements

$$E_1^{e_1} E_2^{e_2} \cdots E_m^{e_m} \tag{37}$$

where each $e_k$ is 0 or 1, and temporarily we adopt the convention that for any $A$

$$A^0 \equiv A' \qquad A^1 \equiv A \tag{37a}$$

We shall see that these $2^m$ elements constitute a basis. In fact, it is clear that the conditions $b$ and $c$ are satisfied. Given any $A$ in $L$, we can reduce it to the form shown on the right of (31) by the following process:

1. Remove all operations other than meet, union, and negation by the quasi definitions.

2. Move negations toward the interior by (18) and (19) until the negated components are elements of $K$ or else contain no meet or join operation.

3. Remove multiple negations by statement (5).

4. Multiply out by the distributive law until we have a union of terms, each of which is a meet of elements which either is in $K$, or is some $E_k$, or is some $E_k'$.

5. Strike out terms which for some $k$ contain both $E_k$ and $E_k'$. [These are 0, by (1).]

6. If a term contains neither $E_i$ nor $E_i'$, multiply it by $E_i \vee E_i'$ [which is 1 by (2)] and continue as before.

At the end of step (6) every term will be a product (i.e., meet) of elements each of which is either an ob of $K$ or some $E_k$ or some $E_k'$ and such that for every $k$ one and only one of $E_k$, $E_k'$ is present. One has then only to collect together, using the distributive law, all terms with the same factors from the $E$'s in order to have an ob of the form appearing on the right in (31). (If any terms are missing, the corresponding $a_i$ is 0.)

The development in such a case is called the *alternative* (or disjunctive)[1] *normal form* of $A$ relative to $E_1, \ldots, E_m$. The dual development is called the *conjunctive normal form*.

An algebra $L$ formed in the manner above described is called a *free (Boolean) extension of K with the m generators* $E_1, \ldots, E_m$; if $K$ consists of 0 and 1 only, $L$ is the *free Boolean algebra with m generators*. The elements of $L$ may be thought of as functions of the "variables" $E_1, \ldots, E_m$ ranging over the set 0, 1. If $f(E_1, \ldots, E_m)$ is such a function, then the coefficient of (37) in the expansion (31), where $A \equiv f(E_1, \ldots, E_m)$, is $f(e_1, \ldots, e_m)$; for the substitution, for all $k$, of $e_k$ for $E_k$ converts (37) into 1 and all other terms of the expansion (31) into 0. This proves the following:

**Theorem 13.** *Let $L$ be a free extension of a Boolean algebra $K$ with respect to indeterminates $E_1, \ldots, E_m$. Then the $2^m$ obs (37) constitute a basis for $L$ relative to $K$ and the coefficient of an ob (37) in the expansion (31) is the result of substituting $e_k$ for $E_k$ in $A$ for all $k = 1, 2, \ldots, m$. If $K$ is a finite Boolean algebra with $k$ elements, then $L$ is a finite Boolean algebra with $k^{2^m}$ elements.*

COROLLARY 13.1.    *The number of elements in a free Boolean algebra with $m$ generators is $2^{2^m}$.*

Since the $f(e_1, \ldots, e_m)$ are precisely the values for a 0-1 valuation of $A$ in which each $E_k$ is assigned the value $e_k$, the completeness theorem, Theorem 6, can be obtained as a corollary of Theorem 13. A stronger type of completeness theorem is the following:

**Theorem 14.** *Let $A \equiv f(E_1, \ldots, E_m)$ be such that $A$ is not a tautology with respect to 0-1 tables. Then if one adjoins*

$$\vdash f(x_1, \ldots, x_m) \tag{38}$$

*to* HK *as a new axiom scheme, the '$x_1$', $\ldots$, '$x_m$' being* U *variables for arbitrary obs, every ob $B$ is assertible in the resulting system.*

*Proof.* Let the conjunctive normal form of $A$ be (32), where for each $i$ there exist $e_1, \ldots, e_m$ such that

$$U_i \equiv E_1^{e_1} \vee E_2^{e_2} \vee \cdots \vee E_m^{e_m} \tag{39}$$

and, by the dual[2] of Theorem 13,

$$a_i = f(e_1, \ldots, e_m) \tag{40}$$

---

[1] The separate terms are actually disjoint by condition (b), and one can replace '$\vee$' by '$+$', hence the term 'disjunctive' is not inappropriate.

[2] In this dualization it is necessary to interchange the exponents 0 and 1 in (37a).

Then we have for every $i = 1, 2, \ldots, n = 2^m$,

$$A \leq a_i \vee U_i \tag{41}$$

Since $A$ is not a tautology, we can choose $i$ so that

$$a_i = 0$$

Then (41) becomes, by virtue of (39) and (40),

$$A \leq E_1^{e_1} \vee E_2^{e_2} \vee \cdots \vee E_m^{e_m}$$

Since the $E_1, \ldots, E_m$ are indeterminates, we have as theorem scheme of HK

$$\vdash f(x_1, \ldots, x_m) \supset x_1^{e_1} \vee x_2^{e_2} \vee \cdots \vee x_m^{e_m} \tag{42}$$

Hence by (38) and modus ponens, we have in the extension

$$\vdash x_1^{e_1} \vee x_2^{e_2} \vee \cdots \vee x_m^{e_m}$$

Here let $x_k \equiv B^{e_k}$; then by repeated applications of VW and (5) we have $\vdash B$, Q.E.D.

Thus, if HK is formulated with a substitution rule, every nondemonstrable elementary statement is absurd. The substitution rule is, however, essential.

**5. Boolean equations.** To the algebraic logicians of the nineteenth century, a problem of central importance was the development of a technique, similar to that which one has in elementary algebra, for manipulating the elementary statements of Boolean algebra. A sketch of this technique will be given here.

To begin with, an elementary statement of Boolean algebra can be expressed in any one of the following forms:

$$A = 0, A = 1, A = B, A \leq B, A = M$$

where $M$ is a fixed ob, given in advance, and $A$, $B$ are arbitrary obs. In fact the first, second, and fifth forms are special cases of the third; the third and fourth forms can be expressed in any form of the first four by means of the equivalences

$$A = B \rightleftarrows A + B = 0 \rightleftarrows A \sim B = 1 \rightleftarrows A \vee B \leq AB \tag{43}$$

$$A \leq B \rightleftarrows AB' = 0 \rightleftarrows A' \vee B = 1 \rightleftarrows A = AB \rightleftarrows B = A \vee B \tag{44}$$

while the first and second forms have the equivalences

$$A = 0 \rightleftarrows A' = 1 \rightleftarrows A \leq 0 \rightleftarrows A + M = M\dagger$$

$$\tag{45}$$

$$A = 1 \rightleftarrows A' = 0 \rightleftarrows 1 \leq A$$

One can therefore take any one of these five forms as fundamental; the analogy with ordinary algebra causes mathematicians, in general, to prefer the first.

In the second place, any set of simultaneous equations is equivalent to a single equation by virtue of the equivalences

$$A_1 = 0 \;\&\; A_2 = 0 \;\&\; \cdots \;\&\; A_n = 0 \rightleftarrows A_1 \vee A_2 \vee \cdots \vee A_n = 0 \tag{46}$$

† This is known as the 'law of forms'.

Suppose now one has an equation of the form

$$A \equiv f(x) = 0 \qquad (47)$$

where $f(x)$ is an ob constructed from $x$ and certain constants. Let these constants be obs in a Boolean algebra $K$. Then $A$ is an ob in the free extension $L$ of $K$ with the single generator $x$. If we put $A$ in normal form, (47) becomes

$$ax \lor bx' = 0 \qquad (48)$$

By (46) and (44), (48) is equivalent to the double inclusion

$$b \leq x \leq a' \qquad (49)$$

Hence (47) will have a solution just when

$$ab = 0$$

and then it will have as solution any $x$ satisfying (49). The solution may be expressed

$$x = b \lor ta' = b + ta'b' \qquad (50)$$

where $t$ is an arbitrary element of $K$.

Another method of treating the equation is to write

$$ax \lor bx' = ax + b(1 + x) = (a + b)x + b$$

Thus the equation is of the form

$$cx + b = 0$$

or

$$cx = b \qquad (51)$$

Hence, by $\Lambda K'$, $P_1$,

$$b \leq x \leq c \supset b = b \lor c' = b \lor a'b' \qquad (52)$$

where the last step comes by taking $c = a + b = ab' \lor a'b$, $c' = ab \lor a'b'$ and absorbing the $ab$ into $b$. The most general $x$ satisfying (52) is given by (50), and this satisfies (51) and hence (48).

Now suppose we have an equation

$$A \equiv f(x_1, \ldots, x_m) = 0 \qquad (53)$$

where $f(x_1, \ldots, x_m)$ is a construction from obs of $K$ and the indeterminates $x_1, \ldots, x_m$. We may suppose that $A$ is in normal form in the free extension of $K$ with respect to $x_1, \ldots, x_m$ as generators. Then we consider the following two questions:

1. (Elimination problem.) What conditions on the coefficients are necessary and sufficient for a solution to exist, i.e., that there be elements of $K$ which, when substituted for $x_1, \ldots, x_m$, satisfy (53)?

2. (Resolution problem.) Assuming the conditions of elimination fulfilled, how can one represent the solutions explicitly in a formula depending on certain parameters?

A solution of both these problems for $m = 1$ has just been given. For $m > 1$ one can, of course, solve for the variables one at a time. This is a laborious process, at least so far as problem 2 is concerned, as one can ascertain by working it out for the case $m = 2$; moreover, the results are quite different in appearance according to the order in which one chooses to solve for the unknowns. We shall go no further here with problem 2.

One can easily show, however, by induction on $m$, that the general solution of the problem 1 is

$$\alpha = 0 \tag{54}$$

where $\alpha$ is the meet of all the coefficients in the alternative normal form of $A$. In fact, this has been shown for $m = 1$. Suppose we have $m + 1$ unknowns $x_1, \ldots, x_m, y$ and that, for $n = 2^m$ (the $U_i$ being a basis (37) with $E_k \equiv x_k$),

$$A = \bigvee_{i=1}^{n} a_i U_i \qquad B = \bigvee_{i=1}^{n} b_i U_i$$

Then any equation (53) for $m + 1$ will be of the form

$$Ay + By' = 0 \tag{55}$$

This will have a solution for $y$ in terms of $x_1, \ldots, x_m$ if and only if

$$AB = 0$$

By Theorem 12 this is equivalent to

$$\bigvee_{i=1}^{m} a_i b_i U_i = 0$$

By the inductive hypothesis this is possible if and only if

$$a_1 a_2 \cdots a_m b_1 b_2 \cdots b_m = 0$$

This is, however, precisely the criterion (54) for the equation (55).

A slight generalization of this result is the following:

**Theorem 15.** *Let*

$$A \equiv f(x_1, \ldots, x_m) \equiv \bigvee_{i=1}^{n} a_i U_i$$

*where $n = 2^m$, be an ob in normal form of the free extension of $K$ with the $m$ generators $x_1, \ldots, x_m$. Let $\alpha$ be the meet and $\beta$ the join of all the coefficients in $A$. Then a necessary and sufficient condition that the equation*

$$f(x_1, \ldots, x_m) = t \tag{56}$$

*have solutions in $K$ is that*

$$\alpha \leq t \leq \beta \tag{57}$$

*Proof.* The equation (56) is, by (45), equivalent to

$$A + t = 0 \tag{58}$$

Let

$$A + t = \bigvee_{i=1}^{n} c_i U_i$$

where the right side is the alternative normal form of $A + t$. Let $\gamma \equiv c_1 c_2 \cdots c_n$. Then by property (a) for (58), a necessary and sufficient condition for the solubility of (58) is that

$$\gamma = 0 \tag{59}$$

But by the distributive law and Theorem 12,

$$c_i = a_i + t = a_i' t \vee a_i t'$$

Let $y$ be an additional indeterminate, and let $L_y$ be the free extension of $K$ formed by adjoining $y$ to $K$.   Let

$$C_i = a_i'y + a_iy'$$

Then by Theorems 12 (relative to $L_y$) and 8,

$$C_1C_2 \cdots C_n = \beta'y + \alpha y'$$

and hence, putting $t$ for $y$,

$$\gamma = \beta't + \alpha t'$$

Thus the condition (59) becomes

$$\beta't + \alpha t' = 0$$

For this, by the discussion of (48), it is necessary and sufficient that (57) hold, Q.E.D.

This discussion gives as much of the Boolean technique as is of general interest.   Special developments must be sought elsewhere.

### EXERCISES

See also Birkhoff [LTh$_2$], chap. 10.

**1.** Let $L$ be a complemented lattice such that if $C$ is any complement of $B$,

$$AB = 0 \to A \le C$$

Show that $L$ is a Boolean algebra.   (The formulation is, essentially, the second set of Huntington [SIP]; cf. the discussion of Sec. 4S1.)

**2.** Show that the axiom schemes of Whitehead and Russell [PMt], cited at the end of Sec. 2, in combination with both parts ($\le$ and $\ge$) of the quasi definitions (11) and (20), are sufficient for HK, but that if one omits either half of (11), it is not possible to derive any elementary theorem in which implication is the only operation. (Contributed by my colleague W. Howard.)

**3.** Show that the schemes of Whitehead and Russell [PMt] (see Exercise 2) admit the matrix interpretation

| V | 0 | 1 | 2 | ($\rceil$) |
|---|---|---|---|---|
| 0 | 0 | 1 | 2 | 1 |
| 1 | 1 | 1 | 1 | 0 |
| 2 | 2 | 1 | 2 | 1 |

with 1 as the only designated value.   How do you reconcile this fact with the theorems of Secs. 1 and 2 and Sec. 5A1?   (Huntington [NSI, p. 297].)

**4.** Huntington's fourth set of postulates for Boolean algebra contains the axiom schemes

$$A \vee B = B \vee A$$
$$(A \vee B) \vee C = A \vee (B \vee C)$$
$$(A' \vee B')' \vee (A' \vee B)' = A$$

Show that with suitable additions (so as to give properties of equality, including Rp, and quasi definitions) these characterize Boolean algebra.   (Huntington [NSI].   For the redundancy of his postulate 4.5 see his [BAC].)

**5.** Show that the axiom schemes (26) and the quasi definitions (21) and (27) are indeed sufficient for HK.   (Cf. Rosser [LMt, chap. 4].)

**6.** Show that the schemes

$$A = AB \rightleftarrows AB' = CC'$$
$$(AB)C = (BC)A$$

together with properties of equality and suitable quasi definitions, generate Boolean algebra. (Byrne [TBF].)

**7.** Show that the system formed by adjoining to HA the quasi definitions (22) and (23) and the schemes

$$\vdash \neg A \supset . A \supset B$$
$$\vdash \neg A \supset B . \supset . A \supset B . \supset B$$

are sufficient for HK (Porte [DSS]; the system is extracted from Tarski [FBM$_1$]).

**8.** Assuming $b \leq a$, show that the double inclusion

$$b \leq x \leq a$$

is equivalent to each of the following:

$$a'x \vee bx' = 0$$
$$a'x + bx' = 0$$
$$x = ax + bx'$$
$$(a' + b)x + b = 0$$
$$x = b \vee ta \qquad \text{for some } t$$
([LLA].) $\qquad x = at + bt' \qquad \text{for some } t$

**9.** Show that in every set of axiom schemes for HK there are at least three distinct U variables for propositions. (Announced in Łukasiewicz and Tarski [UAK, Theorem 15] as due to Wajsberg; see Tarski and Woodger [LSM, p. 47], where references to proofs are given. The most accessible proof is by Diamond and McKinsey [ASA].)

## S. SUPPLEMENTARY TOPICS

**1. Historical and bibliographical comment.** The present chapter extends to negation the general program initiated in Chap. 5. For comments and references concerning the program as a whole, see Secs. 5S1 and 1S5. The reasons for separating negation so drastically from the operations were given there as well as here in Sec. A1. The chapter is based on [TFD], chap. 4, and [LLA], chap. 5.

The possibility of defining two different kinds of negation corresponding to absurdity and refutability arose directly from the semantical approach. The distinction was already present in [PFD]. The name 'absurdity' was taken from the intuitionists because its formal properties coincide with those postulated for intuitionistic negation, e.g., in Heyting [FRI]. The name 'refutability' was taken from Carnap [ISm], who defined it in a way very similar to that given here (in his [LSL] he used the term in a different sense); what are here called "counteraxioms" appeared there as "directly refutable sentences." A similar notion, called "rejection," appears in Łukasiewicz [ASS]; it had appeared earlier in his [SAr]. There was almost certainly no connection between either of these approaches and that of [PFD]. Whether there was any connection between Carnap and Łukasiewicz I do not know.

Of the five types of formalized negation considered here, the J and the K types appeared in Gentzen's [ULS]; all the others, including the positive A and C types of Chap. 5, were introduced later as modifications.  The K type evolved from the work of logicians using the traditional approach (Sec. 5S1).  The J type represented intuitionistic logic, for which the presentations in Heyting [FRI] and Glivenko [PLB] had only recently appeared. Possibly the fact that Gentzen's [ULS] was written under the direction of Weyl had some influence in drawing Gentzen's attention to it.  The M type was the subject of Johansson [MKR]; in part, Kolmogorov [PTN] anticipated him, but I know that work only through remarks about it in Church [IML$_2$] and Feys [MRD].  Apparently the motivation was a feeling of dissatisfaction about efq [ = (6) of Sec. C], but Johansson explicitly mentions the fact that minimal negation can be defined as the property of implying a fixed but completely arbitrary (i.e., indeterminate) proposition $F$.  (There are some similar remarks in Moisil [RAL].)  Johansson also mentioned the system D (without so naming it) and suggested that it might form a system of "strict implication," but this system was not extensively studied until [TFD], where the semantical approach intimated that it might be suitable under circumstances in which we wish to have implication represent deducibility in the same sense that it does in LA, and yet to be platonistic to the extent of assuming a law of excluded middle.  The E type was suggested by Bernays in his review of [SLD] and [DNF], and some of its properties were studied from the standpoint of its H formulation in Kanger [NPP]; an extensive study of it from the standpoint of an L formulation was made by Kripke [SLE].  In correspondence Kripke has described several other systems of intermediate character; these remain for the future (see his abstract [DCn], which appeared while this was in process).

On the singular and multiple formulations the situation for LJ has already been discussed in Sec. 5S1.  For the other systems the question was in doubt for some time, and false leads were followed (e.g., in [SLD], sec. 3). It now seems clear that the singular systems are the more natural semantically; the multiple systems, which are much more natural from the standpoint of convenience in use, have to be justified by translation into the singular, or by auxiliary interpretation.  The results of this book on multiple systems other than LK are for the most part new, but have been anticipated by the Japanese work on LJ$_m$ cited in Sec. 5S1, and by Kripke [DCn].

The idea of defining negation in terms of implication and $F$ goes far back in the history of our subject.  I do not know its origin.  It was used in a paper by Russell in 1906 (cited in Church [IML$_2$, p. 157]).  It was implicit in Gentzen's formulation of his natural rules (here the T rules, see Sec. C1), but he did not use it in his L formulation.  Wajsberg, in his [MBt] and [UAK], studied H systems based on implication and $F$ (his symbol for it was '$O$').  The idea of so defining negation was used in [PFD], but it was not used for the L rules in [TFD] because of certain technical difficulties.  The equivalence between F and N formulations for the L formulations was first shown in [DNF].  The present treatment contains improvements over [DNF]; these were worked out in the spring of 1960.

The results on T and H systems in Sec. C are taken from [TFD] and [LLA]. The H formulations come mostly from the latter.  References to sources are given in both of those publications.  The remarks given here will supplement them.

The intuitionistic system HJ was formulated in Heyting [FRI]; formulations previous to that time, in Glivenko [PLB], Kolmogorov [PTN], etc. (see Sec. 4S2), were somewhat fragmentary. (The references on the positive part of the Heyting formulation in Sec. 5S1 give also some information about negation.) In the Heyting formulation the schemes for negation were separated. There were two such schemes, NS and efq. Johansson simply dropped efq in his formulation of HM. The formulation of HK in Hilbert and Bernays [GLM] had three prime assertion schemes for negation, namely, NB', NI', and $\neg\,\neg\, A \supset A$; it turned out that the omission of the last scheme from this list gave a formulation of HM. In 1938 Scholz proposed the problem of so formulating HK that (1) there should be exactly three prime schemes for negation, such that, in conjunction with HA, the first should give exactly HM and the first two exactly HJ, and (2) the resulting set of prime statements should be independent. Łukasiewicz [LGL] gave the first published solution of this problem; the three prime schemes are, respectively, NC, efq, and Exercise C9b. Bernays, in correspondence (see Hermes and Scholz), proposed as alternatives the schemes described in Exercises C3 and C9a. Wajsberg [UAK] used NW and efq (see Exercise C2) for HJ. Both Łukasiewicz and Wajsberg mention that they received an "Anregung" from Scholz and his collaborators; one should note also the historical remarks in Schröter [UHA], which is a belated publication of results known for a long time. Wajsberg [UAK], apparently misunderstanding a letter from Scholz, stated that NW was sufficient for HM; he corrected the error, substituting NC, in his [MBt.II], p. 139. What is here taken as the standard formulation of HK (and also the formulation of [TFD]) is a solution of Scholz's problem except for condition b of Exercise C9. Hermes and Scholz (loc. cit.) deny the possibility of an independent and separated set of prime assertion schemes for HK; Kanger [NPP] showed the contrary (Exercise C10).

The matters treated in Sec. D are much older than those of Secs. A to C which we have just been discussing, and the literature is much more extensive. The material in these sections was mostly known before 1930; it is taken over here with minor changes from [LLA], chap. 5, secs. 7 to 11. The aim is to present a central core which is of interest to every logician and to show its relation to other approaches. In order to discuss its history it is necessary to backtrack a little.

The development of logic to about 1930 was dominated by the traditional approach. In this development one can distinguish three principal directions. The first of these is the relational algebraic approach, which began (in principle) with Boole and forms the basis of modern lattice theory. The second is that leading to assertional deductive systems (H systems); this began with Frege and was continued by Russell and his [PMt]. The third direction is that of matrix interpretations (Sec. 5A4). These directions, although historically distinct, are not incompatible. From the present point of view they are equivalent, but this equivalence and the reasons for it were not recognized at the start. The inferential direction, with which this book is principally concerned, may be regarded as a fourth direction, which is an offshoot of the second.

From the first direction one is led to Boolean algebras and Boolean rings.

An account of the history in this direction was given in Sec. 4S1. Since it was impossible at that stage to separate negation completely, much that pertains to the present chapter was discussed there. In particular, the general references on lattice theory in Secs. 4S1 and 4S3 should be consulted for the developments of Boolean algebra which go beyond the scope of this book, particularly in those ways in which the developments diverge from the interests of logic. For an elementary introduction to Boolean algebra, the recommendation of Couturat [ALg] in Sec. 4S1 still applies. This book draws its inspiration from the older works cited in Sec. 4S1, but the interest in these older works is now chiefly historical. Another excellent introduction is Rosenbloom [EML, chaps. 1, 2]. This gives an exposition of the proof, by Frink, of the Stone representation theorem. It also has interesting comments on the transition to assertional systems and to relations between the two types. Its historical comments on pp. 194ff. give information of that sort not elsewhere readily available. Boolean algebra has recently found application to the design of electrical networks, and this has caused the appearance of a number of practical handbooks, but I have no detailed information about them.

The second direction gave rise to a great variety of formulations of HK. Since this was the dominant direction in logic, most of the general references in Sec. 1S, especially Sec. 1S5, give information about it. For elementary approaches see Sec. 1S1c. Church [IML$_2$] is rich in historical comments, particularly in regard to systems of prime assertions for HK.

In regard to the third direction, see Church [IML$_2$, pp. 161ff.] for the early history. Some modern authors, e.g., Quine, prefer to treat propositional algebra wholly from the point of view of 0-1 tautologies and show no interest in deductive treatments. For an elaboration of techniques from this standpoint see Quine [MeL], [MLg]. The matrix point of view has suggested several generalizations; for these see Sec. 2 below.

There is a great variety of ways in which Boolean algebra and HK can be formulated. For a survey of formulations from the H point of view see Church [IML$_2$, secs. 23-29]; for those from the E point of view see Birkhoff [LTh$_2$, chap. 10, secs. 3, 4]; Rosenbloom [EML, p. 194]. The formulations of Whitehead [UAl], Huntington [SIP], [NSI] (gives bibliography to 1933), Byrne [TBF] have been referred to rather frequently in the literature. Porte [DSS] gives a survey of systems based on P and N; his [SCP] does this for systems based on Λ and N. The latter type of system has interest on account of the theorem stated in Exercise B9; on this see also Łukasiewicz [ITD], which tells the fate of the Sobocinski system.

According to Huntington [NSI, footnote p. 278], the term 'Boolean algebra' is due to Sheffer [SFI].

On the properties of finite Boolean algebras see also Bernstein [FBA].

The normal form theorems in Sec. D4 are more or less standard theorems which go back, in principle, at least to Schröder [VAL]. Theorem D14 is due to Post [IGT].

The theorems on Boolean equations in Sec. D5 are all in Couturat [ALg] (actually they go back at least to Schröder [VAL]), except those relating to ring addition; for these see Birkhoff [LTh$_2$, sec. X9]. Couturat [ALg] also contains a brief account of a technique due to Poretskiĭ; this technique, which is said to be useful for certain purposes, is further developed in Blake [CEB]. Ledley, in his [CMS], [DCM], [DSF], [MFC], has proposed improvements in the technique with an eye to machine applications.

**2. Further developments.** A few topics will be mentioned here which, although they are related to the subject of this chapter, it was not possible to include.

There has been considerable study of the relations between HJ and HK, in particular to transformations which map the theorems of HK into those of HJ. The theorem of Glivenko (see Exercise B7) is an example of such a transformation; so also is that of Gödel and Kolmogorov (Exercise B9). The last has the further property that it can be extended to certain forms of arithmetic. For other theorems of this nature see Kleene [IMM, sec. 81], Shanin [LPA], Lukasiewicz [ITD].

By virtue of Theorem 5E5, the system HJ has the alternation property. This property generalizes to some extent; see, for example, Harrop [DES]. Łukasiewicz [ITD] conjectured that this property was characteristic for HJ. Kreisel and Putnam [UBM] prove that this conjecture is false in that if one adjoins to HJ the assertion scheme

$$\vdash \neg A \supset B \vee C . \supset . \neg A \supset B . \vee \neg A \supset C$$

one gets a system which is more inclusive than HJ and yet has the alternation property.

This brings up the question of logics intermediate between HJ and HK. Gödel [IAK] showed by the use of the matrix method (Sec. 5A4) that there were infinitely many such intermediate logics. Umezawa has studied these systematically (see his [IPL] and papers there cited).

The matrix method of approach has suggested generalizations going in different directions from those followed here. Post made a thorough study of two-valued matrices in his [TVI]. However, the most striking generalization is that to matrices with more than two elements. The systematic work in this field is Rosser and Turquette [MVL]; accessible pioneering works are Post [IGT] and Łukasiewicz and Tarski [UAK]; for the history see Church [IML$_2$, pp. 161ff.]; for surveys see, for example, Łukasiewicz [LGL], Frink [NAL]. Relational systems ("Post algebras," "Moolean algebras") for this field have also been studied; see, for example, Rosenbloom [PAl], Chang [AAM] and publications there cited.

Gödel [IAK], already cited, showed that HJ is not identical with any logic generated by a matrix with a finite number of values. The proof is expounded in Schmidt [VAL, sec 141]. However, an infinite matrix representation is given in Jaśkowski [RSL].

The equivalence between E and H formulations has been stressed in this book. Rosenbloom [EML] also makes this same point. However, there are jokers about this equivalence. When the conditions of Sec. 5A1 are violated, strange things may happen. Thus Hiż [ESC] exhibits an assertional system in which every proposition is an assertion which is an assertion of HK, yet the system is not Post-complete. The system is not an H system. This sheds light on the significance of the hypotheses in Theorem D14.

The treatment of the technique of Boolean algebra in Sec. D was very brief. The older works, already cited, developed many other topics. One of these was a theory of inequations, which depends on introducing an additional predicate of inequality or nonnullity. This is of some interest because in it one can give a treatment of traditional logic (syllogisms, etc.). For a very brief account of this, see [LLA, chap. 6].

These developments and those mentioned in Sec. 4S are samples of related investigations which have been made. Many others exist. Those which involve nonconstructive semantical considerations are beyond the scope of this book.

In regard to Boolean algebra, a treatise, Sikorski [BA1], has appeared.

# Chapter 7

# QUANTIFICATION

The preceding two chapters have been concerned with propositional connectives, i.e., operations which combine propositions to form other propositions. In these operations the propositions are taken as unanalyzed wholes. When the theory is interpreted[1] in the epitheory of an underlying formalism $\mathfrak{S}$, these propositions are formed from the elementary statements of $\mathfrak{S}$ without regard to how those elementary statements are themselves formed in terms of the formal objects of $\mathfrak{S}$. In other words, $\mathfrak{S}$ can be an arbitrary deductive theory, provided its rules are elementary in the sense (of Sec. 2D3) that they can be formulated in the form of (1) of Sec. 5C.

In this chapter we shall study ways of expressing generality and related notions. This requires operations of rather a different nature from those used in the preceding chapters. Such operations are traditionally called *quantifiers*.[2] They require that the underlying theory $\mathfrak{S}$ be actually a system, and that we consider two types of obs which we shall call *propositions* and *terms*. In the interpretation in the epitheory of an underlying $\mathfrak{S}$, the terms correspond to the obs of $\mathfrak{S}$. Then the quantified propositions

$$(\forall x)A \qquad (\exists x)A$$

are interpreted as epistatements of $\mathfrak{S}$ saying, respectively, that $A$ is true for all terms $x$ and that $A$ is true for some terms $x$.

The treatment in this chapter will parallel, at least partially, that in Chap. 6. In Sec. A we shall inquire more deeply into the meaning and nature of quantifiers, with the aim of arriving at a formulation of L systems of various sorts containing these operations. Since the semantical difficulties in regard to quantification are rather less than in the case of negation, but the formal difficulties are much greater, most of the space will be devoted to formal matters, and we shall end with an actual formalization. In Sec. B the fundamental epitheorems concerning these L systems will be derived. In Sec. C we shall study the relations between the L systems and the more usual systems of predicate calculus—the T and H formulations. The last section, Sec. D, will be devoted to theorems peculiar to the classical systems.

---

[1] More strictly, evaluated.

[2] The term is due to Peirce, the idea to Frege. For details concerning the history, see Church [IML₂, sec. 49].

The superscript '*' will be used henceforth to indicate systems with quantification. Thus LA*, LC*, LK*, HK* will be the systems formed by adjoining suitable postulates for quantification (as described later) to LA, LC, LK, HK, respectively. The L systems of this chapter may be referred to collectively as L* systems. An unspecified one of these L* systems will be called a system LX*; the 'X' can then be replaced by any one of the letters 'A', 'C', 'D', 'E', 'J', 'K', 'M'.

## A. FORMULATION

We shall begin, in Sec. 1, with a semantical study; the purpose will be to specify more definitely the meaning of the quantified propositions

$$(\forall x)A \qquad (\exists x)A \qquad (1)$$

when interpreted as epistatements concerning an underlying formalism $\mathfrak{S}$. The formal difficulties connected with the "bound variables" so introduced will be our next concern. This requires a meticulous formulation of certain details concerning occurrence of variables, substitution, etc.; these details are tedious, but they are necessary for exactness. This will lead up to the precise formulation of various L* systems in Sec. 5.

**1. Semantical study.** The nature of generalization in connection with the epitheory of a system $\mathfrak{S}$ has already been discussed in Sec. 3A3. We are concerned here with the type of generalization which is there called schematic. Thus, in interpretation, the first statement (1) is true just when $A$ is a statement scheme depending on the term parameter $x$ which becomes true whenever a term is substituted for $x$; in other words, whenever $A$ is a theorem of the extension $\mathfrak{S}(x)$ formed by adjoining $x$ to $\mathfrak{S}$ as an adjoined indeterminate.

The second of the propositions (1), interpreted as an epistatement, is to mean that $A$ is true for some term $t$. This we interpret as meaning that $A$ is a statement of the extension $\mathfrak{S}(x)$ for which there is a term $t$ of $\mathfrak{S}$ such that $A$ becomes a true statement $A'$ of $\mathfrak{S}$ when $t$ is substituted for $x$.

Certain technicalities concerning the bound variables will be deferred until later. Apart from these technicalities we have, in principle, an interpretation for the propositions (1) considered as epistatements relative to $\mathfrak{S}$.

Next let us consider the semantics of the propositions (1) when they appear as constituents in a deducibility epistatement of the form

$$\mathfrak{X} \Vdash B \qquad (2)$$

As in Sec. 5C1, this amounts to asking under what circumstances a constituent of the form (1) may be introduced into (2), in other words, when a statement (2) containing a constituent (1) may be introduced into discourse. There are two cases to consider, viz., when the new constituent is on the right and when it is on the left. In both of these cases we shall suppose that $A'$ is the result of substituting $t$ for $x$ in $A$.

Let us first consider introduction on the right. As in Sec. 5C1, we interpret (2) as meaning that $B$ is true in a system $\mathfrak{S}(\mathfrak{X})$ formed by adjoining $\mathfrak{X}$ to $\mathfrak{S}$. If $B$ is $(\forall x)A$, this will be true just when $x$ is an indeterminate for $\mathfrak{S}(\mathfrak{X})$, so that $x$ does not occur in $\mathfrak{X}$, and $A$ is true in the system $\mathfrak{S}(\mathfrak{X},x)$ formed by adjoining $x$ to $\mathfrak{S}(\mathfrak{X})$. If $B$ is $(\exists x)A$, epistatement (2) will be true just when there is a term $t$ such that $A'$ is true in $\mathfrak{S}(\mathfrak{X})$. The formal rules for introduction on the right, viz.,

$$\mathfrak{X} \Vdash A \rightarrow \mathfrak{X} \Vdash (\forall x)A \tag{3}$$

$$\mathfrak{X} \Vdash A' \rightarrow \mathfrak{X} \Vdash (\exists x)A \tag{4}$$

subject to the indicated restrictions, express just these principles.

To treat introduction on the left we have to consider the interpretation of

$$\mathfrak{X}, C \Vdash B$$

where $C$ is one of the propositions (1). This amounts to asking under what circumstances we conclude, in the presence of $\mathfrak{X}$, that $B$ is a consequence of $C$. We can conclude this, of course, if $B$ is the same as $C$ or is such that the inference can be made without regard to the nature of $C$ (i.e., schematically, with $C$ as parameter). In the nontrivial cases we use the method of Secs. 5A3 and 5C1. The general principle of that method is that the rule for introducing an operation on the right determines the meaning of that operation in the following sense: if $C$ is so introduced by a rule $R$, then the consequences of $C$, wherever it occurs, are the same as when it was first introduced and are to be determined by examining the possible premises for $R$.[1] Thus if $C$ is $(\forall x)A$, then $C$ can be introduced into $\mathfrak{S}(\mathfrak{X})$ only when $A$ is true in $\mathfrak{S}(\mathfrak{X};x)$, and in that case there is a proof $\mathfrak{D}_1$ (of some sort)[2] terminating in $A$; if now $B$ is in $\mathfrak{S}(\mathfrak{X},A')$, then we have a proof tree $\mathfrak{D}_2$ terminating in $B$ and having $A'$ as a premise, so that by putting over each occurrence of $A'$ a proof of it obtained from $\mathfrak{D}_1$ by substituting $t$ for $x$ throughout $\mathfrak{D}_1$, we have a proof that $B$ is in $\mathfrak{S}(\mathfrak{X})$. Thus the rule

$$\mathfrak{X}, A' \Vdash B \rightarrow \mathfrak{X}, (\forall x)A \Vdash B \tag{5}$$

is semantically acceptable. On the other hand, if $C \equiv (\exists x)A$, then $C$ is true for $\mathfrak{S}(\mathfrak{X})$ only when there is a tree $\mathfrak{D}_1$ terminating in some $A'$; if now there is proof $\mathfrak{D}_2$ terminating in $B$ and having a premise $A$, and neither $B$ nor any other premise of $\mathfrak{D}_2$ contains $x$, then by substituting $t$ for $x$ throughout $\mathfrak{D}_2$ we have a proof of $B$ from $A'$. Thus the rule

$$\mathfrak{X}, A \Vdash B \rightarrow \mathfrak{X}, (\exists x)A \Vdash B \tag{6}$$

is semantically acceptable, subject to the indicated restrictions on $x$.

These form the basis of the rules given formally in Sec. 5.

**2. Formal difficulties.** Although the interpretations given in Sec. 1 are fairly straightforward, they involve certain complexities due to the fact that quantification requires bound variables. Some of these complexities were discussed in Sec. 3D4. It is desirable to bring them up again here with particular reference to the present context.

In order to iterate the rules of Sec. 1 with respect to introduction of quantifiers, it is necessary to consider cases of (2) where $\mathfrak{X}$ and $B$ contain other

---

[1] Thus, if $R$ is a one-premise rule, it is semantically invertible. This is quite different from the formal invertibility of Theorem 5D1. That theorem shows agreement of the formal theory with the intended interpretation.

[2] This being a discussion of motivation, we do not have to go into the exact specification of such a proof. We can suppose it is like the T proofs of Sec. 5A3.

adjoined indeterminates besides $x$.    In other words, we have to consider (2) with reference to a term extension $\mathfrak{S}(\mathfrak{a})$ where $\mathfrak{a}$ is a set of indeterminates, and we may indicate this dependence by writing (2) as

$$\mathfrak{X} \mid \mathfrak{a} \vdash B$$

The terms $t$ which may be substituted for $x$ may thus contain adjoined indeterminates, and hence we may run into the situation, already commented on in Sec. 3D4, known as confusion of bound variables.    Thus if $\mathfrak{S}$ is a suitable formulation of elementary number theory,

$$(\exists y) . x < y$$

is a theorem scheme in which $x$ is an indeterminate; yet if we were to substitute $y$ for $x$ naively, we should have

$$(\exists y) . y < y$$

which is false.

The phenomenon of confusion of bound variables shows that conditions of substitution have to be formulated with some care.    There are various methods of doing this.    One very drastic method is that of combinatory logic (Sec. 3D5); this eliminates bound variables altogether and shows how expressions involving bound variables can be defined in terms of combinators in such a way that the rules for manipulating bound variables can be inferred from the definitions.[1]    Inasmuch as this answers the questions of principle, we can use a method which is advantageous from the standpoint of convenience.    The method adopted here is essentially that used in Hilbert and Bernays [GLM, vol. 1].    The formal variables used for terms—here called *term variables*—are divided into two classes; in formal developments those of the first class are used for free variables, those of the second for bound variables.    The classes are here called *real* and *apparent variables*, respectively, so as to leave the words 'free' and 'bound' available for describing occurrences of variables; in the discussion of rules it is sometimes necessary to have apparent variables occurring free.[2]    The letters '$a$', '$b$', '$c$', . . . will be used for real variables, and '$x$', '$y$', '$z$' for apparent variables; the letters '$u$', '$v$', '$w$' will be used for variables which may be either real or apparent.

This convention is of some help.    But it is still necessary to formulate notions of substitution, free and bound occurrence, etc., with great care.    After a digression we shall return to this formulation in Sec. 4.

**3. The B language.**    On account of the complexity of the analysis of variables it is necessary to introduce considerable technical terminology into the U language.    This will be explained as we proceed, but it will help to clarify matters if we first survey it as a whole.

The situation in which we find ourselves requires some amplification of the fundamental grammatical conventions of Secs. 2A3 and 2A4.    Whereas in Chaps. 5 and 6 we were talking—to put it naively—about two fundamentally

---

[1] See, for example, [CLg], sec. 6D.

[2] The terms 'real variable' and 'apparent variable' were used in "Principia mathematica," but seem to have fallen out of use.    Hence it is permissible to introduce them in this technical sense.

different kinds of things, viz., propositions (i.e., obs) and statements made about them, we here have to talk—in the same naive sense—about three different kinds, viz., terms, propositions, and statements.    In our A language there will be three basic grammatical categories: term nouns, proposition nouns, and sentences; for the time being we shall call these $\tau$, $\pi$, $\sigma$, respectively.    Thus $\tau$ will contain the *primitive constants*, i.e., names of specific atoms of $\mathfrak{S}$; the *term variables*, which are names of indeterminates to be adjoined to $\mathfrak{S}$ to form *term extensions;* as well as more complex phrases designating specific terms.    The category $\pi$ will contain the names of specific propositions however they may be formed; when our systems are interpreted in the way intended, these will become statements about $\mathfrak{S}$, but this fact is not relevant to the purely formal considerations.    The category $\sigma$ will contain the sentences expressing the statements which we assert, deny, or otherwise consider.

In addition, we need functors of various kinds.    Those which, in the intended interpretation, become the operators of $\mathfrak{S}$ will belong to the categories $F_1\tau\tau$, $F_2\tau\tau\tau$, etc. (depending on their degree); these will be called *term operators*, or *term functors*.[1]    Those which, in the interpretation, become predicators of $\mathfrak{S}$ belong to categories such as $F_1\tau\pi$, $F_2\tau\tau\pi$, etc.; the name '*predicator*' will be reserved henceforth for functors of this type, and '*predicate*' for their designata.[2]    Functors of the type we have considered hitherto —such as the infixes $\supset$, $\wedge$, $\vee$—belong to the category $F_2\pi\pi\pi$ (and negation to $F_1\pi\pi$); these will be called *propositional operators*, the word 'propositional' being generally omitted.    This name will also be applied to quantifiers, although a quantifier strictly belongs to the category $F_2\tau\pi\pi$—more accurately (cf. Sec. 3D4) to $F_1(F_1\tau\pi)\pi$.    Finally, functors which form sentences will be called henceforth *verbs*, or *sentential* (or *statement*) *functors*, and their designata, *verbal* (or *statement*) *functions*, properties, relationships, etc.

All this discussion concerns the A language.    For epitheoretic purposes we need in the U language also U variables ranging over various categories; names for categories and subcategories, and U variables for them; verbs for expressing epitheoretic relationships; etc.    The totality of this technical terminology, which is necessary for expressing not only the elementary statements, but the rules, morphology, and certain epitheoretic properties, constitutes a language which will be known as the *B language*.

The principal symbols of the B language are exhibited in Table 3.    Here column 1 gives the names of various categories into which these symbols are classified.    Column 2 lists the constants, classified according to category, which constitute the A language of the system $\mathfrak{S}*$ formed by adjoining an infinite set of term variables to $\mathfrak{S}$.    Column 3 lists symbols used as proper names of the categories in column 1.    Column 5 lists U variables for unspecified members of those categories, and column 6, U variables for unspecified subclasses of those categories.    Column 4 exhibits ways of indicating specific subclasses, depending on parameters taken from column 6.    All the

---

[1] An alternative name, for which there is some justification in the literature, is 'descriptive functor'.

[2] There is a disagreement in the literature in regard to this usage, which identifies 'predicate' with 'propositional function'.    Some authors, e.g., Church, prefer to confine 'predicate' to statement functions or functors (see Church [IML$_2$, p. 289]).

TABLE 3

| Name of category (1) | U constants | | | U variables | |
|---|---|---|---|---|---|
| | Elements (2) | Classes (3) | Subclasses (4) | Elements (5) | Subclasses or sequences (6) |
| Primitive constants | $e_1, e_2, \ldots$ | $\mathfrak{e}$ | | | |
| Term variables | $q_1, q_2, \ldots$ | q | | $u, v, w$ | $\mathfrak{u, v, w}$ |
| Primitive term operators | $\omega_1, \omega_2, \ldots$ | $\Omega$ | | | |
| Primitive predicators | $\phi_1, \phi_2, \ldots$ | $\Phi$ | | | |
| Real variables | | $\mathfrak{r}$ | $\mathfrak{h}$ | $a, b, c, f, g, h$ | $\mathfrak{a, b, c, g}$ |
| Apparent variables | | $\mathfrak{s}$ | | $x, y, z$ | $\mathfrak{x, y, z}$ |
| Terms (obs of $\mathfrak{S}^*$) | | t | t(u) | $s, t$ | |
| Null class of q | | $\mathfrak{o}$ | | | |
| Elementary propositions | $E_1, E_2, \ldots, F$ | $\mathfrak{E}$ | $\mathfrak{E}(u)$ | | |
| Propositions | | $\mathfrak{P}$ | $\mathfrak{P}(u), \mathfrak{F}$ | $A, B, C, D$ | $\mathfrak{X, Y, Z, U, B, M}$ |
| Axioms | | $\mathfrak{A}$ | $\mathfrak{A}(u)$ | | |
| Elementary theorems | | $\mathfrak{S}$ | $\mathfrak{S}(u)$ | | |
| Theorems | | $\mathfrak{T}$ | $\mathfrak{T}(\mathfrak{X},u), \mathfrak{T}(u)$ | | |
| Null class or prosequence | | $0$ | | | |
| Null system | | $\mathfrak{O}$ | | | |

symbols may be used as nouns, but certain of them may be used, in combination with parentheses and commas, as functors, according to ordinary mathematical usage (cf. Sec. 2A4). The terminology has been chosen so as to be consistent, in so far as possible, with the convention that German letters denote classes or sequences of which the members are denoted by italic letters of the same kind.

The verbs of the B language include those necessary to state the elementary and auxiliary statements of the various systems, the morphological statement forms

$u$ occurs in $t$

$u$ occurs free in $A$

$u$ occurs bound in $A$

as well as relations among classes. There will also be the three-place operations whose closures

$$[s/u]t \qquad [s/u]A$$

designate the result of substituting $s$ for $u$ in $t$ or $A$, respectively. Various connectives from Sec. 2A4, as well as additional phrases necessary to state the rules in Sec. 5, will also be regarded as belonging to the B language, but it is hardly necessary to be explicit about these.

In connection with classes, the following special conventions will be observed. The infixes '$\epsilon$', '$\subset$', '$=$' will be used in their ordinary senses (cf. Sec. 2A4) of membership, inclusion, and class equality, respectively; the infix '$\equiv$' is used, as heretofore, for identity by definition. A name for a prosequence appearing in the position of a class name will be understood as designating the class of propositions having one or more occurrences as constituent in the prosequence, and similarly for other names of structures. The union of two or more classes will be indicated (unless it is necessary to be more explicit in order to avoid confusion) by writing their names in a series separated by commas; also classes containing only one element are not distinguished from those elements themselves. For example, '$\mathfrak{a}$, $\mathfrak{b}$, $c$' will designate the class whose elements consist of $c$ together with the elements of $\mathfrak{a}$ or $\mathfrak{b}$ or both.

The letters '$i$', '$j$', '$k$', '$l$', '$m$', '$n$', and sometimes (when the context prevents confusion) '$p$', '$q$', '$r$', '$s$', are used for natural numbers. The operational and relational symbols of arithmetic will have their ordinary senses in that connection.

The B language is not the same as the A language of any of the systems LA, LJ, TA, etc. The latter is obtained by adding to column 2 phrases sufficient to make particular elementary statements. Although we attempt to be precise as to the use of the B language, yet we do not attempt either to formalize it or to exhaust the possibilities of the U language in it.

**4. Rules for terms and propositions.** These rules relate to the formulation of the system $\mathfrak{S}$ and its various term extensions, to the constitution of the class $\mathfrak{P}$ of propositions in relation to these term extensions, and to the rules for substitution and occurrence of variables.

PRIMITIVE IDEAS OF $\mathfrak{S}^*$. The system $\mathfrak{S}^*$ is that formed by adjoining to $\mathfrak{S}$ an infinite class q of *term variables*. Its primitive ideas are then as follows:

Atoms of $\mathfrak{S}$, (e):   $e_1, e_2, \ldots$ $\Big\}$   Atoms of $\mathfrak{S}^*$
Term variables (q): $q_1, q_2, \ldots$ 
Primitive operations $\omega_1, \omega_2, \ldots$   Of degrees $m_1, m_2, \ldots$, respectively
Primitive predicates $\phi_1, \phi_2, \ldots$   Of degrees $n_1, n_2, \ldots$, respectively

All except the term variables are primitive ideas of $\mathfrak{S}$.

FORMULATION OF $\mathfrak{S}(\mathfrak{u})$. If $\mathfrak{u}$ is a subclass of q, then $\mathfrak{S}(\mathfrak{u})$ is the system obtained by adjoining to $\mathfrak{S}$ just the members of $\mathfrak{u}$. Then $\mathfrak{S}(\mathfrak{q})$ is $\mathfrak{S}^*$ and $\mathfrak{S}(\mathfrak{o})$ is $\mathfrak{S}$. The formulation is as follows:

I. *Terms*

(a) Every element of $e$ is in $t(\mathfrak{u})$.

(b) Every element of $\mathfrak{u}$ is in $t(\mathfrak{u})$.

(c) If $t_1, t_2, \ldots, t_{m_k}$ are in $t(\mathfrak{u})$, so is $\omega_k(t_1, t_2, \ldots, t_{m_k})$.

II. *Elementary propositions* ($\mathfrak{E}(\mathfrak{u})$). [In the interpretation these are the elementary statements of $\mathfrak{S}(\mathfrak{u})$.]

1. The propositions $E_1, E_2, \ldots, F$ are in $\mathfrak{E}(\mathfrak{u})$. (This convention is similar to that in Chaps. 4 and 5, but we are here insisting that these be constants.) The $E_1, E_2, \ldots, F$ are propositions not otherwise specified; they play the role of propositional variables in $\mathfrak{D}$.

2. If $t_1, t_2, \ldots, t_{n_k}$ are in $t(\mathfrak{u})$, then

$$\phi_k(t_1, t_2, \ldots, t_{n_k})$$

is in $\mathfrak{E}(\mathfrak{u})$.†

---

† The $E_i$ may be regarded as the special case where $n_k = 0$.

III. *Auxiliary statements.* The rules and axioms of $\mathfrak{S}$ are specified by statement schemes of the form

$$A_1, \ldots, A_m \,|\, a \vdash_0 B \tag{7}$$

where $a$ is a class of real variables, and all the $A_1, \ldots, A_m$, $B$ are in $\mathfrak{E}(a)$. These statements are called auxiliary statements because they are analogous to the statements so called in Sec. 5C3. If $m = 0$, $B$ will be called an *axiom*.

The definition of a system $\mathfrak{S}$ may also specify counteraxioms $\mathfrak{F}$. These may also depend on a class $a$ of real variables.

In the void system $\mathfrak{O}$ there will be no statements of the form (7) and no counteraxioms.

The following five definitions define formally the notion of proposition, and also notions related to occurrence of variables and substitutions for them. This is a highly technical matter. Some readers will doubtless prefer to take them and the theorems based on them as given intuitively by the notation.

DEFINITION 1. (Occurrence of an atom in a term.) If $u$ is a term variable (or more generally any atom) and $t$ is in $\mathfrak{t}(\mathfrak{q})$, then

$$u \text{ occurs in } t$$

is defined by induction on the structure of $t$ as follows:
  (i) If $t \equiv u$, then $u$ occurs in $t$.
  (ii) If $t$ is an atom distinct from $u$, then $u$ does not occur in $t$.
  (iii) If $t \equiv \omega_k(t_1, \ldots, t_{m_k})$, then $u$ occurs in $t$ just when it occurs in one or more of the $t_i$.

DEFINITION 2. (Substitution in a term.) If $s$ and $t$ are terms and $u$ is a term variable, or more generally an atom, then

$$[s/u]t$$

is defined by induction on the structure of $t$ as follows:
  (i) If $t \equiv u$, $[s/u]t \equiv s$.
  (ii) If $t$ is an atom distinct from $u$, $[s/u]t \equiv t$.
  (iii) If $t \equiv \omega_k(t_1, \ldots, t_{m_k})$, then

$$[s/u]t \equiv \omega_k(t'_1, \ldots, t'_{m_k})$$

where                    $t'_i \equiv [s/u]t_i \qquad i = 1, 2, \ldots, m_k$

DEFINITION 3. (Propositions, and occurrences of variables in them.) If $u$ is a term variable and $\mathfrak{u}$ is a class of such variables, the statement forms

$$A \text{ is in } \mathfrak{P}(\mathfrak{u})$$

$$u \text{ occurs free in } A$$

$$u \text{ occurs bound in } A$$

are defined simultaneously by induction on the structure of $A$ as follows:
  (i) If $A$ is in $\mathfrak{E}(\mathfrak{u})$, then $A$ is in $\mathfrak{P}(\mathfrak{u})$. If $A$ is some $E_i$ or $F$, then no variable occurs, free or bound, in $A$. If

$$A \equiv \phi_k(t_1, \ldots, t_{n_k})$$

then $u$ occurs free in $A$ just when $u$ occurs in one or more of the $t_i$; no variable occurs bound in $A$.

(ii) If $A$ is $B \supset C$, $B \wedge C$, or $B \vee C$, then $A$ is in $\mathfrak{P}(\mathfrak{u})$ just when $B$ and $C$ are both in $\mathfrak{P}(\mathfrak{u})$. The variables which occur free in $A$ are just those which occur free in $B$ or $C$ or both; likewise the variables which occur bound in $A$ are just those which occur bound in $B$ or $C$ or both.

(iii) If $A$ is $\neg B$, then $A$ is in $\mathfrak{P}(\mathfrak{u})$ just when $B$ is, and the free (bound) variables of $A$ are the same as those of $B$.

(iv) If $A$ is $(\forall x)B$ or $(\exists x)B$, then $A$ is in $\mathfrak{P}(\mathfrak{u})$ just when $B$ is in $\mathfrak{P}(\mathfrak{u},x)$ and $x$ does not occur bound in $B$†; we may or may not add the further restriction that $x$ occur free in $B$ (without this restriction we are said to admit *vacuous quantification*, otherwise not). The variables occurring free in $A$ are those which are distinct from $x$ and occur free in $B$; those occurring bound in $A$ are those occurring bound in $B$ together with $x$.

DEFINITION 4.    A *real term* is a term belonging to $\mathfrak{t}(\mathfrak{r})$; a *real proposition* is one belonging to $\mathfrak{P}(\mathfrak{r})$. Likewise a *constant term* is one belonging to $\mathfrak{t}(\mathfrak{o})$, and a *constant proposition* is one belonging to $\mathfrak{P}(\mathfrak{o})$.

DEFINITION 5.    For each $A$ in $\mathfrak{P}(\mathfrak{q})$, $s$ in $\mathfrak{t}(\mathfrak{q})$, and $u$ in $\mathfrak{q}$, we define

$$[s/u]A$$

by induction on the structure of $A$ as follows:

(i) If $A$ is $E_i$ or $F$, $[s/u]A \equiv A$.

(ii) If

$$A \equiv \phi_k(t_1, \ldots, t_{n_k})$$

then

$$[s/u]A \equiv \phi_k(t'_1, \ldots, t'_{n_k})$$

where

$$t'_i \equiv [s/u]t_i \qquad i = 1, 2, \ldots, n_k$$

(iii) If $A$ is $B \circ C$, where '$\circ$' stands for one of the infixes '$\supset$', '$\wedge$', '$\vee$', then

$$A' \equiv B' \circ C'$$

where

$$B' \equiv [s/u]B \qquad C' \equiv [s/u]C$$

(iv) If $A$ is $\neg B$, and $B'$ is as in (iii), then

$$[s/u]A \equiv \neg B'$$

(v) If $A$ is $(\forall x)B$ or $(\exists x)B$, then

$$[s/x]A \equiv A$$

If $u \not\equiv x$, and $s$ is a real term,[1] or if $s$ is an apparent variable distinct from $x$, then

$$[s/u](\forall x)B \equiv (\forall x)B'$$
$$[s/u](\exists x)B \equiv (\exists x)B'$$

where $B'$ is as in (iii).

† This is an optional restriction; we adopt it because it is convenient.

[1] This is the only place, up to the present, where the distinction between real and apparent variables is relevant. If we did not wish to make this distinction, we could introduce at this point the restriction that $x$ should not be free in $s$; this could be supplemented by a provision for automatic changing of bound variables (cf. [CLg], sec. 3E) if we wanted to have substitution always defined. The restriction made here allows some simplifications to be made later.

*Remark* 1.    Note that these conventions may not define $[s/u]A$ if $A$ contains bound variables and $s$ is a composite term containing apparent variables, or if $s$ is a variable bound in $A$.

The following theorems can be proved by structural induction on the $A$. The proofs are omitted; they are rather tedious, but straightforward.[1]

**Theorem 1.**    *The class* $\mathfrak{P}(\mathfrak{u})$ *has the following properties:*

(i) *The truth of any statement of the forms defined in Definitions 1 and 3 is a definite question.*

(ii) *The class* $\mathfrak{P}(\mathfrak{u})$ *is monotone increasing with* $\mathfrak{u}$; *i.e.,*

$$\mathfrak{u} \subseteq \mathfrak{v} \rightarrow \mathfrak{P}(\mathfrak{u}) \subseteq \mathfrak{P}(\mathfrak{v})$$

(iii) *If $A$ is in* $\mathfrak{P}(\mathfrak{q})$ *and* $\mathfrak{u}$ *is the class of all variables which occur free in $A$, then $A$ is in* $\mathfrak{P}(\mathfrak{u})$.

**Theorem 2.**    *The substitution operation has the following properties:*

(i)                           $$[u/u]A \equiv A$$

(ii) *If $u$ does not occur free in $A$,*

$$[s/u]A \equiv A$$

(iii) *If $A$ is in* $\mathfrak{P}(\mathfrak{u},w)$, $s$ *is in* $\mathfrak{t}(\mathfrak{v})$, *then* $[s/w]A$, *if defined, is in* $\mathfrak{P}(\mathfrak{u},\mathfrak{v})$.

(iv) *If $s$ is in* $\mathfrak{t}(\mathfrak{u})$, $t$ *is in* $\mathfrak{t}(\mathfrak{v})$, $u \not\equiv v$, *and neither $u$ is in $v$ nor $v$ in $\mathfrak{u}$, then*

$$[s/u][t/v]A \equiv [t/v][s/u]A$$

(v) *If $s$ is in* $\mathfrak{t}(\mathfrak{a})$, $t$ *is in* $\mathfrak{t}(\mathfrak{b})$, *and* $b$ *is not in* $\mathfrak{a}$,

$$[s/a][t/b]A \equiv [t'/b][s/a]A$$

*where*                        $$t' \equiv [s/a]t$$

These theorems are also true if we substitute $t$ for $A$ and $\mathfrak{t}$ for $\mathfrak{P}$. The theorems so altered will be called Theorems 1' and 2', respectively.

ASSUMPTIONS CONCERNING $\mathfrak{S}$.    Besides the assumptions inherent in the above formulations, the following assumptions will be made:

A1.    *The class* $\mathfrak{e}$ *is not void.*

A2.    *The auxiliary statements and counteraxioms are invariant of substitution;* i.e., when such a statement contains term variables, then a statement formed by substituting real terms for those variables is also an auxiliary statement or a counteraxiom.

The first of these assumptions corresponds to the assumption, usually made in ordinary treatments of the predicate calculus, that the domain of "individuals" is nonvoid.    There has been a certain amount of interest in the removal of this restriction.    Presumably some of the results obtained by those who have studied this matter could be applied to derive a theory without Assumption A1, but that is not investigated here.

*Remark* 2.    The representation of substitution will often be abbreviated as follows.    Let $A \equiv A(u)$, then

$$A(s) \equiv [s/u]A(u)$$

---

[1] The proof of part (iv) of Theorem 2 is given in detail in [TFD], p. 72, footnote 8.

*Remark* 3. The formulation leaves open a number of possibilities in regard to $\mathfrak{S}$, in particular in regard to $\Phi$. Of course, if $\Phi$ were void, no variable would occur free in any proposition and the situation would be trivial. Thus $\Phi$ must be nonvoid. Beyond that requirement, however, there is still some latitude. In some systems certain of the $\phi_i$ are indeterminate, in the sense that the primitive frame says nothing about them explicitly, so that they enter into an axiom or rule instance only by specialization of some U variable. Such a $\phi_i$ is called a *predicate variable;* a $\phi_i$ which is mentioned specifically is called a *predicate constant.* If $\mathfrak{S}$ is $\mathfrak{O}$ there are, of course, no predicate constants in this sense. If $\mathfrak{S}$ is such that there is an unlimited number of predicate variables of every degree, it will be said to have *unrestricted predicate variables.* Where predicate variables exist, there will be a derived rule of substitution in connection with them (see Exercise 4 at the end of this section).[1]

*Remark* 4. Remarks similar to Remark 3 can be made in regard to other constituents of the primitive frame for $\mathfrak{S}$. In particular, if the e are indeterminates, they do not differ in any essential way from those term variables which are not used as characteristic variables,[2] and it is a matter of arbitrary choice whether they are listed as such and included in the range or they are listed in e. Let us call them e *variables.* Likewise, an $E_i$ which is not an indeterminate may be called a *primitive propositional constant.*

*Remark* 5. It is customary to call an *applied predicate calculus* one in which there are primitive term constants, primitive propositional constants, or predicate constants or term operations; a *pure predicate calculus,* one without these features. If we identify constants as in Remarks 3 and 4, then a pure predicate calculus is one in which $\mathfrak{S}$ is $\mathfrak{O}$ and $\Omega$ is void. However, these expressions are sometimes understood in other senses.[3]

**5. L\* systems.** We now consider the formulation of systems, called L\* systems, which are analogous to the L systems of the two preceding chapters, but are modified to allow for the presence of quantifiers and term extensions. These systems will be designated by adding a superscript '\*' to the name for the corresponding system in Chaps. 5 and 6; we thus have systems LA\*, LC\*, LM\*, LJ\*, LD\*, LE\*, LK\* and their various formulations. In discussions related to such systems in general, the letter 'X' will be understood as standing for any of 'A', 'C', 'D', 'E', 'J', 'K', 'M'; thus the systems of this chapter are the systems LX\*, their T forms TX\*, etc.

PROSEQUENCES. The definitions of Sec. 5C3c are carried over with only the obvious changes. We say that $u$ occurs free in $\mathfrak{X}$ just when $u$ occurs free in one or more constituents of $\mathfrak{X}$, and that $u$ occurs bound in $\mathfrak{X}$ just when it occurs bound in one or more such constituents. Likewise, $[s/u]\mathfrak{X}$ will be the prosequence formed from $\mathfrak{X}$ by replacing every constituent $A$ of $\mathfrak{X}$ by $[s/u]A$.

---

[1] Systems of predicate calculus have been proposed in which such a rule of substitution is taken as a primitive rule. Such systems are not considered here, and there seem to be good reasons for avoiding them (see Henkin [BRS]). If they were to be admitted, it would be better to change 'predicate variable' of the text to 'predicate indeterminate'; 'predicate variable' and 'predicate constant' would then have other meanings.

[2] For the terms 'characteristic variable' and 'range', see Sec. 5.

[3] Cf. the preceding footnote.

AUXILIARY STATEMENTS. These are defined in the formulation of $\mathfrak{S}(\mathfrak{u})$ in Sec. 1. They are subject to the restriction A2.

ELEMENTARY STATEMENTS. These are now of the form

$$\mathfrak{X} \mid \mathfrak{a} \vdash \mathfrak{Y} \tag{8}$$

where $\mathfrak{a}$ is a class of real variables, and

$$\mathfrak{X} \subseteq \mathfrak{P}(\mathfrak{a}) \qquad \mathfrak{Y} \subseteq \mathfrak{P}(\mathfrak{a}) \tag{9}$$

Explicitly, if $\mathfrak{X}$ is $A_1, \ldots, A_m$, $\mathfrak{Y}$ is $B_1 \ldots, B_n$, and $\mathfrak{a}$ is $a_1, \ldots, a_p$, then (8) is

$$A_1, \ldots, A_m \mid a_1, \ldots, a_p \vdash B_1, \ldots, B_n \tag{10}$$

The class $(\mathfrak{a})$ will be called the *range* of (8). When it is not necessary for explicitness, indication of the range is frequently omitted.

We extend definitions of occurrence and substitution to elementary statements as follows. We say $u$ occurs free in (8) just when $u$ occurs free in $\mathfrak{X}$ or $\mathfrak{Y}$ or both; likewise $u$ occurs bound in (8) just when it occurs bound in $\mathfrak{X}$ or $\mathfrak{Y}$ or both. If $\Gamma$ is the statement (8) and $s \, \epsilon \, \mathfrak{t}(\mathfrak{b})$, then $[s/a]\Gamma$ is the statement

$$\mathfrak{X}' \mid \mathfrak{a}', \mathfrak{b} \vdash \mathfrak{Y}' \tag{11}$$

where $\mathfrak{a}'$ is the class formed from $\mathfrak{a}$ by deleting $a$, and

$$\mathfrak{X}' \equiv [s/a]\mathfrak{X} \qquad \mathfrak{Y}' \equiv [s/a]\mathfrak{Y} \tag{12}$$

Note that if (9) holds, the analogous condition for (11) is automatically fulfilled by Theorem 2.

PRIME STATEMENTS. These are the same as before, with the additional stipulation that the range $\mathfrak{a}$ is any range satisfying (9).

DERIVATIONAL RULES. The rules for the finite operations, as given in Secs. 5C and 6B, were stated under the supposition that all propositions, auxiliary statements, etc., are related to a fixed system $\mathfrak{S}$. With the understanding that (8) is to be interpreted as stating

$$\mathfrak{X} \Vdash \mathfrak{Y}$$

with reference to $\mathfrak{S}(\mathfrak{a})$, these rules are now postulated for any $\mathfrak{a}$. This means that the range is the same in all the premises as in the conclusion; that all auxiliary statements, counteraxioms, etc., relate to $\mathfrak{S}(\mathfrak{a})$; and that new propositions, such as are introduced by $*K*$, and non-Ketonen forms of $*\Lambda$ and $V*$, are propositions of $\mathfrak{S}(\mathfrak{a})$ [i.e., are in $\mathfrak{P}(\mathfrak{a})$]. These rules will be called henceforth the *algebraic rules* to distinguish them from the new rules presently to be introduced.

In addition, on the supposition that

$$\mathfrak{X}, \mathfrak{Y}, 3 \subseteq \mathfrak{P}(\mathfrak{a})$$
$$t \, \epsilon \, \mathfrak{t}(\mathfrak{a})$$
$$A(c) \, \epsilon \, \mathfrak{P}(\mathfrak{a},c) \qquad B(c) \, \epsilon \, \mathfrak{P}(\mathfrak{a},c)$$
$$c \text{ is not in } \mathfrak{a}$$
$$x \text{ is not bound in } A(c) \text{ or } B(c)$$

we have the following *quantification rules:*

Π    *Universal quantification* (or generalization)

$$*\Pi \quad \frac{\mathfrak{X}, A(t) \mid \mathfrak{a} \vdash \mathfrak{Y}}{\mathfrak{X}, (\forall x)A(x) \mid \mathfrak{a} \vdash \mathfrak{Y}} \qquad \Pi* \quad \frac{\mathfrak{X} \mid \mathfrak{a}, c \vdash B(c), \mathfrak{Z}}{\mathfrak{X} \mid \mathfrak{a} \vdash (\forall x)B(x), \mathfrak{Z}}$$

Σ    *Existential quantification* (or instantiation)

$$*\Sigma \quad \frac{\mathfrak{X}, A(c) \mid \mathfrak{a}, c \vdash \mathfrak{Y}}{\mathfrak{X}, (\exists x)A(x) \mid \mathfrak{a} \vdash \mathfrak{Y}} \qquad \Sigma* \quad \frac{\mathfrak{X} \mid \mathfrak{a} \vdash B(t), \mathfrak{Z}}{\mathfrak{X} \mid \mathfrak{a} \vdash (\exists x)B(x), \mathfrak{Z}}$$

SINGULARITY RESTRICTIONS. In the multiple forms of LA*, LM*, LJ*, and LD*, Π* is restricted to be singular; otherwise there is no restriction.

*Remark* 1. The rules *Π and Σ* could have been formulated without use of the variable c. In the rules Π* and *Σ, however, the variable c plays an essential role; it will be called the *characteristic variable* of the inference.

*Remark* 2. The conventions of Sec. 5C6 apply to these rules, and it is clear that the new rules have the properties (r1) to (r5). However, the rules Π* and *Σ do not have the property (r6), since the new parametric constituent might contain the characteristic variable. Furthermore, the quantification rules do not have the composition property, so that the consequences drawn from that property for systems LX do not necessarily hold for systems LX*—in fact, the analogue of Corollary 5E9.1 is known to be false. We shall see that they do have a generalized form of the composition property, and that from this some of the other results of Sec. 5E follow.

*Remark* 3. In a statement (8), the prosequences $\mathfrak{X}$ and $\mathfrak{Y}$ have only a finite number of constituents, and the range $\mathfrak{a}$ is also finite. It would be possible to admit infinite prosequences and ranges, provided certain restrictions were fulfilled, but this will not be done here.[1] It is supposed, of course, that q, r, s are all infinite.

*Remark* 4. The conventions are such that whenever a statement of form (8) can be derived, then (9) holds. For this it is only necessary to postulate (9) for the prime statements. Since the rules preserve this property, it then follows that all demonstrable statements are elementary statements in the above sense.

*Remark* 5. There is an obvious duality between the rules Π* and *Σ, on the one hand, and *Π and Σ*, on the other. This duality breaks down in certain respects, notably in that Π* may be singular on the right but *Σ is never singular on the left. Nevertheless, it is possible to abbreviate proofs by saying that the proof for one member of these dual pairs is obtained from that for the other by duality.

## EXERCISES

**1.** Show that

$$(\forall x)(\forall y)A \Vdash (\forall y)(\forall x)A$$
$$(\exists x)(\forall y)A \Vdash (\forall y)(\exists x)A$$
$$(\exists x)(\exists y)A \Vdash (\exists y)(\exists x)A$$
$$(\forall x)(\forall y)A(x,y) \Vdash (\forall x)A(x,x)$$
$$(\exists x)A(x,x) \Vdash (\exists x)(\exists y)A(x,y)$$

[1] In [TFD] such possibilities were admitted but hardly used. A "finiteness restriction," stated there on p. 74, was necessary for quantification theory.

**2.** Show that bound variables can be changed arbitrarily, so long as the rules for being members of $\mathfrak{P}$ are followed.

**3.** Show that it is sufficient to have predicates of degrees 1 and 2 in $\Phi$. (This was one of the theorems of Löwenheim [MRK]. Much stronger results of this nature are now known.)

**4.** State and prove a theorem concerning the substitution of a propositional function with certain arguments for a predicate variable with the same arguments, in such a way as to avoid confusion of bound variables. (For the history of this, see Church [rev. HA]. For solutions and further discussion see Hilbert and Bernays [GLM.I], Church [IML₂], Zubieta [SVF], Henkin [BRS].)

## B. THEORY OF THE L* SYSTEMS

The L* systems were defined in Sec. A5. In this section theorems analogous to those in Secs. 5C to 5E and 6B will be proved. This will begin, in Sec. 1, with theorems concerning the extensions and other forms of weakening principles. We then proceed to the inversion theorem in Sec. 2. The elimination theorem, and other theorems of Sec. 5D in which little change occurs in the resulting theorem (though there may be more in the proof), will be treated together in Sec. 3. The deducibility theorems of Secs. 5E1 to 5E4 will be treated in Sec. 4. In Sec. 5 there will be a discussion of classical valuations and the use of them to give constructive proofs of non-demonstrability. The final subsection, Sec. 6, will treat proof tableaux.

**1. Theorems on extensions and substitution.** The first difficulty to be overcome is that $\Pi*$ and $*\Sigma$ do not have the property $(r6)$. This is because the new parameter might contain the characteristic variable, and that would invalidate the inference. It is therefore necessary to establish some results which show that this property can, in principle at least, be restored. These theorems will also show that substitution for the free variables can be made.

The argument begins with two theorems of a preliminary nature. The first is called a lemma because it will be superseded by Theorem 3.

LEMMA 1.    *Let $\Delta$ be a regular demonstration* (Sec. 5C7) *terminating in*

$$\mathfrak{X} \mid \mathfrak{a} \vdash \mathfrak{Y} \tag{1}$$

*and let $a \in \mathfrak{a}$. Let $s \in \mathfrak{t}(\mathfrak{b})$, where $\mathfrak{b}$ is a class of real variables not containing any characteristic variable of $\Delta$. Let $\Delta'$ be the sequence of statements obtained from those of $\Delta$ by substituting $s$ for $a$ throughout. Then $\Delta'$ is a regular demonstration.*

*Proof.*    Let the statements of $\Delta$ be $\Gamma_1, \ldots, \Gamma_n$, where $\Gamma_n$ is (1). Since the range of the conclusion of any rule is never larger than that of every premise, it follows that $a$ must occur in the range of every $\Gamma_k$; moreover, since characteristic variables drop out, $a$ is not a characteristic variable in $\Delta$. It follows that $\Gamma_k$ is of the form

$$\mathfrak{X}_k \mid \mathfrak{a}_k, a \vdash \mathfrak{Y}_k$$

where $a$ is not in $\mathfrak{a}_k$. Let the corresponding statement in $\Delta'$ be $\Gamma'_k$; then $\Gamma'_k$ is

$$\mathfrak{X}'_k \mid \mathfrak{a}_k, \mathfrak{b} \vdash \mathfrak{Y}'_k$$

where

$$\mathfrak{X}'_k \equiv [s/a]\mathfrak{X}_k \qquad \mathfrak{Y}'_k \equiv [s/a]\mathfrak{Y}_k$$

To prove the lemma we have only to show that ($a$) if $\Gamma_k$ is prime, so is $\Gamma_k'$; ($b$) if $\Gamma_k$ follows from $\Gamma_i$, $\Gamma_j$, ... by a rule $R$, then $\Gamma_k'$ follows from $\Gamma_i'$, $\Gamma_j'$, ... by the same $R$. In either case we can ignore the condition (9) of Sec. A by virtue of Remark 4 of Sec. A5.

The proof of statement $a$ is clear if $\Gamma_k$ is of type ($p1$). If $\Gamma_k$ is of type ($p2$), then statement $a$ follows by Assumption A2. It is therefore sufficient to prove statement $b$.

The proof of statement $b$ is clear if $R$ is one of the rules of Chaps. 5 and 6 which do not involve any auxiliary premise, for the transformation from $\Delta$ to $\Delta'$ is a homomorphism with respect to these rules. For rules with an auxiliary premise, namely, $\vdash_*$ and $F_*$, this is still true by Assumption A2.

If $R$ is $\Pi_*$, the inference must be

$$\frac{\mathfrak{X}_i \mid \mathfrak{a}_i, a, c \vdash B(c), \mathfrak{Z}_i}{\mathfrak{X}_i \mid \mathfrak{a}_i, a \vdash (\forall x)B(x), \mathfrak{Z}_i}$$

The transformed inference is

$$\frac{\mathfrak{X}_i' \mid \mathfrak{a}_i, \mathfrak{b}, c \vdash B'(c), \mathfrak{Z}_i'}{\mathfrak{X}_i' \mid \mathfrak{a}_i, \mathfrak{b} \vdash [s/a](\forall x)B(x), \mathfrak{Z}_i'}$$

where

$$B'(c) \equiv [s/a]B(c)$$
$$\mathfrak{Z}_i' \equiv [s/a]\mathfrak{Z}_i$$

Since, by the conventions of Sec. A4 (part v of Definition 5 and Remark 2) and Theorem A2 (part v),

$$[s/a](\forall x)B(x) \equiv [s/a](\forall x)[x/c]B(c)$$
$$\equiv (\forall x)[s/a][x/c]B(c)$$
$$\equiv (\forall x)[x/c][s/a]B(c)$$
$$\equiv (\forall x)B'(x)$$

the inference is a valid application of $R$.

The proof of statement $b$ for the case where $R$ is $*\Sigma$ is dual to this.

If $R$ is $\Sigma_*$, the inference must be

$$\frac{\mathfrak{X}_i \mid \mathfrak{a}_i, a \vdash B(t), \mathfrak{Z}_i}{\mathfrak{X}_i \mid \mathfrak{a}_i, a \vdash (\exists x)B(x), \mathfrak{Z}_i}$$

The transformed inference is

$$\frac{\mathfrak{X}_i' \mid \mathfrak{a}_i, \mathfrak{b} \vdash [s/a]B(t), \mathfrak{Z}_i'}{\mathfrak{X}_i' \mid \mathfrak{a}_i, \mathfrak{b} \vdash [s/a](\exists x)B(x), \mathfrak{Z}_i'}$$

Now let $\mathfrak{b}$ be a real variable not in $\mathfrak{a}_i, \mathfrak{b}$† and not appearing as characteristic variable in $\Delta$.   Let

$$B'(\mathfrak{b}) \equiv [s/a]B(\mathfrak{b})$$
$$t' \equiv [s/a]t$$

Then, by Sec. A4 (Remark 2, part v of Theorem A2, part v of Definition 5),

$$[s/a]B(t) \equiv [s/a][t/b]B(\mathfrak{b}) \equiv [t'/b]B'(\mathfrak{b}) \equiv B'(t')$$
$$[s/a](\exists x)B(x) \equiv [s/a](\exists x)[x/b]B(\mathfrak{b}) \equiv (\exists x)B'(x)$$

† Cf. Sec. A5, Remark 3.

Hence the inference is again a correct application of $R$. For the case where $R$ is $*\Pi$, we proceed dually. This completes the proof of Lemma 1.

**Theorem 1.** *Let $\Gamma$ be an elementary theorem of LX\* and $\mathfrak{g}$ an infinite subclass of $\mathfrak{r}$. Then there exists a derivation $\Delta$ of $\Gamma$ such that the characteristic variables of $\Delta$ are distinct from one another and belong to $\mathfrak{g}$.*

*Proof.* By hypothesis there exists a derivation $\Delta'$ of $\Gamma$. The theorem will be proved by induction on the length of $\Delta'$.

If this length is 1, then $\Gamma$ is prime. Since there are no characteristic variables, $\Delta'$ is itself the desired $\Delta$. This disposes of the basic step of the induction. It therefore suffices to prove the theorem when $\Gamma$ is obtained by a rule $R$ from premises $\Gamma_1, \Gamma_2, \ldots, \Gamma_p$ for which the theorem is already proved.

If $R$ is a rule which does not have any characteristic variable, let $\mathfrak{g}_1$, $\mathfrak{g}_2, \ldots, \mathfrak{g}_p$ be mutually exclusive subsets of $\mathfrak{g}$. By the inductive hypothesis there exist derivations $\Delta_1, \Delta_2, \ldots, \Delta_p$ of $\Gamma_1, \Gamma_2, \ldots, \Gamma_p$, respectively, such that the characteristic variables of each $\Delta_i$ are in the corresponding $\mathfrak{g}_i$. Then $\Delta_1, \Delta_2, \ldots$, followed by $R$ to deduce $\Gamma$ from $\Gamma_1, \ldots, \Gamma_p$, will give the $\Delta$ sought. This takes care of all cases except that where $R$ is $\Pi*$ or $*\Sigma$.

Finally, suppose $R$ is $\Pi*$ or $*\Sigma$. Let the premise be $\Gamma_1$ and $c$ the characteristic variable. Let $g \in \mathfrak{g}$, and let $\mathfrak{g}'$ be an infinite subclass of $\mathfrak{g}$ which does not include $c$ or $g$. By the inductive hypothesis there is a derivation $\Delta'$ terminating in $\Gamma_1$ and having its characteristic variables in $\mathfrak{g}'$. Let $\Delta_1$ be constructed from $\Delta_1'$ as in Lemma 1 with $g$ for $s$, $g$ as the sole member of $\mathfrak{b}$, and $c$ for $a$. Then $\Delta_1$ will terminate in a $\Gamma_1'$, from which $\Gamma$ can be obtained by the same $R$ except that $g$ takes the place of $c$. Such a derivation will be the desired $\Delta$, Q.E.D.

**Theorem 2.** *If (1) holds and $\mathfrak{b}$ is any finite subclass of $\mathfrak{r}$ such that*

$$\mathfrak{X} \subseteq \mathfrak{P}(\mathfrak{b}) \qquad \mathfrak{Y} \subseteq \mathfrak{P}(\mathfrak{b}) \tag{2}$$

*then*

$$\mathfrak{X} \mid \mathfrak{b} \vdash \mathfrak{Y} \tag{3}$$

*Proof.* Let $\Gamma$ be the statement (1) and $\Delta$ a derivation of it. The theorem will be proved by induction on the length of $\Delta$.

If $\Gamma$ is prime and (2) holds, then (3) is also prime. This disposes of the basic step of the induction. It therefore suffices to suppose that $\Gamma$ is obtained by a rule $R$ from premises $\Gamma_1, \Gamma_2, \ldots$ for which the theorem is already proved.

Suppose $R$ is such that all variables which occur free in any premise also occur in the conclusion. This includes all the rules $*C*$, $*K*$, $*W*$, $*P*$, $*\Lambda*$, $*V*$, $*N*$, Fj. In such a case, if (2) holds for the conclusion, the analogous inclusion holds for all the premises. By the inductive hypothesis the premises all hold if the range is changed to $\mathfrak{b}$. Then (3) follows by $R$.

If $R$ is one of the rules $\Pi*$ or $*\Sigma$, then all variables occurring free in the premises, excepting only the characteristic variable, occur free in the conclusion. Let this characteristic variable be $c$. Then if (2) holds for the conclusion, the analogue of (2) will hold for the premise if we take $\mathfrak{b}$, $c$ as the range. By the inductive hypothesis the premise holds if the range is changed to $\mathfrak{b}$, $c$. From the altered premise we again have (3) by $R$.

The remaining possibilities are $\vdash*$, Px, Nx, F*, $*\Pi$, and $\Sigma*$. In all these cases variables may occur free in the premise(s) which do not occur free in the conclusion. By Assumption A1 there is an element $e_1$ of $\mathfrak{e}$. Let $a$ be a variable which occurs free in some premise but not in the conclusion. By Lemma 1 we can derive the premise with $e_1$ substituted for $a$ and $a$ deleted from the range. We can continue in this way until all variables which occur free in any premise also occur free in the conclusion. Since the conclusion can still be derived by the same rule $R$, the theorem follows by the case treated in the second preceding paragraph.

*Remark* 1. The theorem would not be true as stated without Assumption A1. For if $A(a)$ is a proposition in which $a$ and no other variables explicitly appear, then we can easily derive

$$(\forall x)A(x) \mid a \vdash (\exists x)A(x) \tag{4}$$

On the other hand, if $\mathfrak{t}(\mathfrak{o})$ is void, we cannot derive

$$(\forall x)A(x) \mid \mathfrak{o} \vdash (\exists x)A(x) \tag{5}$$

If $\mathfrak{b}$ is not void, one can use some $b$ in $\mathfrak{b}$ instead of $e_1$, but in that case it might be necessary to involve Theorem 1 in order to satisfy the hypotheses of Lemma 1.

*Remark* 2. The theorem allows us to generalize rules which, like $*K*$ and the original forms of $*\Lambda$ and $V*$, introduce a new component $B$ in the conclusion, to cases where $B$ contains variables not in the range $\mathfrak{a}$ of the premises. For if $B \in \mathfrak{P}(\mathfrak{b})$, we use the theorem to change the range of the premises to $\mathfrak{a}, \mathfrak{b}$ and apply the original rules to draw the conclusion with range changed to $\mathfrak{a}, \mathfrak{b}$. A similar remark applies if $B$ is an additional parameter adjoined to rules satisfying $(r6)$.

**Theorem 3.** *If $\Gamma$ is a true statement* (1) *and $s \in \mathfrak{t}(\mathfrak{b})$, then $[s/a]\Gamma$ is also true.*

*Proof.* Let $\Gamma'$ be $[s/a]\Gamma$ (for its definition see the specifications for elementary statements in Sec. A5). If $a$ is not in $\mathfrak{a}$, then $\Gamma'$ is

$$\mathfrak{x} \mid \mathfrak{a}, \mathfrak{b} \vdash \mathfrak{Y}$$

and this follows from $\Gamma$ by Theorem 2. Suppose, then, that $a \in \mathfrak{a}$. By Theorem 1 there is a derivation $\Delta$ of $\Gamma$ such that none of the characteristic variables of $\Delta$ are in $\mathfrak{b}$. The theorem then follows by Lemma 1, Q.E.D.

These theorems show how we may reinstate the property $(r6)$. That property evidently holds for the rules $*\Pi$ and $\Sigma*$, and it holds for $\Pi*$ and $*\Sigma$, provided the characteristic variable does not occur free in the new parameter. If that condition is not met, then Theorem 3 shows that we can replace the characteristic variable by one which satisfies the condition. We thus have the following:

COROLLARY 3.1. *The rules $*\Pi*$ and $*\Sigma*$ have the property $(r6)$, and hence are regular, provided that the characteristic variable, if any, is changed so as not to occur free in the new parameter.*

The quantifier rules also fail to satisfy $(r7)$ in the strict sense. But they do satisfy $(r7)$ in a modified sense, for the $A(x)$ in the conclusion differs from the $A(c)$ or $A(t)$ in the premise only in changes of its terms.

**2. The inversion theorem.** We now study the effect of the introduction of quantification on the inversion theorem. Except for obvious adaptations to the present situation, we use the same notational conventions as in Secs. 5D1 and 6B2.

The inversion theorem itself—Theorem 5D1 and its modification, Theorem 6B1—was stated and proved for a general L system; and consequently, in view of the results of Sec. 1, it holds for L* systems also. Furthermore, since no subaltern in an algebraic operational rule can contain a variable not present in the principal constituent, there can be no conflict with the characteristic variable of a quantification rule under condition $(c)$. Hence the impact of the conditions $(a)$ to $(f)$ of these theorems on the algebraic operational rules is not affected by the presence of quantification rules. We have only to consider what these conditions allow us to say about the quantification rules themselves.

Before we consider these complications there are two preliminary matters to be disposed of. In the first place we may suppose, by virtue of Theorem 1, that the characteristic variables in $\Delta$ are distinct from one another and from the variables in the range of $\Gamma$. In the second place we note that there can be several parametric ancestors of $M$ only when there is a branching of $\Delta_1$ or an application of *W*.

Now suppose that $R$ is $\Pi*$ and that $M$ is $(\forall x)A(x)$. Then the original ancestors of $M$ which are introduced by $R$ are replaced by $A(c_1)$, $A(c_2)$, ..., $A(c_r)$, where the $c_1$, ..., $c_r$ are distinct characteristic variables; we may suppose that those introduced by K* are replaced in the same way with variables suitably chosen. It is evident that condition $(b)$ of Theorem 5D1 is not satisfied because of the possible differences in the characteristic variables. However, suppose we proceed down $\Delta_1$ as in Sec. 5D1. Where two or more branches come together, or *W* is applied to two instances of $M$, we may use Theorem 3 to identify the different $c_i$ in the branches, substituting one of them for each of the others in the appropriate branches. If we carry this all the way down to the bottom, we shall have replaced all the $c_1$, ..., $c_r$ by the same $c$. Condition $(b)$ will then be fulfilled in principle. Since all the variables other than $c$ were present in the original $\Delta_1$, and since the $c_i$ are distinct from the other characteristic variables, condition $(c)$ is also fulfilled. Thus $\Pi*$ will be directly invertible. There may be conflict with condition $(a)$ in those systems where $\Pi*$ is required to be singular, but otherwise it does not make any difference whether the system is absolutely or classically based.

What has been said about $\Pi*$, except the remark about possible failure of condition $(a)$, applies by duality to *$\Sigma$.

Next, suppose that $R$ is $\Sigma*$ and that $M$ is $(\exists x)A(x)$. Then the subalterns which replace the ultimate quasi-parametric ancestors of $M$ introduced by $R$ will be $A(t_1)$, $A(t_2)$, ..., $A(t_r)$. Again condition $(b)$ may not be fulfilled. We can restore condition $(b)$ by using the entire set $A(t_1)$, $A(t_2)$, ..., $A(t_r)$ in place of each $A(t_i)$, and we can use this same set also where $M$ is introduced by K* (including Fj considered as a form of K*). But it may happen that some $t_i$ contains a characteristic variable of $\Delta_1$. In such a case condition $(c)$ may not be fulfilled. An example of this will be given later (Example 1). If condition $(c)$ is fulfilled, then we need at the end an inference of the form

$$\Sigma'* \qquad \frac{\mathfrak{X} \mid \mathfrak{a} \vdash A(t_1), A(t_2), \ldots, A(t_r), \mathfrak{Z}}{\mathfrak{X} \mid \mathfrak{a} \vdash (\exists x)A(x), \mathfrak{Z}}$$

This may be obtained by $r$ applications $\Sigma*$ together with applications of $W*$; it will be a single rule if we replace $\Sigma*$ by $\Sigma'*$, which may be regarded as analogous to the Ketonen form of $V*$.

The case where $R$ is $*\Pi$ is handled dually. The rule analogous to $\Sigma'*$ is $*\Pi'$.

$$*\Pi' \qquad \frac{\mathfrak{X}, A(t_1), \ldots, A(t_r) \mid \mathfrak{a} \vdash \mathfrak{Y}}{\mathfrak{X}, (\forall x)A(x) \mid \mathfrak{a} \vdash \mathfrak{Y}}$$

These considerations prove the following:

**Theorem 4.** *The rules $\Pi*$ and $*\Sigma$ are directly invertible in all systems. If $*\Pi$ and $\Sigma*$ are replaced by $*\Pi'$ and $\Sigma'*$, then a direct inversion may be carried out unless there is conflict with characteristic variables of some $\Pi*$ or $*\Sigma$ in $\Delta_1$. The situation with respect to inversion of the algebraic rules is unchanged.*

*Example 1.* $(\forall x)A(x) \Vdash (\forall x).A(x) \lor B$. This can be established in $LA_m^*$ thus:

$$\frac{\dfrac{A(c) \mid c \vdash A(c) \lor B}{(\forall x)A(x) \mid c \vdash A(c) \lor B} *\Pi}{(\forall x)A(x) \Vdash (\forall x).A(x) \lor B} \Pi*$$

If $R$ is $*\Pi$, condition $(c)$ is not satisfied because $A(c)$ contains the characteristic variable of the following $\Pi*$. Kleene [PIG, p. 25] shows that it cannot be derived, even in $LK*$, with $*\Pi$ last.

*Example 2.* $A(a) \lor A(b) \mid a, b \vdash (\exists x)A(x)$. This can be derived in LA thus:

$$\frac{\dfrac{A(a) \mid a, b \vdash A(a)}{A(a) \mid a, b \vdash (\exists x)A(x)} \Sigma* \qquad \dfrac{A(b) \mid a, b \vdash A(b)}{A(b) \mid a, b \vdash (\exists x)A(x)} \Sigma*}{A(a) \lor A(b) \mid a, b \vdash (\exists x)A(x)} *V$$

Kleene gives this as an example of nonpermutability in $LA_1$. It can, however, be permuted in $LA_m$, thus:

$$\frac{\dfrac{A(a) \mid a, b \vdash A(a), A(b) \qquad A(b) \mid a, b \vdash A(a), A(b)}{A(a) \lor A(b) \mid a, b \vdash A(a), A(b)} *V}{A(a) \lor A(b) \mid a, b \vdash (\exists x)A(x)} \Sigma'*$$

**3. Other basic theorems.** We now consider the analogues of the theorems of Secs. 5C to 5E in which the changes to adapt them to the presence of quantifiers are rather slight. The situation is summed up in the following:

**Theorem 5.** *The theorems of Sec. 5C, the elimination theorem, the equivalence of singular and multiple formulations (Theorems 5D7 and 5D8 and 6B5), and the equivalence of different formulations of negation (Theorem 6B4), all hold for the systems $LX*$.*

*Proof.* Theorems 5C1 to 5C3 hold without change; and as in Sec. 6B1 we pay no further attention to distinctions between Formulations 1 and II. In Theorem 5C4 we need to consider the additional cases

$$\frac{\dfrac{A(b) \mid a, b \vdash A(b)}{(\forall x)(A(x)) \mid a, b \vdash A(b)} *\Pi}{(\forall x)A(x) \mid a \vdash (\forall x)A(x)} \Pi*$$

and the dual of this argument for $(\exists x)A(x)$.

In the proof of Stages 1 and 2 of the elimination theorem, we can avoid conflict due to the possibility of different ranges by Theorem 2. Conflict with respect to characteristic variables can be avoided by Theorem 1. In fact we can suppose, to begin with, that the characteristic variables in either premise of ET are distinct from one another and from all variables appearing in the conclusion. Thus the rules are regular [in the modified sense of ($r7$)]. As for Stage 3, we require two additional cases as follows:

CASE $\Pi$.    $A$ is $(\forall x)B(x)$.    The premises are

$$\mathfrak{X}_1, (\forall x)B(x) \mid \mathfrak{a} \vdash \mathfrak{Y} \tag{6}$$
$$\mathfrak{X}_2 \mid \mathfrak{a} \vdash (\forall x)B(x), \mathfrak{Z} \tag{7}$$

and the conclusion is

$$\mathfrak{X}_1, \mathfrak{X}_2 \mid \mathfrak{a} \vdash \mathfrak{Y}, \mathfrak{Z} \tag{8}$$

By the hypothesis of the stage, (6) and (7) are obtained by $*\Pi$ and $\Pi*$, respectively, from

$$\mathfrak{X}_1, B(t) \mid \mathfrak{a} \vdash \mathfrak{Y} \tag{9}$$
$$\mathfrak{X}_2 \mid \mathfrak{a}, c \vdash B(c), \mathfrak{Z} \tag{10}$$

where the characteristic variable $c$ may be chosen so as not to occur in (9) either. From (10) and Theorem 3,

$$\mathfrak{X}_2 \mid \mathfrak{a} \vdash B(t), \mathfrak{Z}$$

Eliminating $B(t)$ from this and (9), we have (6).

CASE $\Sigma$.    $A$ is $(\exists x)A(x)$.    This may be handled dually.

In principle, no change is necessary in the extension of ET to the mixed systems. We treat all cases where $A$ has the form of the principal constituent of a rule $R$ which is singular on the right in the same way as we treated the case where $A$ was $B \supset C$ in Sec. 5D2. Also, such a rule $R$ plays a role similar to that of P$*$ in the rest of the argument of Sec. 5D2.

In regard to the equivalence between the singular and multiple systems (Theorems 5D7, 5D8, and 6B5), this theorem takes the following form: A necessary and sufficient condition that

$$\mathfrak{X} \mid \mathfrak{a} \vdash_m \mathfrak{Y} \tag{11}$$

is that

$$\mathfrak{X} \mid \mathfrak{a} \vdash_1 C \tag{12}$$

the notations being the same as in Sec. 5D. In view of the results of Sec. 2, the sufficiency proof is not at all affected by the presence of quantifiers. For the necessity proof we have to consider four new cases, namely, $*\Pi*$ and $*\Sigma*$. Of these the rules on the left involve no difficulty since the right sides are parametric. The rule $\Pi*$ is singular in LA$*$, LM$*$, LJ$*$, and LD$*$; hence it needs to be considered only for LE$*$ and LK$*$.

Let us take up first the case of $\Sigma*$. Here $\mathfrak{Y}$ is $(\exists x)A(x)$, $\mathfrak{Z}$, and $C$ is $(\exists x)A(x) \lor D$. By the inductive hypothesis

$$\mathfrak{X} \mid \mathfrak{a} \vdash_1 A(t) \lor D$$

From this we have the desired conclusion by ET, provided we establish

$$A(t) \lor D \mid \mathfrak{a} \vdash (\exists x)A(x) \,.\!\lor D$$

This may be shown as follows:

$$\frac{A(t) \mid \mathfrak{a} \vdash A(t)}{\begin{array}{c}\dfrac{A(t) \mid \mathfrak{a} \vdash (\exists x) . A(x)}{A(t) \mid \mathfrak{a} \vdash (\exists x) . A(x) . \vee D} \text{V}_* \qquad \dfrac{D \mid \mathfrak{a} \vdash D}{D \mid \mathfrak{a} \vdash (\exists x) A(x) . \vee D} \text{V}_*\end{array}}{A(t) \vee D \mid \mathfrak{a} \vdash (\exists x) A(x) \vee D}} \Sigma_* \qquad {}_*\text{V}$$

In the case of $\Pi_*$ we note that we have available all the apparatus of $LC_1$. The following hold in $LA_1^*$:

$$A \supset . (\forall x) B(x) \mid \mathfrak{a} \vdash (\forall x) . A \supset B(x) \qquad (13)$$

$$\mathfrak{X} \Vdash A \vee B \to \mathfrak{X}, A \supset B \Vdash B \qquad (14)$$

while the following holds in $LC_1$:

$$\mathfrak{X}, A \supset B \Vdash B \to \mathfrak{X} \Vdash A \vee B \qquad (15)$$

Using these, we can complete the necessity proof for $\Pi_*$ as follows. The inference in question is

$$\frac{\mathfrak{X} \mid \mathfrak{a}, c \vdash A(c), \mathfrak{Z}}{\mathfrak{X} \mid \mathfrak{a} \vdash (\forall x) A(x), \mathfrak{Z}}$$

By the premise and the inductive hypothesis

$$\mathfrak{X} \mid \mathfrak{a}, c \vdash A(c) \vee D$$

From this we conclude, successively,

| | |
|---|---|
| $\mathfrak{X}, D \supset A(c) \mid \mathfrak{a}, c \vdash A(c)$ | by (14) |
| $\mathfrak{X}, (\forall x) . D \supset A(x) \mid \mathfrak{a}, c \vdash A(c)$ | by $*\Pi$ |
| $\mathfrak{X}, D \supset . (\forall x) A(x) \mid \mathfrak{a}, c \vdash A(c)$ | by (13), ET |
| $\mathfrak{X}, D \supset . (\forall x) A(x) \mid \mathfrak{a} \vdash (\forall x) A(x)$ | by $\Pi_*$ |
| $\mathfrak{X} \mid \mathfrak{a} \vdash (\forall x) A(x) . \vee D$ | by (15) |

This completes the proof of necessity, and thus of the theorem under discussion.

Next we look at the theorem of Sec. 6B4 regarding the different formulations of negation. Inspection of this proof shows that it is not at all affected by the presence of the rules for quantification.

This completes the proof of Theorem 5.

**4. L\* deducibility.** We have already noticed (Sec. A5, Remark 2) that the quantifier rules do not have the composition property, even with respect to compound constituents. But they do have a modified form of this property, viz., that in which we interpret the word 'like' as meaning that the components differ by changes in terms only. This still leaves the number of possibilities infinite, so that the proof of the decidability theorem fails.[1] Yet the modified composition property enables us to establish many of the other theorems of Sec. 5E.

---

[1] The undecidability of LK* was proved in Church [NEP]; see also his correction [CNE]. In principle this proof extends to the other L* systems, but we shall not look into the question of exactly how.

Let us call this form of the composition property the *modified composition property*. Then we have the following:

**Theorem 6.** *If a system* LX *has the composition property, then* LX* *has the modified composition property. In so far as* LX *has the separation, conservation, and, in the singular case, the alternation property,* LX* *does also.*

In the case of the alternation property we have to take account of the fact that $*\Sigma$ is dilemmatic according to the definition of Sec. 5E4 because it fails to satisfy $(r6)$. But the only reason for that failure is that the new parameter may contain the characteristic variable. However, if in climbing the tree we encounter a case where $*\Sigma$ was applied, the characteristic variables cannot be in $A \lor B$, and hence when the latter is replaced by $A$ or $B$, as the case may be, the inference will still be correct.

**COROLLARY 6.1.** *If* $\Gamma$ *is an elementary theorem of* LX* *which does not involve quantifiers, then* $\Gamma$ *is demonstrable in the corresponding* L *system.*

*Proof.* For all except LD* this follows by the separation property. If $\Gamma$ is true in LD*, then $\Gamma$ is true in $\text{LD}_1^*$. But $\text{LD}_1^*$ can be formulated with Nx taken in the form

$$\mathfrak{X}, \ \neg A \Vdash A \to \mathfrak{X} \Vdash A$$

and this does not allow a quantifier to be eliminated. Hence, if a quantification rule were used, a quantifier would appear in $\Gamma$, Q.E.D.

In the case of the alternation property we have the following generalization:

**Theorem 7.** *Let the operations* $\lor$ *and* $\Sigma$ *be regarded as dilemmatic, the others as nondilemmatic. Let the operations in* $\mathfrak{X}$ *be all nondilemmatic, and let*

$$\mathfrak{X} \mid \mathfrak{a} \vdash (\exists x) A(x) \tag{16}$$

*hold in* $\text{LA}_1^*$, $\text{LM}_1^*$, *or* $\text{LJ}_1^*$. *Then for some* t [*obtained constructively from the proof of* (16)],

$$\mathfrak{X} \mid \mathfrak{a} \vdash A(t)$$

*holds in the same system.*

*Proof.* Similar to that of Theorem 5E5. The tree climbing can never reach a quasi-prime statement.

The following theorem is the relevant generalization of Theorem 5E4.

**Theorem 8.** *Let* $\Delta$ *be a demonstration of* (1) *in a system* LX*($\mathfrak{S}$). *Let the auxiliary statements and counteraxioms used in* $\Delta$ *be obtained from auxiliary statement schemes and counteraxiom schemes in which the free* U *variables for terms are* $\mathfrak{b} \equiv \{b_1, \ldots, b_n\}$. *To each elementary statement scheme*

$$A_1, A_2, \ldots, A_m \mid \mathfrak{b} \vdash_0 B$$

*let there be assigned a proposition* $G(b_1, \ldots, b_n)$, *where*

$$G \equiv A_1 \supset. \ A_2 \supset. \ \cdots \supset. \ A_m \supset B$$

*and to each counteraxiom scheme*

$$F_i(b_1, \ldots, b_n)$$

*let there be assigned a*

$$G(b_1, \ldots, b_n)$$

*viz.,*

$$G \equiv \neg F_i$$

*Let* $\mathfrak{M}$ *consist of all propositions of the form*

$$(\forall x_1)(\forall x_2) \cdots (\forall x_n) \, . G(x_1, \ldots, x_n)$$

*(where quantifiers pertaining to variables not actually present can be omitted).*
*Then*

$$\mathfrak{X}, \mathfrak{M} \mid \mathfrak{a} \vdash \mathfrak{Y}$$

*is demonstrable in* LX*($\mathfrak{D}$).

The proof is left as an exercise (Exercise 9). It can be obtained either by analogy with that of Theorem 5E4 or directly by deductive induction.

The theorems on simplification of the structural rules hardly need separate treatment. In Formulation III we need a quasi-principal constituent in $\Sigma *$ and $*\Pi$.

**5. Nonderivability; classical evaluation.** Although the L* systems are in principle undecidable, yet it frequently happens that applications of methods similar to those in Sec. 5C5 will lead to a decision in special cases. This is frequently helped out by the use of other methods, of which one of the most important is contravalidity by classical evaluation. We shall examine the latter method here, and then apply it to a proof of nondemonstrability in LA*.

We shall begin by stating some definitions connected with the notion of *classical valuation.*

DEFINITION 1. If $\mathfrak{a} \subseteq \mathfrak{q}$, a *classical valuation with range* $\mathfrak{a}$ is a mapping $f$ which assigns to every proposition $A$ of an extension $\mathfrak{S}(\mathfrak{a})$ [i.e., to every element of $\mathfrak{P}(\mathfrak{a})$] a value $f(A)$ which is either 0 or 1. This assignment is arbitrary for the elementary propositions [members of $\mathfrak{E}(\mathfrak{a})$],[1] and each such assignment defines a separate valuation. For compound propositions, $f(A)$ is defined recursively thus:[2]

$$f(A \supset B) = 1 - f(A) + f(A)f(B)$$
$$f(A \wedge B) = f(A)f(B)$$
$$f(A \vee B) = f(A) + f(B) - f(A)f(B)$$
$$f(\sqcap A) = 1 - f(A)$$
$$f((\forall x)A(x)) = \min_t\{f(A(t))/t \; \epsilon \; \mathfrak{t}(\mathfrak{a})\}\dagger$$
$$f((\exists x)A(x)) = \max_t\{f(A(t))/t \; \epsilon \; \mathfrak{t}(\mathfrak{a})\}$$

Such a valuation is said to be *constructive* if its value can be determined by an effective process; this means, in the case of $f((\forall x)A(x))$, that we can either determine effectively that $f(A(t)) = 1$ for all $t$ or produce effectively an $s$ such that $f(A(s)) = 0$.

---

[1] Note that an elementary proposition of the form

$$\phi_k(t_1, \ldots, t_{n_k})$$

is distinct from all the $E_i$, and from every other such proposition with a different $\phi_k$, or with the same $\phi_k$ and different $t_1, \ldots, t_{n_k}$. If there are infinitely many distinct members of $\mathfrak{E}(\mathfrak{a})$, the valuations form a nonenumerable set.

[2] The right sides are to be interpreted as arithmetical expressions. Note that the first four cases give the ordinary 0-1 truth tables.

† This means the minimum for all $t$ of the class of values in the braces. Thus the minimum is 1 just when all the values in the braces are 1. Similarly, 'max$_t$' means the maximum of the values in the following braces.

DEFINITION 2. A valuation is *admissible* just when $f(B) = 1$ for all elementary statements $B$ such that

$$A_1, A_2, \ldots, A_m \vdash_0 B$$

where $f(A_1) = f(A_2) = \cdots = f(A_m) = 1$; and $f(B) = 0$ for $B$ a counteraxiom or $B \equiv F$. (In $\mathfrak{D}$ every valuation is admissible if $f(F) = 0$.)

DEFINITION 3. If $f$ is a classical valuation and $\mathfrak{X}$ is a prosequence, we define

$$f(\mathfrak{X}) = 1 \qquad f(\mathfrak{X}) = 0$$

to mean that every constituent of $\mathfrak{X}$ has, respectively, the value 1 or the value 0.

DEFINITION 4. If $f$ is an admissible valuation whose range includes $\mathfrak{a}$, then $f$ shall be said to be a *countervaluation* for the elementary statement (1) just when

$$f(\mathfrak{X}) = 1 \quad \text{and} \quad f(\mathfrak{Y}) = 0 \tag{17}$$

The statement (1) is valid by classical evaluation just when no countervaluation exists.

This notion of validity is evidently indefinite. However, we can sometimes establish invalidity constructively by exhibiting a constructive countervaluation. In such a case the following theorem shows that nondemonstrability can also be constructively established.

**Theorem 9.** *If a countervaluation for* (1) *is constructively defined, then* (1) *is not demonstrable in any* LX*.*

*Proof.* We show, by deductive induction, that if (1) is demonstrable, then the assumption that there is a constructive countervaluation $f$ will lead to a contradiction.[1] This is clear if (1) is prime. The inductive step is completed if we show that if (1) is obtained by a rule $R$ and there is given constructively an $f$ satisfying (17) for the conclusion, then there will be one for at least one of the premises also.

For the algebraic rules this inductive step is clear. (It has already been used in the proof of Theorems 5D6 and 6B9.)

The inductive step for the case where $R$ is $\Pi*$ is as follows. Let the inference be

$$\frac{\mathfrak{X} \mid \mathfrak{a}, c \vdash A(c), \mathfrak{Z}}{\mathfrak{X} \mid \mathfrak{a} \vdash (\forall x)A(x), \mathfrak{Z}}$$

If $f$ satisfies (17) for the conclusion, then

$$f(\mathfrak{X}) = 1 \qquad f((\forall x)A(x)) = 0 \qquad f(\mathfrak{Z}) = 0 \tag{18}$$

From the second of these equations, since $f$ is constructively given, there is an $s \in t(\mathfrak{a})$ such that

$$f(A(s)) = 0$$

Now let $g$ be the valuation over the range $\mathfrak{a}, c$ such that for each elementary $B$

$$g(B) = f([s/c]B) \tag{19}$$

---

[1] This contradiction will consist in the fact that some proposition will be assigned two values.

Then we can show by structural induction on $B$ that (19) holds for all $B \,\epsilon\, \mathfrak{P}(\mathfrak{a})$. For example, if $B \equiv (\forall y)C(y)$, and $C' \equiv [s/c]C$, then

$$g(B) = \min_t\{g(C(t))/t \,\epsilon\, \mathfrak{t}(\mathfrak{a},c)\}$$
$$= \min_t\{f(C'(t'))\}$$

where $t' \equiv [s/c]t$. Since $t' \,\epsilon\, \mathfrak{t}(\mathfrak{a})$ and any $t \,\epsilon\, \mathfrak{t}(\mathfrak{a})$ is such a $t'$,

$$g(B) = \min_t\{f(C'(t))/t \,\epsilon\, \mathfrak{t}(\mathfrak{a})\}$$
$$= f((\forall y)C'(y))$$
$$= f([x/c]B)$$

Proceeding analogously in the other cases, we have (19) for all $B$. From this we have

$$g(\mathfrak{X}) \quad = f(\mathfrak{X}) \quad = 1$$
$$g(\mathfrak{Z}) \quad = f(\mathfrak{Z}) \quad = 0$$
$$g(A(c)) = f(A(s)) = 0$$

so that the premise of $R$ is constructively invalid, which is a contradiction.

The inductive step for the case where $R$ is $*\Sigma$ is dual to this. The other cases do not cause any difficulty. This completes the proof.

As a corollary of this theorem we have the following:

**Theorem 10.** *If* $\mathfrak{a} \subseteq \mathfrak{r}$ *and* $A$ *and* $B$ *are elementary, the statement*

$$(\forall x) .A \vee B(x) \mid \mathfrak{a} \vdash A \vee. (\forall x)B(x) \tag{20}$$

*is not demonstrable in* $\mathrm{LA}*(\mathfrak{D})$.

*Proof.* If (20) is demonstrable, then by Theorem 2 so is the statement formed by extending the range arbitrarily.

The statement (20) is the special case $m = 0$, $n = 0$, of

$$(\forall x) .A \vee B(x), \mathfrak{U}_m, \mathfrak{V}_n \mid \mathfrak{a} \vdash A \vee. (\forall x)B(x) \tag{21}$$

where
$$\mathfrak{U}_m \equiv B(t_1), \ldots, B(t_m)$$
$$\mathfrak{V}_n \equiv A \vee B(s_1), \ldots, A \vee B(s_n)$$

We shall see, using the rules of Formulation IV, that no statement of form (21) is demonstrable.

First, if (21) were obtained by application of $V*$, the premise would be one or the other of

$$(\forall x) .A \vee B(x), \mathfrak{U}_m, \mathfrak{V}_n \mid \mathfrak{a} \vdash A$$
$$(\forall x) .A \vee B(x), \mathfrak{U}_m, \mathfrak{V}_n \mid \mathfrak{a} \vdash (\forall x)B(x)$$

These are both invalid by classical evaluation over $\mathfrak{D}$: the first by taking $f(A) = 0$, $f(B(t)) = 1$ for all $t$; the second by taking $f(A) = 1$, $f(B(t_k)) = 1$ for all $k$, $f(B(t)) = 0$ for some $t$ distinct from $t_1, \ldots, t_m$ (this is possible in a suitable extension of $\mathfrak{a}$). By Theorem 9 neither of these premises can be derived in $\mathrm{LA}*$. Hence (21) cannot be obtained by $V*$.

The only other rules of Formulation IV which can lead to (21) are $*\Pi$, with principal constituent $(\forall x) .A \vee B(x)$, and $*V$, with principal constituent some $A \vee B(t_k)$. In the former case the premise is again of form (21); in the latter case there are two premises, of which one is again of form (21). Thus the search for a proof of (20) will lead to an infinite regress, and no demonstration is possible, Q.E.D.

COROLLARY 10.1.    *The multiple form of* $\Pi^*$ *is not valid in* LA*.

*Proof.*    If it were, we could derive (20) thus:

$$\frac{\dfrac{A \mid a, c \vdash A, B(c) \qquad B(c) \mid a, c \vdash A, B(c)}{A \vee B(c) \mid a, c \vdash A, B(c)} *V}{\dfrac{(\forall x) . A \vee B(x) \mid a, c \vdash A, B(c)}{\dfrac{(\forall x) A \vee B(x) \mid a \vdash A, (\forall x) B(x)}{(\forall x) A \vee B(x) \mid a \vdash A \vee (\forall x) B(x)} V*} \Pi*} *\Pi$$

*Remark.*    The notions used here are related to the ordinary notion of model. There are, however, certain differences. In the present context let us understand a *model* in the following sense. Let there be given a contensive system $\mathfrak{M}$ (which may be another formal system), and let objects of $\mathfrak{M}$ be assigned as values to the terms. We thus have values for the terms as well as for the propositions. It is not necessary that distinct terms have distinct values. We then assign classes or relations (of the values) to the predicates and interpret the elementary propositions as contensive statements; if the elementary statement is

$$\phi_k(t_1, \ldots, t_{n_k})$$

the contensive statement is one to the effect that the values of $t_1, \ldots, t_{n_k}$ are related in the order named by the relation assigned to $\phi_k$; the $E_i$, which may be thought of as special cases of the above where $n_k = 0$, are simply interpreted as contensive statements of some kind concerning $\mathfrak{M}$. When the contensive statement is true, we assign to the proposition the value 1; when it is false, the value 0.† The values of compound propositions are then determined as in Definition 1. Under these circumstances we say that $\mathfrak{M}$ is a *semimodel* (in relation to the correspondence set up); if the interpretants of all asserted propositions are true,[1] $\mathfrak{M}$ is a *model*. Note that a classical valuation, as defined in Definition 1, is a special case, viz., where we take the terms as their own values and the contensive statements as those stating that the proposition has the value 1. In the study of models one takes a platonistic point of view and does not insist on constructiveness; from this standpoint Theorem 9 says that an admissible valuation gives a model.

**6. Proof tableaux.**    Inasmuch as the classical predicate calculus is known to be undecidable, it is not to be expected that any algorithm will give a decision as to demonstrability or nondemonstrability in every case. The

---

† This is then a valuation in the above sense, but two propositions with the same $\phi_k$ whose corresponding terms have the same value must have the same value as propositions.

[1] There are at least two variants to this. On the one hand, we may suppose that only the elementary propositions have interpretants and the truth of the compound propositions, elementary statements, etc., is determined by Definitions 1 to 4. In this case $\mathfrak{M}$ is a model for $\mathfrak{S}$. On the other hand, we may suppose that our operations correspond to connectives in $\mathfrak{M}$ and then compound propositions have interpretants. In such a case it is required that truth as determined by our conventions imply contensive truth of the interpretants.

most that can be expected of a tableau is that its closure will be a necessary and sufficient condition for demonstrability, but the tableau may go on indefinitely without either closing or becoming provably impossible to close. In such a case it is important, when the tableau splits, to consider all the subtableaux simultaneously, so that no possibility of decision may be overlooked if the tableau is carried far enough.

An algorithm satisfying these conditions will now be proposed. This is subject to the general conventions of Sec. 5E8. It is supposed that the system under consideration is $\text{LA}_m^*$, $\text{LC}_m^*$, $\text{LM}_m^*$, $\text{LJ}_m^*$, $\text{LE}_m^*$, or $\text{LK}_m^*$; further, that $\mathfrak{S}$ is $\mathfrak{O}$, $\Omega$ is void, and $\mathfrak{e}$ consists of $e_1$ only. The range of the datum is not indicated; that for the head of the tableau is to include all the term variables which occur free; at any step the range of each statement of the result is to be the same as that of the datum, except that the $c$ in II and III, which is to be added to the range of the datum to give that of the result, is to be the first variable in $\mathfrak{g}$ (Theorem 1) that is not in the range of the datum. In X and XI, $t_1, \ldots, t_r$ are $e_1$ and all the variables in the range (and hence all atoms which appear in the datum, together with $e_1$). The L rule permitting the inference from result to datum is indicated at the extreme right, it being understood that *C* can be used whenever needed.

$$\text{I} \quad \frac{\mathfrak{X} \Vdash \mathfrak{Z}_1,\ A \supset B,\ \mathfrak{Z}_2}{\mathfrak{X},\ A \Vdash \mathfrak{Z}_1,\ B,\ \mathfrak{Z}_2} \qquad (\text{P}*)$$

$$\text{II} \quad \frac{\mathfrak{X} \Vdash \mathfrak{Z}_1,\ (\forall x)A(x),\ \mathfrak{Z}_2}{\mathfrak{X} \mid c \vdash \mathfrak{Z}_1,\ A(c),\ \mathfrak{Z}_2} \qquad (\Pi*)$$

$$\text{III} \quad \frac{\mathfrak{X}_1,\ (\exists x)A(x),\ \mathfrak{X}_2 \Vdash \mathfrak{Y}}{\mathfrak{X}_1,\ A(c),\ \mathfrak{X}_2 \mid c \vdash \mathfrak{Y}} \qquad (*\Sigma)$$

$$\text{IV} \quad \frac{\mathfrak{X} \Vdash \mathfrak{Y}_1,\ A \vee B,\ \mathfrak{Y}_2}{\mathfrak{X} \Vdash \mathfrak{Y}_1,\ A,\ B,\ \mathfrak{Y}_2} \qquad (\text{V}*)$$

$$\text{V} \quad \frac{\mathfrak{X}_1,\ A \wedge B,\ \mathfrak{X}_2 \Vdash \mathfrak{Y}}{\mathfrak{X}_1,\ A,\ B,\ \mathfrak{X}_2 \Vdash \mathfrak{Y}} \qquad (*\Lambda)$$

$$\text{VI} \quad \frac{\mathfrak{X} \Vdash \mathfrak{Y}_1,\ A \wedge B,\ \mathfrak{Y}_2}{\mathfrak{X} \Vdash \mathfrak{Y}_1,\ A,\ \mathfrak{Y}_2\ \&\ \mathfrak{X} \Vdash \mathfrak{Y}_1,\ B,\ \mathfrak{Y}_2} \qquad (\Lambda*)$$

$$\text{VII} \quad \frac{\mathfrak{X}_1,\ A \vee B,\ \mathfrak{X}_2 \Vdash \mathfrak{Y}}{\mathfrak{X}_1,\ A,\ \mathfrak{X}_2 \Vdash \mathfrak{Y}\ \&\ \mathfrak{X}_1,\ B,\ \mathfrak{X}_2 \Vdash \mathfrak{Y}} \qquad (*\text{V})$$

$$\text{VIII} \quad \frac{\mathfrak{X} \Vdash \mathfrak{Y}}{\mathfrak{X} \Vdash \mathfrak{Y},\ F} \qquad (\text{Fj})$$

$$\text{IX} \quad \frac{\mathfrak{X} \Vdash A,\ \mathfrak{Y}}{\mathfrak{X} \Vdash A \quad \text{or} \quad \mathfrak{X} \Vdash \mathfrak{Y},\ A} \qquad (\text{K}*)$$

$$\text{X} \quad \frac{\mathfrak{X} \Vdash \mathfrak{Y}_1,\ (\exists x)A(x),\ \mathfrak{Y}_2}{\mathfrak{X} \Vdash \mathfrak{Y}_1,\ A(t_1), \ldots, A(t_r),\ \mathfrak{Y}_2,\ (\exists x)A(x)} \qquad (\Sigma'*)$$

$$\text{XI} \quad \frac{\mathfrak{X}_1,\ (\forall x)A(x),\ \mathfrak{X}_2 \Vdash \mathfrak{Y}}{\mathfrak{X}_1,\ A(t_1), \ldots, A(t_r),\ \mathfrak{X}_2,\ (\forall x)A(x) \Vdash \mathfrak{Y}} \qquad (*\Pi')$$

$$\text{XII} \quad \frac{\mathfrak{X}_1,\ A \supset B,\ \mathfrak{X}_2 \Vdash \mathfrak{Y}}{\mathfrak{X}_1,\ \mathfrak{X}_2,\ A \supset B \Vdash \mathfrak{Y},\ A\ \&\ \mathfrak{X}_1,\ \mathfrak{X}_2,\ B \Vdash \mathfrak{Y}} \qquad (*\text{P})$$

The convention that the rules are to be taken in the order given is subject to an important exception; viz., as soon as any of the rules X–XII is applied to Γ we must apply these rules and only these rules with the principal constituents originally appearing in Γ until all such possibilities have been exhausted, without regard to the appearance during the process of other possibilities of applying these or any other rules; only after each of these rules has gone through a cycle is the normal course of the algorithm resumed. The purpose of this provision is to insure that every constituent is reached; with new terms being introduced by II and III, it is conceivable that the algorithm might otherwise continue indefinitely without reaching certain constituents.

In the systems based on LC the rule IX is omitted, the quasi-principal constituent in XII is removed, and XII is moved up to come immediately after VII. (For other possible modifications, see Sec. D4.) The rule VIII is omitted in all systems which do not postulate Fj.

**Theorem 11.** *A necessary and sufficient condition that an elementary statement be demonstrable is that its tableau close.*

*Proof.* This follows the pattern of Sec. 5E8, but contains some modifications.[1] It will be expedient to make a few definitions and preliminary remarks before beginning the proof proper.

A standard demonstration is one made in Formulation IV,[2] with prime statements of type $(p1)'$ (Sec. 5C9) or $(p2)$, and applications of ∗K∗ made initially or in a group immediately after an application of a rule which is singular on the right (but without any restriction on the form of the principal constituent).[3] By the theorems already proved, any demonstration can be made standard. The degree of a standard demonstration will be, as in Corollary 5D1.6, the total number of its nonstructural inferences; in this the rule Fj, which simply omits $F$, will be regarded as nonstructural.

A constituent $C$ on a specified side will be called simple in exactly the following cases: (1) $C$ is elementary; (2) $C$ is on the right and has the form of the principal constituent of a rule which is singular on the right; and (3) $C$ has the form of the quasi-principal constituent of a rule which requires such on the same side. Any constituent $C$ is formed from simple constituents by the operations which form the principal constituents of the rules III to VII from their subalterns; the number $m$ of applications of such operations will be called the order of $C$. The K-order of a standard demonstration Δ is defined as the sum total of the orders of all the constituents introduced into Δ by ∗K∗.

After these preliminaries we pass to the proof proper.

The proof of sufficiency follows at once, since the inference from result to datum can be made by the rule indicated at the right. In certain cases, where the two subalterns are equal or a subaltern already occurs in the datum, the special rules of Formulation IV may be needed.

In the proof of necessity we deal first with the case where we have a

---

[1] Some of these modifications were made too late to affect Secs. 5E8 and 6B6.

[2] Due to the restrictions on ∗K∗, certain rules discarded as trivial in Sec. 5E6 will be needed here. The provisions of Sec. 5E8 for omitting repetitions take care of all special rules of Formulation IV automatically. There is no difficulty about extending Theorem 5D1 and its corollaries to Formulation IV.

[3] This is necessary on account of the condition $(d)$ in Theorem 5D1.

system based on LA*, then consider what modifications are suitable for a system based on LC*. The proof proceeds, as in Theorem 5E10, by an induction. Let Γ be the given statement. Let Δ be a standard demonstration of Γ, and let its degree be $n$ and its K-order be $m$. If $n = 0$, then Γ is quasi-prime, and the algorithm closes at once. Otherwise we assume as inductive hypothesis that the theorem holds in all cases where the degree is less than $n$; also, if $m > 0$, in all cases where the degree is $n$ and the K-order is less than $m$.

If one of the rules I to VII is applicable to Γ, then we invoke the inversion theorem as in Theorem 5E10. Let $M$ be the principal constituent of the first step of the algorithm applied to Γ. If $M$ has at least one quasiparametric ancestor which was introduced by the operational rule, then by Corollary 5D1.6 the result(s) of the first step of the algorithm will have standard demonstration(s) of degree less than $n$. If not, $M$ must have positive order; for $M$ is necessarily composite, and the omission of a constituent of either of the types I or II would lead to an elementary theorem with a void consequent, which is impossible in an F formulation. Then, by Corollary 5D1.6, the demonstration(s) of the result(s) of the first step of the algorithm will have degree not greater than $n$ and K-order less than $m$. In either case the algorithm closes by virtue of the inductive hypothesis.

We may, therefore, suppose that Γ contains no constituent having the form of the principal constituent of any of the rules I to VII. No rule previous to VIII is then applicable, and the last inference in Δ must be by one of the rules Fj, K∗, Σ∗, ∗Π, or ∗P.

If the last inference in Δ is by Fj, then the result of VIII will give the premise of the inference. Since the result can be obtained from the datum by K∗, it has likewise a demonstration of degree $n$. We may therefore suppose that Γ contains an instance of $F$, and hence that the algorithm will pass through to IX.

If the last inference in Δ is by K∗, then, since this cannot be an initial instance of K∗, it must be one of a group occurring immediately after application of a rule which is singular on the right. Then one of the alternatives of IX will give the conclusion of that application, and the algorithm will then give the premise by I or II. Since that premise has a demonstration of degree $n - 1$, the algorithm will close by the inductive hypothesis. (Note that we need only consider those alternatives in IX in which the $A$ has the proper form, and that we should exclude all other alternatives if we want the demonstration formed by inverting the algorithm to be standard.) If the last inference in Δ is not by K∗, then it must be by Σ∗, ∗Π, or ∗P. In that case we cycle through IX, and let the algorithm pass through to X.

We continue the proof of the theorem. We recall that we have handled all cases except that where the last inference in Δ is by one of the rules Σ∗, ∗Π, or ∗P, and none of the rules I to VIII is applicable to Γ. At least one of the rules X to XII must then be applicable to Γ. By our special convention we must make a cycle through each of these rules using only principal constituents which were present in Γ to begin with, without regard to the possible principal constituents (for these or any other rules) which may appear during the process. The cycles through X and XI will convert Γ into a Γ′ which is derived from it by weakening. If the constituents of the form $A \supset B$ which were in Γ are those of (14) in Sec. 5E8, then the cycle through XII will convert $\Gamma'$ into a conjunction $\Gamma'_1, \Gamma'_2, \ldots, \Gamma'_p$, where $p = 2^m$ (see Exercise 5E12) and each is derived from $\Gamma'$ by inversion and weakening. The special convention requires that all this be done as if it were a single step of the algorithm.

If the last inference in $\Delta$ is by one of the rules $\Sigma*$, $*\Pi$ from a premise $\Gamma''$, then the principal constituent is in $\Gamma$. By the reasoning used at the end of the proof of Theorem B2, we can suppose that the subaltern is one of the $A(t_i)$. It must therefore appear in $\Gamma'$. Thus $\Gamma'$ can be obtained by weakening from $\Gamma''$. Since the process of inversion does not increase the degree, each of the $\Gamma'_i$ can be obtained by weakening from a statement with a standard demonstration of degree less than $n$. By the inductive hypothesis, the algorithm closes for each of the $\Gamma'_1, \ldots, \Gamma'_p$, and hence for $\Gamma$ in this case.

The only remaining case is that the last inference in $\Delta$ is by $*P$. Let the premises be $\Gamma_a$ and $\Gamma_b$. If we perform on these the same weakening steps as led from $\Gamma$ to $\Gamma'$, we shall obtain statements $\Gamma'_a$ and $\Gamma'_b$ from which $\Gamma'$ follows by a homologous instance of $*P$. Each of the $\Gamma'_i$ can be obtained from one or the other of $\Gamma'_a$, $\Gamma'_b$ by inversion with respect to certain of the $A_j \supset B_j$ and weakenings. Since inversion alters the structure of a proof only in the deletion of certain operational inferences (viz., those by the rule $R$ of Sec. 5D1), each of the $\Gamma'_i$ can be obtained by weakening a statement with a standard demonstration of degree less than $n$. As before, the algorithm therefore closes for each $\Gamma'_i$, and hence for $\Gamma$.

This completes the proof of the theorem for systems based on LA*. For systems based on LC* the situation is simpler. In the first place, all applications of $*K*$ in a standard demonstration are initial; therefore we do not need IX, and have no alternative splittings of a tableau. Again since $*P$ is invertible without a quasi-principal constituent, we can cancel that constituent, move the rule up in front of VIII, and treat it like I to VII. This simplifies matters at the end of the proof considerably.

Theorem 11 is therefore proved for all the systems mentioned in the preliminary discussion.

### EXERCISES

Exercises 1 to 3 are lemmas in the reduction to prenex normal form in Sec. D1. Partial answers may be found there.

**1.** Show that the following hold in LC* and that exactly three of them fail to hold in LA*.

$$(\forall x).A \supset B(x) \Vdash A \supset . (\forall x)B(x)$$

$$A \supset . (\exists x)B(x) \Vdash (\exists x).A \supset B(x)$$

$$(\forall x).A(x) \supset B \Vdash (\exists x)A(x) .\supset B$$

$$(\forall x)A(x) .\supset B \Vdash (\exists x).A(x) \supset B$$

$$(\forall x).A(x) \wedge B \Vdash (\forall x)A(x) .\wedge B$$

$$(\exists x)A(x) .\wedge B \Vdash (\exists x).A(x) \wedge B$$

$$(\forall x).A(x) \vee B \Vdash (\forall x)A(x) .\vee B$$

$$(\exists x)A(x) .\vee B \Vdash (\exists x).A(x) \vee B$$

**2.** Show that one of the following holds in LM*, the other in LK* but not in LJ*:

$$\neg\,(\forall x)A(x) \Vdash (\exists x)\,.\,\neg\,A(x)$$
$$(\forall x)\,.\,\neg\,A(x) \Vdash \neg(\exists x)A(x)$$

In the first of these show that the universal quantifier must be introduced last; in the second one, the existential quantifier.

**3.** Let $M[A]$ be a proposition containing an occurrence of a proposition $A$, such that $M$ does not contain $x$ except in so far as it is in $A$. With the understanding that $M[B]$ is the result of replacing $A$ by $B$ in that particular occurrence, show that if the occurrence of $A$ is positive,

$$M[(\forall x)A(x)] \Vdash (\forall x)M[A(x)]$$
$$(\exists x)M[A(x)] \Vdash M[(\exists x)A(x)]$$

whereas if the occurrence is negative,

$$M[(\exists x)A(x)] \leq (\forall x)M[A(x)]$$
$$(\exists x)M[A(x)] \leq M[(\forall x)A(x)]$$

**4** Determine which of the following hold in LJ* (for $A$, $B$, $C$, etc., elementary):

(a)　　　　　　$\Vdash \neg\,\neg(\forall x)(A(x) \vee \neg A(x))$

(b)　　　　　　$\Vdash \neg\,\neg((\forall x).A \vee B(x)\,.\supset.\,A \vee (\forall x)B(x))$

(c)　　　　　　$\Vdash \neg\,\neg(\forall y)((\forall x)(A(y) \vee B(x)) \supset A(y) \vee (\forall x)B(x))$

(d)　　　　　　$\Vdash \neg\,\neg((\forall x)\neg\,\neg A(x)\,.\supset.\,(\forall x)A(x))$

(e)　　　　　　$\neg\,\neg(\,\neg(\exists x)\,\neg A(x)) \Vdash \neg\,\neg(\forall x)\,\neg\,\neg A(x)$

(Kleene [IMM, Theorem 58], which refers to sources.)

**5.** Show that

$$A(a) \vee A(b) \Vdash (\exists x)A(x)$$

cannot be demonstrated in $LA_1^*$ with the $\Sigma^*$ last (Kleene [IMM, p. 463]).

**6.** Verify the statement in Example 1 that

$$(\forall x)A(x) \Vdash (\forall x).A(x) \vee B$$

cannot be derived even in $LK_m^*$ with $*\Pi$ last (Kleene [PIG, p. 25]).

**7.** Show that

$$(\forall x).A \wedge B(x) \Vdash A \wedge (\forall x)B(x)$$
$$A \vee (\exists x)B(x) \Vdash (\exists x).A \vee B(x)$$

are valid in LA*, but the external quantification must be made before the internal one, even in LC*.

**8.** Show that a necessary and sufficient condition that

$$\mathfrak{X} \Vdash \mathfrak{Y}$$

hold in LK* is that

$$\mathfrak{X} \Vdash \mathfrak{Y},\,F$$

hold in LE* (Suranyi [RTE, Theorem III], credited to H. Thiele).

**9.** Complete the proof of Theorem 8.

**10.** What is the effect on Theorem 7 of not excluding $*\vee$?

**11.** Show (not necessarily constructively) that a demonstrable elementary statement in any L* system is valid in any model. What modifications are necessary to make this constructive?

**\*12.** Revise the algorithm of Sec. 5 so as to be applicable even if $\Omega$ is not void. (For the classical cases such a revision is given in Sec. D4.)

**\*13.** Suppose one were to introduce the Hilbert $\epsilon$ operator with the rules

$$\frac{\mathfrak{X}, A(b) \mid a, b \vdash \mathfrak{Y}}{\mathfrak{X}, A((\epsilon x)A(x)) \mid a \vdash \mathfrak{Y}} \qquad \frac{\mathfrak{X} \mid a \vdash A(t), \mathfrak{Z}}{\mathfrak{X} \mid a \vdash A((\epsilon x)A(x)), \mathfrak{Z}}$$

In what way would the theory of the L\* system be modified? Can one deduce an elimination theorem, and if so what can one infer in regard to the separation of the new rules? Is not the left-hand rule superfluous? (These questions are related to the $\epsilon$ theorems stated in Hilbert and Bernays [GLM.II]. It seems evident that their proofs could be greatly improved by using the Gentzen technique. Cf. also Asser [TLA]; Maehara [PCE], [EAH]; Rasiowa [ETh].)

## C. OTHER FORMS OF QUANTIFICATION THEORY

In this section we discuss other formulations of quantification theory. The T formulation will concern us in Sec. 1, the H formulation in Sec. 2. The lattice-type formulations will not be treated in this volume.

**1. T formulation of quantification.** The T rules for quantification are suggested by the semantical discussion in Sec. A. They are as follows:

$$\Pi e \quad \frac{(\forall x)A(x)}{A(t)} \qquad \qquad \Pi i \quad \frac{A(c)}{(\forall x)A(x)}$$

$$\Sigma e \quad \frac{(\exists x)A(x) \quad B}{B} \qquad \qquad \Sigma i \quad \frac{A(t)}{(\exists x)A(x)}$$

it being understood that $c$ does not occur in any other premise in $\Pi i$, or in $B$ or other premises from which it is obtained in $\Sigma e$. When these rules are adjoined to the formulation of a system TX, the resulting formulation will be called the system TX\*. We thus have formulations TA\*, TC\*, ..., TK\*, etc.

Let the statement

$$\mathfrak{X} \mid a \vdash^T B \tag{1}$$

hold just when there is a tree whose conclusion is $B$ and whose premises are in $\mathfrak{X}$, and the range $a$ includes all variables which occur free in the premises or in $B$.

As before, we use the notation

$$\mathfrak{X} \mid a \vdash^L B \tag{2}$$

when explicitness is necessary, to indicate an elementary statement of an L\* system.

In the following proofs technicalities concerning the range are omitted. They may be taken care of by Theorem B2.

**Theorem 1.** *A necessary and sufficient condition that* (2) *hold in an* L\* *system is that* (1) *hold in the corresponding* T\* *system.*

*Proof of Necessity.* For the algebraic rules this follows from Theorems 5D6 and 6C1. It suffices to add consideration of the induction step for the quantificational rules. We have to show that the L\* rules hold for the T\* system if '$\vdash^L$' is interpreted as '$\vdash^T$'.

This is clear for $\Pi*$ and $\Sigma*$, since these rules, when interpreted as stated, become precisely the rules $\Pi i$ and $\Sigma i$, respectively.

For $*\Pi$ the premise states that there is a T proof of $B$ with premises $\mathfrak{X}$ and $A(t)$. Putting over the last premise its derivation from $(\forall x)A(x)$ by $\Pi e$, we have the derivation postulated by the conclusion of $*\Pi$.

For $*\Sigma$ we argue thus. By the premise of the interpreted $*\Sigma$ there is a T derivation of $B$ from $A(c)$ and $\mathfrak{X}$, with $c$ not occurring in $\mathfrak{X}$ or $B$. By $\Sigma e$ there is then a T derivation of $B$ from $\mathfrak{X}$ and $(\exists x)A(x)$.

This completes the proof of necessity.

*Proof of Sufficiency.* We have now to show that the T rules, when interpreted in the $L^*$ system, are valid as inferences in the latter. The interpretation is that each premise $A$ is interpreted as $\mathfrak{X} \mid \mathfrak{a} \vdash A$ (for suitable $\mathfrak{a}$), and a rule such as

$$\frac{A \quad \overset{[B]}{C}}{D}$$

is interpreted as

$$\frac{\mathfrak{X} \Vdash A \quad \mathfrak{X}, B \Vdash C}{\mathfrak{X} \Vdash D}$$

with suitable range.

This is clear for $\Pi i$ and $\Sigma i$, since, when so interpreted, they become the same as $\Pi *$ and $\Sigma *$. For $\Pi e$ we argue thus:

$$\frac{\mathfrak{X} \mid \mathfrak{a} \vdash (\forall x)A(x) \quad \dfrac{A(t) \mid \mathfrak{a} \vdash A(t)}{(\forall x)A(x) \mid \mathfrak{a} \vdash A(t)} *\Pi}{\mathfrak{X} \mid \mathfrak{a} \vdash A(t)} \text{ET}$$

For $\Sigma e$ the proof is as follows:

$$\frac{\mathfrak{X} \mid \mathfrak{a} \vdash (\exists x)A(x) \quad \dfrac{\mathfrak{X}, A(c) \mid \mathfrak{a}, c \vdash B}{\mathfrak{X}, (\exists x)A(x) \mid \mathfrak{a} \vdash B} *\Sigma}{\mathfrak{X} \mid \mathfrak{a} \vdash B} \text{ET}$$

This completes the proof.

**2. H formulations of quantification.** There are several ways of forming systems of quantification theory analogous to the systems HX. The formulation is different according to whether one does or does not admit vacuous quantification, and whether one insists on having modus ponens as the sole rule or allows also a rule of generalization for free variables. We shall call all these systems $H^*$ systems, and a system equivalent to the systems $LX^*$, $TX^*$ will be called a system $HX^*$.† It would be appropriate to call such a system a propositional calculus, letting the word 'calculus', in contrast to 'algebra', suggest the presence of quantifiers; but traditional usage employs the term 'propositional calculus' for the algebraic systems of Chaps. 5 and 6 and 'predicate calculus' or 'functional calculus' for systems $H^*$.

The predicate calculus is based on the same definitions of term, proposition, occurrence, and substitution as were made in Sec A4. We shall suppose, however, that the system $\mathfrak{S}$ is void and that there are no operations $(\Omega)$; this, however, is not essential, since only minor changes

† The different variants of HX* are analogous to the different formulations of LX*.

are necessary to treat the case where this assumption is not made.    The
systems H are assertional; an elementary statement may be written

$$\vdash A \tag{1}$$

We shall use the notation

$$\mathfrak{X} \mid \mathfrak{a} \vdash^H A \tag{2}$$

to mean that there is a derivation of (1), using the rules of the system from
premises which are either prime statements of the system or assertions of
propositions in $\mathfrak{X}$ and such that all variables free in $\mathfrak{X}$ or $A$ are in $\mathfrak{a}$.    On
occasion we shall ignore the distinction between a proposition $A$ and the
statement (1).

Specific statement schemes, from which the prime statements are chosen,
are assigned names as follows:

$\Pi_0 \quad \vdash (\forall x) A(x) . \supset . A(t)$

$\Pi_1 \quad \vdash (\forall x). C \supset A(x) : \supset. C \supset (\forall x) A(x)$

$\Pi_2 \quad \vdash C \supset (\forall x) C$

$\Pi P \quad \vdash (\forall x). A(x) \supset B(x) : \supset : (\forall x) A(x) . \supset. (\forall x) B(x)$

$\Sigma_0 \quad \vdash A(t) \supset. (\exists x) A(x)$

$\Sigma_1 \quad \vdash (\forall x). A(x) \supset C : \supset : (\exists x) A(x) . \supset. C$

$\Sigma_2 \quad \vdash (\exists x) C . \supset. C$

$\Sigma P \quad \vdash (\forall x). A(x) \supset B(x) : \supset : (\exists x) A(x) . \supset. (\exists x) B(x)$

Here $A(x)$, $B(x) \in \mathfrak{P}(\mathfrak{a},x)$, $C \in \mathfrak{P}(\mathfrak{a})$ in which $x$ is not free, and $t \in \mathfrak{t}(\mathfrak{a})$.    Note
that $\Pi_2$ and $\Sigma_2$ require vacuous quantification.

The rules of the predicate calculuses are the following:

Ph              (modus ponens)

$$\frac{A \supset B \quad A}{B}$$

Пh    (generalization).    If $c$ is an indeterminate,[1]

$$\frac{A(c)}{(\forall x) A(x)}$$

These are to be interpreted as procedural rules in the same sense as in the T
system.

A system HX* which postulates the rules Ph and Пh will be called a
*predicate calculus with generalization* and indicated as $HX_g^*$; one which postulates only the rule Ph will be called a *proper predicate calculus* and indicated
as $HX_p^*$.    We shall consider these two cases separately.

**3. The predicate calculus with generalization.**    The systems $HX_g^*$
will have two forms.    In the first form the axiom schemes are $\Pi_0$, $\Pi_1$, $\Sigma_0$,
$\Sigma_1$; in the second form the schemes $\Pi_1$, $\Sigma_1$ are replaced by $\Pi_2$, $\Pi P$, $\Sigma_2$, $\Sigma P$.
The second form is, of course, only suitable if vacuous quantification is
permitted.    The following theorem is true for either form, but the proof
applies only to the first form; the validity for the second form will follow
from Theorem 3.

---

[1] That is, if the premise of the rule is derivable from premises which do not contain $c$,
the conclusion is derivable from the same premises.

**Theorem 2.** *A necessary and sufficient condition that* (2) *hold in a system* $HX_g^*$ *is that*

$$\mathfrak{X} \mid \mathfrak{a} \vdash^T A \tag{3}$$

*hold in the system* $TX^*$.

*Proof of Necessity.* Under the interpretation of (2) as (3) the rule Ph becomes the rule Pe, and Пh is Пi. In view of Theorem 1, it is an exercise in the techniques of Sec. B to show that the prime statement schemes give rise to true statements of the form (3). The necessity therefore follows by deductive induction.

*Proof of Sufficiency.* We have to show that the rules of $TX^*$, when interpreted in terms of (2) (cf. the interpretation in terms of L statements in Sec. 1), are valid in $HX^*$. For Пe this thesis follows by $\Pi_0$ and Ph, for $\Sigma i$ by $\Sigma_0$ and Ph, and for Пi by Пh. For the algebraic rules, once Pi has been established, this was shown in Sec. 5B2. It suffices therefore to establish the thesis for Pi and $\Sigma e$.

Next we establish Pi. Suppose that we have a derivation $\Delta$ of $B$ from $A$, and let its steps be $B_1, B_2, \ldots, B_n$. The only addition to be made to the proof in Theorem 5B2 is the case that $B_k$ comes from $B_i$ by an application of Пh. In that case $B_i$ is some $D(c)$ and $B_k$ is $(\forall x)D(x)$. By the hypothesis of the induction we have

$$\vdash A \supset D(c)$$

where $c$ does not occur free in $A$.† By Пh

$$\vdash (\forall x) . A \supset D(x)$$

and hence, by $\Pi_1$ and Ph,

$$\vdash A \supset (\forall x)D(x)$$

which is the desired transform of $B_k$. This, in connection with the proofs in Secs. 5B2 and 5B3, completes the proof for Pi and all algebraic inferences.

The proof for $\Sigma e$ now follows. By the right premise and Pi we have

$$\vdash A(c) \supset B$$

and hence by Пh

$$\vdash (\forall x) . A(x) \supset B$$

Hence by $\Sigma_1$ and Ph we have

$$\vdash (\exists x)A(x) . \supset B$$

From this and the left premise of $\Sigma e$ we have the desired conclusion by Ph.

This completes the proof of Theorem 2 for the first form of $HX_g^*$. The following theorem extends the result to the second form.

**Theorem 3.** *If vacuous quantification is admitted and* $\Pi_0$, *Ph,* Пh *are postulated, then the scheme* $\Pi_1$ *is equivalent to the conjunction of* $\Pi_2$ *and* ПP; *further, if* $\Pi_0$, $\Sigma_0$, $\Pi_1$, *Ph,* Пh *hold, the scheme* $\Sigma_1$ *is equivalent to the conjunction of* $\Sigma_2$ *and* $\Sigma P$.

*Proof.* This will consist of (1) a derivation of $\Pi_1$ from $\Pi_2$, ПP; (2) a derivation of $\Sigma_1$ from $\Sigma_2$, $\Sigma P$; (3) a derivation of $\Pi_2$ from $\Pi_1$; (4) a derivation of

---

† By the hypothesis of Пh, $c$ does not occur in any of the premises used in the derivation of $B_i$. If $A$ is not used as premise in deriving $B_i$, $c$ might occur in $A$, but then we can change $c$ to some other variable which does not occur in $A$. This can be shown by an argument similar to the proof of Lemma 1 in Sec. A.

$\Sigma_2$ from $\Sigma_1$; (5) a derivation of $\Pi P$ from $\Pi_0$, $\Pi_1$; and (6) a derivation of $\Sigma P$ from $\Pi_0$, $\Pi_1$, $\Sigma_0$, $\Sigma_1$.

DERIVATION OF $\Pi_1$

$$\vdash (\forall x).\, C \,\supset\, A(x) :\supset: (\forall x)C \,.\supset.\, (\forall x)A(x) \qquad \text{by } \Pi P$$

From this we have $\Pi_1$ by $\Pi_2$ and Rp.

DERIVATION OF $\Sigma_1$

$$\vdash (\forall x).\, A \,\supset\, C :\supset: (\exists x)A(x) \,.\supset.\, (\exists x)C \qquad \text{by } \Sigma P$$

Here again we have $\Sigma_1$ by $\Sigma_2$ and Rp.

DERIVATION OF $\Pi_2$

$$
\begin{array}{ll}
\vdash C \,\supset\, C & \text{by PI} \\
\vdash (\forall x).\, C \,\supset\, C & \text{by } \Pi\text{h} \\
\vdash (\forall x).\, C \,\supset\, C :\supset: C \,\supset\, (\forall x)C & \text{by } \Pi_1 \\
\vdash C \,\supset.\, (\forall x)C & \text{by Ph}
\end{array}
$$

DERIVATION OF $\Sigma_2$

$$
\begin{array}{ll}
\vdash (\forall x).\, C \,\supset\, C & \text{as in previous case} \\
\vdash (\forall x).\, C \,\supset\, C :\supset: (\exists x)C \,.\supset.\, C & \text{by } \Sigma_1 \\
\vdash (\exists x)C \,.\supset.\, C & \text{by Ph}
\end{array}
$$

DERIVATION OF $\Pi P$.  By Theorem 2 it suffices to show the derivability in TX\*, as follows:

$$
\cfrac{
\cfrac{\dfrac{(\forall x).\, A(x) \,\supset\, B(x)}{A(c) \,\supset\, B(c)}\,\Pi\text{e} \qquad \dfrac{(\forall x)A(x)}{A(c)}\,\Pi\text{e}}{B(c)}\,\text{Pe}
}{(\forall x)B(x)}\,\Pi\text{i}
$$

From this we have $\Pi P$ by two applications of Pi.

DERIVATION OF $\Sigma P$.  As before, it is sufficient to show the derivability in HX\*, thus:

$$
\cfrac{
\cfrac{
\cfrac{\dfrac{\overset{1}{(\forall x).\, A(x) \,\supset\, B(x)}}{A(c) \,\supset\, B(c)}\,\Pi\text{e} \qquad \overset{3}{A(c)}}{B(c)}\,\text{Pe}
}{(\exists x)B(x)}\,\Sigma\text{i} \qquad \overset{2}{(\exists x)A(x)}
}{(\exists x)B(x)}\,\Sigma\text{i} - 3
$$

Here again we have $\Sigma P$ by two applications of Pi.

This completes the proof of Theorem 3.

**4. Proper predicate calculus.**  For the calculus $HX_p^*$ some additional conventions are needed.  If $A(a)$ is a proposition containing $a$, we call $(\forall x)A(x)$ the *closure* of $A(a)$ with respect to $a$.  If $\mathfrak{b}$ is a class of variables, a closure of $A$ with respect to $\mathfrak{b}$ is a proposition obtained by starting with $A$ and taking successively the closure with respect to the variables of $\mathfrak{b}$ in some order.

Given a class $\mathfrak{a}$ of real variables, we formulate a *proper predicate calculus* $HX_p^*$ *with range* $\mathfrak{a}$ by taking as prime statements all closures with respect to variables not in $\mathfrak{a}$ of prime statements of the system $HX_g^*$, together with similar closures of instances of $\Pi P$ (if not already postulated).

**Theorem 4.** *A necessary and sufficient condition that* (2) *hold relative to a system* $HX_p^*$ *is that it be derivable in the corresponding system* $HX_g^*$.

*Proof of Necessity.* If (2) holds in $HX_p^*$, then there is a derivation in which the various statements are either (a) assertions of propositions in $\mathfrak{X}$, (b) axioms, or (c) consequences of preceding statements by Ph. In case $a$, (2) holds in either form of the calculus. In case $b$, $B$ is by definition the closure with respect to some $\mathfrak{b}$ of a $B'$ whose assertion is an axiom of the system $HX_g^*$; then in $HX_g^*$,

$$\mathfrak{X} \mid \mathfrak{a}, \mathfrak{b} \vdash B'$$

from which (2) follows by successive applications of $\Pi h$. In case $c$, we assume as inductive hypothesis that our thesis holds for the premises; it then holds for (2) since Ph is a rule of the system with generalization.

*Proof of Sufficiency.* This amounts to showing that $\Pi h$ is an admissible rule of the proper calculus. We prove this, using the method sketched in Sec. 3A2, by showing that an application of $\Pi h$ which has no other application above it in the proof tree can be eliminated. The proof will be carried through for the case where vacuous quantification is admitted with $\Pi_2$ and $\Pi P$ postulated; the remark at the end of the proof will show how the result holds for other formulations of vacuous quantification; but the consideration of nonvacuous quantification is left as an exercise.

Suppose, therefore, that we have a derivation in $HX_p^*$ showing that

$$\mathfrak{X} \mid \mathfrak{a}, b \vdash A(b) \tag{4}$$

where $b$ does not occur in $\mathfrak{X}$. It is to be shown that

$$\mathfrak{X} \mid \mathfrak{a} \vdash (\forall x)A(x) \tag{5}$$

We do this by deductive induction. There are three cases to be considered, as follows:

CASE 1. $A(b)$ is like a constituent of $\mathfrak{X}$. Then $A(b)$ does not contain $b$, and thus there is a $C$ in $\mathfrak{P}(\mathfrak{a})$ such that

$$A(b) \equiv A(x) \equiv C$$

In this case we can pass from (4) to (5) by $\Pi_2$.

CASE 2. $A(b)$ is an axiom. Then $(\forall x)A(x)$ is also an axiom, and thus (5) holds.

CASE 3. The statement $\vdash A(b)$ is obtained by Ph. Let $A(b) \equiv A_2(b)$; then there is an $A_1(b)$ such that

$$\vdash A_1(b)$$
$$\vdash A_1(b) \supset A_2(b)$$

precede $\vdash A_2(b)$ in the proof of (4). By the hypothesis of the induction,

$$\mathfrak{X} \mid \mathfrak{a} \vdash (\forall x)A_1(x) \tag{6}$$

$$\mathfrak{X} \mid \mathfrak{a} \mid (\forall x) .A_1(x) \supset A_2(x) \tag{7}$$

From (7) we have, by ΠP and Ph,

$$\mathfrak{X} \mid \mathfrak{a} \vdash (\forall x)A_1(x) \mathbin{.\supset.} (\forall x)A_2(x)$$

From this and (6) we have (5) by Ph.

Remark 1. If $\Pi_2$ is not postulated, then it can be derived, as shown in the proof of Theorem 3, from

$$\vdash (\forall x)C \supset C \tag{8}$$

and $\Pi_1$. But (8) follows by Cases 2 and 3 only, and these do not require $\Pi_2$. Note that ΠP is postulated in any form of $HX_p^*$.

## EXERCISES

**1.** Show that if one adjoins to HX the axiom schemes

$$\vdash (\forall x)A(x) \mathbin{.\supset.} A(t)$$
$$\vdash A(t) \mathbin{.\supset.} (\exists x)A(x)$$

and the rules

$$\vdash C \supset A(a) \to \vdash C \supset (\forall x)A(x)$$
$$\vdash A(a) \supset C \to \vdash (\exists x)A(x) \supset C$$

where $a$ is not free in $C$, then one obtains a system with the same assertions as HX*. (This is essentially the formulation of Hilbert and Ackermann [GZT₃]. In [GZT₄] a formulation similar to Schütte's is used.)

**2.** In mathematics we frequently use the following type of argument. Starting with certain premises we derive a theorem to the effect that

$$\vdash (\exists x)A(x)$$

We then say, let $y$ be such that

$$A(y)$$

Then, using modus ponens and premises which do not contain $y$, we derive

$$\vdash B$$

Assuming that these arguments can be justified as proofs using HX* and premises which do not contain $x$ or $y$, show that the arguments establish

$$\vdash B$$

as a consequence of those premises. What limitations are there? (Rosser [LMt, pp. 128ff.]; cf. Quine [MeL].)

**3.** Show that if one interprets

$$A_1, \ldots, A_m \mid \mathfrak{a} \vdash B_1, \ldots, B_n$$

as

$$\vdash \ulcorner A_1 \vee \ulcorner A_2 \vee \cdots \vee \ulcorner A_m \vee B_1 \vee \cdots \vee B_n$$

where all free variables which occur are in $\mathfrak{a}$, then the rules of LK* hold. Use these rules to formulate a system with the same assertions as HK*. (Cf. Schütte [SWK].)

**4.** Derive

$$\vdash (\forall x)(\forall y)A(x,y) \supset (\forall y)(\forall x)A(x,y)$$

in $(HA_g^*)$ and $(HA_p^*)$.

**5.** A "predicate calculus with equality" is often defined as one whose assertions are obtained by adjoining to predicate calculus the schemes

$$\vdash a = a$$
$$\vdash a = b \mathbin{.\supset.} A(a) \supset A(b)$$

Show that if one adjoins these as assertion schemes to some HX* ($\mathfrak{S}$), the resulting system is the same as one obtained from LX*($\mathfrak{S}'$), where $\mathfrak{S}'$ is obtained from $\mathfrak{S}$ by adjoining the auxiliary schemes

$$a = a$$
$$a = b, a = c \vdash b = c$$
$$a = b, \psi(a) \vdash \psi(b)$$

in which $\psi$ is any unary term operation obtained by fixing all but one of the arguments of some $\phi$ in $\Phi$. Thus a predicate calculus with equality is a special case of a predicate calculus over $\mathfrak{S}$.

**6.** Show that the mapping of HK in HM described in Exercise 6B9 cannot be extended to include quantification. Are there other mappings which do work? (See, for example, Prawitz, Dag and P. E. Malmnäs, "A Survey of some Connections between Classical, Intuitionistic, and Minimal Logic," in *Contributions to Mathematical Logic*, edited by H. A. Schmidt, K. Schütte, and H. J. Thiele, Amsterdam, 1968, pp. 215–229.)

**\*7.** In what way would the present theory be modified if Assumption A1 were dropped, and how would the resulting theory be related to H theories of quantification over arbitrary (including null) domains? (Cf. Hailperin [QTE], [TRQ]; Jaskowski [RSF, sec. 5]; Mostowski [RPP]; Quine [QED].)

**\*8.** To what extent do the results of Sec. 4 depend on the order of the quantifiers in forming the closures? (Cf. Quine [MLg$_2$, pp. 88–95].)

## D. CLASSICAL EPITHEORY

Among the immense variety of epitheorems relating to the classical predicate calculus, four typical ones, one of them nonconstructive, have been selected for discussion here. These may presumably be extended, with suitable modifications, to other systems—extensions to LC*, LE*, in most cases, are immediate. But that question is not gone into (except in the incidental remarks); throughout the section it is supposed that we are dealing with LK* or some of its variations.

**1. Prenex normal form.** An ob of a system HX* is said to be in *prenex normal form* just when all its quantifiers are on the outside, i.e., when in its construction from the atoms all the algebraic operations are performed first and the quantification afterward. Thus

$$(\forall x)(\exists y)(\forall u)(\forall v)(\exists w)A$$

is in prenex normal form if $A$ contains no quantifiers, but

$$A \vee (\forall x)B(x) \mathbin{.} \supset \mathbin{.} (\exists x)C(x)$$

is not.

It is a standard theorem of HK*, which is also true for HC*, that given any ob $A$, there is an ob $B$ in prenex normal such that

$$\vdash A \supset B \ \& \ \vdash B \supset A$$

We shall prove a somewhat more general theorem which leads to this result, viz., that if $A$ contains a quantifier in its interior—viz., at the beginning of an algebraic component—then that quantifier can be moved to the outside. The new quantifier will be of the same kind as the original one if the occurrence was positive, and of the opposite kind if the occurrence was negative.

We recall the definitions of $\leq$ and $=$ from Sec. 5A1, viz.,

$$A \leq B \rightleftharpoons \vdash A \supset B$$
$$A = B \rightleftharpoons A \leq B \,\&\, B \leq A$$

By virtue of the theorems of Sec. 5D4, we have

$$A \leq B \rightleftharpoons A \Vdash B$$

The symbols '$\leq$' and '$\Vdash$' may thus be used interchangeably.

If $A$ is a proposition, let $M[A]$ be a proposition containing a specific algebraic component $A$ whose occurrence is either positive or negative. Then $M[B]$ will be the proposition obtained by replacing $A$ by $B$ in that particular occurrence. In the following it will be assumed that neither $b$ nor $x$ occurs in $M[A]$ except in so far as it occurs in $A$.

**Theorem 1.** *If the occurrence of $A$ in $M[A]$ is positive, then*

$$(\forall x)M[A(x)] = M[(\forall x)A(x)] \tag{1}$$
$$(\exists x)M[A(x)] = M[(\exists x)A(x)] \tag{2}$$

*If the occurrence of $A$ in $M[A]$ is negative, then*

$$(\forall x)M[A(x)] = M[(\exists x)A(x)] \tag{3}$$
$$(\exists x)M[A(x)] = M[(\forall x)A(x)] \tag{4}$$

*Proof.* A part of this proof is easy. In fact, since

$$(\forall x)A(x) \leq A(b) \leq (\exists x)A(x)$$

we have by the replacement theorem (Sec. 5D3) in the positive case

$$M[(\forall x)A(x)] \leq M[A(b)] \leq M[(\exists x)A(x)] \tag{5}$$

and hence by $\Pi*$ and $*\Sigma$

$$M[(\forall x)A(x)] \leq (\forall x)M[A(x)] \tag{6}$$
$$(\exists x)M[A(x)] \leq M[(\exists x)A(x)] \tag{7}$$

In the negative case the inequalities in (5) run in the opposite direction; hence we have, by the same reasoning,

$$(\exists x)M[A(x)] \leq M[(\forall x)A(x)] \tag{8}$$
$$M[(\exists x)A(x)] \leq (\forall x)M[A(x)] \tag{9}$$

These results hold in all systems LX*.

To establish the converses of (6) to (9) we employ structural induction. The basic step of this induction, where $M(A) \equiv A$, is trivial. The inductive step in this induction is made by the following:

| | |
|---|---|
| $(\forall x).\, B \supset A(x) \Vdash B \supset .\,(\forall x)A(x)$ | (LA*) |
| $B \supset .\,(\exists x)A(x) \Vdash (\exists x).\, B \supset A(x)$ | (LC*) |
| $(\forall x)A(x) .\supset B \Vdash (\exists x).\, A(x) \supset B$ | (LC*) |
| $(\forall x).\, A(x) \supset B \Vdash (\exists x)A(x) .\supset B$ | (LA*) |
| $(\forall x).\, A(x) \wedge B \Vdash (\forall x)A(x) .\wedge B$ | (LA*) |
| $(\exists x)A(x) .\wedge B \Vdash (\exists x).\, A(x) \wedge B$ | (LA*) |
| $(\forall x).\, A(x) \vee B \Vdash (\forall x)A(x) .\vee B$ | (LC*) |
| $(\exists x)A(x) .\vee B \Vdash (\exists x).A(x) \vee B$ | (LA*) |
| $\neg(\forall x)A(x) \Vdash (\exists x) .\neg A(x)$ | (LK*) |
| $(\forall x).\neg A(x) \Vdash \neg (\exists x)A(x)$ | (LM*) |

The proofs of these are left as exercises (see Exercises B1 and B2). The systems in which they hold are indicated at the right.

**2. The Herbrand-Gentzen theorem.** This is the theorem which Gentzen called his extended principal theorem (*"erweiterter Hauptsatz"*). This theorem is, however, not closely related to ET (which is what corresponds to Gentzen's *"Hauptsatz"* in the present connection).[1] It is, however, closely related to the principal theorem of Herbrand [RTD]. It is therefore appropriate to call the new theorem the Herbrand-Gentzen theorem. Its statement is as follows:

**Theorem 2.** *Let $\Gamma$ be an elementary theorem of a system LX\* which is based on LC. Let the constituents of $\Gamma$ be in prenex normal form. Then there exists a $\Gamma'$ which is an elementary theorem of LX (and hence is algebraic) such that $\Gamma$ can be obtained from $\Gamma'$ by quantification and structural rules only.*

*Proof.* By Theorems B5 and B6, 5C2, 5C4, and 5E6 we can suppose that we are dealing with Formulation II, that the prime statements contain no quantifiers, and that the applications of *K* are initial, with principal constituent containing no quantifiers. Let $\Delta$ be a demonstration of $\Gamma$ conforming to these conditions. We show that whenever a quantification inference is followed immediately (except for structural inferences) by an algebraic inference, we can interchange the two. Using the notation of Sec. 5D1, which does not show the sides on which the constituents occur, a quantification rule is a rule $R_1$ of the form

$$\frac{\mathfrak{X}, P}{\mathfrak{X}, Q} \tag{10}$$

in which $Q$ is the principal constituent and $P$ the subaltern; an algebraic rule $R_2$ with two premises will be of the form

$$\frac{\mathfrak{X}_1, \mathfrak{U}_1 \qquad \mathfrak{X}_2, \mathfrak{U}_2}{\mathfrak{X}_1, \mathfrak{X}_2, M} \tag{11}$$

in which $M$ is the principal constituent and $\mathfrak{U}_1$, $\mathfrak{U}_2$ the subalterns; an algebraic rule with one premise will be obtained by simply omitting the right-hand premise and also dropping $\mathfrak{X}_2$ in the conclusion.[2] A situation where $R_2$ immediately follows $R_1$ in $\Delta$ will be an instance of the scheme

$$\frac{\dfrac{\mathfrak{X}_1, P}{\mathfrak{X}_1, Q}}{\mathfrak{X}_2, Q, \mathfrak{U}_1 \qquad (\mathfrak{X}_3, \mathfrak{U}_2)} \tag{12}$$
$$\overline{\mathfrak{X}_2, (\mathfrak{X}_3), Q, M}$$

---

[1] In Gentzen, ET was a postulate and his *Hauptsatz* was to the effect that any proof could be transformed into one in which ET was not used. Thus, in order to prove the theorem analogous to the one at present under discussion, he had to begin by an appeal to the *Hauptsatz* to be sure that there was a proof without a cut. Here, however, that conclusion follows by definition.

[2] If one were to admit as algebraic rules such a rule as ⊢*, which may have more than two premises, the necessary changes would be easily made.

where the double horizontal line indicates a change by (at most) structural rules. By the proof of the inversion theorem this can be replaced by the following scheme:[1]

$$\frac{\overline{\overline{\frac{\mathfrak{X}_1, P}{\mathfrak{X}_2, P, \mathfrak{U}_1}}} \quad (\mathfrak{X}_3, \mathfrak{U}_2)}{\frac{\mathfrak{X}_2, (\mathfrak{X}_3,) \, M, P}{\mathfrak{X}_2, (\mathfrak{X}_3,) \, M, Q}} \tag{13}$$

(If necessary the characteristic variable in $P$ can be changed by Theorem A1.)

Applying this process as long as there are such pairs $R_1$, $R_2$, we must eventually[2] reach a demonstration where there is no quantifier inference over an algebraic one.

Then, since quantification and structural rules are one-premise rules, we can start at $\Gamma$ and work upward in $\Delta$ until we reach a $\Gamma'$ which is the conclusion of an algebraic rule. Then, by the restrictions assumed in the first paragraph, $\Gamma'$ will not contain any quantified constituents.[3] Thus $\Gamma'$ has the properties required in the theorem, Q.E.D.

**3. The Skolem normal form.** An ob of HK* is said to be in *Skolem normal form* when it is in a prenex normal form in which every existential quantifier precedes every universal quantifier.[4] If the system HK* contains unrestricted predicate variables (Sec. A4, Remark 3; see Exercise A4), then we shall see here that, given any proposition $A$, there is an $A^S$ in Skolem normal form such that

$$\vdash A \rightleftarrows \vdash A^S \tag{14}$$

Suppose first that $A$ is $(\forall x)B(x)$. Let

$$A^* = (\forall x). \, B(x) \supset \phi(x) \, . \supset . \, (\forall y)\phi(y)$$

where $\phi$ is a predicate variable not occurring in $A$. Then we have $A \le A^*$ as follows:

$$\frac{\dfrac{\overset{1}{A}}{B(a)}\text{IIe} \qquad \dfrac{\overset{2}{(\forall x).B(x) \supset \phi(x)}}{B(a) \supset \phi(a)}\text{IIe}}{\dfrac{\dfrac{\phi(a)}{\dfrac{(\forall y)\phi(y)}{\dfrac{A^*}{A \supset A^*}\text{Pi-1}}\text{Pi-2}}\text{IIi}}{}}\text{Pe}$$

---

[1] We can suppose that the structural inferences shown do not include an application of *W* with $Q$ as principal constituent. Such a *W* can be passed below the algebraic inference before we begin; for the latter already admits $Q$ as parameter, and all rules are such that if they admit one parameter they will admit any number of repetitions of it.

[2] If we define the order of $\Delta$ as the total number of pairs $R_1$, $R_2$ such that $R_1$ is a quantification inference, $R_2$ is an algebraic inference, and $R_1$ is (not necessarily immediately) over $R_2$, then the process diminishes the order and must eventually reduce it to zero.

[3] If we were to omit the restrictions on *K*, one could invoke Theorem 5C1 to eliminate any such constituents.

[4] The case where either or both of these kinds of quantifiers are totally missing is to be included.

Conversely, if we substitute $B(a)$ for $\phi(a)$ in $A^*$, we get an $A'$ such that $A' \leq A$.  If we define

$$A^S \equiv (\exists x)(\forall y) : B(x) \supset \phi(x) \mathbin{.\supset.} \phi(y)$$

then $A^S = A^*$, and consequently (14) is satisfied.

The relation between $A$ and $A^S$ so defined is more specific than (14).  In fact, if we define

$$A \,(=)\, B \tag{15}$$

as meaning that

$$A \leq B$$

and that there is a $B'$, obtained from $B$ by substituting for the predicate variables, such that

$$B' \leq A$$

then we have shown that $A \,(=)\, A^S$.

The relation (15) is not an equivalence, but it is reflexive and transitive, and it has the properties

$$A = B \rightarrow A \,(=)\, B \rightarrow. \ \vdash A \rightleftarrows \vdash B \tag{16}$$
$$A \,(=)\, B \rightarrow f(A) \,(=)\, f(B) \tag{17}$$

where $f$ is any directly monotone unary operation.  Thus we can use the process just described to prove the following:

**Theorem 3.**  *Let $A$ be formed by prefixing $p$ universal and $q$ existential quantifiers to a proposition $M$, and let the system admit unrestricted predicate variables.  Let $r = p + q$.  Then there exists an $A^S$ of the form*

$$(\exists \overline{x_1}) \cdots (\exists x_r)(\forall y_1) \cdots (\forall y_p)M^*$$

*where $M^*$ is formed by algebraic operations from $M$ and predicate variables, such that*

$$A \,(=)\, A^S$$

*Proof.*  If $A$ is already in Skolem normal form, there is nothing to prove.  We show that, using the method of the introductory discussion, and an induction on the number of universal quantifiers which precede an existential quantifier, we can find $A^S$.  In this we understand that, for any $C$, $C^P$ is the result of substituting the apparent variables $x_i$, $y$, $z$, indicated in the quantifiers preceding $C^P$, for corresponding real variables $a_i$, $b$, $c$.

Now suppose that

$$A \equiv (\exists x_1) \cdots (\exists x_m) A_1^P \qquad (m < p)$$
$$A_1 \equiv (\forall z) B^P$$
$$A_2 \equiv (\forall z) \mathbin{.} B^P \supset \phi(z) \mathbin{.\supset.} (\forall y)\phi(y)$$

where $\phi$ is a predicate variable depending on the $a_1, \ldots, a_m$ and not appearing in $B$.  Let $A_3$ be obtained from $A_2$ by replacing $B$ by a $B^*$ such that

$$B \,(=)\, B^* \tag{18}$$

Let                          $$A^* \equiv (\exists x_1) \cdots (\exists x_m) A_3^P$$

Then we have by the preliminary discussion

$$A_1 \ (=) \ A_2$$

Since $B$ has a positive position in $A_2$, it follows by (17) that

$$A_2 \ (=) \ A_3$$

Hence by ($\tau$)

$$A_1 \ (=) \ A_3$$

and thus by (17)

$$A \ (=) \ A^*$$

If we take $B^*$ to be $B^S$ [which is defined and satisfies (18) by the inductive hypothesis], then when we transfer the quantifiers of $A^*$ to the beginning by Theorem 1, we shall have, by (16), an $A^S$ which satisfies the conditions of the theorem, Q.E.D.

Note that there is some latitude in the choice of $B^*$, and this allows a more specific characterization of $A^S$ to be made. This will not be gone into here.

**4. The completeness theorem.** Gödel [VAL] showed that HK* is complete in the sense that a proposition which is valid in every enumerable model is demonstrable. This is necessarily a nonconstructive result, because Church [NEP] proved that HK* is recursively undecidable; and if there were a constructive method of deciding whether or not a proposition $A$ is valid in every model, then it would furnish a constructive decision method. As such a nonconstructive result, it lies outside the scope of this book; but there are two reasons for including it nevertheless. The first of these is the fundamental nature of the theorem in relation to large areas of currently active investigation in mathematical logic; the second is that Rasiowa and Sikorski [GTh] have presented a proof which ties in very closely with the methods of this book. It thus forms an ideal transition from the foundations of mathematical logic to its superstructure.

Let us recall the proof tableau of Sec. B6. We modify it so as to apply without the assumption that $\Omega$ is void. We do this as follows. Since in LK the rule *P is reversible, we can move XII forward until it is immediately after VII. Then the rules I to VII and XII will remove all algebraic outside operations, and VIII will simply introduce an $F$ on the right. The algorithm will either close or get through to X. Now the terms of $t(q)$ form an enumerable set; let $t_1, t_2, \ldots$ be a fixed enumeration of them. Suppose we modify X and XI by dropping $A(t_1), A(t_2), \ldots, A(t_{r-1})$ and letting $t_r$ be the first term in the fixed sequence of terms which has not been tried as subaltern for that particular $(\forall x)A(x)$ or $(\exists x)A(x)$ in the branch between that node and the top of the tableau. In order to prevent stalling on the wrong alternative, we provide that X and XI be applied alternately.

Now suppose we start the algorithm with an elementary statement $\Gamma_0$ which we call the head of the tableau. If the algorithm leads to a closed tableau, then we have constructively a proof of $\Gamma_0$. We shall have a proof of the Gödel completeness theorem if we deduce from the fact that the tableau does not close a direct interpretation over an enumerable model in which $\Gamma_0$ is invalid.  In view of the Remark of Sec. B5, we can conclude this from the following converse to Theorem B9:

**Theorem 4.  (Nonconstructive.)**  *The statement* $\Gamma_0$ *is demonstrable in* LK($\mathfrak{D}$) *unless there is a countervaluation for* $\Gamma_0$.

*Proof.*  For the system LK($\mathfrak{D}$) the algorithm has the following characteristics.  There is always an applicable rule unless the constituents of the datum are entirely elementary.  Second, there are no alternative splittings; if the tableau splits, it splits conjunctively and into at most two subtableaux. Consequently, each branch of the tableau will determine a sequence $i_1$, $i_2$, $i_3$, . . . , where $i_k = 0$ if at the $k$th step there was only one result, or if there were two, the left-hand result was taken; and $i_k = 1$ if at the $k$th step there were two results and the right-hand one was taken.  If the tableau does not close, then there are two possibilities.  It may be that in the $k$th step of some branch we reach a $\Gamma_k$ which has all its constituents elementary but is not quasi-prime; or there may be a branch which continues indefinitely without ever reaching a quasi-prime statement.[1]  Let $\Delta$ be a branch satisfying one of these conditions, and let its statements in order be $\Gamma_0$, $\Gamma_1$, $\Gamma_2$, . . . . The sequence will be finite in the first possibility, infinite in the second.

Next we determine two classes $\mathfrak{M}$, $\mathfrak{N}$ of elementary propositions.  Let $\mathfrak{M}$ consist of all elementary propositions which appear as left constituents of some $\Gamma_k$ in $\Delta$, and let $\mathfrak{N}$ be those which appear as right constituents in some $\Gamma_k$.  Then we shall show that (*a*) the classes $\mathfrak{M}$ and $\mathfrak{N}$ are nonoverlapping, and (*b*) any valuation in which all the members of $\mathfrak{M}$ have the value 1 and all those in $\mathfrak{N}$ the value 0 is a countervaluation for $\Gamma_0$.

As to statement (*a*), the algorithm has the property that when an elementary constituent appears on either side in any $\Gamma_i$, it remains on the same side of all $\Gamma_j$ for $j > i$.  Hence, if $\mathfrak{M}$ and $\mathfrak{N}$ had a member in common, some $\Gamma_k$ would be quasiprime, which is a contradiction.

To take care of statement (*b*), let us call a constituent of $\Gamma_k$ which is a right constituent with value 1 or a left constituent with value 0 a positive constituent; one with the opposite character, a negative constituent.  By our construction all elementary constituents are negative.  Let us define the order of a constituent as the number of operational steps used in its construction.  If $\Gamma_0$ is valid, there will be a positive constituent in it; let us call it $A$.  The algorithm is such that $A$ will remain on the same side until it becomes the principal constituent.  Let this happen at $\Gamma_k$.  If $A$ is of any of the forms $B \supset C$, $B \wedge C$, $B \vee C$, $\neg B$, $(\forall x)B(x)$ on the right, or $(\exists x)B(x)$ on the left, then the subaltern in $\Gamma_{k+1}$ will be of lower order.  If $A$ is of the form $(\exists x)B(x)$ on the right, then, since such a constituent is never eliminated, $\Delta$ is infinite.  Since $(\exists x)B(x)$ is positive, there will be a $B(t_i)$ which is positive, and eventually there will be a $\Gamma_k$, where $A$ is principal constituent and $B(t_i)$ is a subaltern.  The situation is dual to this if $A$ is $(\forall x)B(x)$ on the left.  Thus, in all cases, if there is a positive constituent in $\Gamma_0$ of order $n$, there will be one of order $n - 1$ further down, then of order $n - 2$ below

---

[1] We can even use an argument similar to that of usual proof of the Bolzano-Weierstrass theorem to determine such a sequence uniquely.  For having determined $i_1$, . . . , $i_k$, we take $i_{k+1} = 1$ if the tableau splits at that point and the left-hand subtableau closes; otherwise we take $i_{k+1} = 0$.  The nonconstructive element enters because the closing of the subtableau is an indefinite question.

that, and so on. Eventually there will be a positive elementary constituent, which is impossible. This contradiction has come from assuming that $\Gamma_0$ was valid. Hence $\Gamma_0$ is invalid, Q.E.D.

### EXERCISES

**\*1.** What sort of normal form can replace the prenex normal form in systems based on LA?

**2.** Show that the Herbrand-Gentzen theorem is true for $LA_m$, $LM_m$, $LJ_m$ provided the rule \*V is deleted, but that if \*V is admitted, the following is a counterexample:

$$B \vee C, (\forall x). \; B \overset{.}{\supset} A(x) \Vdash (\forall x)A(x), C$$

(Kleene [IMM, pp. 460–463]. His counterexample, which is Example 1 of Sec. B1, does not work for $LA_m$.)

**3.** Show that the Skolem normal form can be further specialized to a "normal alternation," i.e., a union of propositions each of which is in Skolem normal form with only one universal quantifier (Hilbert and Bernays [GLM.I, p. 159]).

### S. SUPPLEMENTARY TOPICS

**1. General historical and bibliographical comment.** This chapter is a revision of chap. 3 of [TFD]. The main part of it extends to the theory of quantification, the general method established in the previous chapters. The references given Secs. 1S5, 5S1, and 6S apply in large part to the present chapter.

The sources in regard to the method of treating bound variables have been given in the main text, Sec. A2. Various alternative ways of doing this are discussed, with critical comments, in Church [IML$_2$, p. 290]. See also Quine [MLg], [MeL]. Quine is especially clear in explaining the translation of quantification into ordinary discourse.

The present treatment (Sec. B5) of the classical evaluation contains improvements over that of [TFD], sec. III6. The idea of model, as described in the remark at the end of Sec. B5, is fundamental in modern semantical epitheory (Sec. 3S3), and the pioneers in that development have formulated it very precisely. There is thus no novelty in Sec. B5. However, in current work, the term 'valuation' is not always used in exactly the same sense as in Sec. B5.

The result of Theorem B10 was obtained independently by several different persons. The first publication of the result, obtained by the use of the Gentzen technique, was in Kleene [ILg]; at about the same time Mostowski [PND] derived the results by another method. Both of these refer to previous proofs of nondeducibility using arithmetic models. The present proof was taken from [TFD], where it was in turn taken, after correction of some errors, from a draft originally prepared for [PFD]. For other examples of intuitionistic nondeducibility, see Exercise B4.

For the history of proof tableaux see Sec. 1S5. The present treatment merely adjoins rules for quantification, based on those given by Beth [SCI], to those given earlier. For alternative procedures, intended to accomplish the same purpose as the Beth algorithm, see Stanley [EPQ], Quine [PPQ].

The systems of predicate calculus appearing in current treatments of mathematical logic are mostly systems HX* and usually $HK_g^*$. Information on the history of these systems is given in Church [IML$_2$, chaps. 3 and 4, especially sec. 49]. Systematic presentations of the system $HK_g^*$ are found, for example, in Church [*loc. cit.*], Hilbert and Ackermann [GZT], Hilbert and Bernays [GLM], and Kleene [IMM].

For system $HK_p^*$ see especially Quine [MLg$_2$], where there is a sketch of the history on pp. 88ff. This history goes back to 1935. The idea that free variables are, semantically considered, to be regarded as bound by a universal quantifier, and that therefore free variables, in principle, should not appear in the elementary theorems of a logic, was widely, if not generally, recognized before that time. Thus one finds the idea in Jaśkowski [RSF, sec. 5]; free variables are excluded from the elementary theorems in Church [SPF]; and one of the announced purposes of early work in combinatory logic (even in Schönfinkel [BML]) was to show that one can form a logic in which one does not admit—in the elementary statements, of course, for it is not advocated that they be abolished as an epitheoretical device—variables of any kind. The present treatment is an outgrowth of this point of view. Thus the name 'ΠP' comes from [UQC], and 'Π$_0$' from [AVS], and the proof of admissibility of Πh was modeled on the proof in [PEI]. That there is a great resemblance of the resulting theory to that of Quine is due to a process of convergence (in the biological sense); the exact relation between the two approaches is left as an exercise (Exercise C8).

The prenex normal form is a standard technique of predicate calculus. An account of it will be found in any treatise. For the history see Church [IML$_2$, p. 292].

The term 'Herbrand-Gentzen theorem' was proposed by Craig [LRN]. In that paper and others he has proved a number of extensions of the theorem. The sources of the theorem are given in the text.

On the Skolem normal form see Hilbert and Bernays [GLM.I, p. 159] and Church [IML$_2$, sec. 42]; these give references to sources.

On the Gödel completeness theorem see Sec. 3S1.

**2. Further developments.** The literature connected with the predicate calculus is so extensive that it is futile to do more than comment on a sample of the developments; the selection is necessarily somewhat arbitrary. For those involving semantical or nonconstructive methods see Sec. 3S3.

For a long time the study of the predicate calculus was dominated by the decision problem (*Entscheidungsproblem*). This is the problem of finding a constructive process which, when applied to a given proposition of HK*, will determine whether or not it is an assertion. (This is the syntactical form of the problem; there is also a semantical form where assertibility is replaced by validity in any evaluation. The two are equivalent by the Gödel completeness theorem.) Since a great variety of mathematical problems can be formulated in the system, this would enable many important mathematical problems to be turned over to a machine. Much effort in the early 1930's went into special results bearing on this problem. In 1936, Church showed in his [NEP] and [CNE] that the problem was unsolvable. Inasmuch as the other systems can be mapped in HK* (see Exercise C6), the decision problem is unsolvable for all the systems HX*.

Since Church's result, the decision problem has become rather a specialized domain of investigation. Two kinds of results continue to be obtained: first, solutions of special cases (such as the case where only unary predicates are present or where, in the Skolem normal form, $r \leq 2$), and second, reductions of the general problem to cases where the proposition to be investigated is of a special form (e.g., the Skolem normal form with $r = 3$ and a single binary predicate). Monographs on both of these aspects now exist: for the first aspect see Ackermann [SCD], for the second Suranyi [RTE]. Considerable information about the problem is given in Church [IML$_2$, chap. 4] and in Hilbert and Ackermann [GZT$_4$, sec. IV 11].

Systems formulated in the predicate calculus as basis have been the subject of systematic study, notably by Tarski and others. On these see Sec. 3S3.

Another question connected with the predicate calculus is the eliminability of descriptive functions. We have admitted here the possibility of term operations $\Omega$. Theories exist in which there are functions from propositions to terms, or rather, since these operations are like quantifiers in that they bind a variable, from propositional functions to terms. Among these are the description operator; this forms from $A(b)$ the term

$$(\iota x)A(x)$$

which is interpreted as the unique object $a$ such that $\vdash A(a)$. Another example is the Hilbert $\epsilon$ operator, where

$$(\epsilon x)A(x)$$

is a term interpreted as some $a$ for which $\vdash A(a)$ (if any exists). For the $\epsilon$ operations one postulates

$$A(t) \leq A((\epsilon x)A(x))$$

and then in the classical system one would have the quasi definitions

$$(\forall x)A(x) \equiv A((\epsilon x)\neg A(x)) \qquad (\exists x)A(x) \equiv A((\epsilon x)A(x))$$

These operations are related. Theorems of eliminability of these various notions are of some importance. Presumably the Gentzen technique would help a great deal in their proofs. For the $\epsilon$ operator this is proposed in Exercise B13. For descriptions see Hilbert and Bernays [GLM.I, sec 7]; Asser [AFP]; Hailperin [RID], [TRQ]; Johansson [CCA]; LeBlanc and Hailperin [NDS]; Montague and Kalish [RDN]; Rosser [CQN], [LMt, chap. 8]; Schröter [TBA]; Schütte [EBA]. In the reduction theory of the decision problem there are results regarding the elimination of descriptive functions generally.

A number of persons have been interested in formulations of predicate calculus in which there are several different sorts of terms, or more generally, variables ranging over domains which are restricted in some way. On this topic see Herbrand [RTD, sec. III 3]; Schmidt [DTM], [ZBM]; Wang [LMS]; Hintikka [RTT]; Quine [UUS]; Hailperin [TRQ]; Lightstone and Robinson [STr]; Gilmore [ALM].

It has occurred to several persons to ask whether we could not have lattice formulations of quantification. Of course we can; the cylindrical algebras of Tarski and Henkin and the polyadic algebras of Halmos are examples of such systems. On the former see Tarski [NMB]; Henkin [ACQ], [SAT], [RTC]; Henkin and Tarski [CA1]. On the latter a general exposition is in Halmos [BCA], a technical summary in Halmos [PBA], and a detailed account in Halmos [ALg]. See also Galler [CPA].

# Chapter 8

# MODALITY

Since the dawn of logic, logicians have noticed that there is a distinction between truths which seem to come to us as a matter of necessity and others which are so just because they happen to be. Some logicians, from Aristotle down, have attempted to take account of this difference. There has thus arisen a branch of logic, called *modal logic*, in which such distinctions are made.

Modal logic is a rather specialized branch of logic. It therefore lies for the most part outside the scope of this book. But there is some interest, even for the foundations, in the fact that it can be treated mathematically. Moreover, its treatment by the methods of this book enables one to gain a deeper insight into the significance of those methods. This chapter is devoted to as much of a discussion of modal logic as helps to attain that objective.

## A. FORMULATION OF NECESSITY

In this section we examine the meaning of necessity, and then set up an L system conforming to the semantic analysis.

**1. The analysis of necessity.** We must begin by divesting ourselves of any feelings of metaphysical or psychological compulsion that we have in regard to logical necessity. Such feelings may exist, but we must regard them as just as irrelevant here as in the previous chapters. We seek here, as there, a fully objective notion.

To attain this, we recall that we have a formal theory whenever we have a category $\mathfrak{E}$ of elementary statements and some means of generating a subclass $\mathfrak{T}$ of $\mathfrak{E}$ which we call the elementary theorems. We have a formal theory with negation if we have a second subclass $\mathfrak{F}$ of $\mathfrak{E}$ satisfying conditions which we studied in Chap. 6. In the same way conventions which determine not only $\mathfrak{T}$ but a subclass $\mathfrak{N}$ of $\mathfrak{T}$ give us a theory with necessity. If the theory is a deductive theory, this means that we have two sets of deductive rules, an inner set which determines $\mathfrak{N}$ and an outer set which determines $\mathfrak{T}$, these being such that whenever the inner rules apply, the outer ones do also. It does not make any difference what metaphysical or psychological notions we associate with $\mathfrak{N}$ and $\mathfrak{T}$.

If the necessary statements are identified with the $\mathfrak{N}$ in such a situation, then there are several potential applications. If we were formalizing an experimental science, we might consider the necessary statements to be those obtained by the theory alone, while $\mathfrak{X}$ might contain statements obtained by certain experiments.   Again, if we were studying biophysics, we might consider $\mathfrak{N}$ to be the statements of the basic physics we are accepting, while $\mathfrak{X}$ was based on the addition to these of certain biological principles. If we were studying theoretical physics, we might consider the statements established on purely mathematical grounds to be in $\mathfrak{N}$.   These are all applications in the field of experimental science.   But we can conceive of applications to mathematics itself.   Thus, if we were studying elementary geometry, we might put in $\mathfrak{N}$ those statements which are obtained from the axioms of incidence alone, or those invariant of an arbitrary projective transformation.   Finally, we might put in $\mathfrak{N}$ those statements obtained in a strictly constructive manner, while in $\mathfrak{X}$ a classical logic was admitted.

These examples not only show that one may conceive of necessity in an objective manner, but they suggest certain generalizations.   For one thing, there may be cases where distinctions of more than two levels exist.   This, however, does not appear to introduce anything new in principle.   More interesting is the question of whether one could get an analysis of possibility. Evidently, if negation is present, one could define a possible statement as one whose negation is not necessary.   But one should be able to define possibility directly.   Thus in relation to Euclidean geometry one could define the necessary statements to be those of absolute geometry, whereas our intuition would say that the possible statements should include those of hyperbolic and elliptic noneuclidean geometry.   This example suggests that we might define necessity and possibility with reference to a set $\mathfrak{X}_1$, $\mathfrak{X}_2, \ldots$ of theories by saying that the necessary statements are those that hold in all the $\mathfrak{X}_i$ and the possible statements are those that hold in some $\mathfrak{X}_i$.   In such a case necessity has something of the character of a universal quantifier, possibility that of an existential one.   As the example indicates, possible statements may be incompatible with one another.

In this chapter we shall ignore these generalizations[1] and pass on to formalize the notion of necessity described in the third preceding paragraph.

**2. The formalization of necessity.**   We proceed in the manner of Sec. 6A.   It is not necessary to dwell on the preliminary steps, which are the same here as there; we can suppose that we are dealing with propositions, rather than statements, and that we have a unary operation, indicated by the prefix '$\square$',[†] such that $\square A$ is an assertion just when $A$ is necessary.

In order to formulate L rules, we distinguish two levels of the system, an inner level and an outer level.   As elementary statements we take the following two forms:

$$\mathfrak{X} \Vdash \mathfrak{Y} \qquad \mathfrak{X} \Vdash' \mathfrak{Y} \qquad (1)$$

with the understanding that the rules of the inner level are to apply to the left-hand statement (1), those of both levels to the right-hand statement.

---

[1] The suggestion of the preceding paragraph in regard to possibility has never been satisfactorily worked out.   Conjectures about possibility in [TFD] have turned out to be false, and the possibility proposed in [ETM] is merely that in which the possible statements form a subclass of $\mathfrak{E}$ more inclusive than $\mathfrak{X}$.   Possibility is treated as dual of necessity in Ohnishi and Matsumoto [GMM].

[†] In [TFD] the prefix '$\#$' was used, but '$\square$' is more in agreement with standard practice. Confusion with the infix '$\square$' used in Sec. 2D1 is hardly likely.

We shall use the infix '⊩' in cases where it is not specified whether the inner or the outer system (i.e., level) is intended; in any context all occurrences of '⊩' are to be replaced by the same one of '⊩′', '⊩″'.

In seeking to find a justification for the rules relating to (1), we confine attention to the singular case, because we have found that that is the case which one justifies semantically; the justification in the multiple case is then to be sought by interpretation of the multiple system in the singular one.

We then inquire under what circumstances we can conclude

$$A_1, A_2, \ldots, A_m \Vdash \Box B \tag{2}$$

According to the interpretation we associate with (2), this will mean that there is a deduction of some kind leading from the premises

$$\vdash A_i \tag{3}$$

to

$$\vdash \Box B \tag{4}$$

But by the interpretation in Sec. 1, (4) will mean that there is a deduction, valid in the inner system, leading to

$$\vdash B \tag{5}$$

But the deduction is valid in the inner system only if (a) it is conducted according to the rules of the inner system and (b) its premises are true in the inner system.   Now the $i$th premise (3) does not say that $A_i$ is an assertion of the inner system, but only that it is an assertion of the outer system.   Thus a deduction from (3) to (5) conducted according to the rules of the inner system will be a valid deduction of (5) in the inner system only if each $A_i$ is of the form $\Box A_i'$.   Thus the rule for the introduction of $\Box B$ on the right should be of the form

$$\Box \mathfrak{X} \Vdash B \to \Box \mathfrak{X} \Vdash \Box B \tag{6}$$

where $\Box \mathfrak{X}$, in analogy with the $\neg \mathfrak{Y}$ of Chap. 6,[1] is the prosequence formed from $\mathfrak{X}$ by changing every constituent $A$ to $\Box A$.†

Now for rules of introduction on the left, we ask what inference we can draw from $\vdash \Box A$.   The basic semantical principle is that this has the same meaning as if it had been just introduced.   But in that case the premise would be $\vdash A$; hence one can draw any inference from $\vdash \Box A$ that one can draw from $\vdash A$, leading to the rule

$$\mathfrak{X}, A \Vdash B \to \mathfrak{X}, \Box A \Vdash B \tag{7}$$

Thus (6) and (7) are suitable singular rules.   But the multiple form of (6), even in the form

$$\Box \mathfrak{X} \Vdash B, \Box 3 \to \Box \mathfrak{X} \Vdash \Box B, \Box 3 \tag{8}$$

---

[1] See Exercise 6B7.

† Note that if the '$\Box$' were omitted before the '$\mathfrak{X}$' in (6), we should derive

$$B \Vdash \Box B$$

which is contrary to the interpretation we have for $\Box B$.   One should note, by examining the above discussion, why such a conclusion is a fallacy.

would not be acceptable; for by its aid we should have

$$\frac{A \Vdash A, \ \Box B \qquad \Box B \Vdash A, \ \Box B}{A \vee \Box B \Vdash A, \ \Box B}$$
$$\overline{\Box(A \vee \Box B) \Vdash A, \ \Box B}$$
$$\overline{\Box(A \vee \Box B) \Vdash \Box A, \ \Box B}$$
$$\overline{\Box(A \vee \Box B) \Vdash \Box A \vee \Box B}$$

The conclusion of this proof would not be derivable in the singular system,[1] and thus the interpretation of the multiple system in the singular one would be invalid. On the other hand, there is no objection to the multiple form of (7). Thus we find the following rules appropriate:

Y    *Rules for necessity*

$$*Y \quad \frac{\mathfrak{X}, A \Vdash \mathfrak{Y}}{\mathfrak{X}, \Box A \Vdash \mathfrak{Y}} \qquad Y* \quad \frac{\Box \mathfrak{X} \Vdash B}{\Box \mathfrak{X} \Vdash \Box B}$$

The system formed by adjoining these rules to the system LX will be called the *system* LXY. If quantifiers are present, this will be indicated in the usual fashion; they do not appear to cause any essential difficulty.

## B. THE L THEORY OF NECESSITY

This section will be devoted to a brief study of the L theory of necessity. The study will be confined to the inversion and elimination theorems, on the one hand, and a theorem on representation of the outer system in the inner system, on the other. One could go on to deduce decidability theorems, etc., but this will not be done here.

**1. The inversion and elimination theorems.** Our first inquiry will be as to the extent to which the inversion and elimination theorems are affected by the new rules.

In the case where the $M$ of the inversion theorem or the $A$ of the elimination theorem are nonmodal, i.e., not of the form $\Box B$, I maintain that the proofs of these theorems are not affected at all. For, since the rule Y* does not admit any nonmodal parameters, it will be impossible for this rule to occur in the $\Delta_1$ of the inversion theorem or the $\Delta_2$ of Stages 1 and 2 of ET.[2] The rule *Y is regular. Thus there is no difficulty in these stages of the proof. This part of the argument is true even if Y* is replaced by (8).

If the rule $R$ of the inversion theorem is Y*, then there can be no singular inferences applied in $\Delta_1$ (since $M$ is itself parametric on the right) and hence, in particular, no instances of Y*. Thus the replacement of $M$ (that is, $\Box B$) by $B$ will invalidate no inferences in $\Delta_1$. Therefore Y* is directly invertible, although the direct inversion may not be completed on account of a violation of condition $(a)$. If the rule $R$ is *Y, then replacement of $\Box B$ by $B$ may invalidate any inferences by Y* which there may be in $\Delta_1$. Thus

---

[1] One can show this either directly or by using a valuation over a Boolean algebra with two points $\alpha$ and $\beta$, taking $\Box 0 = \Box \alpha = 0$, $\Box \beta = \beta$, $\Box 1 = 1$, and interpreting $\Vdash$ as meaning $\leq$ between the Boolean meet of the left constituents and the right constituent. (Take $A = \alpha$, $B = \beta$.)

[2] In Stage 2 of ET either $\mathfrak{U}_k$ or $\mathfrak{V}_k$ is nonvoid, and then constituents are nonmodal.

*Y is not directly invertible.   We have the same difficulty on both sides if
Y* is replaced by (8) (but not if arbitrary parameters are allowed on the
right).

Now let us look at ET.   Stages 1 and 2 where $A$ is nonmodal have been
taken care of.   Suppose $A$ is $\square B$.   The situation is then analogous to the
treatment of $A \supset B$ in $\mathrm{LA}_m$.   We consider Stage 2 first.   There can be no
instance of Y*, except one which introduces $A$, in $\Delta_2$; for either there is a
parametric instance of $A$ (in $\mathfrak{B}_k$) or we have a singular system and a nonmodal constituent on the left (in $\mathfrak{U}_k$).   For the same reason either we have a
singular system or no other singular rule can be used in $\Delta_2$.   Thus all rules
in $\Delta_2$ satisfy (r6), and the proof of Stage 2 goes through.   We can therefore
suppose that $\mathfrak{X}'$ (in ET, Sec. 5D2) is of the form $\square\,\mathfrak{W}$ and $\mathfrak{Z}$ is void.   Then
the replacements we have to make in the proof of Stage 1 do not invalidate
any of the inferences of Stage 1, and the proof of Stage 1 also goes through.

It remains to consider Stage 3.   If the hypotheses of Stages 1 and 2 are
satisfied, we have an argument of the following sort:

$$\frac{\dfrac{\square\,\mathfrak{W} \Vdash B}{\square\,\mathfrak{W} \Vdash \square B} \qquad \dfrac{\mathfrak{X}, B \Vdash \mathfrak{Y}}{\mathfrak{X}, \square B \Vdash \mathfrak{Y}}}{\mathfrak{X}, \square\,\mathfrak{W} \Vdash \mathfrak{Y}}\ \text{ET}$$

This can be replaced as follows:

$$\frac{\dfrac{\square\,\mathfrak{W} \Vdash B}{\square\,\mathfrak{W} \Vdash B} \qquad \mathfrak{X}, B \Vdash \mathfrak{Y}}{\mathfrak{X}, \square\,\mathfrak{W} \Vdash \mathfrak{Y}}\ \text{ET}$$

Here the first step on the left is necessary only when the conclusion is in the
outer system; in that case it follows by deductive induction since all the
postulates of the inner system hold in the outer.   We thus have the following:

**Theorem 1.**   *The elimination theorem holds for all the systems* LXY; *further,
the rule* Y* *is directly invertible.*

*Remark.*   The proof fails if Y* is replaced by (8); this is because we cannot
be sure, when $A \equiv \square B$, that $A$ does not occur in the $\square\mathfrak{Z}$ of some other
application of Y*.

**2. Representation of the outer system in the inner.**   The theorem
which we shall now prove contains Theorems 5E4, 6B9, and 7B8 and Exercises 6B9 and 7B9 (with some auxiliary argument) as special cases.

**Theorem 2.**   *Suppose* ET *holds in the inner system, and the outer system is
obtained by adjoining to the inner system rules of the forms*

$$\frac{\mathfrak{X} \Vdash' A_i, \mathfrak{Z} \qquad i = 1, 2, \ldots, m}{\mathfrak{X} \Vdash' B, \mathfrak{Z}} \tag{1}$$

$$\frac{\mathfrak{X}, B \Vdash' \mathfrak{Y}}{\mathfrak{X}, A_1, \ldots, A_m \Vdash' \mathfrak{Y}} \tag{2}$$

*Let* $\Delta$ *be a proof in the outer system that*

$$\mathfrak{X} \Vdash' \mathfrak{Y} \tag{3}$$

*If* (1) *or* (2) *is applied at a node of* $\Delta$, *let*

$$C \equiv A_1 \supset.\ A_2 \supset.\ \cdots \supset.\ A_m \supset B$$

*and let $C^*$ be obtained by closing $C$ with respect to all characteristic variables appearing in $\Delta$ below that node. Let $\mathfrak{M}$ be a prosequence consisting of all such $C^*$. Then*

$$\mathfrak{X}, \mathfrak{M} \Vdash \mathfrak{Y} \tag{4}$$

*Proof.* Let $\Delta \equiv \Gamma_1, \Gamma_2, \ldots, \Gamma_n$ be the given proof of (3), and let $\mathfrak{M}_k$ be the $C^*$ which have been introduced in $\Delta$ in the nodes at or above $\Gamma_k$. If $\Gamma_k$ is

$$\mathfrak{X}_k \Vdash' \mathfrak{Y}_k$$

let $\Gamma'_k$ be

$$\mathfrak{X}_k, \mathfrak{M}_k \Vdash \mathfrak{Y}_k$$

We show by a deductive induction that $\Gamma'_k$ holds for all $k$.

If $\Gamma_k$ is quasi prime of type $(p1)$, or of type $(p2)$ where the axiom is an axiom of the inner system, then $\Gamma'_k$ is also quasi prime, $\mathfrak{M}_k$ is void. If $\Gamma_k$ is quasi prime of type $(p2)$, using an axiom $E$ of the outer system, then we have a case of (1) with $m = 0$, and $\Gamma'$ can be derived by successive applications of $*\Pi$ from the quasi-prime statement

$$\mathfrak{X}_k, E \Vdash E, 3_k$$

Here $\mathfrak{M}_k$ consists solely of $E^*$. This also applies whenever $\Gamma_k$ is derived by (1) with $m = 0$.

If $\Gamma_k$ is derived by a rule $R$ which is nonmodal and valid in the inner system, or if $R$ is $*Y$, let the premise(s) be $\Gamma_{i_1}, \Gamma_{i_2}, \ldots$. Then $\Gamma'_k$ can be obtained from $\Gamma'_{i_1}, \Gamma'_{i_2}, \ldots$ by the same rule $R$. The addition of $\mathfrak{M}_k$ as additional parameters does not invalidate the inference, since none of them can contain the characteristic variable, if any, of $R$.

If $\Gamma_k$ is obtained by rule $Y*$, then $\Gamma_k$ is obtained thus:

$$\frac{\mathfrak{X}_k \Vdash D}{\mathfrak{X}_k \Vdash' \Box D}$$

The inference is valid if we change $\Vdash'$ to $\Vdash$. From this we obtain $\Gamma'_k$ by $*K$.

Suppose $\Gamma_k$ is obtained by (1). Then by the inductive hypothesis and $*K$

$$\mathfrak{X}_k, \mathfrak{M}_{k-1} \Vdash A_i, 3 \qquad i = 1, 2, \ldots m \tag{5}$$

Let

$$C_i \equiv A_{i+1} \supset. A_{i+2} \supset. \cdots. \supset A_m \supset B$$

Then we shall see that for all $i$,

$$\mathfrak{X}_k, \mathfrak{M}_{k-1}, C_i \Vdash B, 3 \tag{6}$$

In fact, since $C_m \equiv B$, this statement is quasi prime for $i = m$; supposing it true for a given $i$, we have

$$\frac{\mathfrak{X}_k, \mathfrak{M}_{k-1} \Vdash A_i, 3 \qquad \mathfrak{X}_k, \mathfrak{M}_{k-1}, C_i \Vdash B, 3}{\mathfrak{X}_k, \mathfrak{M}_{k-1}, C_{i-1} \Vdash B, 3} *P$$

Thus (6) is true for all $i$, in particular for $i = 0$. Hence, by successive applications of $*\Pi$,

$$\mathfrak{X}_k, \mathfrak{M}_{k-1}, C_0^* \Vdash B, 3$$

This is $\Gamma'_k$.

Suppose $\Gamma_k$ is obtained by (2); then the inference is

$$\frac{\mathfrak{X}_k, B \Vdash' \mathfrak{Y}}{\mathfrak{X}_k, A_1, \ldots, A_m \Vdash' \mathfrak{Y}}$$

By the inductive hypothesis

$$\mathfrak{X}_k, \mathfrak{M}_{k-1}, B \Vdash \mathfrak{Y}$$

But in LA, hence in the inner system, we have

$$A_1, \ldots, A_m, C \Vdash B$$

From the last two statements and ET for the inner system we have

$$\mathfrak{X}_k, \mathfrak{M}_{k-1}, C, A_1, \ldots, A_m \Vdash \mathfrak{Y}$$

From this we have $\Gamma_k'$ by applications of $*\Pi$. This completes the proof of the theorem. Note that ET is not needed in the outer system.

The following corollaries are obtained by taking LM* as the inner system and LJ*, LD* as the outer system. These corollaries could be obtained from H formulations by the deduction theorem, but their relation to modal logic is not without interest.

COROLLARY 2.1.    *If (3) holds in* LJ*, *then (4) holds with* LM* *taken as the inner system if* $\mathfrak{M}$ *consists of suitable closures of all* $F \supset A_i$, *where* $A_1, \ldots, A_n$ *are all the principal constituents of the various instances of* Fj.

COROLLARY 2.2.    *If (3) holds in* LD*, *then (4) holds with* LM* *taken as the inner system, provided* $\mathfrak{M}$ *contains suitable closures of instances of the law of excluded middle.*

For an inference by Nx can be obtained thus:

$$\frac{\dfrac{\mathfrak{X}, A \Vdash \mathfrak{Y} \qquad \mathfrak{X}, \neg A \Vdash \mathfrak{Y}}{\mathfrak{X}, A \vee \neg A \Vdash \mathfrak{Y}}}{\mathfrak{X} \Vdash \mathfrak{Y}} \ (2)$$

In case the inner system is taken to be $LA_1^*$, we have also the following:

COROLLARY 2.3.    *If (3) holds in* $LC_1^*$, *then (4) holds as stated in* $LA_1^*$, *with* $\mathfrak{M}$ *consisting of suitable closures of instances of Peirce's law.*

For an inference by Px can be obtained thus:

$$\frac{\dfrac{\dfrac{\mathfrak{X}, A \supset B \Vdash' A}{\mathfrak{X} \Vdash' A \supset B . \supset A} \qquad \mathfrak{X}, A \Vdash' A}{\mathfrak{X}, A \supset B . \supset A . \supset A \Vdash' A}}{\mathfrak{X} \Vdash' A} \ \overset{*P}{(2)}$$

An analogue of Corollary 2.2 can, of course, be obtained for LK*.

## C. THE T AND H FORMULATIONS OF NECESSITY

**1. The T formulation.** The T rules for necessity, which are obtained from the L rules in a manner analogous to that used in the previous chapters, are as follows:

$$\text{Ye} \quad \frac{\Box A}{A} \qquad \text{Yi} \quad \frac{(\Box)}{A} \\ \frac{A}{\Box A}$$

Here the '($\Box$)' over the '$A$' in Yi indicates that $A$ is derived from necessary premises by the rules of the inner system.   It is convenient to call TXY the system formed by adjoining these rules to TX.

It is hardly necessary to give a formal proof that these rules are equivalent to the L rules in the same sense as in the corresponding places in the previous chapters.   Instead we shall state a definition and derive some particular results.

DEFINITION 1.   We define strict implication, denoted by the infix '$\prec$', thus:

$$A \prec B \equiv \Box(A \supset B)$$

**Theorem 1.**   *The following hold, whatever the underlying system* $\mathfrak{S}$.

(a)                                        $\vdash \Box A \prec A$

(b)                                        $\vdash \Box A \prec \Box\Box A$

(c)                                        $\vdash A \prec B . \prec . \Box A \prec \Box B$

*Further, we have, if* $\mathfrak{S}$ *is* $\mathfrak{D}$,

(d)                              *If* $\vdash A$, *then* $\vdash \Box A$

*Proof.*   We have (a) by Ye and Pi and Yi; (d) by Yi.   To derive (b) and (c) we use schemes as follows:

$$\frac{\dfrac{\dfrac{\Box A}{\Box\Box A}\;\text{Yi}}{\Box A \supset \Box\Box A}\;\text{Pi-1}}{\Box A \prec \Box\Box A}\;\text{Yi}$$

$$\cfrac{\cfrac{\cfrac{\cfrac{\cfrac{A \prec B}{A \supset B}\;\text{Ye} \qquad \Box A \quad \overset{2}{\phantom{x}} }{\cfrac{B}{\Box B}\;\text{Yi}}\;\cfrac{}{A}\;\text{Pe}}{\Box A \supset \Box B}\;\text{Pi-2}}{\Box A \prec \Box B}\;\text{Yi}}{\cfrac{A \prec B \supset . \Box A \prec \Box B}{A \prec B . \prec . \Box A \prec \Box B}\;\text{Pi-1}}\;\text{Yi}$$

An H formulation having the same assertions as TXY will be called a system HXY.   It is convenient to extend this term so as to include not only the system based on modus ponens (Ph) alone, but also on a rule Yh, analogous to IIh, as follows:

Yh                                     $\dfrac{\vdash A}{\vdash \Box A}$

We treat first the case where there is no rule but Ph, because previous proofs then apply without revision.

**Theorem 2.**   *With respect to Ph as sole rule, a set of prime assertions for* HXY *consists of those of the form* $\Box A$, *where* $A$ *is a prime assertion of* HX *together with those in the schemes* (a) *to* (c) *of Theorem* 1 *and the scheme*

(a')                                        $\vdash \Box A \supset A$

*Proof.* Let the system described in the theorem be called $H_2$. Then all the prime propositions of $H_2$ are assertions of HXY by Theorem 1. [Note that $(a')$ follows directly from Ye and Pi.] Since Ph is the same as Pe, it is an admissible rule for HXY. By deductive induction every assertion of $H_2$ is also one for HXY.

To prove the converse we must show that the rules of TXY are admissible for $H_2$. This is clear for Pe, since it is the same as Ph. For the nonmodal rules it was shown in the previous chapters. For Ye it follows immediately from $(a')$ and Ph.

It remains to consider Yi. To establish this, note first that the rule

Ph′
$$\frac{\vdash A \prec B \qquad \vdash A}{\vdash B}$$

is admissible by Ph and $(a')$. Suppose then that $A$ is deduced in $H_2$ from premises of the form $\square C$. Let $B_1, B_2, \ldots, B_n$ be the stages in such a deduction. We shall see that, for every $k$, $\square B_k$ is an assertion of $H_2$. If $B_k$ is of the form $\square C$, then $\vdash \square B_k$ follows by $(b)$ and Ph′. This includes all cases where $B_k$ is a premise or a prime assertion other than $(a')$. If $B_k$ is an instance of $(a')$, then $\square B$ is an instance of $(a)$. If $B_k$ is obtained by Pe, let the premises be $B_i$ and $B_j \equiv B_i \supset B_k$. Then we have $\vdash \square B_k$ from $\vdash \square B_i$ and $\vdash \square B_j$ by $(c)$ and Ph′, Q.E.D.

**Theorem 3.** *With respect to the rules Ph and Yh, a set of prime assertions for HXY consists of those of HX together with*

$(a')$          $\vdash \square A \supset A$

$(b')$          $\vdash \square A \supset \square\square A$

$(c')$          $\vdash A \prec B . \supset : \square A . \prec . \square B$

*Proof.* Let the present system be $H_3$, that of Theorem 2 be $H_2$. Then the prime assertions of $H_2$ follow in $H_3$ by Yh. Those of $H_3$ are valid in $H_2$ by $(a')$ and Pe; further, Yh, which is a special case of Yi, is admissible in $H_2$.

## S. Supplementary Topics

**1. Historical and bibliographical comment.** The immediate sources for this chapter are [TFD], chap. 5, and [ETM]. I have purposely refrained from discussing further developments of the subject on the ground that the interest is rather too special for a book of this sort.

For ancient modal logic see the historical works mentioned in Sec. 1S3, particularly that of Bocheński.

Modern modal logic was initiated by Lewis, although there is said to have been some anticipation by MacColl. (For references see under these authors in Church [BSL].) Lewis presented one version of his system in his [SSL]. Further developments were in his part of Lewis and Langford [SLg]. At the end of that book he presented five systems, S1 to S5, which, in their numerical order, were of increasing strength. He expressed himself as unable to decide which of his systems "expresses the acceptable principles of deduction."

After Lewis, work on modal logic was done by a number of scholars, including Becker, Parry, Wajsberg, Feys, McKinsey, Tang, Ruth Barcan, Moh, von Wright, H. A. Schmidt, Hallden, Lemmon, and Anderson. Hallden has a whole series of systems going beyond Lewis, and he and some others have interpolated additional ones in the series. All these systems are based on HK. The only nonclassical modal logic that I know of is Fitch [IML]. For a general survey of what might be called the classical approach to modality see Prior [FLg, chap. III.1] (the postulate lists in the appendix are particularly useful); Lemmon [NFL]. Schmidt [VAL, secs. 161–192] gives an extensive detailed treatment of modal logic; he does not, however, establish contact with work of other authors. Becker [ELg] may also still be useful.

Gödel [IIA] pointed out that there was a close relation between Lewis's system S4 and the system HJ. He showed that HJ could be interpreted in HKY; and the schemes (a) to (c) of Theorems C2 and C3 are due to him. For further studies along this line see McKinsey and Tarski [TSC], Maehara [DIL]. Topological connections with closure algebras have been made by Tang [APG], McKinsey and Tarski [ATp], and others.

The system arrived at here by a semantical approach is equivalent to the Lewis system S4. Other semantical approaches have also arrived at S4, or in some cases S5. Thus McKinsey [SCS] arrives at S4; Carnap [MNc] at S5.

The Gentzen methods in connection with modal logics have been studied also in Ohnishi and Matsumoto [GMM]. According to them, the system in which one replaces Y∗ by (8) of Sec. A leads to the Lewis system S5. They formulate Gentzen rules for several modal systems, including possibility as well as necessity; establish equivalence with the more usual H formulations; and prove ET for a number of them. Their second paper claims a proof of ET for S5.

Kripke has recently made a profound study of modal logics using techniques allied to those of Gentzen and Beth. See his [CTM], [SAM].

Lattice systems related to modal logics have been considered by a few authors, e.g., Porte [RLM]; Rasiowa and Sikorski [ALL], [ETN]; Rubin [RCA]; and the papers cited above under topological applications.

There are several vexatious questions in regard to the interpretations of quantifiers in modal logic. There is a fairly extensive literature relating to this, of which the following are samples: Hintikka [MRM]; Kanger [MSP], [NQM]; Kripke [SAM]; Myhill [PAF]; Quine [PIM].

# BIBLIOGRAPHY

This bibliography is intended to list only items cited in the text. Citations are made by author's name and abbreviated title in brackets. These abbreviated titles are listed here alphabetically under each author, except that abbreviations beginning with "rev," which refer to reviews, are listed after the other works by the same author.

In the case of joint authors, the group of authors is considered as a new author. When it is not desirable to mention all the authors, the abbreviation "*et al.*" is used; this means that the work referred to is by the author mentioned and one or more others. Cross references to the additional authors will be made through the index.

A citation without an author's name (expressed or implied) is to a work by Curry, or in the case of [CLg], Curry and Feys. In the case of cross references in this bibliography, however, the reference is to another work by the same author, unless otherwise indicated.

Ordinarily the initials of the given names of the authors are not given in making citations. They may be found from this bibliography. In the cases where the initials do not suffice for identification, the given names are stated.

When a work contains two or more volumes, or otherwise detached parts, reference to a particular volume or part is made by putting the volume or part number inside the brackets after a period. Different editions of the same work are distinguished by numerical subscripts. Thus, under Whitehead and Russell, [PMt.I$_2$] is the second edition of the first volume of [PMt]. Occasionally this device is used to distinguish works which have the same title, even though they are scarcely different editions of the same work.

References to parts of a work are made either by page numbers or the natural divisions of the work cited. Such references are frequently included in the brackets.

Journals are abbreviated according to the practice of *Mathematical Reviews* (see vol. 19, pp. 1417–1430, and similar lists in earlier volumes). When *Mathematical Reviews* does not list a journal, an abbreviation is used that conforms to their practice in principle.

In referring to parts of journals, the volume number is written first, then a colon, then the page numbers, then (in parentheses) the presumed date of publication. Everyone knows that the actual date of publication is sometimes difficult to determine. In such cases the best information available has been used, but an exhaustive inquiry has not been made.

In transliterating Russian, the system described in *Math. Rev.*, 19:1433, and ascribed to *Science Abstracts*, has been followed. This requires fewer special letters than other systems.

**Ackermann, W.**

[BSI]   Begründung einer strengen Implikation, *J. Symb. Logic*, **21**:113–128 (1956).

[SCD]   "Solvable Cases of the Decision Problem," Amsterdam, 1954.

**Ajdukiewicz, K.**
[TCD]   Three Concepts of Definition, *Logique et analyse*, n.s., 1ᵉ Année: 115–126 (1958).

**American Mathematical Society**
[SLM]   Structure of Language and its Mathematical Aspects, *Proc. 12th Symposium in Applied Mathematics*, New York, 1960 (published 1961).

**Anderson, A. R.**
[CTS]   Completeness Theorems for the Systems E of Entailment and EQ of Entailment with Quantification, *Tech. Rep.* 6, Contract SAR/Nonr—609 (16) (Problem Solving and Social Interaction) for U.S. Office of Naval Research, New Haven, Conn., 1959.

**Anderson, A. R., and N. D. Belnap, Jr.**
[PCE]   "The Pure Calculus of Entailment," ditto manuscript prepared at Yale University, May 5, 1960. See also Anderson [CTS].

**Asser, G.**
[AFP]   Über die Ausdrucksfähigkeit des Prädikatenkalküls der ersten Stufe mit Funktionalen, *Z. Math. Logik Grundlagen Math.*, 2:250–264 (1956).
[NPA]   Normierte Postsche Algorithmen, *Z. Math. Logik Grundlagen Math.*, 5:323–333 (1959).
[TLA]   Theorie der logischen Auswahlfunktionen, *Z. Math. Logik Grundlagen Math.*, 3:30–68 (1957).
[TMM]   Turing-Maschinen und Markowsche Algorithmen, *Z. Math. Logik Grundlagen Math.*, 5:346–365 (1959).

**Bachmann, H.**
[TZh]   "Transfinite Zahlen," Berlin, 1955.

**Becker, O.**
[ELg]   "Einführung in die Logistik," Meisenheim am Glan, 1951.

**Behmann, H.**
[PKL]   Der Prädikatenkalkül mit limitierten Variablen. Grundlegung einer natürlichen exakten Logik, *J. Symb. Logic.* 24:112–140 (1959).
[WLM]   Zu den Widersprüchen der Logik und der Mengenlehre, *Jber. Deutsch. Math. Verein*, 40:37–48 (1931).

**Bennett, A. A.**
[SSO]   Semi-serial Order, *Amer. Math. Monthly*, 37:418–423 (1930).

**Bergmann, G.**
[FHN]   Frege's Hidden Nominalism, *Philos. Rev.*, 67:437–459 (1958).

**Bernays, P.**
[AUA]   Axiomatische Untersuchung des Aussagenkalkuls der "Principia mathematica," *Math. Z.*, 25:305–320 (1926).
[BSB]   Zur Beurteilung der Situation in der beweistheoretischen Forschung, *Theoria* (Madrid), 1:153–154 (1952).
[HGG]   Über Hilberts Gedanken zur Grundlegung der Arithmetik, *Jber. Deutsch Math. Verein.*, 31:10–19 (1922).
[LCl]   "Logical Calculus," notes by Bernays with assistance of F. A. Ficken at Institute for Advanced Study, Princeton, N.J., 1935–1936.
[PMt]   Sur le platonisme dans les mathématiques, *Enseignement Math.*, 34:52–69 (1935–1936).
[QMA]   Sur les questions méthodologiques actuelles de la théorie hilbertienne de la démonstration, *Les Entretiens de Zurich sur les fondements et la méthode des sciences mathématiques* (December 6–9, 1938), 1941, pp. 144–152, Discussion, pp. 153–161.
[rev. C]   Review of Curry [DNF] and [SLD], *J. Symb. Logic*, 18:266–268 (1953).

**Bernays, P., and A. Fraenkel**
[AST]   "Axiomatic Set Theory," Amsterdam, 1958.

**Berstein, B. A.**

[FBA] On Finite Boolean Algebras, *Amer. J. Math.*, **57**:733–742 (1935).

**Berry, G. D. W.**

[OSL] Symposium: On the Ontological Significance of the Löwenheim-Skolem Theorem, *Academic Freedom, Logic, and Religion* (American Philosophical Association, Eastern Division), Philadelphia, 1953, pp. 39–55. See also Myhill [OSL].

**Beth, E. W.**

[CSL] Construction sémantique de la logique intuitionniste, *Le raisonnement en mathématiques et en sciences expérimentales* (Colloques internationaux du Centre National de la Recherche Scientifique), vol. 70, 1958, pp. 77–84.

[CTL] Some Consequences of the Theorem of Löwenheim-Skolem-Gödel-Malcev, *Nederl. Akad. Wetensch. Proc.*, ser. A, **56**(= *Indag. Math.*, 15):66–71 (1953).

[FMt] "The Foundations of Mathematics," Amsterdam, 1959.

[SCI] Semantic Construction of Intuitionistic Logic, *Meded. Kon. Nederl. Akad. Wetensch., Letterkunde*, n.s., **19**:357–388 (1956).

[SEF] Semantic Entailment and Formal Derivability, *Meded. Kon. Nederl. Akad. Wetensch. Letterkunde*, n.s., **18**:309–342 (1955).

[SLG] Symbolische Logik und Grundlegung der exakten Wissenschaften, "Bibliographische Einführungen in das Studium der Philosophie," 1948.

[TPT] A Topological Proof of the Theorem of Löwenheim-Skolem-Gödel, *Nederl. Akad. Wetensch.*, ser. A., **54**(= *Indag. Math.*, 13):436–444 (1951).

[VIn] Verstand en Intuitie, *Algemeen Nederl. Tijdschr. Wijsbegeerte en Psychologie*, **46**:213–224 (1953–1954).

**Birkhoff, Garrett**

[CSA] On the Combination of Subalgebras, *Proc. Cambridge Philos. Soc.*, **29**:441–464 (1933).

[LTh] "Lattice Theory," American Mathematical Society, 1st ed., 1940; 2d ed., 1948.

**Birkhoff, G., and J. von Neumann**

[LQM] The Logic of Quantum Mechanics, *Ann. of Math.*, (2) **37**:823–842 (1936).

**Black, M.**

[NMt] "The Nature of Mathematics: A Critical Survey," New York, 1934.

[RMP] The Relevance of Mathematical Philosophy to the Teaching of Mathematics, *Math. Gaz.*, **22**:149–163 (1938).

**Blake, A.**

[CEB] "Canonical Expressions in Boolean Algebra," thesis, University of Chicago, 1938.

**Blanché, Robert**

[ILC] "Introduction à la logique contemporaine," Paris, 1957.

**Bocheński, I. M.**

[AFL] "Ancient Formal Logic," Amsterdam, 1951.

[FLg] "Formale Logik," Munich, 1956. An English translation by Ivo Thomas has been published by the University of Notre Dame, Notre Dame, Ind.

[NLL] "Nove lezioni di logica simbolica," Rome, 1938.

[PLM] "Précis de logique mathématique," Bussum, 1949. For German translation and revision 1954, see I. M. Bocheński and A. Menne [GRL]; an English translation by O. Bird has been announced.

**Bocheński, I. M., and A. Menne**

[GRL] "Grundriss der Logistik" (German translation and revision of I. M. Bocheński [PLM]), Paderborn, 1954.

**Boehner, P.**

[MLg] "Medieval Logic: An Outline of Its Development from 1250–c. 1400," Chicago, 1952.

**Borkowski, L., and J. Słupecki**
[LWJ]   The Logical Works of J. Łukasiewicz, *Studia Logica,* **8**:7–56 (1958).
**Brouwer, L. E. J.**
[CPM]   Consciousness, Philosophy, and Mathematics, *Proc.* 10*th Internat. Congress of Philosophy,* Amsterdam, 1948, pp. 1235–1249.
[GLW]   "Over de grondslagen der wiskunde," Amsterdam, 1907. (Dutch.)
[HBP]   Historical Background, Principles and Methods of Intuitionism, *South African J. Sci.,* **49**:139–146 (1952).
[IBF]   Intuitionistische Betrachtungen über den Formalismus, *Nederl. Akad. Wetensch. Proc.,* **31**:374–379 (1931); also *S.-B. Preuss. Akad. Wiss. Phys.- Math. Kl.,* **1928**:48–52.
[IZM]   Zur intuitionistischen Zerlegung mathematischer Grundbegriffe, *Jber. Deutsch. Math. Verein.,* **36**:127–129 (1927).
[OLP]   De onbetrouwbaarheid der logische principes, *Tijdschrift wijsbegeerte,* **2**:152–158 (1908). (Dutch.)
**Brown, R.**
[WTh]   "Words and Things: An Introduction to Language," Glencoe, Ill., 1959.
**Burali-Forti, C.**
[LMt]   "Logica matematica," Milan, 1894; 2d. ed., 1919.
**Byrne, L.**
[TBF]   Two Brief Formulations of Boolean Algebra, *Bull. Amer. Math. Soc.,* **52**:269–272 (1946).
**Cantor, G.**
[GAb]   "Gesammelte Abhandlungen," edited by E. Zermelo, Berlin, 1932.
**Carnap, R.**
[ESL]   "Einführung in die symbolische Logik," Vienna, 1954.   For English edi- tion see [ISL].
[FLg]   "Formalization of Logic," Cambridge, Mass., 1943.
[FLM]   Foundations of Logic and Mathematics, "International Encyclopedia of Unified Science," vol. I, no. 3, Chicago, 1939.
[ISL]   "Introduction to Symbolic Logic and Applications," New York, 1958. English edition of [ESL].
[ISm]   "Introduction to Semantics," Cambridge, Mass., 1942.
[LSL]   "The Logical Syntax of Language" (translation of [LSS] by Amethe Smeaton), London and New York, 1937.
[LSS]   "Logische Syntax der Sprache," Vienna, 1934.   For English translation, see [LSL].
[MNc]   "Meaning and Necessity: A Study in Semantics and Modal Logic," Chicago, 1947.
**Chang, C. C.**
[AAM]   Algebraic Analysis of Many Valued Logic, *Trans. Amer. Math, Soc.,* **88**:467–490 (1958).
**Chernyavskiĭ, V. S.**
[KNA]   Ob odnom klasse normal'nykh algorifmov Markova (On a class of normal algorithms of Markov), "Logicheskie issledovaniya," Academy of Sciences, U.S.S.R., Moscow, 1959, pp. 263–299.
**Chin, L. H., and A. Tarski**
[DML]   Distributive and Modular Laws in the Arithmetic of Relation Algebras, *Univ. California Publ. Math.,* **1**:341–384 (1951).
**Choudhury, A. C.**
[BNr]   On Boolean Narings, *Bull. Calcutta Math. Soc.,* **46**:41–45 (1954.)
**Church, A.**
[BBF]   Brief Bibliography of Formal Logic, *Proc. Amer. Acad. Arts Sci.,* **80**:155– 172 (1952).

[BSL]  A Bibliography of Symbolic Logic, *J. Symb. Logic*, 1:121–218 (1936). Additions and corrections, *ibid.*, 3:178–212 (1938). Supplements have been published in *ibid.* from time to time since.

[CLC]  The Calculi of Lambda-conversion, *Ann. Math. Studies*, no. 6, Princeton, N.J., 1941; 2d ed., 1951.

[CNE]  Correction to "A Note on the Entscheidungsproblem," *J. Symb. Logic*, 1:101–102 (1936).

[Dfn]  Definition (article), in Runes [DPh], pp. 74–75.

[FST]  A Formulation of the Simple Theory of Types, *J. Symb. Logic*, 5:56–68 (1940).

[IML]  "Introduction to Mathematical Logic, Part I" (*Ann. Math. Studies*, no. 13), Princeton, N.J., 1944; 2d ed. (greatly expanded as vol. I of a two-volume work in the Princeton Mathematical Series), Princeton, N.J., 1956.

[LEM]  On the Law of the Excluded Middle, *Bull. Amer. Math. Soc.*, 34:75–78 (1928).

[MiL]  Minimal Logic (abstract), *J. Symb. Logic*, 16:239 (1951).

[NEP]  A Note on the Entscheidungsproblem, *J. Symb. Logic*, 1:40–41 (1936).

[PLg]  Paradoxes, Logical (article), in Runes [DPh], pp. 224ff.

[SAS]  Schröder's Anticipation of the Simple Theory of Types, *J. Unified Sci.*, 9:149–152 (1939).

[SPF]  A Set of Postulates for the Foundation of Logik, part I, *Ann. of Math.*, (2)33:346–366 (1932); part II, (2)34:839–864 (1933).

[UPE]  An Unsolvable Problem of Elementary Number Theory, *Amer. J. Math.*, 58:345–363 (1936).

[WTI]  The Weak Theory of Implication, *Kontrolliertes Denken, Untersuchungen zum Logikkalkul und zur Logik der Einzelwissenschaften*, edited by A. Menne, A. Wilhelmy, and Helmut Angsil (Festgabe zum 60 Geburtstag von Prof. W. Britzelmayr), Munich, 1951, pp. 22–37.

[rev. C]  Review of Curry [RDN], *J. Symb. Logic.* 22:85–86 (1957).

[rev. HA]  Review of D. Hilbert and W. Ackermann [GZT], *J. Symb. Logic.*, 15:59 (1950).

**Cogan, E.**

[FTS]  A Formalization of the Theory of Sets from the Point of View of Combinatory Logic, *Z. Math. Logik Grundlagen Math.*, 1:198–240 (1955).

**Copeland, A. H., Sr.**

[IBA]  Implicative Boolean Algebra, *Math. Z.*, 53:285–290 (1950).

[POP]  Probabilities, Observations, and Predications, *Proc. 3d Berkeley Symposium on Math. Statist. and Probability*, Berkeley, Calif., 2:41–47 (1955).

**Copi, I. M.**

[BFP]  The Burali-Forti Paradox, *Philos. Sci.*, 25:281–286 (1958).

**Couturat, L.**

[ALg]  "L'algèbre de la logique," 1st ed., Paris, 1905; 2d ed., 1914. English translation by L. G. Robinson, "The Algebra of Logic," Chicago and London, 1914.

**Craig, W.**

[LRN]  Linear Reasoning. A New Form of the Herbrand-Gentzen Theorem, *J. Symb. Logic*, 22:250–268 (1957).

**Curry, H. B.**

[ALS]  An Analysis of Logical Substitution, *Amer. J. Math.*, 51:363–384 (1929).

[APM]  Some Aspects of the Problem of Mathematical Rigor, *Bull. Amer. Math. Soc.*, 47:221–241 (1941).

[AVS]  Apparent Variables from the Standpoint of Combinatory Logic, *Ann. of Math.*, (2)34:381–404 (1933).

[CFS]   Calculuses and Formal Systems, *Dialectica*, **12**:249–273 (1958).

[DFS]   Definitions in Formal Systems, *Logique et analyse*, n.s., 1ᵉ Année **1**:105–114 (1958).

[DNF]   On the Definition of Negation by a Fixed Proposition in the Inferential Calculus, *J. Symb. Logic*, **17**:98–104 (1952).

[DSR]   On the Definition of Substitution, Replacement, and Allied Notions in an Abstract Formal System, *Rev. philos. Louvain*, **50**:251–269 (1952).

[DTC]   The Deduction Theorem in the Combinatory Theory of Restricted Generality, *Logique et analyse*, n.s., 3ᵉ Année :15–39 (1959).

[ETM]   The Elimination Theorem when Modality Is Present, *J. Symb. Logic*, **17**:249–265 (1952).

[GDT]   Generalizations of the Deduction Theorem, *Proc. Internat. Congress of Mathematicians*, 1954, pp. 399–400.

[GKL]   Grundlagen der kombinatorischen Logik, *Amer. J. Math.*, **52**:509–536, 789–834 (1930).

[IAL]   The Inferential Approach to Logical Calculus, *Logique et analyse*, n.s., part 1, 3ᵉ Année :119–136 (1960); part 2, 4ᵉ Année :5–22 (1961).

[IFI]   The Interpretation of Formalized Implication, *Theoria* (Sweden), **25**:1–26 (1959).

[LAG]   Some Logical Aspects of Grammatical Structure, in American Mathematical Society [SLM], pp. 56–68.

[LFS]   Languages and Formal Systems, *Proc. 10th Internat. Congress of Philosophy*, Amsterdam, 1948, pp. 770–772 (abstract of [LMF]).

[LLA]   "Leçons de logique algébrique," Paris and Louvain, 1952.

[LMF]   Language, Metalanguage, and Formal System, *Philos. Rev.*, **59**:346–353 (1950). For abstract see [LFS].

[LSF]   L-semantics as a Formal System, *Actualités Sci. Ind.*, **1134**:19–29 (1951).

[LSG]   The Logical Structure of Grammar, lecture delivered November, 1948. Unfinished manuscript mimeographed at University of Chicago and distributed to a number of persons. Subsequently incorporated in [LAG].

[MSL]   Mathematics, Syntactics and Logic, *Mind*, **62**:172–183 (1953).

[NAL]   A Note on the Associative Law in Logical Algebras, *Bull. Amer. Math. Soc.*, **42**:523–524 (1936).

[NRG]   A Note on the Reduction of Gentzen's Calculus LJ, *Bull. Amer. Math. Soc.*, **45**:288–293 (1939).

[OFP]   "Outlines of a Formalist Philosophy of Mathematics," Amsterdam, 1951.

[PBP]   Philosophische Bemerkungen zu einigen Problemen der mathematischen Logik, *Arch. Philos.*, **4**:147–156 (1951).

[PEI]   Some Properties of Equality and Implication in Combinatory Logic, *Ann. of Math.*, (2)**35**:849–860 (1934).

[PFD]   Some Properties of Formal Deducibility (abstract), *Bull. Amer. Math. Soc.*, **43**:615 (1937).

[PKR]   The Paradox of Kleene and Rosser, *Trans. Amer. Math. Soc.*, **50**:454–516 (1941).

[PRC]   The Permutability of Rules in the Classical Inferential Calculus, *J. Symb. Logic*, **17**:245–248 (1952).

[RDN]   Remarks on the Definition and Nature of Mathematics, *J. Unified Sci.*, **9**:164–169 (1939). (All copies of this issue were destroyed during World War II, but "preprints" were distributed at the International Congress for the Unity of Science in September, 1939.) Reprinted without certain minor corrections in *Dialectica*, **8**:228–333 (1954).

[SFL]   Les Systèmes formels et les langues, *Les méthodes formelles en axiomatique*

(Colloques internationaux du Centre National de la Recherche Scientifique, vol. 36), 1953, pp. 1–10.

[SLD] The System LD, *J. Symb. Logic*, 17:35–42 (1952).

[STC] A Simplification of the Theory of Combinators, *Synthese*, 7:391–399 (1949).

[TCm] La Théorie des combinateurs, *Rend. Mat. e Appl.*, 5th ser., 10:347–359 (1951).

[TEA] "Two Examples of Algorithms," to be published.

[TFD] "A Theory of Formal Deducibility," Notre Dame Mathematical Lectures, no. 6, Notre Dame, Ind., 1950; 2d ed., 1957.

[UDB] On the Use of Dots as Brackets in Logical Expressions, *J. Symb. Logic*, 2:26–28 (1937).

[UQC] The Universal Quantifier in Combinatory Logic, *Ann. of Math.*, (2)32:154–180 (1931).

[rev. C] Review of Church [IML$_2$], *J. Franklin Inst.*, 264:244–246 (1957).

[rev. R] Review of Rosser [LMt], *Bull. Amer. Math. Soc.*, 60:266–272 (1954).

**Curry, H. B., and R. Feys**

[CLg] "Combinatory Logic," Amsterdam, 1958.

**Davis, M.**

[CUn] "Computability and Unsolvability," New York and London, 1958.

**Dedekind, R.**

[GMW] "Gesammelte mathematische Werke," edited by R. Fricke, E. Noether, and O. Ore, 3 vols., Braunschweig, 1930–1932.

[WSW] "Was sind und was sollen die Zahlen?," 6th ed., Braunschweig, 1930.

[ZZG] Über Zerlegungen von Zahlen durch ihre grössten gemeinsamen Teiler, *Festschrift Tech. Hoch. Braunschweig*, 1897; reprinted in [GMW], vol. 2, pp. 103–148.

**Detlovs, V. K.**

[NAR] Normal'nye algorifmy̅ i rekursivnye funktsii (Normal algorithms and recursive functions), *Doklady Akad. Nauk SSSR*, 90:723–725 (1953).

**Diamond, A. H., and J. C. C. McKinsey**

[ASA] Algebras and Their Subalgebras, *Bull. Amer. Math. Soc.*, 53:959–962 (1947).

**Dilworth, R. P.**

[ARL] Abstract Residuation over Lattices, *Bull. Amer. Math. Soc.*, 44:262–268 (1938).

**Dopp, J.**

[VSD] Les variétés syntaxiques de la définition dans les langages rigoureux, *Études philos.*, 1956, no. 2, pp. 209–225.

**Dubislav, W.**

[Dfn] "Über die Definition," Berlin, 1926; 3d ed., Leipzig, 1931.

[PMG] "Die Philosophie der Mathematik in der Gegenwart," Berlin, 1932.

**Dubreil-Jacotin, M. L., L. Lesieur, and R. Croisot**

[LTT] "Leçons sur la théorie des treillis, des structures algébriques ordonnées, et des treillis géométriques," Paris, 1953.

**Dyson, V. H., and G. Kreisel**

[ABS] Analysis of Beth's Construction of Intuitionistic Logic, *Appl. Math. and Statist. Lab. Tech. Rep.* 3, Stanford University, Calif., 1961.

**Exner, R. M., and M. F. Rosskopf**

[LEM] "Logic in Elementary Mathematics," New York, 1959.

**Feys, R.**

[Lgs] Logistique, *Philosophie* (Chronique des années d'après-guerre 1946–1948), XIII, *Philosophie des sciences*, l'Institut International de Philosophie (Actualités scientifiques et industrielles, no. 1105), Paris, 1950, pp. 19–32.

[MRD]  Les méthodes récentes de déduction naturelle, *Rev. Philos. Louvain,* **44**:370–400 (1946).
[NCM]  Note complémentaire sur les méthodes de déduction naturelle, *Rev. Philos. Louvain,* **1**:60–72 (1947).
[SLP]  Systèmes L propositionnels avec soustraction, *Logique et analyse,* no. 6, pp. 3–16, 1956.
**Findlay, J.**
[GSN]  Gœdelian Sentences: A Non-numerical Approach, *Mind,* n.s., **51**:259–265 (1942).
**Fitch, F. B.**
[EVE]  An Extensional Variety of Extended Basic Logic, *J. Symb. Logic,* **23**:13–21 (1958).
[IML]  Intuitionistic Modal Logic with Quantifiers, *Portugal. Math.,* **7**:113–118 (1948).
[QCF]  Quasi-constructive Foundation for Mathematics, in Heyting [CMT], pp. 26–36.
[SLg]  "Symbolic Logic: An Introduction," New York, 1952.
**Fowler, H. W.**
[DME]  "A Dictionary of Modern English Usage," Oxford, England, 1926.
**Fraenkel, A. A.**
[AST]  "Abstract Set Theory," Amsterdam, 1953.
[EML]  "Einleitung in die Mengenlehre," 3d ed., Berlin, 1928.
**Fraenkel, A. A., and Y. Bar-Hillel**
[FST]  "Foundations of Set Theory," Amsterdam, 1958.
**Frege, G.**
[GGA]  "Grundgesetze der Arithmetik, begriffsschriftlich abgeleitet," vol. I, Jena, 1893, vol. II, Jena, 1903.
**Frink, O.**
[NAL]  New Algebras of Logic, *Amer. Math. Monthly,* **45**:210–219 (1938).
**Galler, B. A.**
[CPA]  Cylindric and Polyadic Algebras, *Proc. Amer. Math. Soc.,* **8**:176–183 (1957).
**Gentzen, G.**
[EUA]  Über die Existenz unabhängiger Axiomensysteme zu unendlichen Satzsystemen, *Math. Ann.,* **107**:329–350 (1932).
[NFW]  Neue Fassung des Widerspruchsfreiheitsbeweises für die reine Zahlentheorie, *Forsch. Logik und Grundlegung exakten Wiss.,* n.s., **4**:19–44 (1938).
[RDL]  "Recherches sur la déduction logique" (translation, with supplementary notes by R. Feys and J. Ladrière, of [ULS]), Paris, 1955.
[ULS]  Untersuchungen über das logische Schliessen, *Math. Z.,* **39**:176–210, 405–431 (1934). For French translation, see [RDL].
[WFR]  Die Widerspruchsfreiheit der reinen Zahlentheorie, *Math. Ann.,* **112**: 493–565 (1936).
**Gilmore, P. C.**
[ALM]  An Addition to Logic of Many-sorted Theories, *Compositio Math.,* **13**: 277–281 (1958).
[AST]  "An Alternative to Set Theory," mimeographed manuscript, 1959.
**Glivenko, M. V.**
[LBr]  Sur la logique de M. Brouwer, *Acad. Roy. Belg. Bull. Cl. Sci.,* **14**:225–228 (1928).
[PLB]  Sur quelques points de la logique de M. Brouwer, *Acad. Roy. Belg. Bull. Cl. Sci.,* **15**:183–188 (1929).
[TGS]  "Theorie genérale des structures," Paris, 1938.

**Gödel, K.**

[BNB] Über eine bisher noch nicht benützte Erweiterung des finiten Standpunktes, *Dialectica*, **12**:280–287 (1958).

[CAC] "The Consistency of the Axiom of Choice and of the Generalized Continuum-hypothesis with the Axioms of Set Theory" (*Ann. Math. Studies*, no. 3), Princeton, N.J., 1940.

[FUS] Über formal unentscheidbare Sätze der Principia mathematica und verwandter Systeme, I, *Monatsh. Math. Phys.*, **38**:173–198 (1931).

[IAK] Zum intuitionistischen Aussagenkalkül, *Akad. Wiss. Anzeiger*, **69**:65–66 (1932).

[IAZ] Zur intuitionistischen Arithmetik und Zahlentheorie, *Erg. math. Kolloq.*, **4**:34–38 (1933).

[IIA] Eine Interpretation des intuitionistischen Aussagenkalküls, *Erg. math. Kolloq.*, **4**:39–40 (1933).

[RML] Russell's Mathematical Logic, in "The Philosophy of Bertrand Russell," edited by P. A. Schlipp, Chicago, 1944, pp. 123–153.

[UPF] On Undecidable Propositions of Formal Mathematical Systems, mimeographed report of lectures by Gödel at the Institute for Advanced Study, Princeton, N.J., 1934.

[VAL] Die Vollständigkeit der Axiome des logischen Funktionenkalküls, *Monatsh. Math. Phys.*, **37**:349–360 (1930).

[WIC] What Is Cantor's Continuum Problem?, *Amer. Math. Monthly*, **44**:515–525 (1947).

**Gonseth, F.**

[MRI] "Les mathématiques et la réalité: Essai sur la méthode axiomatique," Paris, 1936.

**Goodstein, R. L.**

[CFr] "Constructive Formalism," Leicester, England, 1951.

[MLg] "Mathematical Logic," Leicester, England, 1957.

[NMS] On the Nature of Mathematical Systems, *Dialectica*, **12**:296–316 (1958).

[RNT] "Recursive Number Theory," Amsterdam, 1957.

**Grzegorczyk, A.**

[SLR] The Systems of Leśniewski in Relation to Contemporary Logical Research, *Studia Logica*, **3**:77–95 (1955).

**Hailperin, T.**

[QTE] Quantification Theory and Empty Individual-domains, *J. Symb. Logic*, **18**:197–200 (1953).

[RID] Remarks on Identity and Description in First-order Axiom Systems, *J. Symb. Logic.* **19**:14–20 (1954).

[TRQ] Theory of Restricted Quantification, *J. Symb. Logic.* **22**:19–35, 113–129 (1957).

**Halmos, P. R.**

[ALg] Algebraic Logic, Part I, Monadic Boolean Algebras, *Compositio Math.*, **12**:217–249 (1955); Part II, Homogeneous Locally Finite Polyadic Boolean Algebras of Infinite Degree, *Fund. Math.*, **43**:255–325 (1955); Part III, Predicates, Terms, and Operations in Polyadic Algebras, *Trans. Amer. Math. Soc.*, **83**:430–470 (1956); Part IV, Equality in Polyadic Algebras, *Trans. Amer. Math. Soc.*, **86**:1–27 (1957).

[BCA] The Basic Concepts of Algebraic Logic, *Amer. Math. Monthly*, **63**:363–387 (1956).

[PBA] Polyadic Boolean Algebras, *Proc. Nat. Acad. Sci. U.S.A.*, **40**:296–301 (1954).

**Hardy, G. H.**

[MPr] Mathematical Proof, *Mind.* **38**:1–25 (1929).

**Harrop, R.**
[DES]  On Disjunctions and Existential Statements in Intuitionistic Systems of Logic, *Math. Ann.*, **132**:347–361 (1956).
**Hausdorff, F.**
[GZM]  "Grundzüge der Mengenlehre," Leipzig, 1914. Reprinted New York, 1949.
**Henkin, L.**
[ACQ]  An Algebraic Characterization of Quantifiers, *Fund. Math.*, **37**:63–74 (1950).
[ASM]  The Algebraic Structure of Mathematical Theories, *Bull. Soc. Math. Belg.*, **7**:131–136 (1955).
[BRS]  Banishing the Rule of Substitution for Functional Variables, *J. Symb. Logic*, **18**:201–208 (1953).
[CTT]  Completeness in the Theory of Types, *J. Symb. Logic*, **15**:81–91 (1950).
[RTC]  The Representation Theorem for Cylindrical Algebras, in Skolem *et al.* [MIF], pp. 85–97.
[SAT]  "La structure algébrique des théories mathématiques," Paris, 1956.
**Henkin, L., P. Suppes, and A. Tarski**
[AMS]  "The Axiomatic Method with Special Reference to Geometry and Physics," Amsterdam, 1959.
**Henkin, L., and A. Tarski**
[CAl]  Cylindrical Algebras, Summaries of talks presented at the summer institute for symbolic logic, Cornell University, 1957, pp. 332–340.
**Herbrand, J.**
[RTD]  "Recherches sur la théorie de la démonstration" (Travaux de la Société des Sciences et des Lettres de Varsovie, Classe III, Sciences Mathématiques et Physiques), Warsaw, 1930.
**Hermes, H.**
[EVT]  "Einführung in die Verbandstheorie," Berlin, 1955.
[GLM]  Über die gegenwärtige Lage der mathematischen Logik und Grundlagenforschung, *Jber. Deutsch. Math. Verein.*, **59**:49–69 (1956).
[IPO]  Zur Inversionsprinzip der operativen Logik, in Heyting [CMt], pp. 62–68.
[SMt]  Semiotik, *Forsch. Logik und Grundlegung exakten Wiss.*, n.s., **5**:5–22 (1938).
**Hermes, H., and G. Köthe**
[TVr]  Theorie der Verbände, "Enzyklopädie der mathematischen Wissenshaften," vol. I, no. 13, 1938, 28 pp.
**Hermes, H., and W. Markwald**
[GLM]  Grundlagen der Mathematik, *Grundzüge Math. für Lehrer an Gymnasien sowie für Math. in Industrie und Wirtschaft*, **1**:1–89 (1958).
**Hermes, H., and H. Scholz**
[MLg]  Mathematische Logik, in "Enzyklopädie der mathematischen Wissenschaften," vol. I, no. 1 Leipzig, 1952, 82 pp.
[NVB]  Ein neuer Vollständigkeitsbeweis für das reduzierte Fregeschen Axiomensystem des Aussagenkalküls, *Forsch. Logik und Grundlegung exakten Wiss.*, n.s., **1**:1–40 (1937).
**Hertz, P.**
[ASB]  Über Axiomensysteme für beliebige Satzsysteme, *Math. Ann.*, **101**:457–514 (1929).
**Heyting, A.**
[CIL]  La conception intuitionniste de la logique, *Études philos.*, n.s., **11**:226–233 (1956).
[CMt]  "Constructivity in Mathematics," Amsterdam, 1959.
[FMI]  "Les fondements des mathématiques, intuitionnisme, théorie de la démonstration," Paris, 1955. (Translation and revision of [MGL].)

[FRI] Die formalen Regeln der intuitionistischen Logik, *S.-B. Preuss. Akad. Wiss. Phys.-Math. Kl.*, 1930, pp. 42–56.

[Int] "Intuitionism: An Introduction," Amsterdam, 1956.

[MGL] "Mathematische Grundlagenforschung Intuitionismus: Beweistheorie," Berlin, 1934. (For French translation and revision see [FMI].)

**Hilbert, D.**

[GAb] "Gesammelte Abhandlungen," 3 vols., Berlin, 1932–1935.

[GLG] "Grundlagen der Geometrie," 7th ed., Leipzig and Berlin, 1930 (appeared first in 1899).

[GLM] "Die Grundlagen der Mathematik" (Hamburger mathematische Einzelschriften, no. 5), Leipzig, 1928..

[NBM] Neubegründung der Mathematik, *Abh. Math. Sem. Univ. Hamburg*, 1:157–177 (1922). Reprinted in [GAb], 3:157–177.

**Hilbert, D., and W. Ackermann**

[GZT] "Grundzüge der theoretischen Logik," Berlin, 1928; 2d ed., 1938; 3d ed., 1949; 4th ed., 1959. (For English translation of 2d ed., see [PML].)

[PML] "Principles of Mathematical Logic" (English translation of [GZT$_2$] by L. M. Hammond, G. G. Leckie, and F. Steinhardt, edited and with notes by Robert E. Luce), New York, 1950.

**Hilbert, D., and P. Bernays**

[GLM] "Grundlagen der Mathematik," vol. 1, Berlin, 1934; vol. 2, 1939.

**Hintikka, K. J. J.**

[FCQ] Form and Content in Quantification Theory, *Acta Philos. Fenn.*, 8:11–55 (1955).

[MRM] Modality as Referential Multiplicity, *Ajatus*, 20:49–63 (1957).

[RTT] Reductions in the Theory of Types, *Acta Philos. Fenn.*, 8:57–115 (1955).

**Hiż, H.**

[ESC] Extendible Sentential Calculus, *J. Symb. Logic*, 24:193–202 (1959).

**Huntington, E. V.**

[BAC] Boolean Algebra: A Correction, *Trans. Amer. Math.Soc.*, 35:557–558 (1933).

[NSI] New Sets of Independent Postulates for the Algebra of Logic with Special Reference to Whitehead and Russell's Principia mathematica, *Trans. Amer. Math. Soc.*, 35:274–304 (1933).

[NSP] A New Set of Postulates for Betweenness, with Proof of Complete Independence, *Trans. Amer. Math. Soc.*, 26:257–282 (1924).

[SIP] Sets of Independent Postulates for the Algebra of Logic, *Trans. Amer. Math. Soc.*, 5:288–309 (1904).

**Jaśkowski, S.**

[RSF] On the Rules of Suppositions in Formal Logic, *Studia Logica*, No. 1:5–32 (1934).

[RSL] Recherches sur le système de la logique intuitioniste, *Internat. Congress Philos. Sci.*, 6:58–61 (1936).

**Johansson, I.**

[CCA] Sur le concept de "le" (ou de "ce qui") dans le calcul affirmatif et dans les calculs intuitionistes, *Les méthodes formelles en axiomatique* (Colloques internationaux du Centre National de la Recherche Scientifique, Paris, vol. 36), 1953, pp. 65–72.

[MKR] Der Minimalkalkül, ein reduzierter intuitionistischer Formalismus, *Compositio Math.*, 4:119–136 (1936).

**Johnson, W. E.**

[Lgc] "Logic," London, part I, 1921; part II, 1922; part III, 1924.

**Jonsson, B., and A. Tarski**

[BAO] Boolean Algebras with Operators, part I, *Amer. J. Math.*, 73:891–939 (1951); part II, 74:127–162 (1952).

**Jordan, P.**
[TSV] Die Theorie der Schrägverbände, *Abh. Math. Sem. Univ. Hamburg*, 21:127–138 (1957).
**Jordan, Z.**
[DML] "The Development of Mathematical Logic and of Logical Positivism in Poland between the Two Wars," London, 1945.
**Jørgensen, J.**
[TFL] "A Treatise of Formal Logic," 3 vols., Copenhagen and London, 1931.
**Kamke, E.**
[MLh] "Mengenlehre," Berlin and Leipzig, 1928.
**Kanger, S.**
[MSP] The Morning Star Paradox, *Theoria* (Sweden), 23:1–11 (1957).
[NPP] A Note on Partial Postulate Sets for Propositional Logic, *Theoria* (Sweden), 21:99–104 (1955).
[NQM] A Note on Quantification and Modalities, *Theoria* (Sweden), 23:133–134 (1957).
**Kemeny, J. G.**
[NAS] A New Approach to Semantics, Part I, *J. Symb. Logic*, 21:1–27 (1956).
**Kempner, A. J.**
[PCS] "Paradoxes and Common Sense," New York, 1959.
**Ketonen, D.**
[UPK] Untersuchungen zum Prädikatenkalkül, *Ann. Acad. Sci. Fenn.*, ser. A, 1944, 71 pp.
**Kleene, S. C.**
[GRF] General Recursive Functions of Natural Numbers, *Math. Ann.*, 112:727–742 (1936).
[ILg] On the Intuitionistic Logic, *Proc. 10th Internat. Congress of Philosophy*, Amsterdam, 1948, pp. 741–743.
[IMM] "Introduction to Metamathematics," Amsterdam and Groningen, 1952.
[PIG] Permutability of Inferences in Gentzen's Calculi LK and LJ, *Mem. Amer. Math. Soc.*, 10:1–26 (1952).
[rev. C] Review Curry [APM], *J. Symb. Logic*, 6:100–102 (1941).
**Kneale, W.**
[PLg] The Province of Logic, in H. D. Lewis [CBP], pp. 235–261.
**Kolmogorov, A. N.**
[PTN] O principe tertium non datur, *Mat. Sb.*, 32:646–667 (1924–1925).
**Kotarbinski, T.**
[JLW] Jan Łukasiewicz's Works on the History of Logic, *Studia Logica*, 8:57–62 (1958).
[LPO] La Logique en Pologne, son originalité et les influences étrangères, *Acad. Polacca Sci. Lett. Biblioteca Roma*, vol. 7, 24 pp. (1959).
**Kreisel, G.**
[HPr] Hilbert's Programme, *Dialectica*, 12:346–372 (1958).
[INF] On the Interpretation of Non-finitist Proofs, Part I, *J. Symb. Logic*, 16:241–267 (1951).
**Kreisel, G., and H. Putnam**
[UBM] Eine Unableitbarkeitsbeweismethode für den intuitionistischen Aussagenkalkül, *Arch. Math. Logik Grundlagenforsch.*, 3:74–78 (1957).
**Kripke, S. A.**
[CTM] A Completeness Theorem in Modal Logic, *J. Symb. Logic*, 24:1–14 (1959).
[DCn] Distinguished Constituents (abstract), *J. Symb. Logic*, 24:323 (1959).
[PEn] The Problem of Entailment (abstract), *J. Symb. Logic*, 24:324 (1959).
[SAM] Semantical Analysis of Modal Logic (abstract), *J. Symb. Logic*, 24:323ff. (1959).

[SLE] "The System LE." Submitted to Westinghouse Science Talent Search, February, 1958; not yet published.

**Ladrière, J.**
[LIF] "Les limitations internes des formalismes," Paris and Louvain, 1957.

**Landau, E.**
[FAn] "The Foundations of Analysis: The Arithmetic of Whole, Rational, Irrational, and Complex Numbers" (English translation of [GLA] by F. Steinhart, New York, 1951).
[GLA] "Grundlagen der Analysis," Leipzig, 1930; 3d ed., New York, 1960. For English translation see [FAn].

**Leblanc, H.**
[IDL] "An Introduction to Deductive Logic," New York and London, 1955.

**Leblanc, H., and T. Hailperin**
[NDS] Nondesignating singular terms, *Philos. Rev.*, **68**:239–243 (1959).

**Ledley, R. S.**
[CMS] "Computational Methods in Symbolic Logic: A New Methodology in Operations Research," Johns Hopkins University Operations Research Office, 1955.
[DCM] Digital Computational Methods in Symbolic Logic with Examples in Biochemistry, *Proc. Nat. Acad. Sci. U.S.A.*, **41**:498–511 (1955).
[DSF] A Digitalization, Systematization and Formulation of the Theory and Methods of the Propositional Calculus, Nat. Bureau Standards Rep. 3363, 1954.
[MFC] Mathematical Foundations and Computational Methods for a Digital Logic Machine, *J. Operations Res. Sci. Amer.*, **2**:249–274 (1954).

**Leggett, H. W.**
[BRP] "Bertrand Russell, O.M.: A Pictorial Biography," New York, 1950.

**Lemmon, E. J.**
[NFL] New Foundations for Lewis Modal Systems, *J. Symb. Logic*, **22**:176–186 (1957).

**Leśniewski, S.**
[GZN] Grundzüge eines neuen Systems der Grundlagen der Mathematik, *Fund. Math.*, **14**:1–81 (1929).

**Lewis, C. I.**
[ASL] Alternative Systems of Logic, *Monist*, **42**:481–507 (1932).
[SSL] "A Survey of Symbolic Logic," Berkeley, Calif., 1918.

**Lewis, C. I., and C. Langford**
[SLg] "Symbolic Logic," New York and London, 1932; reprinted New York, 1959.

**Lewis, H. D.**
[CBP] "Contemporary British Philosophy," 3d ser., London and New York, 1956.

**Lightstone, A. H., and A. Robinson**
[STr] Syntactical Transforms, *Trans. Amer. Math. Soc.*, **86**:220–245 (1957).

**Lorenzen, P.**
[ALU] Algebraische und logistische Untersuchungen über freie Verbände, *J. Symb. Logic*, **16**:81–106 (1951).
[EOL] "Einführung in die operative Logik und Mathematik," Berlin-Göttingen-Heidelberg, 1955.
[FLg] "Formale Logik," Berlin, 1958.
[KBM] Konstruktive Begründung der Mathematik, *Math. Z.*, **53**:162–202 (1950).
[LRF] Logical Reflection and Formalism, *J. Symb. Logic*, **23**:241–249 (1958).

**Löwenheim, L.**
[MRK] Über Möglichkeiten im Relativkalkül, *Math. Ann.*, **76**:447–470 (1915).

**Łukasiewicz, J.**
[ASS]   "Aristotle's Syllogistic from the Standpoint of Modern Formal Logic," Oxford, England, 1951.
[ITD]   On the Intuitionistic Theory of Deduction, *Nederl. Akad. Wetensch. Proc.*, ser A, **55**( = *Indag. Mag.*, 14):202–212 (1952).
[LGL]   Die Logik und das Grundlagenproblem, *Les Entretiens de Zurich sur les fondements et la méthode des sciences mathématiques*, 1941, pp. 82–100; Discussion, pp. 100–108.
[SAr]   O sylogistyce Arystotelesa (On Aristotle's syllogistic), *Sprawozdania z czynnosci i posiedzeń Polskiej Akad. Umiejętności*, **44**:220–227 (1939).

**Łukasiewicz, J., and A. Tarski**
[UAK]   Untersuchungen über den Aussagenkalkül, *Soc. Sci. Lett. Varsovie, C. R.*, Cl. III, **23**:30–50 (1935).

**McKinsey, J. C. C.**
[PIP]   Proof of the Independence of the Primitive Symbols of Heyting's Calculus of Propositions, *J. Symb. Logic*, **4**:155–158 (1939).
[SCS]   On the Syntactical Construction of Systems of Modal Logic, *J. Symb. Logic*, **10**:83–94 (1945).

**McKinsey, J. C. C., and A. Tarski**
[ATp]   The Algebra of Topology, *Ann. of Math.*, (2)**45**:141–191 (1944).
[CEC]   On Closed Elements in Closure Algebras, *Ann. of Math.*, (2)**47**:122–162 (1946).
[TSC]   Some Theorems about the Sentential Calculi of Lewis and Heyting, *J. Symb. Logic*, **13**:1–15 (1948).

**MacLane, S.**
[ABL]   Abgekürzte Beweise im Logikkalkül, (inaugural dissertation), Göttingen, 1934.

**Maehara, S.**
[DIL]   Eine Darstellung der intuitionistischen Logik in der Klassischen, *Nagoya Math. J.*, **7**:45–64 (1954).
[EAH]   Equality Axiom on Hilbert's ε-symbol, *J. Faculty Sci., Univ. Tokyo*, sec. I, **7**:419–435 (1957).
[PCE]   The Predicate Calculus with ε-symbol, *J. Math. Soc. Japan*, **7**:323–344 (1955).

**Markov, A. A.**
[TAl₁]   Teoriya algorifmov (Theory of Algorithms), *Trudy Mat. Inst. Steklov*, **38**:176–189 (1951). English translation by Edwin Hewitt in American Mathematical Society Translations, ser. 2, **15**:1–14 (1960).
[TAl₂]   Teoriya algorifmov (Theory of Algorithms), *Trudy Mat. Inst. Steklov*, vol. 42 (1954).

**Martin, R.**
[SPr]   "Toward a Systematic Pragmatics," Amsterdam, 1959.
[TDn]   "Truth and Denotation: A Study in Semantical Theory," Chicago, 1958.

**Menger, K.**
[ATF]   An Axiomatic Theory of Functions and Fluents, in Henkin et al. [AMS], pp. 454–473 (1959).
[GAp]   Gulliver in Applyland, *Eureka*, October, 1960.
[GLO]   Gulliver in the Land without One, Two, Three, *Math. Gaz.*, **43**:241–250 (1959).
[GRL]   Gulliver's Return to the Land without One, Two, Three, *Amer. Math. Monthly*, **67**:641–648 (1960).
[MMi]   Multiderivatives and Multi-integrals, *Amer. Math. Monthly*, **64**:58–70 (1957).

[RVP]  Random Variables from the Point of View of a General Theory of Variables, *Proc. Berkeley Symposium on Math. Statist. and Probability*, 1955, pp. 215–229.

**Meredith, C. A.**
[SAP]  A Single Axiom of Positive Logic, *J. Computing Systems*, 1:169–170 (1953).

**Moisil, G.**
[LPs]  Sur la logique positive, *An. Univ. C.I. Parhon-Bucuresti, seria Acta Logica*, 1958, no. 1, pp. 149–171.
[RAL]  Recherches sur l'algèbre de la logique, *Ann. Sci. Univ. Jassy*, 22:1–118 (1936).
[RPI]  Recherches sur le principe d'identité, *Ann. Sci. Univ. Jassy*, 23:7–56 (1937).

**Montague, R., and D. Kalish**
[RDN]  Remarks on Descriptions and Natural Deduction, *Arch. Math. Logik Grundlagenforsch.*, 3:50–64, 65–73, (1957).

**Monteiro, A. A.**
[AIA]  Axiomes independants pour les algèbres de Brouwer, *Rev. Un. Mat. Argentina*, 17:149–160 (1955).

**Monteiro, A. A., and O. Varsavsky**
[AHM]  Algebras de Heyting monadicas, *Actas X Jornadas Un. Mat. Argentina, Bahia Blanca*, 1957, pp. 52–62.

**Moody, E.**
[TCM]  "Truth and Consequence in Medieval Logic," Amsterdam, 1953.

**Morris, C. W.**
[FTS]  Foundations of the Theory of Signs, "International Encyclopedia of Unified Science," vol. 1, no. 2, Chicago, 1938.
[SLB]  "Signs, Language and Behavior," New York, 1946.

**Mostowski, A.**
[LMt]  Logika matematyczna. Kurs uniwersytecki (Mathematical Logic. University course), Warsaw, 1948.
[OSJ]  L'oeuvre scientifique de Jan Łukasiewicz dans le domain de la logique mathématique, *Fund. Math.*, 44:1–11 (1957).
[OUM]  Quelques observations sur l'usage des méthodes non-finitistes dans la méta-mathématique, *Le raisonnement en mathématiques et en sciences experimentales* (Colloques internationaux du Centre National de la Recherche Scientifique, vol. 70), 1958, pp. 19–28.
[PND]  Proofs of Non-deducibility in Intuitionistic Functional Calculus, *J. Symb. Logic*, 13:204–207 (1948).
[PSI]  The Present State of Investigations on the Foundations of Mathematics, *Rozprawy Mat.*, vol. 9, 48 pp. (1955). This paper was written in collaboration with several other authors. It also appeared in Russian, German, and Polish. For citations see *Math. Rev.*, 16:552 (1955).
[RPP]  On the Rules of Proof in the Pure Functional Calculus of the First Order, *J. Symb. Logic*, 16:107–111 (1951).
[SUF]  "Sentences Undecidable in Formalized Arithmetic: An Exposition of the Theory of Kurt Gödel," Amsterdam, 1952.

**Myhill, J.**
[OSL]  Symposium: On the Ontological Significance of the Löwenheim-Skolem Theorem, *Academic Freedom, Logic, and Religion* (American Philosophical Association, Eastern Division), Philadelphia, 1953, pp. 57–70. See also Berry [OSL].
[PAF]  Problems Arising in the Formalization of Intensional Logic, *Logique et analyse*, n.s., 1:74–83 (1958).

[PIM]    Some Philosophical Implications of Mathematical Logic. I. Three Classes of Ideas, *Rev. Metaphys.*, **6**:165–198 (1952).

**Myhill, J., and J. C. E. Dekker**

[RET]    "Recursive Equivalence Types," Berkeley and Los Angeles, 1960.

**Nagel, E.**

[FMC]    The Formation of Modern Conceptions of Formal Logic in the Development of Geometry, *Osiris*, **7**:142–222 (1939).

**Nagel, E., and J. R. Newman**

[GPr]    "Gödel's Proof," New York, 1958.

**Nagorniĭ, N. M.**

[UTP]    K usileniyu teoremy privedeniya teorii algorifmov (On the Strengthening of the Reduction Theorem of the Theory of Algorithms), *Doklady Akad. Nauk SSSR*, n.s., **90**:341–342 (1953).

**Neumann, J. von**

[HBT]    Zur Hilbertschen Beweistheorie, *Math. Z.*, **26**:1–46 (1927).

**Novikov, P. S.**

[ANP]    Ob algoritmicheskoĭ nerazreshimosti problemȳ tozhdestva slov v teorii grupp (On the Algorithmic Unsolvability of the Word Problem in the theory of Groups), *Trudy Mat. Inst. Steklov.*, Vol 44, 1955.

[EML]    "Elementȳ matematicheskoĭ logiki" (Elements of Mathematical Logic). Moscow, 1959.    An English translation by L. Boron has been announced.

**Ogasawara, T.**

[RIL]    Relations between Intuitionistic Logic and Lattices, *J. Sci. Hiroshima Univ.*, ser. A, **9**:157–164 (1939).

**Ogden, C. K., and I. A. Richards**

[MMn]    "The Meaning of Meaning," New York and London, 1927.

**Ohnishi, M. and K. Matsumoto**

[GMM]    Gentzen Method in Modal Calculi, part I, *Osaka Math. J.*, **9**:113–129 (1957); part II, **11**:115–120 (1959).

**Peano, G.**

[APN]    "Arithmetices principia, nova methodo exposita," Turin, 1889. Reprinted in [OSc], vol. II, pp. 20–56.

[CNm]    Sul concetto di numero, *Riv. di. mat.*, **1**:87–102, 256–267 (1891). Reprinted in [OSc], vol. III, pp. 80–109.

[FMt]    "Formulaire de mathématiques," Turin, vol. I, 1895; vol. II, 1898; vol. III, 1901; vol. IV, 1902; vol. V, 1908.

[OSc]    "Opere scelte: A cura dell'Unione Matematica Italiana e col contributo del Consiglio Nazionale delle Ricerche," vol. I, Rome, 1957; vol. II, 1958; vol. III, 1959.

**Peirce, C.**

[ALg$_1$]    On the Algebra of Logic, *Amer. J. Math.*, **3**:15–57 (1880).    Reprinted with corrections in [CPC], **3**:104–157.

[ALg$_2$]    On the Algebra of Logic. A Contribution to the Philosophy of Notation, *Amer. J. Math.*, **7**:180–202 (1885).    Reprinted in [CPC], **3**:210–249.

[CPC]    "Collected papers of Charles Sanders Peirce," vols. I-VIII, edited by Charles Hartshorne, Paul Weiss, and Arthur Burks, Cambridge, Mass., 1931–1958.

**Péter, R.**

[RFn]    "Rekursive Funktionen," Budapest, 1951; 2d ed., 1957.

**Porte, J.**

[DSS]    Deux systèmes simples pour le calcul des propositions, *Publ. Sci. Univ. Alger.*, ser. A, **5**:1–16 (1958).

[PCP]   Une propriété du calcul propositionnel intuitionniste, *Nederl. Akad. Wetensch. Proc.*, ser. A, **60**(= *Indag. Math.*, 20):362–365 (1958).

[RLM]   Recherches sur les logiques modales, *Le raisonnement en mathématiques et en sciences expérimentales* (Colloques internationaux du Centre National de la Recherche Scientifique, vol. 70), 1958, pp. 117–126.

[SCP]   Schémas pour le calcul des propositions fondé sur la conjonction et la négation, *J. Symb. Logic*, **23**:421–431 (1958).

[SPA]   Systèmes de Post, Algorithmes de Markov, *Cybernetica*, 1958, no. 2, pp. 1–35.

**Post, E. L.**

[FRG]   Formal Reductions of the General Combinatorial Decision Problem, *Amer. J. Math.*, **65**:197–215 (1943).

[IGT]   Introduction to a General Theory of Elementary Propositions, *Amer. J. Math.*, **43**:163–185 (1921).

[RES]   Recursively Enumerable Sets of Positive Integers and Their Decision Problems, *Bull. Amer. Math. Soc.*, **50**:284–316 (1944).

[TVI]   "The Two Valued Iterative Systems of Mathematical Logic," Princeton, N.J., 1941.

**Prior, A. N.**

[ECr]   Epimenides the Cretan, *J. Symb. Logic*, **23**:261–266 (1958).

[FLg]   "Formal Logic," Oxford, England, 1955.

[PAP]   Peirce's Axioms for Propositional Calculus, *J. Symb. Logic*, **23**:135–136 (1958).

**Quine, W. V.**

[CBA]   Concatenation as a Basis for Arithmetic, *J. Symb. Logic*, **11**:105–114 (1946).

[CQT]   Completeness of Quantification Theory: Löwenheim's Theorem, appendix to Quine [MeL], pp. 253–260.

[ELg]   "Elementary Logic," Boston, Mass. 1941.

[ISC]   Interpretations of Sets of Conditions, *J. Symb. Logic*, **19**:97–102 (1954).

[LPV]   "From a Logical Point of View, 9 Logico-Philosophical Essays," Cambridge, Mass., 1951.

[MeL]   "Methods of Logic," New York, 1950; revised edition, 1959.

[MLg]   "Mathematical Logic," New York, 1940; 2d ed., Cambridge, Mass., 1951.

[NDd]   On Natural Deduction, *J. Symb. Logic*, **15**:93–102 (1950).

[NFM]   New Foundations for Mathematical Logic, *Amer. Math. Monthly*, **44**:70–80 (1937). Reprinted with supplementary remarks in [LPV], pp. 80–101.

[PIM]   The Problem of Interpreting Modal Logic, *J. Symb. Logic*, **12**:43–48 (1947).

[PPQ]   A Proof Procedure for Quantification Theory, *J. Symb. Logic*, **20**:141–149 (1955).

[QED]   Quantification and the Empty Domain, *J. Symb. Logic*, **19**:177–179 (1954).

[SLg]   "A System of Logistic," Cambridge, Mass., 1934.

[SNL]   O sentido da nova lógica, São Paulo, 1944.

[UUS]   Unification of Universes in Set Theory, *J. Symb. Logic*, **21**:267–279 (1956).

[WOb]   "Word and Object," New York and Cambridge, Mass., 1960.

**Ramsey, F. P.**

[FML]   "The Foundations of Mathematics and Other Logical Essays," edited by R. B. Braithwaite, London, 1931.

[FMt]   The Foundations of Mathematics, *Proc. London Math. Soc.*, 2d ser., **25**:338–384 (1926). Reprinted in [FML], pp. 1–61.

**Rasiowa, H.**
[ASP]  Axiomatisation d'un système partiel ed la théorie de la déduction, *Soc. Sci. Lett. Varsovie, C.R.*, Cl. III *Sci. Math. Phys.* 40:22–37 (for 1947, published 1948).
[ETh]  On the ε-theorems, *Fund. Math.*, 43:156–165 (1956).
**Rasiowa, H., and R. Sikorski**
[ALL]  An Application of Lattices to Logic, *Fund. Math.*, 42:83–100 (1955).
[ETN]  On Existential Theorems in Non-classical Functional Calculi, *Fund. Math.*, 41:21–28 (1954).
[GTh]  On the Gentzen Theorem, *Fund. Math.*, 48:57–69 (1960).
[PSL]  A Proof of the Löwenheim-Skolem Theorem, *Fund. Math.*, 38:230–232 (1951, published 1952).
**Robinson, A.**
[CTh]  "Complete Theories," Amsterdam, 1956.
[MMA]  "The Metamathematics of Algebra," Amsterdam, 1951.
[OIM]  Outline of an Introduction to Mathematical Logic, I, *Canad. Math. Bull.*, 1:41–54 (1958); II, 1:113–127 (1958); III, 1:193–208 (1958); IV, 2:33–42 (1959).
**Rogers, H., Jr.**
[TRF]  "Theory of Recursive Functions and Effective Computability," vol. I, mimeographed copy from Massachusetts Institute of Technology, Cambridge, Mass., 1957.
**Rosenbloom, P. C.**
[EML]  "The Elements of Mathematical Logic," New York, 1950.
[PAl]  Post Algebras. I. Postulates and General Theory, *Amer. J. Math.*, 44:167–188 (1942).
**Rosser, J. B.**
[BFP]  The Burali-Forti Paradox, *J. Symb. Logic*, 7:1–17 (1942).
[CQN]  On the Consistency of Quine's "New Foundations for Mathematical Logic," *J. Symb. Logic*, 4:15–24 (1939).
[ETG]  Extensions of Some Theorems of Gödel and Church, *J. Symb. Logic*, 1:87–91 (1936).
[IEP]  An Informal Exposition of Proofs of Gödel's Theorems and Church's Theorem, *J. Symb. Logic*, 4:53–60 (1939).
[LMt]  "Logic for Mathematicians," New York, 1953.
[MLV]  A Mathematical Logic without Variables, *Ann. of Math.*, (2)36:127–150 (1935), and *Duke Math. J.*, 1:328–355 (1935).
**Rosser, J. B., and A. R. Turquette**
[MVL]  "Many-valued Logics," Amsterdam, 1952.
**Rubin, J. E.**
[RCA]  Remarks about a Closure Algebra in which Closed Elements Are Open, *Proc. Amer. Math. Soc.*, 7:30–34 (1956).
**Runes, D.**
[DPh]  "The Dictionary of Philosophy," New York, 1942.
**Russell, B.**
[IMP]  "Introduction to Mathematical Philosophy," London, 1919.
[PMt]  "Principles of Mathematics," Cambridge, England, 1903; 2d ed., 1938.
**Schilpp, P. A.**
[PBR]  "The Philosophy of Bertrand Russell," Chicago, 1944.
**Schmidt, H. A.**
[DTM]  Über deduktive Theorien mit mehreren Sorten von Grunddingen, *Math. Ann.*, 115:485–506 (1938).
[MGL]  Mathematische Grundlagenforschung, "Enzyklopädie der mathematischen Wissenschaften," vol. I, no. 2, 1950, 48 pp.

[VAL] "Vorlesungen über Aussagenlogik" (same as "Mathematische Gesetze der Logik, I") Berlin-Göttingen-Heidelberg, 1960.

[ZBM] Die Zulässigkeit der Behandlung mehrsortiger Theorien mittels der üblichen einsortigen Prädikatenlogik, *Math. Ann.*, **123**:187–200 (1951).

**Scholz, H.**

[GZM] See [VGZ].

[MJL] In Memoriam Jan Lukasiewicz, *Arch. Math. Logik Grundlagenforsch.*, **3**:3–18 (1957).

[VGZ] "Vorlesungen über Grundzüge der mathematischen Logik," parts 1, 2; 1st ed., Münster, 1949; 2d ed., 1950.

**Schönfinkel, M.**

[BML] Über die Bausteine der mathematischen Logik, *Math. Ann.*, **92**:305–316 (1924).

**Schröder, E.**

[VAL] "Vorlesungen über die Algebra der Logik," 3 vols., Leipzig, 1890–1905.

**Schröter, K.**

[AKB] Ein allgemeiner Kalkülbegriff, *Forsch. Logik und Grundlegung exakten Wiss.*, n.s., vol. 6 (1941).

[TBA] Theorie des bestimmten Artikels, *Z. Math. Logik Grundlagen Math.*, **2**:37–56 (1956).

[UHA] Eine Umformung des Heytingschen Axiomensystems für den intuitionistischen Aussagenkalkül, *Z. Math. Grundlagen Math.*, **3**:18–29 (1957).

[VIE] Die Vollständigkeit der die Implikation enthaltenden zweiwertigen Aussagenkalküle und Prädikatenkalküle der ersten Stufe, *Z. Math. Logik Grundlagen Math.*, **3**:81–107 (1957).

[WIM] Was ist eine mathematische Theorie?, *Jber. Deutsch. Math. Verein.*, **53**:69–82 (1943).

**Schütte, K.**

[BTh] "Beweistheorie," Berlin, 1960.

[EBA] Die Eliminierbarkeit des bestimmten Artikels in Kodifikaten der Analysis, *Math. Ann.*, **123**:166–186 (1951).

[SVS] Ein System des verknüpfenden Schliessens, *Arch. Math. Logik Grundlagenforsch.*, **2**:55–67 (1956).

[SWK] Schlussweisen-Kalküle der Prädikatenlogik, *Math. Ann.*, **122**:47–65 (1950).

**Shanin, N. A.**

[LPA] O nekotorȳkh logicheskikh problemakh aritmetiki (On some Logical Problems of Arithmetic), *Trudy Mat. Inst. Steklov*, no. 43, Moscow, 1955.

**Sheffer, H. M.**

[SFI] A Set of Five Independent Postulates for Boolean Algebras with Application to Logical Constants, *Trans. Amer. Math. Soc.*, **14**:481–488 (1913).

**Sierpinski, W.**

[CON] "Cardinal and Ordinal Numbers," Warsaw, 1958.

[LNT] "Leçons sur les nombres transfinis," Paris, 1928.

**Sikorski, R.**

[BAl] "Boolean Algebra," Berlin, 1960.

**Skolem, T.**

[BEA] Begründung der elementaren Arithmetik durch die rekurrierende Denkweise ohne Anwendung scheinbarer Veränderlichen mit unendlichem Ausdehnungsbereich, *Videnskapsselskapets Skrifter, I. Mat.-Nat. Klasse*, no. 6, 1923, 38 pp.

[CRF] Some Considerations Concerning Recursive Functions, *Math. Scand.*, **1**:213–221 (1953).

[MIF] "Mathematical Interpretations of Formal Systems," Amsterdam, 1955.

[PTL]   Sur la portée du theorème de Löwenheim-Skolem, *Les Entretiens de Zurich sur les fondements et la méthode des sciences mathématiques* (December 6–9, 1938), 1941, pp. 25–47, Discussion, pp. 47–52.

[UAK]   Untersuchungen über die Axiome des Klassenkalküls und über Produktations- und Summations-probleme, welche gewisse Klassen von Aussagen betreffen, *Videnskapsselskapets Skrifter, I. Mat.-Nat. Klasse*, 1919, no. 3, 37 pp.

[VLt]   Über gewisse "Verbände" oder "lattices," *Avhandlinger utgitt av det Norske Videnskaps-Akademi i Oslo, I. Mat. naturv. kl.*, 1936, no. 7, 16 pp.

**Słupecki, J.**

[GML]   Toward a Generalized Mereology of Leśniewski, *Studia Logica*, **8**:131–154 (1958).

[SLC]   S. Leśniewski's Calculus of Names, *Studia Logica*, **3**:7–71 (1955).

[SLP]   S. Leśniewski's Protothetics, *Studia Logica*, **1**:44–112 (1954).

**Smullyan, R. M.**

[TFS]   "Theory of Formal Systems" (*Ann. Math. Studies*, no. 47), Princeton, N.J., 1961.

**Sobociński, B.**

[MJL]   In Memoriam Jan Łukasiewicz, *Philos. Studies*, **6**:3–49 (1956).

**Specker, E. P.**

[ACQ]   The Axiom of Choice in Quine's New Foundations for Mathematical Logic, *Proc. Nat. Acad. Sci. U.S.A.*, **39**:972–975 (1953).

[AML]   Die Antinomien der Mengenlehre, *Dialectica*, **8**:234–244 (1954).

**Stanley, R.**

[EPQ]   An Extended Procedure in Quantificational Logic, *J. Symb. Logic*, **18**:97–104 (1953).

**Stegmüller, W.**

[WPl]   "Das Wahrheitsproblem und die Idee der Semantik: Eine Einführung in die Theorien von A. Tarski und R. Carnap," Vienna, 1957.

**Stenius, E.**

[PLA]   Das Problem der logischen Antinomien, *Soc. Sci. Fenn. Comment. Phys.-Math.*, vol. 14 (1949).

**Stone, M. H.**

[NFL]   Note on Formal Logic, *Amer. J. Math.*, **59**:506–514 (1937).

[RBA]   The Representation of Boolean Algebras, *Bull. Amer. Math. Soc.*, **44**:807–816 (1938).

[TRB]   The Theory of Representations for Boolean Algebras, *Trans. Amer. Math. Soc.*, **40**:37–111 (1936).

**Strawson, P. F.**

[ILT]   "Introduction to Logical Theory," London, 1952.

**Suppes, P.**

[AST]   "Axiomatic Set Theory," Princeton, N.J., 1960.

[ILg]   "Introduction to Logic," New York, 1957.

**Suranyi, J.**

[RTE]   "Reduktionstheorie des Entscheidungsproblems im Prädikatenkalkül der ersten Stufe," Budapest, 1959.

**Takeuti, G.**

[GLC]   On a Generalized Logic Calculus, *Jap. J. Math.*, **23**:39–96 (1953).

**Tang, T. C.**

[APG]   Algebraic Postulates and a Geometric Interpretation for the Lewis Calculus of Strict Implication, *Bull. Amer. Math. Soc.*, **44**:737–744 (1938).

**Tarski, A.**
[BBO]  Einige Betrachtungen über die Begriffe der ω-Widerspruchsfreiheit und der ω-Vollständigkeit, *Monatsh. Math. Phys.*, **40**:97–112 (1933). Reprinted in Tarski and Woodger [LSM], pp. 279–295.
[FBM]  Fundamentale Begriffe der Methodologie der deduktiven Wissenschaften, I, *Monatsh. Math. Phys.*, **37**:361–404 (1930). Reprinted in Tarski and Woodger [LSM], pp. 60–109.
[GTP]  A General Theorem Concerning Primitive Notions of Euclidean Geometry, *Nederl. Akad. Wetensch. Proc.*, ser. A, **59**(= *Indag. Math.*, 18):468–474 (1956).
[GZS]  Grundzüge des Systemenkalküls, Part I, *Fund. Math.*, **25**:503–526 (1935); Part II, **26**:283–301 (1936). Reprinted in Tarski and Woodger [LSM], pp. 342–383.
[ILM]  "Introduction to Logic and to the Methodology of the Deductive Sciences," translated by Olaf Helmer, New York, 1941. (Appeared originally in Polish, 1936, and was translated into German, 1937.)
[NMB]  Some Notions and Methods on the Borderline of Algebra and Metamathematics, *Proc. Internat. Congress of Mathematicians, Cambridge, Mass.*, 1950, **1**:705–720 (1952).
[SCT]  The Semantic Conception of Truth and the Foundations of Semantics, *Philos. and Phenomenol. Res.*, **4**:341–376 (1944).
[WBF]  Der Wahrheitsbegriff in den formalisierten Sprachen, *Studia Philos.*, **1**:261–405 (1936). Reprinted in Tarski and Woodger [LSM], pp. 152–278.
**Tarski, A., A. Mostowski, and R. Robinson**
[UDT]  Undecidability and Essential Undecidability in Arithmetic, in [UTh], pp. 39–74.
[UTh]  "Undecidable Theories," Amsterdam, 1953.
**Tarski, A., and J. H. Woodger**
[LSM]  "Logic, Semantics, and Metamathematics," translation by J. H. Woodger of selected papers by Tarski, Oxford, England, 1956.
**Turing, A. M.**
[CNA]  On Computable Numbers with an Application to the Entscheidungsproblem, *Proc. London Math. Soc.*, **42**:230–265 (1936).
**Umezawa, T.**
[IPL]  On Intermediate Propositional Logics, *J. Symb. Logic*, **24**:20–36 (1959).
**Vaught, R. L.**
[ALS]  Applications of the Löwenheim-Skolem-Tarski Theorem to Problems of Completeness and Decidability, *Nederl. Akad. Wetensch. Proc.*, ser. A, **57**(= *Indag. Math.* 16):467–472 (1954).
**Waisman, F.**
[EMD]  "Einführung in das mathematische Denken: Die Begriffsbildung der modernen Mathematik," Vienna, 1936; 2d ed., 1947.
**Wajsberg, M.**
[MLB]  "Metalogische Beiträge," *Wiadomosci Mat.*, part I, **43**:131–168 (1936); part II, **47**:119–139 (1939).
[UAK]  Untersuchungen über den Aussagenkalkül von A. Heyting, *Wiadomosci Mat.*, **46**:45–101 (1938).
**Wang, H.**
[FMt]  The Formalization of Mathematics, *J. Symb. Logic*, **19**:241–266 (1954).
[LMS]  The Logic of Many-sorted Theories, *J. Symb. Logic*, **17**:105–116 (1952).
**Wang, H., and R. MacNaughton**
[SAT]  "Les systèmes axiomatiques de la théorie des ensembles," Paris, 1953.

**Weyl, H.**

[DHM]  David Hilbert and his Mathematical Work, *Bull. Amer. Math. Soc.*, **50**:612–654 (1944).

[PMN]  "Philosophie der Mathematik und Naturwissenschaft," Munich and Berlin, 1927. (English translation by Olaf Helmer, revised and augmented, "Philosophy of Mathematics and Natural Science," Princeton, N.J., 1949.)

**Whitehead, A. N.**

[UAl]  "A Treatise on Universal Algebra with Applications, I," Cambridge, England, 1898.

**Whitehead, A. N., and B. Russell**

[PMt]  "Principia mathematica," 3 vols., Cambridge, England, 1910–1913; 2d ed., 1925–1927.

**Wilder, R. L.**

[IFM]  "Introduction to the Foundations of Mathematics," New York and London, 1952.

**Yanovskaya, S. A.**

[MLO]  Matematicheskaya logika i osnovaniya matematiki (Mathematical Logic and the Foundations of Mathematics), "Matematika v SSSR za sorok let 1917–1957," Moscow, 1959, pp. 13–120.

[OMM]  Osnovania matematiki i matématicheskaya logika (Foundations of Mathematics and Mathematical Logic), "Matématiki v SSSR za tridtsat let 1917–1947," Moscow and Leningrad, 1948, pp. 9–50.

**Zermelo, E.**

[UGL]  Untersuchungen über die Grundlagen der Mengenlehre, I, *Math. Ann.*, **65**:261–281 (1908).

**Zhegalkin, I. I.**

[ASL]  Arifmetizatsiya simbolicheskoĭ logiki (Arithmetization of Symbolic Logic), *Mat. Sb.*, **35**:311–377 and **36**:205–338 (1928–1929).

**Zubieta, R. G.**

[SVF]  Sobre la substitucion de les variables funcionales en el calculo funcional de primer orden, *Bol. Soc. Mat. Mexicano*, **7**:1–21 (1950).

# INDEX

# LIST OF ADDITIONAL SYMBOLS

Symbols, operators, or expressions that do not lend themselves to alphabetizing are placed here, in approximate order of first appearance in the text. Their meaning is usually explained on first mention.

| | | | | | | | |
|---|---|---|---|---|---|---|---|
| . | 36n., 70 | ⊢* | 188, 193, 200, 214, 276 | + | 75, 153, 162 | | |
| → | 35, 68, 70, 97 | | | − | 144 | | |
| ⇄ | 35 | *⊢ | 188 | $(-)_1$ | 144 | | |
| & | 35, 96 | ⊩ | 185, 360 | $(-)_2$ | 144 | | |
| ≡ | 35, 108–109, 317 | $⊩^L$ | 217 | ∩ | 161–162 | | |
| | | $⊩^T$ | 217, 280 | ∪ | 162 | | |
| = | 35, 126, 162, 316 | $⊩_m$ | 219 | * | 312 | | |
| | | $⊩_1$ | 219 | [ ] | 104, 114, 117 | | |
| (=) | 353 | ⊩′ | 360 | [/] | 318 | | |
| ≤ | 35, 126, 162, 166 | ⊣ | 35, 257–258 | ∀ | 116, 311–312 | | |
| ⊆ | 35, 162 | ⊃ | 35, 139, 162, 165, 172 | ∃ | 116, 311–312 | | |
| ⊂ | 162 | | | 0 | 316 | | |
| \| | 294 | $⊃_x$ | 116 | ⊂ | 316 | | |
| \|...⊢, \|a⊢ | 314, 322 | ∽ | 35, 158 | ° | 319 | | |
| \|...$⊢^L$ | 342 | ∨ | 35, 126, 161–162, 172 | ⧽ | 366 | | |
| \|...$⊢^T$ | 342 | | | ι | 358 | | |
| \|...$⊢^H$ | 344 | ∧ | 35, 51, 55, 126, 162, 172 | $\overset{n}{\underset{i=1}{\bigvee}}$ | 297 | | |
| ⊢ | 35, 65, 84, 165 | ⌐ | 35, 55, 257–258, 279 | | | | |
| $⊢_0$ | 185, 191, 318 | □ | 35, 65, 360–361, 366 | $\overset{n}{\underset{i=1}{\bigwedge}}$ | 298 | | |